CALCULUS

AND ITS APPLICATIONS

SECOND EDITION

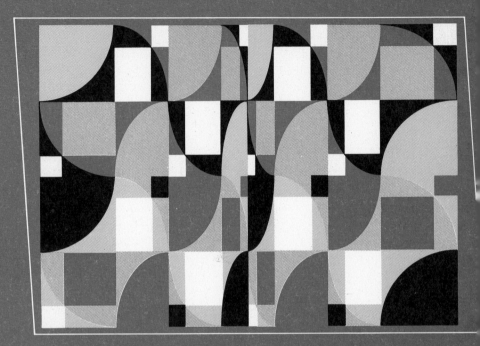

SECOND EDITION

PRENTICE-HALL, INC., Englewood Cliffs, New Jersey 07632

CALCULUS

AND ITS APPLICATIONS

LARRY J. GOLDSTEIN

DAVID C. LAY

DAVID I. SCHNEIDER

Department of Mathematics
University of Maryland

Library of Congress Cataloging Publication Data

Goldstein, Larry Joel.
    Calculus and its applications.

    Includes indexes.
      1. Calculus. I. Lay, David C., joint author.
II. Schneider, David I., joint author.
III. Title.
QA303.G625 1980b   515   79–23789
ISBN 0-13-111963-X

Calculus and Its Applications, Second Edition

Larry J. Goldstein, David C. Lay, David I. Schneider

10 9 8 7 6 5 4

Editorial/production supervision by Nancy Milnamow
Interior design and cover design by Walter A. Behnke
Manufacturing buyer: Ed Leone
Photo researcher: Frances L. Orkin

PRENTICE-HALL INTERNATIONAL, INC., *London*
PRENTICE-HALL OF AUSTRALIA PTY. LIMITED, *Sydney*
PRENTICE-HALL OF CANADA, LTD., *Toronto*
PRENTICE-HALL OF INDIA PRIVATE LIMITED, *New Delhi*
PRENTICE-HALL OF JAPAN, INC., *Tokyo*
PRENTICE-HALL OF SOUTHEAST ASIA PTE. LTD., *Singapore*
WHITEHALL BOOKS LIMITED, *Wellington, New Zealand*

PHOTO CREDITS

Chapter 0 (page 4) Daniels/Photo Researchers, Inc.

Chapter 1 (page 36) Woolfitt/Woodfin Camp & Associates.

Chapter 2 (page 102) Clark Equipment.

Chapter 3 (page 164) NASA.

Chapter 4 (page 180) Kinne/Photo Researchers, Inc.

Chapter 5 (page 200) Carmichael/Bruce Coleman, Inc.

Chapter 6 (page 218) © ARCH. PHOT., Paris/ S.P.A.D.E.M.

Chapter 7 (page 248) Getsug/Photo Researchers, Inc.

Chapter 8 (page 314) Goldman/Photo Researchers, Inc.

Chapter 9 (page 352) Massar/ Black Star.

Chapter 10 (page 386) Courtesy of PPG Industries.

Chapter 11 (page 426) Photo by Florita Botts/ Nancy Palmer.

Chapter 12 (page 476) Photo by Dr. Harold Edgerton, MIT, Cambridge, MA.

Mathematics and Its Applications

This volume is one of a collection of texts for freshman and sophomore college mathematics courses. Included in this collection are the following.

Calculus and Its Applications, Second Edition by L. Goldstein, D. Lay, and D. Schneider. A text designed for a two-semester course in calculus for students of business and the social and life sciences. Emphasizes an intuitive approach and integrates applications into the development. Much expanded from the highly successful first edition.

Calculus and Its Applications, Brief Edition by L. Goldstein, D. Lay, and D. Schneider. Consists of Chapters zero through eight of the above book. Suitable for shorter courses.

Finite Mathematics and Its Applications by L. Goldstein and D. Schneider. A traditional finite mathematics text for students of business and the social and life sciences. Allows courses to begin with either linear mathematics (linear programming, matrices) or probability and statistics.

Modern Mathematics and Its Applications by L. Goldstein, D. Lay, and D. Schneider. A text for a two-semester course covering finite mathematics, precalculus, and calculus.

CONTENTS

Preface   *xiii*

Index of Applications   *xvii*

Introduction   *2*

0. Functions   *5*

0.1. Functions and Their Graphs   *5*

0.2. Some Important Functions   *15*

0.3. The Algebra of Functions   *21*

0.4. Zeros of Functions—The Quadratic Formula and Factoring   *27*

1. The Derivative   *37*

    1.1.   The Slope of a Straight Line   *38*

    1.2.   The Slope of a Curve at a Point   *49*

    1.3.   The Derivative   *55*

    1.4.   Limits and the Derivative   *64*

    1.5.   Differentiability and Continuity   *73*

    1.6.   Some Rules for Differentiation   *80*

    1.7.   More About Derivatives   *86*

    1.8.   The Derivative as a Rate of Change   *91*

2. Applications of the Derivative   *103*

    2.1.   Describing Graphs of Functions   *103*

    2.2.   The First and Second Derivative Rules   *112*

    2.3.   Curve Sketching (Introduction)   *119*

    2.4.   Curve Sketching (Conclusion)   *126*

    2.5.   Optimization Problems   *133*

    2.6.   Further Optimization Problems   *141*

    2.7.   Applications of Calculus to Business and Economics   *150*

3. Techniques of Differentiation   *165*

    3.1.   The Product and Quotient Rules   *166*

    3.2.   The Chain Rule   *173*

4. The Exponential Function   *181*

    4.1.   Properties of Exponents   *182*

    4.2.   Graphs of Exponential Functions   *185*

    4.3.   Differentiation of Exponential Functions   *192*

5. The Natural Logarithm Function   *201*

    5.1.   The Natural Logarithm Function   *201*

    5.2.   The Derivative of ln x *207*

    5.3.   Properties of the Natural Logarithm Function   *211*

6. Applications of the Exponential and Natural Logarithm Functions   *219*

    6.1.   Exponential Growth and Decay   *220*

    6.2.   Compound Interest   *230*

    6.3.   Further Exponential Models   *236*

7. Integration   *249*

    7.1.   Antidifferentiation   *249*

    7.2.   Definite Integrals   *258*

    7.3.   Areas in the x-y Plane   *266*

    7.4.   Riemann Sums   *276*

8. Functions of Several Variables   *293*

    8.1.   Examples of Functions of Several Variables   *293*

    8.2.   Partial Derivatives   *299*

    8.3.   Maxima and Minima of Functions of Several Variables   *309*

    8.4.   Lagrange Multipliers and Constrained Optimization   *319*

    8.5.   Total Differentials and Their Applications   *330*

    8.6.   The Method of Least Squares   *336*

    8.7.   Double Integrals   *343*

9. The Trigonometric Functions   *353*

    9.1.   Radian Measure of Angles   *354*

    9.2.   The Sine and the Cosine   *358*

    9.3.   Differentiation of sin t and cos t *368*

    9.4.   The Tangent and Other Trigonometric Functions   *377*

10. Techniques of Integration   *387*

10.1.   Integration by Substitution   *389*

10.2.   Integration by Parts   *395*

10.3.   Evaluation of Definite Integrals   *400*

10.4.   The Trapezoidal Rule   *405*

10.5.   Some Applications of the Integral   *413*

10.6.   Improper Integrals   *418*

11. Differential Equations   *427*

11.1.   Solutions of Differential Equations   *428*

11.2.   Separation of Variables   *432*

11.3.   Numerical Solution of Differential Equations   *441*

11.4.   Qualitative Theory of Differential Equations   *447*

11.5.   Applications of Differential Equations   *458*

11.6.   The Lotka-Volterra Model for the Competition Between Two Species   *470*

12. Probability and Calculus   *477*

12.1.   Discrete Random Variables   *477*

12.2.   Continuous Random Variables   *486*

12.3.   Expected Value and Variance   *496*

12.4.   Exponential and Normal Random Variables   *504*

13. Taylor Polynomials and Infinite Series   *523*

13.1.   Taylor Polynomials   *524*

13.2.   The Newton-Raphson Algorithm   *532*

13.3.   Infinite Series   *538*

13.4.   Infinite Series and Probability Theory   *547*

Appendix

Tables   *A1*

Table 1   The Exponential Function   *A2*

Table 2   The Natural Logarithm Function   *A4*

Table 3   Trigonometric Functions in Radians   *A7*

Table 4   Areas Under the Standard Normal Curve   *A8*

Answers to Odd-Numbered Exercises   *A9*

Index   *A23*

PREFACE

We have been pleased with the enthusiastic response accorded the first edition of *Calculus and its Applications* by teachers and students alike. The present work is a revision that includes additional topics and incorporates changes suggested by classroom use.

Although the changes to the first edition have been substantial, we have preserved the approach and the flavor. Our goals remain the same: to begin the actual calculus as soon as possible; to present calculus in an intuitive yet intellectually satisfying way; and to illustrate the many applications of calculus to the biological, social, and management sciences. We have tried to achieve these goals while paying close attention to students' real and potential problems in learning calculus. Our main concern, as always, is: Will it work for the students? Listed on the following pages are some of the features that illustrate various aspects of this student-oriented approach.

Applications We provide realistic applications that illustrate the uses of calculus in other disciplines. The reader may survey the variety of applications by turning to the Index of Applications on page xvii. Wherever possible, we have attempted to use applications to motivate the mathematics. For example, the integral is introduced in Chapter 7 via a discussion of world oil consumption.

Examples We have included many more worked examples than is customary (421). Furthermore, we have included computational details to enhance readability by students whose basic skills are weak.

Exercises The 2443 exercises comprise about one-quarter of the text—the most important part of the text in our opinion. The exercises at the end of sections are usually arranged in the order in which the text proceeds, so that homework assignments may easily be made after only part of a section is discussed. Interesting applications and more challenging problems tend to be located near the end of the exercise sets. Supplementary exercises at the end of each chapter expand the other exercise sets and provide cumulative exercises that require skills from earlier chapters.

Practice Problems A new feature in this edition, the practice problems are carefully selected exercises that are located at the end of each section, just before the exercise set. Complete solutions are given following the exercise set. The practice problems often focus on points that are potentially confusing or are likely to be overlooked. We recommend that the reader seriously attempt the practice problems and study their solutions before moving on to the exercises. In classroom use over the past three years, several thousand students have found practice problems to be a valuable learning aid.

Minimal Prerequisites In Chapter 0 we review those facts that the reader needs to study calculus. A few important topics such as laws of exponents are reviewed again when they are used in a later chapter. A reader familiar with the content of Chapter 0 should begin with Chapter 1 and use Chapter 0 as a reference, whenever needed.

New in This Edition Among the many changes in this edition, the most significant are as follows:

1. *Limits.* In the first edition, all the important differentiation and integration facts were explained by informal limiting arguments, but the usual limit notation was avoided. At the request of many instructors, we have added an intuitive discussion of limits (Section 1.4). The limiting arguments in later sections are essentially the same as before but now use the limit notation. Thus it is possible to omit Section 1.4 except for a brief look at the limit notation on pages 65 and 68. Section 1.5, Differentiability and Continuity, is also an optional section.

2. *Rules for Differentiation.* The product, quotient and chain rules (Chapter 3) have been placed earlier to allow those on a quarter system to easily cover them in the first quarter of the course. If desired, this material may be covered immediately after Chapter 1.

3. *Natural Logarithm Function.* The chapter on the natural logarithm function (Chapter 5) has been placed immediately after the chapter on the exponential function. This allows for the use of the natural logarithm to solve growth and decay problems.

4. *Additional Material.* We have added chapters on probability and calculus (Chapter 12) and Taylor polynomials and infinite series (Chapter 13), as well as additional topics in integration (Sections 8.7 and 10.6). These topics will allow for greater curriculum flexibility in two-semester and three-quarter courses.

The following chapter dependence chart should aid in curriculum planning.

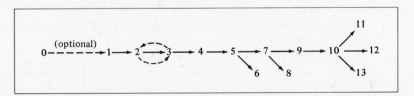

Answers to the odd-numbered exercises are included at the back of the book. Answers to the even-numbered exercises are contained in the Instructor's Manual. A computer manual is also available.

LARRY J. GOLDSTEIN
DAVID C. LAY
DAVID I. SCHNEIDER

College Park, Maryland

ACKNOWLEDGMENTS

While writing this book, we have received assistance from many persons. And our heartfelt thanks goes out to them all. Especially, we should like to thank the following reviewers, who took the time and energy to share their ideas, preferences, and often their enthusiasm, with us.

Reviewers of the first edition: Russell Lee, Allan Hancock College; Donald Hight, Kansas State College of Pittsburgh; Ronald Rose, American River College; W. R. Wilson, Central Piedmant Community College; Bruce Swenson, Foothill College; Samuel Jasper, Ohio University; Carl David Minda, University of Cincinnati; H. Keith Stumpff, Central Missouri State University; Claude Schochet, Wayne State University; and James E. Huneycutt, North Carolina University.

Reviewers of the second edition: Charles Himmelberg, University of Kansas; James A. Huckaba, University of Missouri; Joyce Longman, Villanova University; T. Y. Lam, University of California, Berkeley; W. T. Kyner, University of New Mexico; Shirley A. Goldman, University of California, Davis; Dennis White, University of Minnesota; Dennis Bertholf, Oklahoma State University; Wallace A. Wood, Bryant College; James L. Heitsch, University of Illinois, Chicago Circle; John H. Mathews, California State University, Fullerton; Arthur J. Schwartz, University of Michigan; Gordon Lukesh, University of Texas, Austin.

Our sincere thanks to Peter Schumer and Susanne Corr for their assistance in proofreading and to Patricia Berg for her patience and consummate skill in typing what seemed (to us) to be an infinite number of manuscript pages. In addition, we should like to thank our students and our colleagues around the country who shared with us their opinions of the previous edition.

The staff of Prentice-Hall, Inc. has done a magnificent job of transforming a somewhat dog-eared, often ill-prepared manuscript into a beautiful bound volume. Especially, we should like to thank Susan Formilan for directing the traffic between the authors, reviewers, and publisher, Nancy Milnamow for her expertise as production editor, Robin Bartlett and Robyn Juback for their help in promotion, and Walter Behnke for his imaginative and beautiful design work. Finally, we would like to express our gratitude to Harry Gaines, Executive Editor of Prentice-Hall, for favors too numerous to mention. His partnership, friendship, and constant encouragement have had a profound, positive effect on the ultimate form of this work.

INDEX OF APPLICATIONS

Biology and Medicine

Age distribution of cells, 486, 488, 491, 495
Allometric equation, 215
Autocatalytic enzyme reaction, 467
Basal metabolic rate, 376
Blood pressure, 376
Calcium metabolism, 93
Cardiac output, 332
Chemical reactions, 440, 467, 475
Concentration of a drug in the body, 246, 290
Contraction of the trachea during coughing, 144
Drug time-concentration curve, 13–14, 104–105, 206
E. Coli infection, 228
Elimination time of a drug, 246
Enzyme kinetics, 20
Epidemic model, 218, 241–43
Exponential decay, 223–26
Exponential growth, 220–23, 227, 228, 247, 284
Fish population of a lake, 241
Flu epidemic, 92, 97, 243, 257
Gene frequency, 464, 469
Gestation period, 515
Gompertz growth curve (or equation), 197, 440, 456
Growth of bacteria, 220–21, 227, 228, 247, 467
Growth of a tumor, 96
Half-life and decay constants, 223–26, 227, 228
Incidence of bacterial infection, 554
Intramuscular injection, 104–105, 206
Intravenous infusion of glucose, 239, 245
Iodine in the blood, 469
Laboratory testing of blood samples, 518
Logistic differential equation, 459
Logistic growth, 239–44
Lung cancer and smoking, 339–41
Mitosis in a cell population, 495
Movement of solutes through a cell membrane, 467
One-compartment problems, 463, 469, 470
Population genetics, 464–66
Protozoa in water samples, 553
Relief times of arthritic patients, 520
Renal clearance (kidney function) 335–36

Response of heart muscle to acetylcholine, 12
Restricted organic growth, 111, 239–41
Surface area of a human body, 308
Urinary tract infection, 228
Weight of an animal, 91–92
White blood cell count, 557

Business and Economics

Advertising, 502
Average cost, 168
Bidding on a construction project, 481, 502
Call frequency at a telephone switchboard, 552–53
Capital investment model, 468
Cobb-Douglas production function, 297, 304, 308, 322–24, 328, 333–34
Compound interest, 230–35, 247, 282, 286, 287
Consumers' surplus, 281–82, 285–86
Consumption of food, as a function of income and retail prices, 308
Consumption of fuel oil, 111, 250, 254, 259–60, 272–73, 274, 287
Continuous annuity, 461, 468, 475
Continuous stream of income, 413–15, 418, 468
Cost of electricity, 160
Cost functions, 20, 26, 40, 55, 93, 111, 150, 151–52, 159, 255, 260, 264, 265
Demand for beer (as a function of income, retail prices, and strength of the beer), 307
Demand equation, 152–57, 159, 160, 163, 211, 281–82, 285–86, 342
Demand functions (several variables), 298, 305–306, 307–308
Distribution of revenue between labor and capital, 308
Effect of advertising on sales, 90, 96, 149, 226, 335, 349
Effect of an excise tax on sales price, 157, 160
Effect of a second shift on a factory's cost function, 75
Evans price adjustment model, 468
Fire insurance claims, 555
Growth rates of U.S. population and national debt per person, 105–106
Interarrival times, 502, 508, 513, 514
Inventory problems, 142–44, 146, 148, 163
Least squares line, 336ff
Lifetimes of manufactured products, 494, 502, 505, 506, 513, 514, 515, 516, 520
Marginal cost, 93–94, 96, 98, 151–52, 158, 159, 168, 255, 257, 260, 264, 265, 273
Marginal cost of production (as a Lagrange multiplier), 335
Marginal productivity of labor and capital, 304, 324, 329
Marginal productivity of money (as a Lagrange multiplier), 324, 333–34
Marginal profit, 97, 155–58, 257, 265, 274
Marginal revenue, 97, 98, 152–55, 158, 255
Maximizing production, 148, 322–24, 329–30, 333–34
Minimum average cost, 168
Monopoly, profit of, 155–58, 160, 163, 312, 318, 329
Mortgage payment, 306
Multiplier effect, 541–42
National debt, 105
Optimal airline fares, 141–42
Optimal reorder quantity, 142–44, 146, 148, 163
Output, as a function of time, 28, 96, 101, 149, 257, 265

Present value, 234, 235, 298
Present value of an income stream, 287, 413–15, 418
Price discrimination, 312
Producers' surplus, 285–86
Production functions, 297, 304, 307, 308, 322–24, 328–29, 333–34
Production possibilities curve, 329
Profitability of new franchised restaurants, 489, 502
Profit functions, 146, 147, 148, 155–58, 159–60, 163, 211, 265
Profit functions (2 variables), 317, 318, 328–29, 335
Quality control, 483, 550
Rate of compliance with government regulations, 445
Rate of decrease of sales with respect to price, 439
Rate of increase of income, 111
Rate of net investment, 432
Ratio of marginal productivities, 324, 329
Reliability of equipment, 514
Revenue functions, 141–42, 147–48, 152–55, 159, 255
Sales decay curve, 226
Sales function (2 variables), 305–306, 335, 349
Sales trends, 342
Service contracts, 519, 520
Service time, 513
Wage per unit of labor, capital, 308
Waiting times, 494, 503

Ecology   Average world population, 284
Capacity of a lake, 241
Contamination of hay by radioactive iodine-131, 224–25
Depletion of natural resources, 257 (*see also* World consumption of oil)
Food chain, 24
Litter accumulation in a forest, 467, 469
Logistic growth, 239–44
Lotka-Volterra predator-prey model, 470ff
Net primary production of nutrients, 17
Oxygen content of a lake, 96, 98, 111, 149
Population growth, 220–23, 227, 228, 286
Predator-prey model, 372, 470ff
Protozoa in water samples, 553
Radioactive fallout and waste products, 224, 228, 286
Temperature of streams, 341–42
Water pollution, 16, 96, 111
Wind velocity, 207
World consumption of oil, 250, 254, 259–60, 272–73, 274, 287
World's need for arable land, 228–29

Social   Crime rate, as a function of several variables, 349
Sciences Demographic model, 415–16, 418
Diffusion of information by mass media, 237–39, 244, 245
Diffusion of an innovation among physicians, 244
Distribution of incomes, 519

Distribution of IQ's, 409
Ebbinghaus model for forgetting, 245
Growth rate of population, 475
Learning curve, 237
Population density of a city, 416–17, 418
Potassium dating of fossils, 227
Radiocarbon dating, 225–26, 228
Reaction to a stimulus, 439
Social diffusion, 467
Spread of a rumor, 37–38, 243–44
Spread of a technological innovation, 459
U.S. Supreme Court vacancies, 514
Voting model (in political science), 6–7, 171
War fever, 468

General Interest

Carbon dioxide level in a room, 469
Carbon monoxide level in the air, 35
Citrus grower's response to predicted freezing weather, 485
Conversion from Celsius to Fahrenheit, 6
Design of an athletic field, 147
Design of a building to minimize heat loss, 296, 312–14, 325–26
Design of a rain gutter or trough, 147, 370
Design of a wind shelter, 138, 140, 328
Dimensions of a Norman window, 140
Distance traveled, 410, 412
Error in measurement, 331, 335, 350
Estimating area of real estate, 412
Evaporation of water, 446, 467
Flow of water into a tank, 287
Galileo's conjecture about gravity, 467
Height of a building or tree, 382, 384
Maximum height of a projectile, 95, 134, 139
Newton's law of cooling, 430, 432
Optimal design of containers and mailing cartons, 137–38, 139, 140, 146, 147, 162, 317, 350
Optimal design of rectangular enclosures, 134–36, 140, 146, 147, 350
Pitch of a roof, 384
Principle of optimal design in architecture, 326
Probability of having an accident, 440
Roulette, 481
R-rating of insulation, 21
Strength of a beam, 163
Terminal velocity of a skydiver, 111, 236
Velocity and acceleration, 94–95, 97, 98, 100, 254–55, 257, 260, 264, 265
Volume of a gas (as a function of pressure and temperature), 308, 335
Width of a river, 382

CALCULUS

AND ITS APPLICATIONS

SECOND EDITION

INTRODUCTION

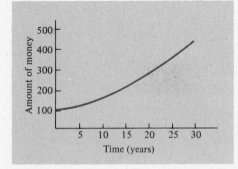

FIGURE 1

Often it is possible to give a succinct and revealing description of a situation by drawing a graph. For example, Fig. 1 describes the amount of money in a bank account drawing 5% interest, compounded daily. The graph shows that as time passes, the amount of money in the account grows. In Fig. 2 we have drawn a graph that depicts the weekly sales of a breakfast cereal at various times after advertising has ceased. The graph shows that the longer the time since the last advertisement, the fewer the sales. Figure 3 shows the size of a bacteria culture at various times. The culture grows larger as time passes. But there is a maximum size that the culture cannot exceed. This maximum size reflects the restrictions imposed by food supply, space, and similar factors. The graph in Fig. 4 describes the

2

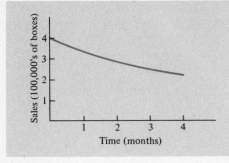

decay of the radioactive isotope iodine 131. As time passes, less and less of the original radioactive iodine remains.

Each of the graphs in Figs. 1 to 4 describes a change that is taking place. The amount of money in the bank is changing, as are the sales of cereal, the size of the bacteria culture, and the amount of the iodine. Calculus provides mathematical tools to study each of these changes in a quantitative way.

FIGURE 3

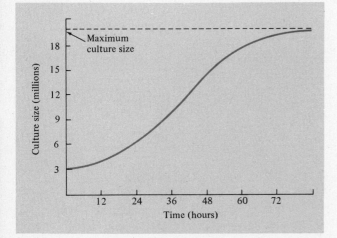

FIGURE 4

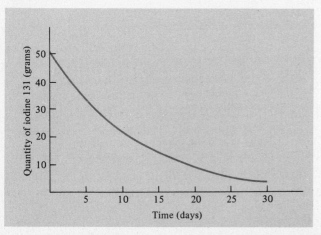

chapter 0

FUNCTIONS

Each of the graphs in Figs. 1 to 4 of the Introduction depicts a relationship between two quantities. For example, Fig. 4 illustrates the relationship between the quantity of iodine (measured in grams) and time (measured in days). In this chapter we develop the concept of a *function*, which is the basic quantitative tool for describing such relationships.

0.1. Functions and Their Graphs

A *function* of a variable x is a *rule* f that assigns to each value of x a number $f(x)$, called *the value of the function at x*. [We read "$f(x)$" as "f of x."]

The functions we shall meet in this book will usually be defined by algebraic formulas. For example, the function

$$f(x) = 3x - 1$$

is the rule that takes a number, multiplies it by 3, and then subtracts 1. If we specify a value of x, say $x = 2$, then we find the value of the function at 2 by substituting 2 for x in the formula:

$$f(2) = 3(2) - 1 = 5.$$

EXAMPLE 1 Let f be the function defined by

$$f(x) = 3x^3 - 4x^2 - 3x + 7.$$

Find $f(2)$ and $f(-2)$.

Solution To find $f(2)$ we substitute 2 for every occurrence of x in the formula for $f(x)$:

$$f(2) = 3(2)^3 - 4(2)^2 - 3(2) + 7$$
$$= 3(8) - 4(4) - 3(2) + 7$$
$$= 24 - 16 - 6 + 7$$
$$= 9.$$

The calculation of $f(-2)$ is similar.

$$f(-2) = 3(-2)^3 - 4(-2)^2 - 3(-2) + 7$$
$$= 3(-8) - 4(4) - 3(-2) + 7$$
$$= -24 - 16 + 6 + 7$$
$$= -27.$$

EXAMPLE 2 If x represents the temperature of an object in degrees Celsius, then the temperature in degrees Fahrenheit is a function of x, given by $f(x) = \frac{9}{5}x + 32$.

(a) Water freezes at 0°C (C = Celsius) and boils at 100°C. What are the corresponding temperatures in degrees Fahrenheit?

(b) Aluminum melts at 660°C. What is its melting point in degrees Fahrenheit?

Solution (a) $f(0) = \frac{9}{5}(0) + 32 = 32$. Water freezes at 32°F.
$f(100) = \frac{9}{5}(100) + 32 = 180 + 32 = 212$. Water boils at 212°F.

(b) $f(660) = \frac{9}{5}(660) + 32 = 1188 + 32 = 1220$. Aluminum melts at 1220°F.

EXAMPLE 3 (*A Voting Model*) Let x be the proportion of the total popular vote which a Democratic candidate for president receives in a U.S. national election (so x is a number between 0 and 1). Political scientists have observed that a good estimate of the proportion of seats in the House of Representatives going to Democratic candidates is given by

$$f(x) = \frac{x^3}{x^3 + (1 - x)^3} \qquad (0 \le x \le 1).$$

This formula is called the "cube law." Compute $f(.6)$ and interpret the result.

Solution   We must substitute .6 for every occurrence of x in $f(x)$:

$$f(.6) = \frac{(.6)^3}{(.6)^3 + (1 - .6)^3} = \frac{(.6)^3}{(.6)^3 + (.4)^3}$$

$$= \frac{.216}{.216 + .064} = \frac{.216}{.280} \approx .77.$$

This calculation shows that the "cube law" function predicts that if .6 (or 60%) of the total popular vote is for the Democratic candidate for president, then approximately .77 (or 77%) of the seats in the House of Representatives will be won by the Democratic candidates; that is, about 335 of the 435 seats will be won by Democrats.

In calculus, it is often necessary to substitute an algebraic expression for x and simplify the result.

EXAMPLE 4   If $f(x) = (4 - x)/(x^2 + 3)$, what is $f(a)$? $f(a + 1)$?

Solution   Here a represents some number. To find $f(a)$, we substitute a for x wherever x appears in the formula defining $f(x)$:

$$f(a) = \frac{4 - a}{a^2 + 3}.$$

To evaluate $f(a + 1)$, we substitute $a + 1$ for each occurrence of x in the formula for $f(x)$:

$$f(a + 1) = \frac{4 - (a + 1)}{(a + 1)^2 + 3}.$$

The expression for $f(a + 1)$ may be simplified, using the fact that $(a + 1)^2 = (a + 1)(a + 1) = a^2 + 2a + 1$:

$$f(a + 1) = \frac{4 - (a + 1)}{(a + 1)^2 + 3} = \frac{4 - a - 1}{a^2 + 2a + 1 + 3} = \frac{3 - a}{a^2 + 2a + 4}.$$

The Domain of a Function   When defining a function, it is necessary to specify the set of possible values of the variable. This set is called the *domain* of the function. For instance, the cube law function of Example 3 was given by

$$f(x) = \frac{x^3}{x^3 + (1 - x)^3} \qquad (0 \le x \le 1).$$

The domain of this function consists of those x for which $0 \le x \le 1$.

If no domain is specified, we will understand that the intended domain consists of all numbers for which the formula defining the function makes sense. For example, consider the function

$$f(x) = \frac{1}{x}.$$

Here x may be any number except zero. (Division by zero is not permissible.) So the domain intended is the set of nonzero numbers. Similarly, when we write

$$f(x) = \sqrt{x}$$

we understand the domain of $f(x)$ to be the set of all nonnegative numbers, since the square root of a number x is defined if and only if $x \geq 0$.

Graphs of Functions   Often it is helpful to describe a function f geometrically, using a rectangular x-y coordinate system. Given any x in the domain of f, we can plot the point $(x, f(x))$. This is the point in the x-y plane whose y-coordinate is the value of the function at x. The set of *all* such points $(x, f(x))$ usually forms a curve in the x-y plane and is called the *graph of the function $f(x)$*.

It is possible to approximate the graph of $f(x)$ by plotting the points $(x, f(x))$ for a representative set of values of x and joining them by a smooth curve. (See Fig. 1.) The more closely spaced the values of x, the closer the approximation.

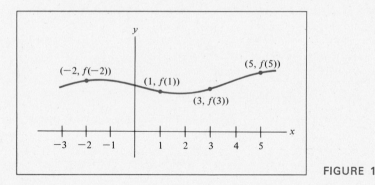

FIGURE 1

EXAMPLE 5   Sketch the graph of the function $f(x) = x^3$.

Solution   The domain consists of all numbers x. We choose some representative values of x and tabulate the corresponding values of $f(x)$. We then plot the points $(x, f(x))$ and sketch the graph indicated. (See Fig. 2.)

EXAMPLE 6   Sketch the graph of the function $f(x) = 1/x$.

Solution   The domain of the function consists of all numbers except zero. The table in Fig. 3 lists some representative values of x and the corresponding values of $f(x)$. A function often has interesting behavior for x near a number not in the domain. So when we chose representative values of x from the domain, we included some

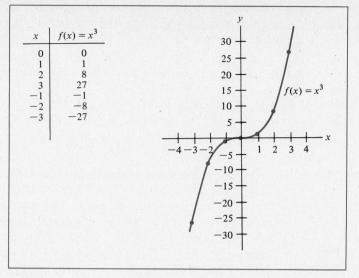

FIGURE 2

values close to zero. The points $(x, f(x))$ are plotted and the graph sketched in Fig. 3.

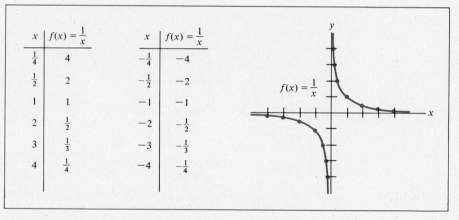

FIGURE 3

EXAMPLE 7 Suppose f is the function whose graph is given in Fig. 4. Notice that the point $(x, y) = (3, 2)$ is on the graph of f.

(a) What is the value of the function when $x = 3$?

(b) Find $f(-2)$.

(c) What is the domain of f?

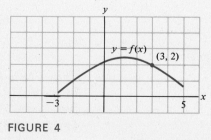

FIGURE 4

Solution   (a) Since (3, 2) is on the graph of *f*, the *y*-coordinate 2 must be the value of *f* at the *x*-coordinate 3. That is, $f(3) = 2$.

(b) To find $f(-2)$ we look at the *y*-coordinate of the point on the graph where $x = -2$. From Fig. 4 we see that $(-2, 1)$ is on the graph of *f*. Thus $f(-2) = 1$.

(c) The points on the graph of $f(x)$ all have *x*-coordinates between -3 and 5 inclusive; and for each value of *x* between -3 and 5 there is a point $(x, f(x))$ on the graph. So the domain consists of those *x* for which $-3 \leq x \leq 5$.

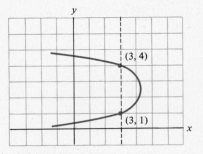

FIGURE 5   A curve that is *not* the graph of a function.

Not *every* curve is the graph of a function. To see this, refer first to the curve in Fig. 4, which *is* the graph of a function. It has the following important property: For each *x* between -3 and 5 inclusive there is a *unique* y such that (x, y) is on the curve. Now refer to the curve in Fig. 5. It cannot be the graph of a function, because a function *f* must assign to each *x* in its domain a *unique* value $f(x)$. However, for the curve of Fig. 5 there corresponds to $x = 3$ (for example) more than one *y*-value, namely, $y = 1$ and $y = 4$.

The essential difference between the curves in Figs. 4 and 5 leads us to the following test.

The vertical line test: A curve in the *x-y*-plane is the graph of a function if and only if each vertical line cuts or touches the curve at no more than one point.

EXAMPLE 8   Which of the following curves are graphs of functions?

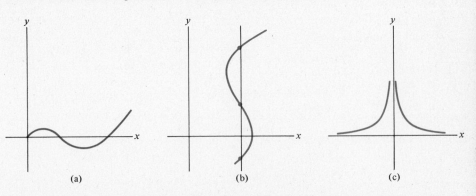

Solution   The curve in (a) is the graph of a function. It appears that vertical lines to the left of the *y*-axis do not touch the curve at all. This simply means that the function represented in (a) is defined only for $x \geq 0$. The curve in (b) is *not* the graph of a function because some vertical lines cut the curve in three places. The curve in (c) is *not* the

is the graph of a function whose domain is all nonzero x. (There is no point on the curve in (c) whose x-coordinate is 0.)

There is another notation for functions which we will find useful. Suppose that $f(x)$ is a function. When $f(x)$ is graphed on an x-y coordinate system, the values of $f(x)$ give the y-coordinates of points of the graph. For this reason, the function is often abbreviated by the letter y and we find it convenient to speak of "the function $y = f(x)$." Thus, for example, the function $y = 2x^2 + 1$ refers to the function $f(x)$ for which $f(x) = 2x^2 + 1$. The graph of a function $f(x)$ is often called *the graph of the equation* $y = f(x)$.

The equations arising in connection with functions are all of the form

$$y = \text{[an expression in } x\text{]}.$$

However, not all equations connecting the variables x and y are of this sort. For example consider these equations:

$$2x + 3y = 5$$
$$x = 3$$
$$x^2 + y^2 = 1.$$

It is possible to graph an equation by plotting points just as for functions. The only difference is that the resulting graph may not satisfy the vertical line test. For example, the graphs of the three equations above are shown in Fig. 6. Only the first graph is the graph of a function.

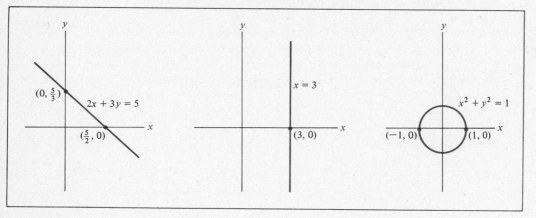

FIGURE 6

PRACTICE PROBLEMS 1

Letters other than f can be used to denote functions.

1. Is the point $(3, 12)$ on the graph of the function $g(x) = x^2 + 5x - 10$?

2. Sketch the graph of the function $h(x) = x^2 - 2$.

EXERCISES 1

1. If $f(x) = x^2 - 3x$, find $f(0)$, $f(5)$, $f(3)$, and $f(-7)$.
2. If $f(x) = 9 - 6x + x^2$, find $f(0)$, $f(2)$, $f(3)$, and $f(-13)$.
3. If $f(x) = x^3 + x^2 - x - 1$, find $f(1)$, $f(-1)$, $f(\frac{1}{2})$, and $f(a)$.
4. If $f(x) = x^3 - 3x^2 + x$, find $f(2)$, $f(-\frac{1}{2})$, $f(\frac{2}{3})$, and $f(a)$.
5. If $f(x) = x/(1 + x)$, find $f(\frac{1}{2})$, $f(-\frac{3}{2})$, and $f(a + 1)$.
6. If $f(x) = x^3/(x^3 - 1)$, find $f(\frac{1}{2})$, $f(-\frac{1}{2})$, and $f(a + 1)$.
7. An office supply firm finds that the number of Remington typewriters sold in year x is approximately given by the function $f(x) = 50 + 4x + \frac{1}{2}x^2$, where $x = 0$ corresponds to 1980.

    (a) What does $f(0)$ represent?

    (b) Find the number of Remington typewriters sold in 1982.
8. When a solution of acetylcholine is introduced into the heart muscle of a frog, it diminishes the force with which the muscle contracts. The data from experiments of A. J. Clark are closely approximated by a function of the form

    $$R(x) = \frac{100x}{b + x}, \qquad x \ge 0,$$

    where x is the concentration of acetylcholine (in appropriate units), b is a positive constant that depends on the particular frog, and $R(x)$ is the response of the muscle to the acetylcholine, expressed as a percentage of the maximum possible effect of the drug.

    (a) Suppose $b = 20$. Find the response of the muscle when $x = 60$.

    (b) Determine the value of b if $R(50) = 60$—that is, if a concentration of $x = 50$ units produces a 60% response.

    In Exercises 9 to 12 describe the domain of the function.

9. $f(x) = \dfrac{8x}{(x - 1)(x - 2)}$

10. $f(x) = \dfrac{1}{\sqrt{x}}$

11. $g(x) = \dfrac{1}{\sqrt{3 - x}}$

12. $g(x) = \dfrac{4}{x(x + 2)}$

    In Exercises 13 to 18 decide which curves are graphs of functions.

13.

14.

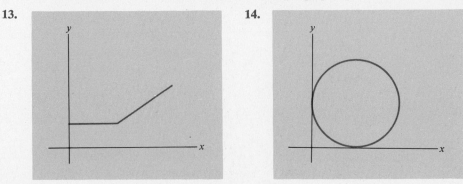

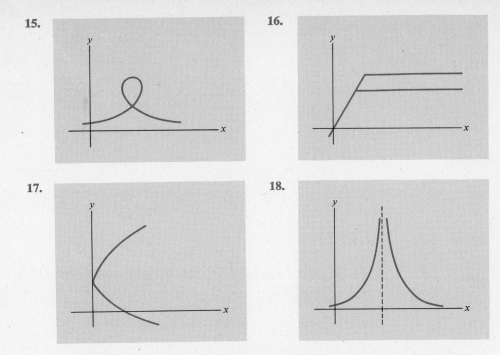

Exercises 19 to 28 relate to the function whose graph is sketched in Fig. 7.

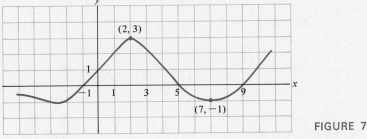

FIGURE 7

19. Find $f(0)$

20. Find $f(7)$

21. Find $f(2)$

22. Find $f(-1)$

23. Is $f(4)$ positive or negative?

24. Is $f(6)$ positive or negative?

25. Is $f(-\frac{1}{2})$ positive or negative?

26. Is $f(1)$ greater than $f(6)$?

27. For what values of x is $f(x) = 0$?

28. For what values of x is $f(x) \geq 0$?

Exercises 29 to 32 relate to Fig. 8. When a drug is injected into a person's muscle tissue, the concentration y of the drug in the blood is a function of the time elapsed since the injection. The graph of a typical time-concentration function f is given in Fig. 8, where $t = 0$ corresponds to the time of the injection.

29. What is the concentration of the drug when $t = 1$?

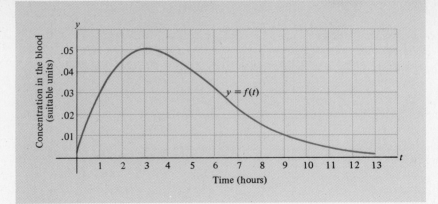

FIGURE 8

30. What is the value of the time-concentration function f when $t = 6$?

31. Find $f(5)$.

32. At what time does $f(t)$ attain its largest value?

33. Is the point $(3, 12)$ on the graph of the function $f(x) = (x - \frac{1}{2})(x + 2)$?

34. Is the point $(-2, 12)$ on the graph of the function $f(x) = x(5 + x)(4 - x)$?

35. Is the point $(\frac{1}{2}, \frac{2}{5})$ on the graph of the function $g(x) = (3x - 1)/(x^2 + 1)$?

36. Is the point $(\frac{2}{3}, \frac{5}{3})$ on the graph of the function $g(x) = (x^2 + 4)/(x + 2)$?

   In Exercises 37–42, sketch the graph of the given function.

37. $f(x) = -x^2 + 2x + 3$ 38. $g(x) = x^3 - 3x$ 39. $h(x) = \dfrac{2x}{1 + x^2}$

40. $h(x) = \dfrac{1}{x + 1}$ 41. $g(x) = x + \dfrac{1}{x}$ 42. $f(x) = (1 - x)^2$

SOLUTIONS TO PRACTICE PROBLEMS 1

1. If $(3, 12)$ is on the graph of $g(x) = x^2 + 5x - 10$, then we must have $g(3) = 12$. This is not the case, however, because

$$g(3) = 3^2 + 5(3) - 10$$
$$= 9 + 15 - 10 = 14.$$

   Thus $(3, 12)$ is *not* on the graph of $g(x)$.

2. Choose some representative values for x, say $x = 0, \pm 1, \pm 2, \pm 3$. For each value of x, calculate $h(x)$ and plot the point $(x, h(x))$.

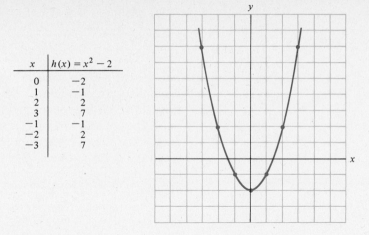

x	$h(x) = x^2 - 2$
0	−2
1	−1
2	2
3	7
−1	−1
−2	2
−3	7

0.2. Some Important Functions

In this section, we introduce some of the functions that will play a prominent role in our discussion of calculus.

Linear Functions As we shall see in the next chapter, a knowledge of the algebraic and geometric properties of straight lines is essential for the study of calculus. Every straight line is the graph of a linear equation of the form

$$cx + dy = e$$

where c, d, and e are given constants, with c and d not both zero. If $d \neq 0$, then we may solve the equation for y to obtain an equation of the form

$$y = mx + b, \tag{1}$$

FIGURE 1

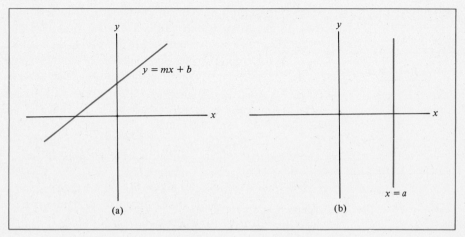

for appropriate numbers m and b. If $d = 0$, then we may solve the equation for x to obtain an equation of the form

$$x = a, \tag{2}$$

for an appropriate number a. So every straight line is the graph of an equation of type (1) or (2). The graph of an equation of the form (1) is a nonvertical line (Fig. 1(a)), whereas the graph of (2) is a vertical line (Fig. 1(b)).

The straight line of Fig. 1(a) is the graph of the function $f(x) = mx + b$. Such a function, which is defined for all x, is called a *linear function*. Note that the straight line of Fig. 1(b) is not the graph of a function since the vertical line test is violated.

An important special case of a linear function occurs if the value of m is zero, that is, if $f(x) = b$ for some number b. In this case, $f(x)$ is called a *constant function* since it assigns the same number, b, to every value of x. Its graph is the horizontal line whose equation is $y = b$. (See Fig. 2.)

Linear functions often arise in real-life situations, as the next example shows.

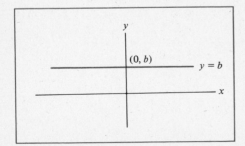

FIGURE 2 The graph of the constant function $f(x) = b$.

EXAMPLE 1   When the U.S. Environmental Protection Agency found a certain company dumping sulfuric acid into the Mississippi River, it fined the company $125,000, plus $1000 per day until the company complied with federal water pollution regulations. Express the total fine as a function of the number x of days the company continued to violate the federal regulations.

Solution   The variable fine for x days of pollution, at $1000 per day, is $1000x$ dollars. The total fine is therefore given by the function

$$f(x) = 125{,}000 + 1000x.$$

Since the graph of a linear function is a line, we may sketch it by locating any two points on the graph and drawing the line through them. For example, to sketch the graph of the function $f(x) = -\frac{1}{2}x + 3$, we may select two convenient values of x, say 0 and 4, and compute $f(0) = -\frac{1}{2}(0) + 3 = 3$ and $f(4) = -\frac{1}{2}(4) + 3 = 1$. The line through the points $(0, 3)$ and $(4, 1)$ is the graph of the function. See Fig. 3.

The point at which the graph of a linear function intersects the y-axis is called the *y-intercept* of the graph. The point at which the graph intersects the x-axis is called the *x-intercept*. The next example shows how to determine the intercepts of a linear function.

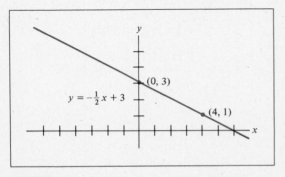

FIGURE 3

EXAMPLE 2 Determine the intercepts of the graph of the linear function $f(x) = 2x + 5$.

Solution Since the y-intercept is on the y-axis, its x-coordinate is 0. The point on the line with x-coordinate zero has y-coordinate

$$f(0) = 2(0) + 5 = 5.$$

So the y-intercept is $(0, 5)$. Since the x-intercept is on the x-axis, its y-coordinate is 0. Since $f(x)$ gives the y-coordinate, we must have

$$2x + 5 = 0$$

$$2x = -5$$

$$x = -\tfrac{5}{2}.$$

So $(-\tfrac{5}{2}, 0)$ is the x-intercept. See Fig. 4.

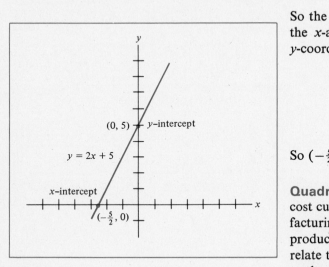

FIGURE 4

Quadratic Functions Economists utilize average-cost curves which relate the average unit cost of manufacturing a commodity to the number of units to be produced. (See Fig. 5.) Ecologists use curves which relate the net primary production of nutrients in a plant to the surface area of the foliage. (See Fig. 6.) Each of the curves is bowl-shaped, opening either up or down. The simplest functions whose graphs resemble these curves are the quadratic functions.

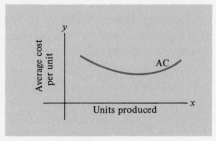

FIGURE 5 An average cost curve.

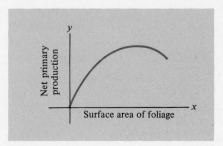

FIGURE 6 Production of nutrients.

A *quadratic function* is a function of the form

$$f(x) = ax^2 + bx + c,$$

where a, b, and c are constants and $a \neq 0$. The domain of such a function consists of all numbers. The graph of a quadratic function is called a *parabola*. Two typical parabolas are drawn in Figs. 7 and 8. We shall develop techniques for sketching

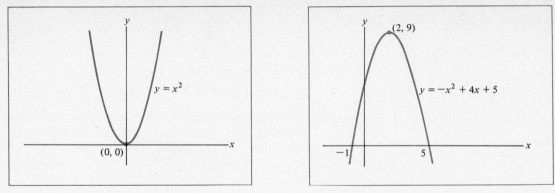

FIGURE 7   The graph of $f(x) = x^2$.

FIGURE 8   The graph of $f(x) = -x^2 + 4x + 5$.

graphs of quadratic functions after we have the properties of the derivative at our disposal.

Polynomial and Rational Functions   A *polynomial function* $f(x)$ is one of the form

$$f(x) = a_n x^n + a_{n-1} x^{n-1} + \ldots + a_0,$$

where n is a nonnegative integer and $a_0, a_1, \ldots, a_n$ are given numbers. Some examples of polynomial functions are

$$f(x) = 5x^3 - 3x^2 - 2x + 4$$

$$g(x) = x^4 - x + 1.$$

Of course, linear and quadratic functions are special cases of polynomial functions. The domain of a polynomial function consists of all numbers.

A function expressed as the quotient of two polynomials is called a *rational function*. Some examples are

$$h(x) = \frac{x^2 + 1}{x}$$

$$k(x) = \frac{x + 3}{x^2 - 4}.$$

The domain of a rational function excludes all values of x for which the denominator is zero. Thus, for example, the domain of $h(x)$ excludes $x = 0$, whereas the domain of $k(x)$ excludes $x = 2$ and $x = -2$. As we shall see, both polynomial and rational functions arise in applications of calculus.

Power Functions   Functions of the form $f(x) = x^r$ are called *power functions*. Table 1 gives the definition and domain for certain special cases. Many familiar functions are power functions. For instance, $\sqrt{x}$ is $x^{1/2}$, and $1/x$ is x^{-1}.

TABLE 1

r	x^r	Domain
n (n is a positive integer)	x^n	all x
$\dfrac{1}{n}$ (n is a positive integer)	$\sqrt[n]{x}$	$x \geq 0$ if n even, all x if n odd
$\dfrac{m}{n}$ (m, n positive integers)	$(\sqrt[n]{x})^m$	
$-s$ (s a positive number)	$\dfrac{1}{x^s}$	same as for x^s except for $x = 0$, which is excluded

EXAMPLE 3   Use Table 1 to evaluate the following power functions at $x = 125$.

(a) $f(x) = x^{1/3}$

(b) $g(x) = x^{2/3}$

(c) $h(x) = x^{-(2/3)}$.

Solution   (a) Here $r = \frac{1}{3}$. By the second line of the table, $x^{1/3}$ is $\sqrt[3]{x}$. Therefore,

$$f(125) = 125^{1/3} = \sqrt[3]{125} = 5.$$

(b) By the third line of the table, $x^{2/3}$ is $(\sqrt[3]{x})^2$. Therefore,

$$g(125) = 125^{2/3} = (\sqrt[3]{125})^2 = 5^2 = 25.$$

(c) By the last line of the table, $x^{-(2/3)}$ is $\dfrac{1}{x^{2/3}}$. Therefore,

$$h(125) = 125^{-(2/3)} = \frac{1}{125^{2/3}} = \frac{1}{25}.$$

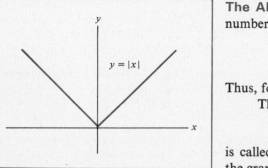

FIGURE 9

The Absolute Value Function   The absolute value of a number x is denoted by $|x|$ and is defined by

$$|x| = \begin{cases} x & \text{if } x \text{ is positive or zero,} \\ -x & \text{if } x \text{ is negative.} \end{cases}$$

Thus, for example, $|5| = 5$, $|0| = 0$, and $|-3| = -(-3) = 3$.

The function defined for all numbers x by

$$f(x) = |x|$$

is called the *absolute value function*. Its graph coincides with the graph of the equation $y = x$ for $x \geq 0$ and with the graph of the equation $y = -x$ for $x < 0$. See Fig. 9.

1. A photocopy service has a fixed cost of $2000 per month (for rent, depreciation of equipment, etc.), and variable costs of $.04 for each page it reproduces for customers. Express its total cost as a (linear) function of the number of pages copied per month.

2. Consider the power function $f(x) = x^{2/5}$. Evaluate (a) $f(32)$, (b) $f(-1)$.

EXERCISES 2

Graph the following functions.

1. $f(x) = 2x - 1$ 2. $f(x) = 3$ 3. $f(x) = 3x + 1$

4. $f(x) = -\frac{1}{2}x - 4$ 5. $f(x) = -2x + 3$ 6. $f(x) = \frac{1}{4}$

Determine the x- and y-intercepts of the graphs of the following functions.

7. $f(x) = 9x + 3$ 8. $f(x) = -\frac{1}{2}x - 1$ 9. $f(x) = 5$

10. $f(x) = 14$ 11. $f(x) = -\frac{1}{4}x + 3$ 12. $f(x) = 6x - 4$

13. In biochemistry, such as in the study of enzyme kinetics, one encounters a linear function of the form $f(x) = (K/V)x + 1/V$, where K and V are constants.

(a) If $f(x) = .2x + 50$, find K and V so that $f(x)$ may be written in the form $f(x) = (K/V)x + 1/V$.

(b) Find the x-intercept and y-intercept of the line $y = (K/V)x + 1/V$ (in terms of K and V).

14. The constants K and V in Exercise 13 are often determined from experimental data. Suppose a line is drawn through data points and has x-intercept $(-500, 0)$ and y-intercept $(0, 60)$. Determine K and V so that the line is the graph of the function $f(x) = (K/V)x + 1/V$. [Hint: Use Exercise 13(b).]

15. In some cities you can still rent a car for $12 per day and $.15 per mile.

(a) Find the cost of renting the car for one day and driving 200 miles.

(b) Suppose the car is to be rented for one day, and express the total rental expense as a function of the number x of miles driven. (Assume that for each fraction of a mile driven, the same fraction of $.15 is charged.)

16. A gas company will pay a property owner $5000 for the right to drill on the land for natural gas, and $.10 for each thousand cubic feet of gas extracted from the land. Express the amount of money the landowner will receive as a function of the amount of gas extracted from the land.

17. In 1978, Blue Cross paid $135 per day for a semiprivate hospital room and $150 for an appendectomy operation. Express the basic amount Blue Cross paid for an appendectomy operation as a function of the number of days of hospital confinement.

18. The R-rating of fiberglass home insulation is directly proportional to the thickness of the insulation. If the R-rating of a 6-inch batt of insulation is 19, express the R-rating of the insulation as a function of the thickness (in inches).

Each of the quadratic functions in Exercises 19–24 has the form $y = ax^2 + bx + c$. Identify a, b, and c.

19. $y = 3x^2 - 4x$
20. $y = \dfrac{x^2 - 6x + 2}{3}$
21. $y = 3x - 2x^2 + 1$

22. $y = 3 - 2x + 4x^2$
23. $y = 1 - x^2$
24. $y = \frac{1}{2}x^2 + \sqrt{3}x - \pi$

Evaluate each of the functions in Exercises 25–36 at the given value of x.

25. $f(x) = x^{-3}, x = 2$
26. $f(x) = x^3, x = -\frac{1}{5}$
27. $f(x) = x^{1/4}, x = 16$

28. $f(x) = x^{3/2}, x = 100$
29. $f(x) = x^{2/3}, x = 8$
30. $f(x) = x^{-4}, x = -1$

31. $f(x) = x^{-1/2}, x = 25$
32. $f(x) = x^{1/3}, x = \frac{1}{27}$
33. $f(x) = |x|, x = -2.5$

34. $f(x) = |x|, x = \pi$
35. $f(x) = |x|, x = 10^{-2}$
36. $f(x) = |x|, x = -\frac{2}{3}$

SOLUTIONS TO PRACTICE PROBLEMS 2

1. If x represents the number of pages copied per month, then the variable cost is $.04x$ dollars. Now [total cost] = [fixed cost] + [variable cost]. If we define

$$f(x) = 2000 + .04x,$$

then $f(x)$ gives the total cost per month.

2. By Table 1, $x^{2/5}$ is $(\sqrt[5]{x})^2$.

   (a) $f(32) = (\sqrt[5]{32})^2 = (2)^2 = 4.$
   (b) $f(-1) = (\sqrt[5]{-1})^2 = (-1)^2 = 1.$

0.3. The Algebra of Functions

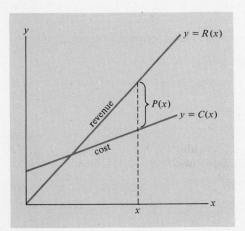

FIGURE 1

Many functions we shall encounter later in the text may be viewed as combinations of other functions. For example, let $P(x)$ represent the profit a company makes on the sale of x units of some commodity. If $R(x)$ denotes the revenue received from the sale of x units, and if $C(x)$ is the cost of producing x units, then

$$P(x) = R(x) - C(x).$$
$$[\text{profit}] = [\text{revenue}] - [\text{cost}]$$

Writing the profit function in this way makes it possible to predict the behavior of $P(x)$ from properties of $R(x)$ and $C(x)$. (See Fig. 1.)

The first four examples below review the algebraic techniques needed to combine functions by addition, subtraction, multiplication, and division.

EXAMPLE 1   Let $f(x) = 3x + 4$, $g(x) = 2x - 6$. Find $f(x) + g(x)$, $f(x) - g(x)$, $\dfrac{f(x)}{g(x)}$, and $f(x)g(x)$.

Solution   For $f(x) + g(x)$ and $f(x) - g(x)$ we add or subtract corresponding terms:

$$f(x) + g(x) = (3x + 4) + (2x - 6) = 3x + 4 + 2x - 6 = 5x - 2.$$

$$f(x) - g(x) = (3x + 4) - (2x - 6) = 3x + 4 - 2x + 6 = x + 10.$$

To compute $\dfrac{f(x)}{g(x)}$ and $f(x)g(x)$, we first substitute the formulas for $f(x)$ and $g(x)$.

$$\frac{f(x)}{g(x)} = \frac{3x + 4}{2x - 6}.$$

$$f(x)g(x) = (3x + 4)(2x - 6).$$

The expression for $\dfrac{f(x)}{g(x)}$ is already in simplest form. To simplify the expression for $f(x)g(x)$, we carry out the multiplication indicated in $(3x + 4)(2x - 6)$. We must be careful to multiply each term of $3x + 4$ by each term of $2x - 6$. A common order for multiplying such expressions is: (1) the First terms, (2) the Outer terms, (3) the Inside terms, and (4) the Last terms. (This procedure may be remembered by the word FOIL.)

$$f(x)g(x) = (3x + 4)(2x - 6) = 6x^2 - 18x + 8x - 24$$
$$= 6x^2 - 10x - 24.$$

EXAMPLE 2   Let $g(x) = \dfrac{2}{x}$, $h(x) = \dfrac{3}{x - 1}$. Express $g(x) + h(x)$ as a rational function.

Solution   First we have

$$g(x) + h(x) = \frac{2}{x} + \frac{3}{x - 1}.$$

In order for us to add two fractions, their denominators must be the same. A common denominator for $\dfrac{2}{x}$ and $\dfrac{3}{x - 1}$ is $x(x - 1)$. If we multiply $\dfrac{2}{x}$ by $\dfrac{x - 1}{x - 1}$, we obtain an equivalent expression whose denominator is $x(x - 1)$. Similarly, if we multiply $\dfrac{3}{x - 1}$ by $\dfrac{x}{x}$, we obtain an equivalent expression whose denominator is

$x(x - 1)$. Thus

$$\frac{2}{x} + \frac{3}{x - 1} = \frac{2}{x} \cdot \frac{x - 1}{x - 1} + \frac{3}{x - 1} \cdot \frac{x}{x}$$

$$= \frac{2(x - 1)}{x(x - 1)} + \frac{3x}{x(x - 1)}$$

$$= \frac{2(x - 1) + 3x}{x(x - 1)}$$

$$= \frac{5x - 2}{x(x - 1)}.$$

So

$$g(x) + h(x) = \frac{5x - 2}{x(x - 1)}.$$

EXAMPLE 3   Find $f(t)g(t)$, where

$$f(t) = \frac{t}{t - 1} \quad \text{and} \quad g(t) = \frac{t + 2}{t + 1}.$$

Solution   To multiply rational functions, multiply numerator by numerator and denominator by denominator:

$$f(t)g(t) = \frac{t}{t - 1} \cdot \frac{t + 2}{t + 1}$$

$$= \frac{t(t + 2)}{(t - 1)(t + 1)}.$$

An alternate way of expressing $f(t)g(t)$ is obtained by carrying out the indicated multiplications:

$$f(t)g(t) = \frac{t^2 + 2t}{t^2 + t - t - 1} = \frac{t^2 + 2t}{t^2 - 1}.$$

The choice of which expression to use for $f(t)g(t)$ depends on the particular application involving $f(t)g(t)$.

EXAMPLE 4   Find $\dfrac{f(x)}{g(x)}$, where

$$f(x) = \frac{x}{x - 3} \quad \text{and} \quad g(x) = \frac{x + 1}{x - 5}.$$

Solution   To divide $f(x)$ by $g(x)$, we multiply $f(x)$ by the reciprocal of $g(x)$:

$$\frac{f(x)}{g(x)} = \frac{x}{x - 3} \cdot \frac{x - 5}{x + 1} = \frac{x(x - 5)}{(x - 3)(x + 1)} = \frac{x^2 - 5x}{x^2 - 2x - 3}.$$

Composition of Functions Another important way of combining two functions $f(x)$ and $g(x)$ is to substitute the function $g(x)$ for every occurrence of the variable x in $f(x)$. The resulting function is called the *composition* (or *composite*) of $f(x)$ and $g(x)$, and is denoted by $f(g(x))$.

EXAMPLE 5 Let $f(x) = x^2 + 3x + 1$, $g(x) = x - 5$. What is $f(g(x))$?

Solution We substitute $g(x)$ in place of each x in $f(x)$.

$$f(g(x)) = [g(x)]^2 + 3g(x) + 1$$
$$= (x - 5)^2 + 3(x - 5) + 1$$
$$= (x^2 - 10x + 25) + (3x - 15) + 1$$
$$= x^2 - 7x + 11.$$

Later in the text we shall need to study expressions of the form $f(x + h)$, where $f(x)$ is a given function and h represents some number. The meaning of $f(x + h)$ is that $x + h$ is to be substituted for each occurrence of x in the formula for $f(x)$. In fact, $f(x + h)$ is just a special case of $f(g(x))$, where $g(x) = x + h$.

EXAMPLE 6 If $f(x) = \dfrac{x^3 - 4x^2}{x^2 + 1}$, find $f(x + h)$.

Solution
$$f(x + h) = \frac{(x + h)^3 - 4(x + h)^2}{(x + h)^2 + 1}.$$

EXAMPLE 7 In a certain lake, the bass feed primarily on minnows and the minnows feed on plankton. Suppose the size of the bass population is a function $f(n)$ of the number n of minnows in the lake, and the number of minnows is a function $g(x)$ of the amount x of plankton in the lake. Express the size of the bass population as a function of the amount of plankton, if $f(n) = 50 + \sqrt{n/150}$ and $g(x) = 4x + 3$.

Solution We have $n = g(x)$. Substituting $g(x)$ for n in $f(n)$, we find that the size of the bass population is given by

$$f(g(x)) = 50 + \sqrt{\frac{g(x)}{150}} = 50 + \sqrt{\frac{4x + 3}{150}}.$$

When two power functions are combined by multiplication, division, or composition the resulting function can always be written as a power function. The rules for such combinations are as follows.

$$x^r \cdot x^s = x^{r+s} \tag{1}$$

$$\frac{x^r}{x^s} = x^{r-s} \tag{2}$$

$$(x^s)^r = x^{rs}. \tag{3}$$

EXAMPLE 8 Let $f(x) = x^2$, $g(x) = x^{1/2}$. Express $f(x)g(x)$, $f(x)/g(x)$, and $f(g(x))$ as power functions.

Solution
$$f(x)g(x) = x^2 \cdot x^{1/2} = x^{2+1/2} = x^{5/2}$$

$$\frac{f(x)}{g(x)} = \frac{x^2}{x^{1/2}} = x^{2-1/2} = x^{3/2}$$

$$f(g(x)) = (x^{1/2})^2 = x^{2(1/2)} = x^1 = x.$$

PRACTICE PROBLEMS 3

1. Let $f(x) = x^5$, $g(x) = x^3 - 4x^2 + x - 8$.

    (a) Find $f(g(x))$ (b) Find $g(f(x))$.

2. Let $f(x) = x^2$, $g(x) = x^{1/3}$. Express $f(x)g(x)$, $f(x)/g(x)$, and $f(g(x))$ as power functions.

EXERCISES 3

Let $f(x) = x^2 + 1$, $g(x) = 9x$, $h(x) = 5 - 2x^2$. Calculate the following functions.

1. $f(x) + g(x)$ 2. $f(x) - h(x)$ 3. $f(x)g(x)$

4. $g(x)h(x)$ 5. $f(t)/g(t)$ 6. $g(t)/h(t)$

In Exercises 7–12, express $f(x) + g(x)$ as a rational function. Carry out all multiplications.

7. $f(x) = \dfrac{2}{x - 3}$, $g(x) = \dfrac{1}{x + 2}$ 8. $f(x) = \dfrac{3}{x - 6}$, $g(x) = \dfrac{-2}{x - 2}$

9. $f(x) = \dfrac{x}{x - 8}$, $g(x) = \dfrac{-x}{x - 4}$ 10. $f(x) = \dfrac{-x}{x + 3}$, $g(x) = \dfrac{x}{x + 5}$

11. $f(x) = \dfrac{x + 5}{x - 10}$, $g(x) = \dfrac{x}{x + 10}$ 12. $f(x) = \dfrac{x + 6}{x - 6}$, $g(x) = \dfrac{x - 6}{x + 6}$

Let $f(x) = \dfrac{x}{x - 2}$, $g(x) = \dfrac{5 - x}{5 + x}$, $h(x) = \dfrac{x + 1}{3x - 1}$. Express the following as rational functions.

13. $f(x) - g(x)$ 14. $f(t) - h(t)$ 15. $f(x)g(x)$

16. $g(x)h(x)$ 17. $f(x)/g(x)$ 18. $h(s)/f(s)$

In Exercises 19–22, express $f(x)g(x)$, $f(x)/g(x)$, and $f(g(x))$ as power functions.

19. $f(x) = x^{-2}$, $g(x) = x^3$ 20. $f(x) = x^{10}$, $g(x) = x^{5/2}$

21. $f(x) = x^{4/3}$, $g(x) = x^{3/4}$ 22. $f(x) = x^{-3}$, $g(x) = x^{-6}$

Let $f(x) = x^6$, $g(x) = \dfrac{x}{1-x}$, $h(x) = x^3 - 5x^2 + 1$. Calculate the following functions.

23. $f(g(x))$ 24. $h(f(t))$ 25. $h(g(x))$

26. $g(f(x))$ 27. $g(h(t))$ 28. $f(h(x))$

29. If $f(x) = x^2$, find $f(x + h) - f(x)$ and simplify.

30. If $f(x) = 1/x$, find $f(x + h) - f(x)$ and simplify.

31. If $g(t) = 4t - t^2$, find $g(t + h) - g(t)$ and simplify.

32. If $g(t) = t^3 + 5$, find $g(t + h) - g(t)$ and simplify.

33. After t hours of operation, an assembly line has assembled $A(t) = 20t - \frac{1}{2}t^2$ power lawn mowers, $0 \le t \le 10$. Suppose that the factory's cost of manufacturing x units is $C(x)$ dollars, where $C(x) = 3000 + 80x$.

   (a) Express the factory's cost as a (composite) function of the number of hours of operation of the assembly line.

   (b) What is the cost of the first two hours of operation?

34. During the first half-hour, the employees of a machine shop prepare the work area for the day's work. After that, they turn out 10 precision machine parts per hour, so that the output after t hours is $f(t)$ machine parts, where $f(t) = 10(t - \frac{1}{2})$ $= 10t - 5$, $\frac{1}{2} \le t \le 8$. The total cost of producing x machine parts is $C(x)$ dollars, where $C(x) = .1x^2 + 25x + 200$.

   (a) Express the total cost as a (composite) function of t.

   (b) What is the cost of the first four hours of operation?

SOLUTIONS TO PRACTICE PROBLEMS 3

1. (a) $f(g(x)) = [g(x)]^5 = (x^3 - 4x^2 + x - 8)^5$.

   (b) $g(f(x)) = [f(x)]^3 - 4[f(x)]^2 + f(x) - 8$

   $= (x^5)^3 - 4(x^5)^2 + x^5 - 8$

   $= x^{15} - 4x^{10} + x^5 - 8$.

2. $f(x)g(x) = x^2 \cdot x^{1/3} = x^{2+(1/3)} = x^{7/3}$

   $\dfrac{f(x)}{g(x)} = \dfrac{x^2}{x^{1/3}} = x^{2-(1/3)} = x^{5/3}$

   $f(g(x)) = (x^{1/3})^2 = x^{2/3}$.

0.4. Zeros of Functions—The Quadratic Formula and Factoring

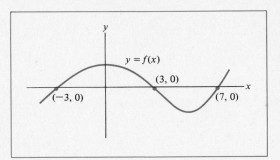

FIGURE 1

A *zero* of a function $f(x)$ is a value of x for which $f(x) = 0$. For instance, the function $f(x)$ whose graph is shown in Fig. 1 has $x = -3$, $x = 3$, and $x = 7$ as zeros. Throughout this book we shall need to determine zeros of functions or, what amounts to the same thing, to solve the equation $f(x) = 0$.

In Section 2 we found zeros of linear functions. In this section our emphasis is on zeros of quadratic functions.

The Quadratic Formula Consider the quadratic function $f(x) = ax^2 + bx + c$, $a \neq 0$. The zeros of this function are precisely the solutions of the quadratic equation

$$ax^2 + bx + c = 0.$$

One way of solving such an equation is via the *quadratic formula*.

> The solutions of the equation $ax^2 + bx + c = 0$ are
> $$x = \frac{-b \pm \sqrt{b^2 - 4ac}}{2a}.$$

The $\pm$ sign tells us to form two expressions, one with $+$ and one with $-$. The quadratic formula is derived at the end of the section.

EXAMPLE 1 Solve the quadratic equation $3x^2 - 6x + 2 = 0$.

Solution Here $a = 3$, $b = -6$, $c = 2$. Substituting these values into the quadratic formula, we find

$$\sqrt{b^2 - 4ac} = \sqrt{(-6)^2 - 4(3)(2)} = \sqrt{36 - 24} = \sqrt{12}$$
$$= \sqrt{4 \cdot 3} = 2\sqrt{3},$$

and

$$x = \frac{-b \pm \sqrt{b^2 - 4ac}}{2a} = \frac{-(-6) \pm 2\sqrt{3}}{2(3)} = \frac{6 \pm 2\sqrt{3}}{6} = 1 \pm \frac{\sqrt{3}}{3}.$$

Therefore, the solutions of the equation are $1 + \sqrt{3}/3$ and $1 - \sqrt{3}/3$. (See Fig. 2.)

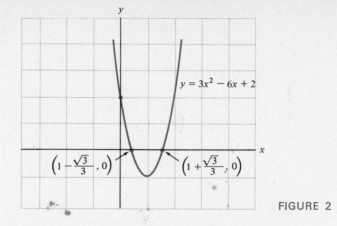

$$y = 3x^2 - 6x + 2$$

$$\left(1 - \frac{\sqrt{3}}{3}, 0\right) \qquad \left(1 + \frac{\sqrt{3}}{3}, 0\right)$$

FIGURE 2

EXAMPLE 2 Find the zeros of the following quadratic functions.

(a) $f(x) = 4x^2 - 4x + 1$ (b) $f(x) = \frac{1}{2}x^2 - 3x + 5$.

Solution (a) We must solve $4x^2 - 4x + 1 = 0$. Here $a = 4$, $b = -4$, $c = 1$, so that

$$\sqrt{b^2 - 4ac} = \sqrt{(-4)^2 - 4(4)(1)} = \sqrt{0} = 0.$$

Thus there is only one zero, namely,

$$x = \frac{-(-4) \pm 0}{2(4)} = \frac{4}{8} = \frac{1}{2}.$$

The graph of $f(x)$ is sketched in Fig. 3.

(b) We must solve $\frac{1}{2}x^2 - 3x + 5 = 0$. Here $a = \frac{1}{2}$, $b = -3$, $c = 5$, so that

$$\sqrt{b^2 - 4ac} = \sqrt{(-3)^2 - 4(\tfrac{1}{2})(5)} = \sqrt{9 - 10} = \sqrt{-1}.$$

The square root of a negative number is undefined, so we conclude that $f(x)$ has no zeros. The reason for this is clear from Fig. 4. The graph of $f(x)$ lies entirely above the x-axis and has no x-intercepts.

FIGURE 3

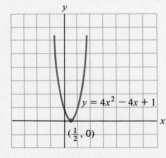

$$y = 4x^2 - 4x + 1$$

$$(\tfrac{1}{2}, 0)$$

FIGURE 4

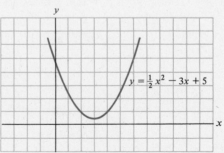

$$y = \frac{1}{2}x^2 - 3x + 5$$

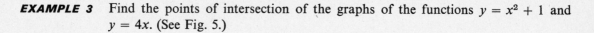

In the following example the quadratic formula is used to solve a type of problem that will arise in another context in our discussion of the definite integral.

EXAMPLE 3 Find the points of intersection of the graphs of the functions $y = x^2 + 1$ and $y = 4x$. (See Fig. 5.)

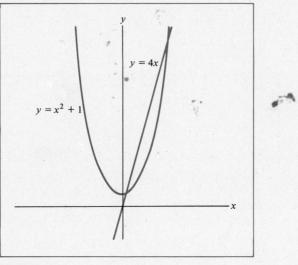

FIGURE 5

Solution If a point (x, y) is on both graphs, then its coordinates must satisfy both equations. That is, x and y must satisfy $y = x^2 + 1$ and $y = 4x$. Equating the two expressions for y, we have

$$x^2 + 1 = 4x.$$

To use the quadratic formula, we rewrite the equation in the form

$$x^2 - 4x + 1 = 0.$$

From the quadratic formula,

$$x = \frac{4 \pm \sqrt{16 - 4}}{2} = \frac{4 \pm \sqrt{12}}{2} = \frac{4 \pm 2\sqrt{3}}{2} = 2 \pm \sqrt{3}.$$

Thus the x-coordinates of the points of intersection are $2 + \sqrt{3}$ and $2 - \sqrt{3}$. To find the y-coordinates, we substitute these values of x into either equation, $y = x^2 + 1$ or $y = 4x$. The second equation is simpler. We obtain $y = 4(2 + \sqrt{3}) = 8 + 4\sqrt{3}$ and $y = 4(2 - \sqrt{3}) = 8 - 4\sqrt{3}$. Thus the points of intersection are $(2 + \sqrt{3}, 8 + 4\sqrt{3})$ and $(2 - \sqrt{3}, 8 - 4\sqrt{3})$.

Factoring If $f(x)$ is a polynomial, we can often write $f(x)$ as a product of linear factors (i.e., factors of the form $ax + b$). If this can be done, then the zeros of $f(x)$

can be determined by setting each of the linear factors equal to zero and solving for x. (The reason is that the product of numbers can be zero only when one of the numbers is zero.)

EXAMPLE 4   Factor the quadratic functions:

(a) $x^2 + 7x + 12$ (b) $x^2 - 13x + 12$ (c) $x^2 - 4x - 12$

(d) $x^2 + 4x - 12$.

Solution   Note first that for any numbers c and d,

$$(x + c)(x + d) = x^2 + (c + d)x + cd.$$

In the quadratic on the right, the constant term is the product cd, while the coefficient of x is the sum $c + d$.

(a) Think of all integers c and d such that $cd = 12$. Then choose the pair that satisfies $c + d = 7$; that is, take $c = 3$, $d = 4$. Thus

$$x^2 + 7x + 12 = (x + 3)(x + 4).$$

(b) We want $cd = 12$. Since 12 is positive, c and d must be both positive or both negative. We must also have $c + d = -13$. These facts lead us to

$$x^2 - 13x + 12 = (x - 12)(x - 1).$$

(c) We want $cd = -12$. Since -12 is negative, c and d must have opposite signs. Also, they must combine to give -4. We find that

$$x^2 - 4x - 12 = (x - 6)(x + 2).$$

(d) This is almost the same as (c).

$$x^2 + 4x - 12 = (x + 6)(x - 2).$$

EXAMPLE 5   Factor the polynomials:

(a) $x^2 - 6x + 9$ (b) $x^2 - 25$ (c) $3x^2 - 21x + 30$

(d) $20 + 8x - x^2$.

Solution   (a) We look for $cd = 9$ and $c + d = -6$. The solution is $c = d = -3$, and

$$x^2 - 6x + 9 = (x - 3)(x - 3) = (x - 3)^2.$$

In general,

$$x^2 - 2cx + c^2 = (x - c)(x - c) = (x - c)^2.$$

(b) We use the identity

$$x^2 - c^2 = (x + c)(x - c).$$

Hence

$$x^2 - 25 = (x + 5)(x - 5).$$

(c) We first factor out a common factor of 3, and then use the method of Example 4.

$$3x^2 - 21x + 30 = 3(x^2 - 7x + 10)$$
$$= 3(x - 5)(x - 2).$$

(d) We first factor out a -1 in order to make the coefficient of x^2 equal to $+1$.

$$20 + 8x - x^2 = (-1)(x^2 - 8x - 20)$$
$$= (-1)(x - 10)(x + 2).$$

EXAMPLE 6 Factor the polynomials:

(a) $x^2 - 8x$ (b) $x^3 + 3x^2 - 18x$ (c) $x^3 - 10x$.

Solution In each case we first factor out a common factor of x.

(a) $x^2 - 8x = x(x - 8)$.

(b) $x^3 + 3x^2 - 18x = x(x^2 + 3x - 18) = x(x + 6)(x - 3)$.

(c) $x^3 - 10x = x(x^2 - 10)$. To factor $x^2 - 10$, we use the identity $x^2 - c^2 = (x + c)(x - c)$, where $c^2 = 10$ and $c = \sqrt{10}$. Thus $x^3 - 10x = x(x^2 - 10) = x(x + \sqrt{10})(x - \sqrt{10})$.

The main use of factoring in this text will be to solve equations.

EXAMPLE 7 Solve the equations:

(a) $x^2 - 2x - 15 = 0$ (b) $x^2 - 20 = x$ (c) $\dfrac{x^2 + 10x + 25}{x + 1} = 0$.

Solution (a) The equation $x^2 - 2x - 15 = 0$ may be written in the form

$$(x - 5)(x + 3) = 0.$$

The product of two numbers is zero only if one or the other of the numbers (or both) is zero. Hence

$$x - 5 = 0 \quad \text{or} \quad x + 3 = 0.$$

That is,

$$x = 5 \quad \text{or} \quad x = -3.$$

(b) First we must rewrite the equation $x^2 - 20 = x$ in the form $ax^2 + bx + c = 0$; that is,

$$x^2 - x - 20 = 0$$
$$(x - 5)(x + 4) = 0.$$

We conclude that

$$x - 5 = 0 \quad \text{or} \quad x + 4 = 0;$$

that is,

$$x = 5 \quad \text{or} \quad x = -4.$$

(c) A rational function will be zero only if the numerator is zero. Thus

$$x^2 + 10x + 25 = 0,$$

$$(x + 5)^2 = 0,$$

$$x + 5 = 0; \quad \text{that is,} \quad x = -5.$$

Since the denominator is not 0 at $x = -5$, we conclude that $x = -5$ is a zero.

Derivation of the Quadratic Formula

$$ax^2 + bx + c = 0$$

$$ax^2 + bx = -c$$

$$4a^2x^2 + 4abx = -4ac \qquad \text{(both sides were multiplied by } 4a),$$

$$4a^2x^2 + 4abx + b^2 = b^2 - 4ac \quad (b^2 \text{ added to both sides}).$$

Now note that $4a^2x^2 + 4abx + b^2 = (2ax + b)^2$. To check this, simply multiply out the right-hand side! Therefore

$$(2ax + b)^2 = b^2 - 4ac$$

$$2ax + b = \pm\sqrt{b^2 - 4ac}$$

$$2ax = -b \pm \sqrt{b^2 - 4ac}$$

$$x = \frac{-b \pm \sqrt{b^2 - 4ac}}{2a}.$$

PRACTICE PROBLEMS 4

1. Solve the equation $x - 14/x = 5$.

2. Use the quadratic formula to solve $7x^2 - 35x + 35 = 0$.

EXERCISES 4

Use the quadratic formula to find the zeros of the functions in Exercises 1–6.

1. $f(x) = 2x^2 - 7x + 6$

2. $f(x) = 3x^2 + 2x - 1$

3. $f(x) = 4x^2 - 12x + 9$

4. $f(x) = \frac{1}{4}x^2 + x + 1$

5. $f(x) = -2x^2 + 3x - 4$

6. $f(x) = 11x^2 - 7x + 1$

Use the quadratic formula to solve the equations in Exercises 7–12.

7. $5x^2 - 4x - 1 = 0$

8. $x^2 - 4x + 5 = 0$

9. $15x^2 - 135x + 300 = 0$

10. $x^2 - \sqrt{2}x - \frac{5}{4} = 0$

11. $\frac{3}{2}x^2 - 6x + 5 = 0$

12. $9x^2 - 12x + 4 = 0$

Factor the polynomials in Exercises 13–24.

13. $x^2 + 8x + 15$

14. $x^2 - 10x + 16$

15. $x^2 - 16$

16. $x^2 - 1$

17. $3x^2 + 12x + 12$

18. $2x^2 - 12x + 18$

19. $30 - 4x - 2x^2$

20. $15 + 12x - 3x^2$

21. $3x - x^2$

22. $4x^2 - 1$

23. $6x - 2x^3$

24. $16x + 6x^2 - x^3$

Find the points of intersection of the pairs of curves in Exercises 25–32.

25. $y = 2x^2 - 5x - 6, y = 3x + 4$

26. $y = x^2 - 10x + 9, y = x - 9$

27. $y = x^2 - 4x + 4, y = 12 + 2x - x^2$

28. $y = 3x^2 + 9, y = 2x^2 - 5x + 3$

29. $y = x^3 - 3x^2 + x + 8, y = x^2 - 3x + 8$

30. $y = \frac{1}{2}x^3 - 2x^2, y = 2x$

31. $y = \frac{1}{2}x^3 + x^2 + 5, y = 3x^2 - \frac{1}{2}x + 5$

32. $y = 30x^3 - 3x^2, y = 16x^3 + 25x^2$

Solve the equations in Exercises 33–38.

33. $\dfrac{21}{x} - x = 4$

34. $x + \dfrac{2}{x - 6} = 3$

35. $x + \dfrac{14}{x + 4} = 5$

36. $1 = \dfrac{5}{x} + \dfrac{6}{x^2}$

37. $\dfrac{x^2 + 14x + 49}{x^2 + 1} = 0$

38. $\dfrac{x^2 - 8x + 16}{1 + \sqrt{x}} = 0$

SOLUTIONS TO PRACTICE PROBLEMS 4

1. Multiply both sides of the equation by x. Then

$$x^2 - 14 = 5x.$$

Now, take the $5x$ term to the left side of the equation and solve by factoring.

$$x^2 - 5x - 14 = 0$$

$$(x - 7)(x + 2) = 0$$

$$x = 7 \quad \text{or} \quad x = -2.$$

2. In this case, each coefficient is a multiple of 7. To simplify the arithmetic, we divide both sides of the equation by 7 before using the quadratic formula.

$$x^2 - 5x + 5 = 0$$

$$\sqrt{b^2 - 4ac} = \sqrt{(-5)^2 - 4(1)(5)} = \sqrt{5}$$

$$x = \frac{-b \pm \sqrt{b^2 - 4ac}}{2a} = \frac{5 \pm \sqrt{5}}{2 \cdot 1} = \frac{5}{2} \pm \frac{1}{2}\sqrt{5}$$

Chapter 0: CHECKLIST

☐ Function
☐ Value of function at x
☐ Domain of a function
☐ Graph of a function
☐ Vertical line test
☐ Graph of an equation
☐ Linear function
☐ Constant function
☐ x- and y-intercepts
☐ Quadratic function
☐ Polynomial and rational functions
☐ Power function
☐ Absolute value function
☐ Addition, subtraction, multiplication, and division of functions
☐ Composition of functions
☐ Rules for combining power functions
☐ Zero of a function
☐ Quadratic formula
☐ Factoring polynomials

Chapter 0: SUPPLEMENTARY EXERCISES

1. Let $f(x) = x^3 + 1/x$. Evaluate $f(1)$, $f(3)$, $f(-1)$, $f(-\frac{1}{2})$, $f(\sqrt{2})$.

2. Let $f(x) = 2x + 3x^2$. Evaluate $f(0)$, $f(-\frac{1}{4})$, $f(1/\sqrt{2})$.

3. Let $f(x) = x^2 - 2$. Evaluate $f(a - 2)$.

4. Let $f(x) = [1/(x + 1)] - x^2$. Evaluate $f(a + 1)$.

   Determine the domains of the following functions.

5. $f(x) = \dfrac{1}{x(x + 3)}$ 6. $f(x) = \sqrt{x - 1}$

7. $f(x) = \sqrt{x^2 + 1}$ 8. $f(x) = 1/\sqrt{3x}$

9. Is the point $(\frac{1}{2}, -\frac{3}{5})$ on the graph of the function $h(x) = (x^2 - 1)/(x^2 + 1)$?

10. Is the point $(1, -2)$ on the graph of the function $k(x) = x^2 + (2/x)$?

Factor the polynomials in Exercises 11–14.

11. $5x^3 + 15x^2 - 20x$ 12. $3x^2 - 3x - 60$

13. $18 + 3x - x^2$ 14. $x^5 - x^4 - 2x^3$

15. Find the zeros of the quadratic function $y = 5x^2 - 3x - 2$.

16. Find the zeros of the quadratic function $y = -2x^2 - x + 2$.

17. Find the points of intersection of the curves $y = 5x^2 - 3x - 2$ and $y = 2x - 1$.

18. Find the points of intersection of the curves $y = -x^2 + x + 1$ and $y = x - 5$.

Let $f(x) = x^2 - 2x$, $g(x) = 3x - 1$, $h(x) = \sqrt{x}$. Find the following functions.

19. $f(x) + g(x)$ 20. $f(x) - g(x)$ 21. $f(x)h(x)$

22. $f(x)g(x)$ 23. $f(x)/h(x)$ 24. $g(x)h(x)$

Let $f(x) = x/(x^2 - 1)$, $g(x) = (1 - x)/(1 + x)$, $h(x) = 2/(3x + 1)$. Express the following as rational functions.

25. $f(x) - g(x)$ 26. $f(x) - g(x + 1)$ 27. $g(x) - h(x)$

28. $f(x) + h(x)$ 29. $g(x) - h(x - 3)$ 30. $f(x) + g(x)$

Let $f(x) = x^2 - 2x + 4$, $g(x) = 1/x^2$, $h(x) = 1/(\sqrt{x} - 1)$. Determine the following functions.

31. $f(g(x))$ 32. $g(f(x))$ 33. $g(h(x))$

34. $h(g(x))$ 35. $f(h(x))$ 36. $h(f(x))$

37. Simplify $(81)^{3/4}$, $8^{5/3}$, $(.25)^{-1}$ 38. Simplify 5^{-2}, $(100)^{3/2}$, $(.001)^{1/3}$.

39. The population of a city is estimated to be $750 + 25t + .1t^2$ thousand people t years from the present. Ecologists estimate that the average level of carbon monoxide in the air above the city will be $1 + .4x$ ppm (parts per million) when the population is x thousand people. Express the carbon monoxide level as a function of the time t.

40. The revenue $R(x)$ (in thousands of dollars) a company receives from the sale of x thousand units is given by $R(x) = 5x - x^2$. The sales level x is in turn a function $f(d)$ of the number d of dollars spent on advertising, where

$$f(d) = 6[1 - 200/(d + 200)].$$

Express the revenue as a function of the amount spent on advertising.

□ **chapter 1**

THE DERIVATIVE

The *derivative* is a mathematical tool used to measure change. To illustrate the sort of change we have in mind, consider the following example. Suppose that a rumor spreads through a town of 10,000 people.* Denote by $N(t)$ the number of people who have heard the rumor after t days. We have tabulated the values for $N(t)$ corresponding to $t = 0, 1, \ldots, 10$ in the chart in Fig. 1. Thus, for example, the rumor starts at $t = 0$ with one person [$N(0) = 1$]; after one day, the number having heard the rumor increases to 6; after two days, to 40; and so forth. Each day the number of people who have heard the rumor increases. We may describe the increase geometrically by plotting the points corresponding to $t = 0, 1, \ldots, 10$ and drawing a smooth curve through them. (See Fig. 1.)

---

* The graph used in this example is derived from a mathematical model used by sociologists. See J. Coleman, *An Introduction to Mathematical Sociology* (London: Collier-Macmillan Ltd., 1964), p. 43.

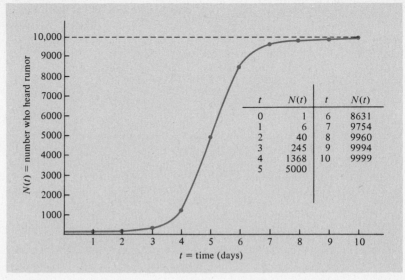

FIGURE 1

Let us describe the change that is taking place in this example. Initially, the number of *new* people to hear the rumor each day is rather small (five the first day, 34 the second, 205 the third). As we shall see later, this situation is reflected in the fact that the graph is not very steep at $t = 1, 2, 3$. (It is almost flat.) On the fourth day, 1123 additional people hear the rumor; on the fifth day, 3632. The more rapid spread of the rumor on the fourth and fifth days is reflected in the increasing steepness of the graph at $t = 4$ and $t = 5$. After the fifth day the rumor spreads less rapidly, since there are fewer people to whom it can spread. Correspondingly, after $t = 5$, the graph becomes less steep. Finally, at $t = 9$ and $t = 10$, the graph is practically flat, indicating little change in the number who heard the rumor.

In the example above, there is a correlation between the rate at which $N(t)$ is changing and the steepness of the graph. This illustrates one of the fundamental ideas of calculus, which may be put roughly as follows: Measure rates of change in terms of steepness of graphs. This chapter is devoted to the *derivative*, which provides a numerical measure of the steepness of a curve at a particular point. By studying the derivative we will be able to deal numerically with the rates of change which occur in applied problems.

1.1. The Slope of a Straight Line

As we shall see later on, the study of straight lines is crucial for the study of the steepness of curves. So this section is devoted to a discussion of the geometric and algebraic properties of straight lines.

A nonvertical line* L has an equation of the form $y = mx + b$. The number m is called the *slope* of L and the point $(0, b)$ is called the *y-intercept*.

---

* Throughout we confine our discussion of straight lines to those that are not vertical.

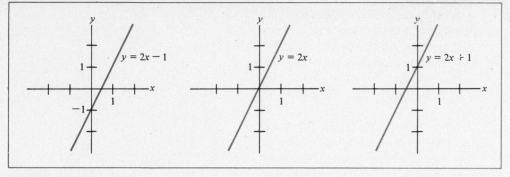

FIGURE 1

If we set $x = 0$, we see that $y = b$, so that $(0, b)$ is on the line L. Thus the y-intercept tells us where the line L crosses the y-axis. The slope measures the steepness of the line. In Fig. 1 we give three examples of lines with slope $m = 2$. In Fig. 2 we give three examples of lines with slope $m = -2$.

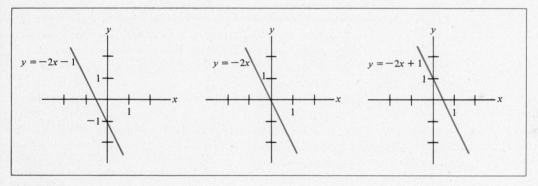

FIGURE 2

FIGURE 3

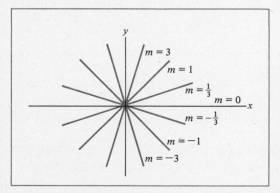

To conceptualize the meaning of slope, think of walking along a line from left to right. On lines of positive slope we will be walking uphill; the greater the slope, the steeper the ascent. On lines of negative slope we will be walking downhill; the more negative the slope, the steeper the descent. Walking on lines of zero slope corresponds to walking on level ground. In Fig. 3 we have graphed lines with $m = 3, 1, \frac{1}{3}, 0, -\frac{1}{3}, -1, -3$, all having $b = 0$. The reader can readily verify our conceptualization of slope for these lines.

The slope and y-intercept of a straight line often have physical interpretations, as the following two examples illustrate.

EXAMPLE 1 A manufacturer finds that the total cost of producing x units of a commodity is $2x + 1000$ dollars. What is the economic significance of the y-intercept and the slope of the line $y = 2x + 1000$? (See Fig. 4.)

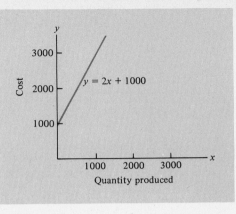

Solution The y-intercept is $(0, 1000)$. The number 1000 represents the *fixed costs* of the manufacturer—those overhead costs, such as rent and insurance, that must be paid no matter how many items are produced. Thus when $x = 0$ (no units produced), the cost is still $y = 1000$ dollars.

FIGURE 4

The slope of the line is 2. This number represents the cost of producing each additional unit. To see this, we can calculate some typical costs.

Quantity produced	Total cost
$x = 1500$	$y = 2(1500) + 1000 = 4000$
$x = 1501$	$y = 2(1501) + 1000 = 4002$
$x = 1502$	$y = 2(1502) + 1000 = 4004$

Each time x is increased by 1, the value of y increases by 2.

EXAMPLE 2 An apartment complex has a storage tank to hold its heating oil. The tank was filled on January 1, but no more deliveries of oil will be made until some time in March. Let t denote the number of days after January 1 and let y denote the number of gallons of fuel oil in the tank. Current records of the apartment complex show that y and t are related approximately by the equation

$$y = 30,000 - 400t. \qquad (1)$$

What interpretation can be given to the y-intercept and slope of this line? (See Fig. 5.)

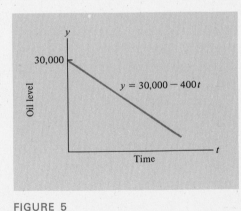

FIGURE 5

Solution The y-intercept is $(0, 30,000)$. This value of y corresponds to $t = 0$, so there were 30,000 gallons of oil in the tank on January 1. Let us examine how fast the oil is removed from the tank.

Days after January 1	Gallons of oil in the tank
$t = 0$	$y = 30{,}000 - 400(0) = 30{,}000$
$t = 1$	$y = 30{,}000 - 400(1) = 29{,}600$
$t = 2$	$y = 30{,}000 - 400(2) = 29{,}200$
$t = 3$	$y = 30{,}000 - 400(3) = 28{,}800$
$\vdots$	$\vdots$

The oil level in the tank drops by 400 gallons each day; that is, the oil is being used at the rate of 400 gallons per day. The slope of the line (1) is -400. Thus the slope gives the rate at which the oil is being used. The negative sign on the -400 indicates that the oil level is decreasing rather than increasing.

Properties of the Slope of a Line Let us now examine several useful properties of the slope of a straight line. At the end of the section we shall explain why these properties are valid.

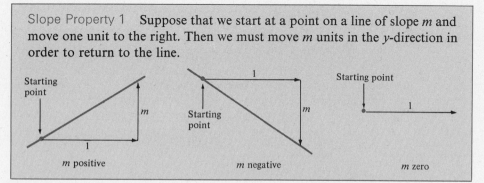

Slope Property 1   Suppose that we start at a point on a line of slope m and move one unit to the right. Then we must move m units in the y-direction in order to return to the line.

m positive m negative m zero

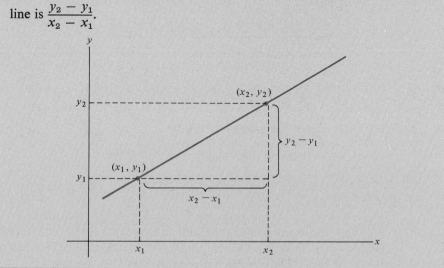

Slope Property 2   We can compute the slope of a line by knowing two points on the line. If (x_1, y_1) and (x_2, y_2) are on the line, then the slope of the line is $\dfrac{y_2 - y_1}{x_2 - x_1}$.

As we move from (x_1, y_1) to (x_2, y_2), the change in the y-coordinates is $y_2 - y_1$ and the change in the x-coordinates is $x_2 - x_1$. Thus the slope of the line is simply the ratio of the change in y to the change in x. This interpretation of the slope is also illustrated by the diagrams given above for Slope Property 1. The slope of a line equals the change in y per unit change in x. We say that the slope gives the *rate of change of y with respect to x*.

Slope Property 3   The equation of a line can be obtained if we know the slope and one point on the line. If the slope is m and if (x_1, y_1) is on the line, then the equation of the line is

$$y - y_1 = m(x - x_1).$$

This equation is called the *point-slope form* of the equation of the line.

Slope Property 4   Distinct lines of the same slope are parallel. Conversely, if two lines are parallel, they have the same slope.

Slope Property 5   When two lines are perpendicular, the product of their slopes is -1.

Calculations Involving Slope of a Line

EXAMPLE 3   Find the slope and the y-intercept of the line whose equation is $2x + 3y = 6$.

Solution   We solve for y in terms of x.

$$3y = -2x + 6$$
$$y = -\tfrac{2}{3}x + 2.$$

The slope is $-\tfrac{2}{3}$ and the y-intercept is $(0, 2)$.

EXAMPLE 4   Sketch the graph of the line

(a) passing through $(2, -1)$ with slope 3,

(b) passing through $(2, 3)$ with slope $-\tfrac{1}{2}$.

Solution   We use Slope Property 1. (See Fig. 6.) In each case, we begin at the given point, move one unit to the right, and then m units in the y-direction (upward for positive m, downward for negative m). The new point reached will also be on the line. Draw the straight line through these two points.

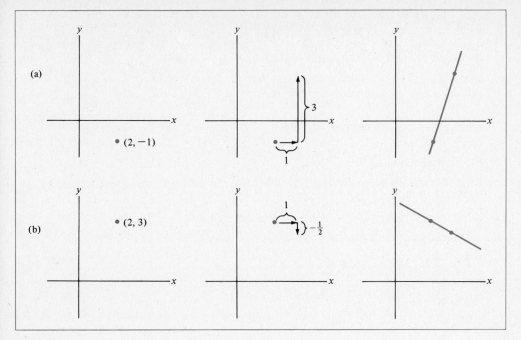

FIGURE 6

EXAMPLE 5   Find the slope of the line passing through the points $(6, -2)$ and $(9, 4)$.

Solution   We apply Slope Property 2 with $(x_1, y_1) = (6, -2)$ and $(x_2, y_2) = (9, 4)$. Then

$$\frac{y_2 - y_1}{x_2 - x_1} = \frac{4 - (-2)}{9 - 6} = \frac{6}{3} = 2.$$

Thus the slope is 2. [We would have reached the same answer if we had let $(x_1, y_1) = (9, 4)$ and $(x_2, y_2) = (6, -2)$.] The slope is just the difference of the y-coordinates divided by the difference of the x-coordinates, with each difference formed in the same order.

EXAMPLE 6   Find an equation of the line passing through $(-1, 2)$ with slope 3.

Solution   We let $(x_1, y_1) = (-1, 2)$ and $m = 3$, and we use Slope Property 3. The equation of the line is

$$y - 2 = 3[x - (-1)]$$

or

$$y - 2 = 3(x + 1).$$

If desired, this equation can be put into the form $y = mx + b$,

$$y - 2 = 3(x + 1) = 3x + 3$$
$$y = 3x + 5.$$

EXAMPLE 7   Find an equation of the line passing through the points $(1, -2)$ and $(2, -3)$.

Solution   By Slope Property 2, the slope of the line is

$$\frac{-3 - (-2)}{2 - 1} = \frac{-3 + 2}{1} = -1.$$

Since $(1, -2)$ is on the line, we can use Property 3 to get the equation of the line:

$$y - (-2) = (-1)(x - 1)$$
$$y + 2 = -x + 1$$
$$y = -x - 1.$$

EXAMPLE 8   Find an equation of the line passing through $(5, 3)$ parallel to the line $2x + 5y = 7$.

Solution   We first find the slope of the line $2x + 5y = 7$.

$$2x + 5y = 7$$
$$5y = 7 - 2x$$
$$y = -\tfrac{2}{5}x + \tfrac{7}{5}.$$

The slope of this line is $-\tfrac{2}{5}$. By Slope Property 4, any line parallel to this line will also have slope $-\tfrac{2}{5}$. Using the given point $(5, 3)$ and Slope Property 3, we get the desired equation

$$y - 3 = -\tfrac{2}{5}(x - 5).$$

This equation can also be written as

$$y = -\tfrac{2}{5}x + 5.$$

EXAMPLE 9   The equation $y = \sqrt{1 - x^2}$ describes a semicircle. (See Fig. 7.) Find the slope of the tangent line l to this semicircle at the point where $x = .6$.

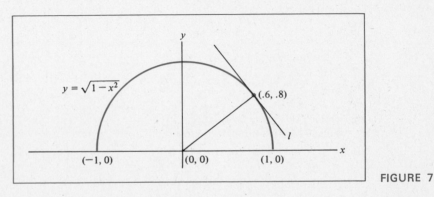

FIGURE 7

Solution First note that if $x = .6$, then $y = \sqrt{1 - (.6)^2} = \sqrt{.64} = .8$. Next, let us compute the slope of the radius from $(0, 0)$ to $(.6, .8)$.

$$[\text{Slope of radius}] = \frac{.8 - 0}{.6 - 0} = \frac{8}{6} = \frac{4}{3}.$$

From geometry we know that the tangent line l is perpendicular to the radius. Therefore, by Property 5, the slope of the radius times the slope of the tangent line is -1. So if we let m denote the slope of the tangent line, then

$$\tfrac{4}{3}m = -1$$

$$m = -\tfrac{3}{4}.$$

Thus l has slope $-\tfrac{3}{4}$.

Remark: By using a calculation identical to that used in Example 9, one can deduce that the slope of the tangent line to the semicircle at the point where $x = a$ is equal to $-a/\sqrt{1 - a^2}$. (Here we must take $a \neq \pm 1$ to avoid vertical lines.) (See Fig. 8.)

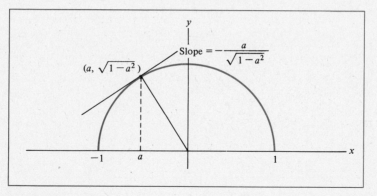

FIGURE 8

Verification of the Properties of Slope It is convenient to verify the properties of slope in the order 2, 3, 1. Verifications of properties 4 and 5 are outlined in Exercises 46 and 47.

Verification of Property 2 Suppose that the equation of the line is $y = mx + b$. Then, since (x_2, y_2) is on the line, we have $y_2 = mx_2 + b$. Similarly, since (x_1, y_1) is on the line, we have $y_1 = mx_1 + b$. Subtracting, we see that

$$y_2 - y_1 = mx_2 - mx_1$$

$$y_2 - y_1 = m(x_2 - x_1),$$

so that

$$\frac{y_2 - y_1}{x_2 - x_1} = m.$$

This is the formula for the slope m stated in Property 2.

Verification of Property 3 The equation $y - y_1 = m(x - x_1)$ may be put into the form

$$y = mx + \underbrace{(y_1 - mx_1)}_{b}.\qquad (2)$$

This is the equation of a line with slope m. Furthermore, the point (x_1, y_1) is on this line, because the equation (2) remains true when we plug in x_1 for x and y_1 for y. Thus the equation $y - y_1 = m(x - x_1)$ corresponds to the line of slope m passing through (x_1, y_1).

Verification of Property 1 Let $P = (x_1, y_1)$ be on a line l and let y_2 be the coordinate of the point on l obtained by moving P one unit to the right and then moving vertically to return to the line. (See Fig. 9.) By Property 2, the slope m of this line l satisfies

$$m = \frac{y_2 - y_1}{(x_1 + 1) - x_1} = \frac{y_2 - y_1}{1} = y_2 - y_1.$$

Thus the difference of the y-coordinates of R and Q is m. This is Property 1.

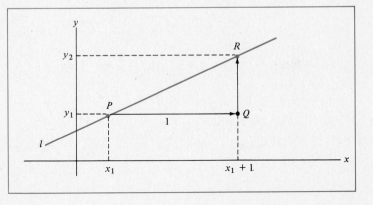

FIGURE 9

PRACTICE PROBLEMS 1

Find the slopes of the following lines.

1. The line whose equation is $x = 3y - 7$.

2. The line passing through the points (2, 5) and (2, 8).

3. The line that is tangent to the semicircle $y = \sqrt{1 - x^2}$ at the point $(\frac{8}{17}, \frac{15}{17})$. (Use the Remark following Example 9.)

EXERCISES 1

Find the slopes of the following lines.

1. $y = 2 - 5x$ 2. $y = -5x$ 3. $y = 2$

4. $y = \frac{1}{3}(x + 2)$ 5. $y = (2x - 1)/7$ 6. $y = \frac{1}{4}$

7. $2x + 3y = 6$ 8. $x - y = 2$

Find the equations of the following lines.

9. Slope is 3; y-intercept is $(0, -1)$ 10. Slope is $-\frac{1}{2}$; y-intercept is $(0, 0)$

11. Slope is 1; $(1, 2)$ on line   12. Slope is $-\frac{1}{3}$; $(6, -2)$ on line

13. Slope is -7; $(5, 0)$ on line   14. Slope is $\frac{1}{2}$; $(2, -3)$ on line

15. Slope is 0; $(7, 4)$ on line   16. Slope is $-\frac{2}{5}$; $(0, 5)$ on line

17. $(2, 1)$ and $(4, 2)$ on line   18. $(5, -3)$ and $(-1, 3)$ on line

19. $(0, 0)$ and $(1, -2)$ on line   20. $(2, -1)$ and $(3, -1)$ on line

21. Parallel to $y = -2x + 1$; $(\frac{1}{2}, 5)$ on line.

22. Parallel to $3x + y = 7$; $(-1, -1)$ on line.

23. Parallel to $3x - 6y = 1$; $(1, 0)$ on line.

24. Parallel to $5x + 2y = -4$; $(0, 17)$ on line.

In each of Exercises 25 to 28 we specify a line by giving the slope and one point on the line. We give the first coordinate of some points on the line. Without deriving the equation of the line, find the second coordinate of each of the points.

25. Slope is 2, $(1, 3)$ on line; $(2, \;)$; $(0, \;)$; $(-1, \;)$.

26. Slope is -3, $(2, 2)$ on line; $(3, \;)$; $(4, \;)$; $(1, \;)$.

27. Slope is $-\frac{1}{4}$, $(-1, -1)$ on line; $(0, \;)$; $(1, \;)$; $(-2, \;)$.

28. Slope is $\frac{1}{3}$, $(-5, 2)$ on line; $(-4, \;)$; $(-3, \;)$; $(-2, \;)$.

For each pair of lines in the figures below, determine the one with the greater slope.

29.   30.

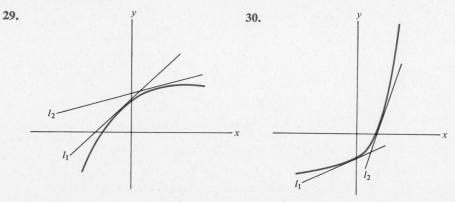

Find the equation and sketch the graph of the following lines.

31. With slope -2 and y-intercept $(0, -1)$. **32.** With slope $\frac{1}{3}$ and y-intercept $(0, 1)$.

33. Through $(2, 0)$ with slope $\frac{4}{5}$. **34.** Through $(-1, 3)$ with slope 0.

Find the slope of the tangent line to the semicircle $y = \sqrt{1 - x^2}$ at the given point. (Use the Remark following Example 9.)

35. $(\frac{4}{5}, \frac{3}{5})$ **36.** $(-\frac{5}{13}, \frac{12}{13})$ **37.** $(-\frac{12}{13}, \frac{5}{13})$ **38.** $(-.6, .8)$

Write the equation of the tangent line to the semicircle $y = \sqrt{1 - x^2}$ at the given point. (In each case use a result from Exercises 35 to 38.)

39. $(\frac{4}{5}, \frac{3}{5})$ **40.** $(-\frac{5}{13}, \frac{12}{13})$ **41.** $(-\frac{12}{13}, \frac{5}{13})$ **42.** $(-.6, .8)$

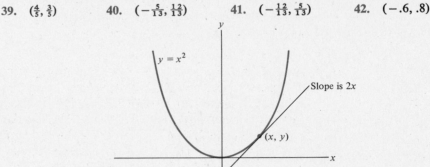

Later in the chapter we shall show that the tangent line to the parabola $y = x^2$ passing through the point with coordinates (x, y) has slope $2x$. Thus the slopes of the tangent lines at the points $(1, 1), (0, 0)$, $(-\frac{1}{2}, \frac{1}{4})$ are, respectively, $2, 0$, and -1. Find the equation of the tangent lines through each of the following points.

43. $(1, 1)$ **44.** $(0, 0)$ **45.** $(-\frac{1}{2}, \frac{1}{4})$

46. Prove Property 4 of straight lines. [Hint: If $y = mx + b$ and $y = m'x + b'$ are two lines, then they have a point in common if and only if the equation $mx + b = m'x + b'$ has a solution x.]

47. Prove Property 5 of straight lines. [Hint: Without loss of generality, assume that both lines pass through the origin. Use Slope Property 1 and the Pythagorean theorem. See the accompanying figure.]

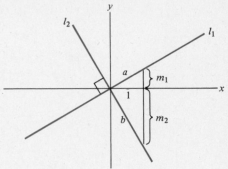

SOLUTIONS TO PRACTICE PROBLEMS 1

1. We solve for y in terms of x.
$$y = \tfrac{1}{3}x + \tfrac{7}{3}.$$
The slope of the line is the coefficient of x, that is, $\frac{1}{3}$.

2. The line passing through these two points is a vertical line; therefore its slope is undefined.

3. We are asked for the slope of the tangent line at the point for which $x = \frac{8}{17}$. Therefore, substitute $\frac{8}{17}$ for a in the formula $-a/\sqrt{1 - a^2}$. (Since $y = \frac{15}{17}$ when $x = \frac{8}{17}$, we know that $\sqrt{1 - (\frac{8}{17})^2} = \frac{15}{17}$.) We obtain

$$[\text{slope at } (\tfrac{8}{17}, \tfrac{15}{17})] = \frac{-\frac{8}{17}}{\sqrt{1 - (\frac{8}{17})^2}} = \frac{-\frac{8}{17}}{\frac{15}{17}} = -\frac{8}{15}.$$

1.2. The Slope of a Curve at a Point

In order to extend the concept of slope from straight lines to more general curves, we must first discuss the notion of the tangent line to a curve at a point.

We have a clear idea of what is meant by the tangent line to a circle at a point P. It is the straight line that touches the circle at just the one point P. Let us focus on the region near P, designated by the dotted rectangle shown in Fig. 1. The enlarged portion of the circle looks almost straight, and the straight line that it resembles is the tangent line. Further enlargements would make the circle near P look even straighter and have an even closer resemblance to the tangent line. In this sense, the tangent line to the circle at the point P is the straight line through P that best approximates the circle near P. In particular, the tangent line at P reflects the steepness of the circle at P. Thus it seems reasonable to define the *slope* of the circle at P to be the slope of the tangent line at P.

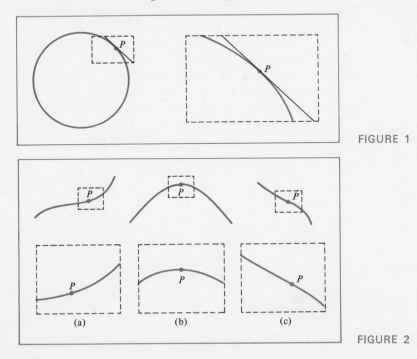

FIGURE 1

FIGURE 2

Similar reasoning leads us to a suitable definition of slope for an arbitrary curve at a point P. Consider the three curves drawn in Fig. 2. We have drawn an enlarged version of the dotted box around each point P. Notice that the portion of each curve lying in the boxed region looks almost straight. In fact, if we took the boxes to be further enlargements, the curves would appear even straighter. This observation suggests that we can approximate each curve near P by a straight line, as in Fig. 3. Let us call the straight line that best approximates a curve near a point P the *tangent line* to the curve at P. Then we define the *slope of a curve at a point P* to be the slope of the tangent line to the curve at P.

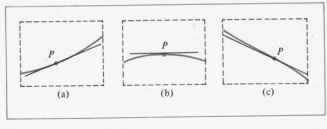

FIGURE 3

The portion of the curve near P can be, at least to within an approximation, replaced by the tangent line. Therefore, the slope of the curve at P—that is, the slope of the tangent line at P—measures the rate of increase or decrease of the curve as it passes through P.

EXAMPLE 1 In Example 2 of Section 1 the apartment complex used approximately 400 gallons of oil per day. Suppose that we keep a continuous record of the oil level in the storage tank. The graph for a typical two-day period appears in Fig. 4. What is the physical significance of the slope of the graph at the point P?

Solution The curve near P is closely approximated by its tangent line. So think of the curve as replaced by its tangent line near P. Then the slope at P is just the rate of decrease of the oil at 7 A.M. on March 5.

FIGURE 4

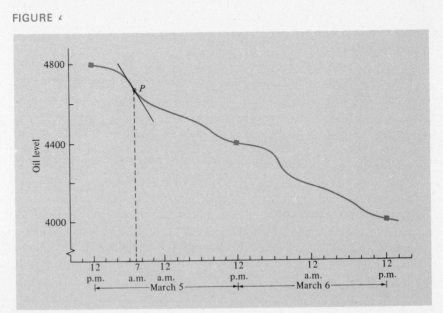

Notice that during the entire day of March 5 the graph in Fig. 4 seems to be the steepest at 7 A.M. That is, the oil level is falling the fastest at that time. This corresponds to the fact that most people awake around 7 A.M., turn up their thermostats, take showers, and so on. Example 1 provides a typical illustration of the manner in which slopes can be interpreted as rates of change. We shall return to this idea in Section 8.

In calculus, we can usually compute slopes by using formulas. For instance, in Section 3 we shall show that the tangent line to the graph of $y = x^2$ at the point $(1, 1)$ has slope 2; the tangent line at $(3, 9)$ has slope 6; and the tangent line at $(-\frac{5}{2}, \frac{25}{4})$ has slope -5. The various tangent lines are shown in Fig. 5. Notice that the slope at each point is two times the x-coordinate of the point. This is a general fact. In Section 3 we shall derive this simple formula (see Fig. 6.):

$$[\text{slope of the graph of } y = x^2 \text{ at the point } (x, y)] = 2x.$$

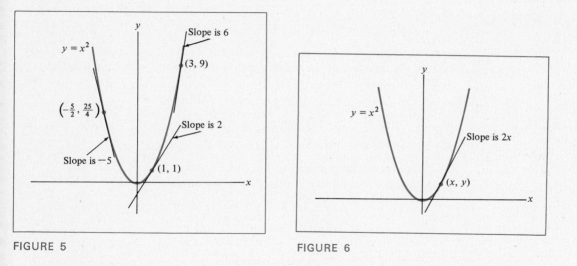

FIGURE 5

FIGURE 6

EXAMPLE 2   (a) What is the slope of the graph of $y = x^2$ at the point $(\frac{3}{4}, \frac{9}{16})$?

(b) Write the equation of the tangent line to the graph of $y = x^2$ at the point $(\frac{3}{4}, \frac{9}{16})$.

Solution   (a) The x-coordinate of $(\frac{3}{4}, \frac{9}{16})$ is $\frac{3}{4}$, so the slope of $y = x^2$ at this point is $2(\frac{3}{4}) = \frac{3}{2}$.

(b) We shall write the equation of the tangent line in the point-slope form. The point is $(\frac{3}{4}, \frac{9}{16})$, and the slope is $\frac{3}{2}$ by part (a). Hence the equation is

$$y - \tfrac{9}{16} = \tfrac{3}{2}(x - \tfrac{3}{4}).$$

1. Refer to the accompanying graph.

    (a) What is the slope of the curve at $(3, 4)$?

    (b) What is the equation of the tangent line at the point where $x = 3$?

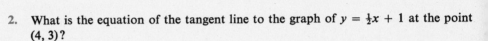

2. What is the equation of the tangent line to the graph of $y = \frac{1}{4}x + 1$ at the point $(4, 3)$?

EXERCISES 2

Trace the curves in Exercises 1–6 onto another piece of paper and sketch the tangent line in each case at the designated point P.

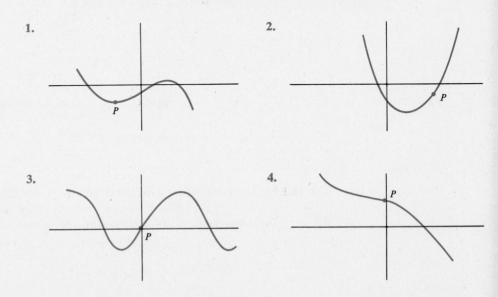

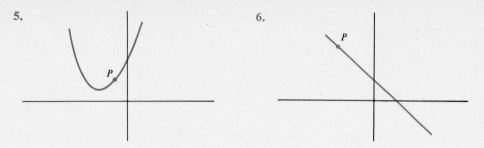

Estimate the slope of each of the following curves at the designated point P.

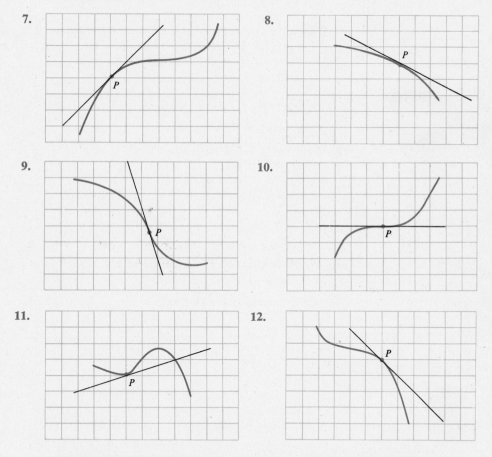

In Exercises 13–15 find the slope of the tangent line at the graph of $y = x^2$ at the point indicated, and then write the corresponding equation of the tangent line.

13. $(-2, 4)$ 14. $(-.4, .16)$ 15. $(\frac{4}{3}, \frac{16}{9})$

16. Find the slope of the tangent line to the graph of $y = x^2$ at the point where $x = -\frac{1}{2}$.

17. Write the equation of the tangent line to the graph of $y = x^2$ at the point where $x = 1.5$.

18. Write the equation of the tangent line to the graph of $y = x^2$ at the point where $x = .6$.

19. Find the point on the graph of $y = x^2$ where the curve has slope $\frac{5}{3}$.

20. Find the point on the graph of $y = x^2$ where the curve has slope -4.

21. Find the point on the graph of $y = x^2$ where the tangent line is parallel to the line $x + 2y = 4$.

22. Find the point on the graph of $y = x^2$ where the tangent line is parallel to the line $3x - y = 2$.

   In the next section we shall see that the tangent line to the graph of $y = x^3$ at the point (x, y) has slope $3x^2$. Using this result, find the slope of the curve at the following points.

23. $(2, 8)$

24. $(\frac{3}{2}, \frac{27}{8})$

25. $(-\frac{1}{2}, -\frac{1}{8})$

26. Find the slope of the curve $y = x^3$ at the point where $x = \frac{1}{4}$.

27. Write the equation of the line tangent to the graph of $y = x^3$ at the point where $x = -1$.

28. Write the equation of the line tangent to the graph of $y = x^3$ at the point where $x = \frac{1}{2}$.

SOLUTIONS TO PRACTICE PROBLEMS 2

1. (a) The slope of the curve at the point $(3, 4)$ is, by definition, the slope of the tangent line at $(3, 4)$. Note that the point $(4, 6)$ is also on the line. Therefore, the slope is

$$\frac{6 - 4}{4 - 3} = \frac{2}{1} = 2.$$

   (b) Use the point-slope formula. The equation of the line passing through the point $(3, 4)$ and having slope 2 is

$$y - 4 = 2(x - 3)$$

   or

$$y = 2x - 2.$$

2. The tangent line at (4, 3) is, by definition, the line that best approximates the curve at (4, 3). Since the "curve" in this case is itself a line, the curve and its tangent line at (4, 3) (and at every other point) must be the same. Therefore, the equation is $y = \frac{1}{2}x + 1$.

1.3. The Derivative

Suppose that a curve is the graph of a function $f(x)$. It is usually possible to obtain a formula that gives the slope of the curve $y = f(x)$ at any point. This slope formula is called the *derivative* of $f(x)$ and is written $f'(x)$. For each value of x, $f'(x)$ gives the slope of the curve $y = f(x)$ at the point with first coordinate x.* (See Fig. 1.)

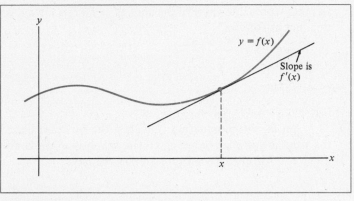

FIGURE 1

The process of computing $f'(x)$ for a given function $f(x)$ is called *differentiation*.

As we shall see later, the concept of a derivative occurs frequently in applications. In economics, the derivatives are often described by the adjective "marginal." For instance, if $C(x)$ is a cost function (the cost of producing x units of a commodity), then the derivative $C'(x)$ is called the *marginal cost function*. The derivative $P'(x)$ of a profit function $P(x)$ is called the *marginal profit function*; the derivative of a revenue function is called the *marginal revenue function*; and so on. We shall discuss the economic meaning of these marginal concepts in Section 8.

For the remainder of this section as well as the next few sections we will concentrate on calculating derivatives.

The case of a linear function $f(x) = mx + b$ is particularly simple. The graph of $y = mx + b$ is a straight line L of slope m. The tangent line to L (at any point) is just L itself, and so the slope of the graph is m at every point. (See Fig. 2.)

---

* As we shall see, there are curves which do not have tangent lines at every point. At values of x corresponding to such points, the derivative $f'(x)$ is not defined. For the sake of the current discussion, which is designed to develop an intuitive feeling for the derivative, let us assume that the graph of $f(x)$ has a tangent line for each x in the domain of f.

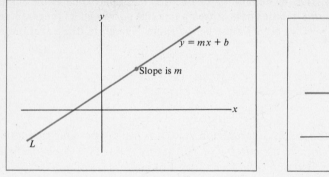

FIGURE 2

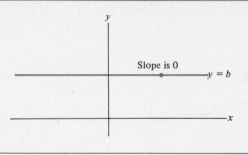

FIGURE 3

In other words, the value of the derivative $f'(x)$ is always equal to m. We summarize this fact as follows:

$$\text{If } f(x) = mx + b, \text{ then } f'(x) = m. \tag{1}$$

Set $m = 0$ in equation (1). Then the function becomes $f(x) = b$, which has the value b for each value of x. The graph is a horizontal line of slope 0, and so $f'(x) = 0$ for all x. [See Fig. 3.] Thus, we have:

$$\text{The derivative of a constant function } f(x) = b \text{ is } f'(x) = 0. \tag{2}$$

Next, consider the function $f(x) = x^2$. As we stated in Section 2 (and will prove at the end of this section), the slope of the graph of $y = x^2$ at the point (x, y) is equal to $2x$. That is, the value of the derivative $f'(x)$ is $2x$:

$$\text{If } f(x) = x^2, \text{ then } f'(x) = 2x. \tag{3}$$

In Exercises 23–28 of Section 2 we made use of the fact that the slope of the graph of $y = x^3$ at the point (x, y) is $3x^2$. This can be restated in terms of derivatives as follows:

$$\text{If } f(x) = x^3, \text{ then } f'(x) = 3x^2. \tag{4}$$

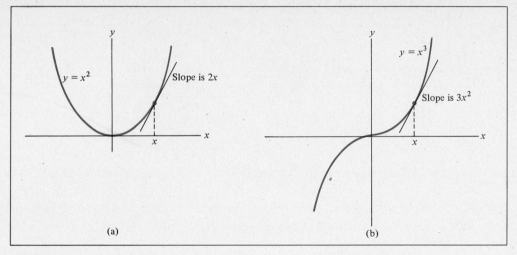

FIGURE 4

We should, at this stage at least, keep the geometric meaning of these formulas clearly in mind. Figure 4 shows the graphs of x^2 and x^3 together with the interpretations of formulas (3) and (4) in terms of slope.

One of the reasons calculus is so useful is that it provides general techniques which can be easily used to determine derivatives. One such general rule, which contains formulas (3) and (4) as special cases, is the so-called power rule.

> *Power Rule:* Let r be any number and let $f(x) = x^r$. Then $f'(x) = rx^{r-1}$.

Indeed, if $r = 2$, then $f(x) = x^2$ and $f'(x) = 2x^{2-1} = 2x$, which is formula (3). If $r = 3$, then $f(x) = x^3$ and $f'(x) = 3x^{3-1} = 3x^2$, which is (4).

EXAMPLE 1   Let $f(x) = \sqrt{x}$. What is $f'(x)$?

Solution   Recall that $\sqrt{x} = x^{1/2}$. We may apply the power rule with $r = \frac{1}{2}$.

$$f(x) = x^{1/2}$$

$$f'(x) = \tfrac{1}{2}x^{1/2-1}$$

$$= \tfrac{1}{2}x^{-1/2}$$

$$= \frac{1}{2} \cdot \frac{1}{x^{1/2}}$$

$$= \frac{1}{2\sqrt{x}}.$$

Another important special case of the power rule occurs for $r = -1$, corresponding to $f(x) = x^{-1}$. In this case, $f'(x) = (-1)x^{-1-1} = -x^{-2}$. However, since $x^{-1} = 1/x$ and $x^{-2} = 1/x^2$, the power rule for $r = -1$ may also be written as follows*:

$$\text{If } f(x) = \frac{1}{x}, \text{ then } f'(x) = -\frac{1}{x^2} \ (x \neq 0).$$

(5)

EXAMPLE 2 Find the slope of the curve $y = 1/x$ at $(2, \frac{1}{2})$.

Solution Set $f(x) = 1/x$. The point $(2, \frac{1}{2})$ corresponds to $x = 2$, so in order to find the slope at this point, we compute $f'(2)$. From formula (5) we find that

$$f'(2) = -\frac{1}{2^2} = -\frac{1}{4}.$$

Thus the slope of $y = 1/x$ at the point $(2, \frac{1}{2})$ is $-\frac{1}{4}$. (See Fig. 5.)

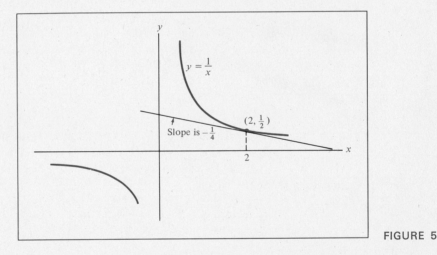

FIGURE 5

Warning: Do not confuse $f'(2)$, the value of the derivative at 2, with $f(2)$, the value of the y-coordinate at the point on the graph at which $x = 2$. In Example 2 we have $f'(2) = -\frac{1}{4}$, whereas $f(2) = \frac{1}{2}$. The number $f'(2)$ gives the *slope* of the graph at $x = 2$; the number $f(2)$ gives the *height* of the graph at $x = 2$.

The Secant-Line Calculation of the Derivative So far, we have said nothing about how to derive differentiation formulas, such as (3), (4), or (5). Let

---

* The formula gives $f'(x)$ for $x \neq 0$. The derivative of $f(x)$ is not defined at $x = 0$ since $f(x)$ itself is not defined there.

us remedy that omission now. The derivative gives the slope of the tangent line, so we must describe a procedure for calculating that slope.*

The fundamental idea for calculating the slope of the tangent line at a point P is to approximate the tangent line very closely by *secant lines*. A secant line at P is a straight line passing through P and a nearby point Q on the curve. (See Fig. 6.) By taking Q very close to P, we can make the slope of the secant line approximate the slope of the tangent line to any desired degree of accuracy. Let us see what this amounts to in terms of calculations.

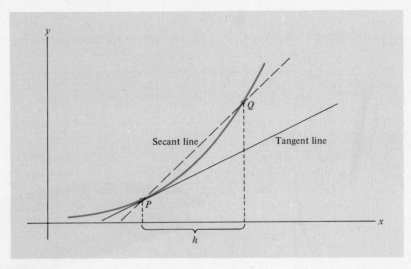

FIGURE 6

Suppose that the point P is $(x, f(x))$. Suppose also that Q is h horizontal units away from P. Then Q has x-coordinate $x + h$ and y-coordinate $f(x + h)$. The slope of the secant line through the points $P = (x, f(x))$ and $Q = (x + h, f(x + h))$ is simply

$$[\text{slope of secant line}] = \frac{f(x + h) - f(x)}{(x + h) - x} = \frac{f(x + h) - f(x)}{h}.$$

(See Fig. 7.)

In order to move Q close to P along the curve, we let h approach zero. Then the secant line approaches the tangent line, and so

[slope of secant line]   approaches   [slope of tangent line];

that is,

$$\frac{f(x + h) - f(x)}{h} \quad \text{approaches} \quad f'(x).$$

---

* The following discussion is designed to develop geometric intuition for the derivative. It will also provide the basis for a formal definition of the derivative in terms of limits in Sections 4 and 5.

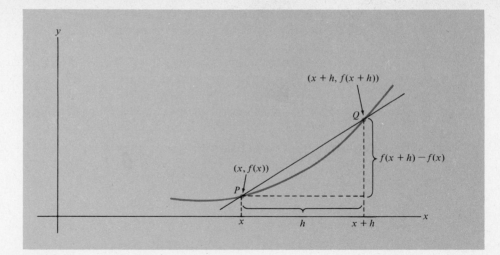

FIGURE 7

Since we can make the secant line as close to the tangent line as we wish, by taking h sufficiently small, the quantity $[f(x + h) - f(x)]/h$ can be made to approximate $f'(x)$ to any desired degree of accuracy. And we arrive at the following method to compute the derivative $f'(x)$.

> To calculate $f'(x)$:
>
> 1. First calculate $\dfrac{f(x + h) - f(x)}{h}$.
>
> 2. Then let h approach zero.
>
> 3. The quantity $\dfrac{f(x + h) - f(x)}{h}$ will approach $f'(x)$.

Let us use this method to verify the differentiation formulas (3) and (5), which as we saw were special cases of the power rule for $r = 2$ and $r = -1$, respectively.

Verification of the Power Rule for r = 2   Here $f(x) = x^2$, so that the slope of the secant line is

$$\frac{f(x + h) - f(x)}{h} = \frac{(x + h)^2 - x^2}{h}.$$

However, by multiplying out, we have $(x + h)^2 = x^2 + 2xh + h^2$, so that

$$\frac{f(x + h) - f(x)}{h} = \frac{\cancel{x^2} + 2xh + h^2 - \cancel{x^2}}{h}$$

$$= 2x + h.$$

As h approaches zero (that is, as the secant line approaches the tangent line), the quantity $2x + h$ approaches $2x$. Thus we have

$$f'(x) = 2x,$$

which is formula (3).

Verification of the Power Rule for r = −1 Here $f(x) = x^{-1} = 1/x$, so that the slope of the secant line is

$$\frac{f(x + h) - f(x)}{h} = \frac{1}{h}\left[\frac{1}{x + h} - \frac{1}{x}\right]$$

$$= \frac{1}{h}\left[\frac{x - (x + h)}{(x + h)x}\right]$$

$$= \frac{1}{h}\left[\frac{-h}{(x + h)x}\right]$$

$$= -\frac{1}{x(x + h)}.$$

As h approaches zero, $-\dfrac{1}{x(x + h)}$ approaches $-\dfrac{1}{x^2}$. Hence

$$f'(x) = -\frac{1}{x^2},$$

which is formula (5).

Similar arguments can be used to verify the power rule for other values of r. The cases $r = 3$ and $r = \frac{1}{2}$ are outlined in Exercises 45 and 46, respectively.

Notation The operation of forming a derivative $f'(x)$ from a function $f(x)$ is also indicated by the symbol $\dfrac{d}{dx}$ (read: the derivative with respect to x). Thus

$$\frac{d}{dx} f(x) = f'(x).$$

For example,

$$\frac{d}{dx}(x^6) = 6x^5, \qquad \frac{d}{dx}(x^{5/3}) = \frac{5}{3}x^{2/3}, \qquad \frac{d}{dx}\left(\frac{1}{x}\right) = -\frac{1}{x^2}.$$

When working with an equation of the form $y = f(x)$, we often write $\dfrac{dy}{dx}$ as a symbol for the derivative $f'(x)$. For example, if $y = x^6$, we may write

$$\frac{dy}{dx} = 6x^5.$$

PRACTICE PROBLEMS 3

1. Consider the curve $y = f(x)$ in the accompanying sketch.

    (a) Find $f(5)$.

    (b) Find $f'(5)$.

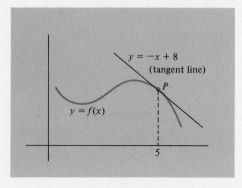

2. Let $f(x) = 1/x^4$.

    (a) Find its derivative.

    (b) Find $f'(2)$.

EXERCISES 3

Find the derivatives of the following functions.

1. $f(x) = 2x - 5$ **2.** $f(x) = 3 - \frac{1}{2}x$ **3.** $f(x) = x^8$

4. $f(x) = x^{75}$ **5.** $f(x) = x^{5/2}$ **6.** $f(x) = x^{4/3}$

7. $f(x) = x^{1/3}$ **8.** $f(x) = x^{3/4}$ **9.** $f(x) = x^{-2}$

10. $f(x) = 5$ **11.** $f(x) = x^{-1/4}$ **12.** $f(x) = x^{-3}$

13. $f(x) = \frac{3}{4}$ **14.** $f(x) = x^{-1/3}$ **15.** $f(x) = 1/x^3$

16. $f(x) = 1/x^5$

In Exercises 17–24, find the derivative of $f(x)$ at the designated value of x.

17. $f(x) = x^6$, at $x = -2$ **18.** $f(x) = x^3$, at $x = \frac{1}{4}$

19. $f(x) = 1/x$, at $x = 3$ **20.** $f(x) = 5x$, at $x = 2$

21. $f(x) = 4 - x$, at $x = 5$ **22.** $f(x) = x^{2/3}$, at $x = 1$

23. $f(x) = x^{3/2}$, at $x = 9$ **24.** $f(x) = 1/x^2$, at $x = 2$

25. Find the slope of the curve $y = x^4$ at $x = 3$.

26. Find the slope of the curve $y = x^5$ at $x = -2$.

27. Find the slope of the curve $y = \sqrt{x}$ at $x = 9$.

28. Find the slope of the curve $y = x^{-3}$ at $x = 3$.

29. If $f(x) = x^2$, compute $f(-5)$ and $f'(-5)$.

30. If $f(x) = x + 6$, compute $f(3)$ and $f'(3)$.

31. If $f(x) = 1/x^5$, compute $f(2)$ and $f'(2)$.

32. If $f(x) = 1/x^2$, compute $f(5)$ and $f'(5)$.

33. If $f(x) = x^{4/3}$, compute $f(8)$ and $f'(8)$.

34. If $f(x) = x^{3/2}$, compute $f(16)$ and $f'(16)$.

35. Find the slope of the tangent line to the curve $y = x^3$ at the point $(4, 64)$, and write the equation of this line.

36. Find the slope of the tangent line to the curve $y = \sqrt{x}$ at the point $(25, 5)$, and write the equation of this line.

In Exercises 37–44, find the indicated derivative.

37. $\dfrac{d}{dx}(x^8)$ 38. $\dfrac{d}{dx}(x^{-3})$ 39. $\dfrac{d}{dx}(x^{3/4})$

40. $\dfrac{d}{dx}(x^{-1/3})$ 41. $\dfrac{dy}{dx}$, if $y = 1$ 42. $\dfrac{dy}{dx}$, if $y = x^{-4}$

43. $\dfrac{dy}{dx}$, if $y = x^{1/5}$ 44. $\dfrac{dy}{dx}$, if $y = \dfrac{x - 1}{3}$

45. Use an argument like those used to verify formulas (3) and (5) to show that the derivative of $f(x) = x^3$ is $3x^2$. [Hint: Recall that $(x + h)^3 = x^3 + 3x^2h + 3xh^2 + h^3$.]

46. Use an argument like those used to verify formulas (3) and (5) to show that the derivative of $f(x) = \sqrt{x}$ is $1/(2\sqrt{x})$. [Hint: After forming the expression for the slope of a secant line, eliminate the square roots from the numerator by multiplying both the numerator and denominator by the quantity $\sqrt{x + h} + \sqrt{x}$.]

SOLUTIONS TO PRACTICE PROBLEMS 3

1. (a) The number $f(5)$ is the y-coordinate of the point P. Since the tangent line passes through P, the coordinates of P satisfy the equation $y = -x + 8$. Since its x-coordinate is 5, its y-coordinate is $-5 + 8 = 3$. Therefore, $f(5) = 3$.

   (b) The number $f'(5)$ is the slope of the tangent line at P, which is readily seen to be -1.

2. (a) The function $1/x^4$ can be written as the power function x^{-4}. Here $r = -4$. Therefore,

$$f'(x) = (-4)x^{(-4)-1} = -4x^{-5} = \frac{-4}{x^5}.$$

   (b) $f'(2) = -4/2^5 = -4/32 = -\frac{1}{8}$.

1.4. Limits and the Derivative

The notion of a limit is one of the fundamental notions of calculus. Indeed, any "theoretical" development of calculus rests on an extensive use of the theory of limits. Even in this book, where we have adopted an intuitive viewpoint, limit arguments are used throughout (although in an informal way). In this section, we give a brief introduction to limits and their role in calculus. As we shall see, the limit concept will allow us to define the notion of a derivative independently of our geometric intuition.

Actually, we have already dealt with limits in our discussion of the derivative! Using geometric reasoning, we arrived at the following procedure for calculating the derivative of a function $f(x)$ at $x = a$. First we calculate the quotient

$$\frac{f(a + h) - f(a)}{h}.$$

Next, we allow h to approach zero. The value of the quotient then approaches the value of the derivative $f'(a)$. Let us recall this calculation in one particular case—namely, for the function $f(x) = x^2$ at $x = 2$. To determine $f'(2)$, we first calculate the quotient

$$\frac{f(2 + h) - f(2)}{h} = \frac{(2 + h)^2 - 2^2}{h}.$$

Next, we allow h to become smaller and smaller. The accompanying table gives values of the quotient for progressively smaller values of h, both positive and negative. It is clear that, as h approaches zero, the value of the quotient approaches 4. We express this fact by saying that the *limit of the quotient as h approaches zero is 4*.

h	$\dfrac{(2 + h)^2 - 2^2}{h}$	h	$\dfrac{(2 + h)^2 - 2^2}{h}$
1	$\dfrac{(2 + 1)^2 - 2^2}{1} = 5$	-1	$\dfrac{(2 + (-1))^2 - 2^2}{-1} = 3$
.1	$\dfrac{(2 + .1)^2 - 2^2}{.1} = 4.10$	$-.1$	$\dfrac{(2 + (-.1))^2 - 2^2}{-.1} = 3.9$
.01	$\dfrac{(2 + .01)^2 - 2^2}{.01} = 4.01$	$-.01$	$\dfrac{(2 + (-.01))^2 - 2^2}{-.01} = 3.99$
.001	$\dfrac{(2 + .001)^2 - 2^2}{.001} = 4.001$	$-.001$	$\dfrac{(2 + (-.001))^2 - 2^2}{-.001} = 3.999$
.0001	$\dfrac{(2 + .0001)^2 - 2^2}{.0001} = 4.0001$	$-.0001$	$\dfrac{(2 + (-.0001))^2 - 2^2}{-.0001} = 3.9999$

A special (and convenient) notation saves excess writing. Write $h \to 0$ to mean that h approaches zero. Further, write

$$\lim_{h \to 0} \frac{(2 + h)^2 - 2^2}{h}$$

to mean "the limit of the quotient as h approaches zero." From our above discussion, the value of this limit is 4. That is,

$$\lim_{h \to 0} \frac{(2 + h)^2 - 2^2}{h} = 4.$$

The discussion above suggests the following definition: Let $g(x)$ be a function, a a number. We say that the number L is the limit of $g(x)$ as x approaches a provided that, as x gets arbitrarily close (but not equal) to a, the values of $g(x)$ approach L. In this case we write

$$\lim_{x \to a} g(x) = L.$$

If, as x approaches a, the values $g(x)$ do *not* approach a specific number, then we say that the limit of $g(x)$ as x approaches a *does not exist*. Let us give some further examples of limits.

EXAMPLE 1 Determine $\lim_{x \to 2}(3x - 5)$.

Solution Let us make a table of values of x approaching 2 and the corresponding values of $3x - 5$:

x	$3x - 5$	x	$3x - 5$
2.1	1.3	1.9	.7
2.01	1.03	1.99	.97
2.001	1.003	1.999	.997
2.0001	1.0003	1.9999	.9997

As x approaches 2, we see that $3x - 5$ approaches 1. In terms of our notation,

$$\lim_{x \to 2}(3x - 5) = 1.$$

EXAMPLE 2 For each of the following functions determine if $\lim_{x \to 2} g(x)$ exists. (The circles drawn on the graphs are meant to represent breaks in the graph, indicating that the functions under consideration are not defined at $x = 2$.)

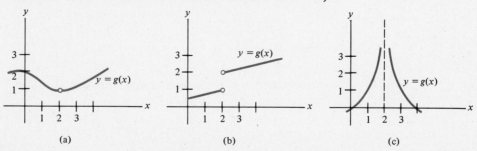

(a) (b) (c)

Solution   (a) $\lim\limits_{x\to 2} g(x) = 1$. We can see that as x gets closer and closer to 2, the values of $g(x)$ get closer and closer to 1. This is true for values of x to both the right and the left of 2.

(b) $\lim\limits_{x\to 2} g(x)$ does not exist. As x approaches 2 from the right, $g(x)$ approaches 2. However, as x approaches 2 from the left, $g(x)$ approaches 1. In order for a limit to exist, the function must approach the *same* value from each direction.

(c) $\lim\limits_{x\to 2} g(x)$ does not exist. As x approaches 2, the values of $g(x)$ become larger and larger and do not approach a fixed number.

The following limit theorems, which we cite without proof, allow us to reduce the computation of limits for combinations of functions to computations of limits involving the constituent functions.

Limit Theorems   Suppose that $\lim\limits_{x\to a} f(x)$ and $\lim\limits_{x\to a} g(x)$ both exist. Then we have the following results.

I. If k is a constant, then $\lim\limits_{x\to a} k \cdot f(x) = k \cdot \lim\limits_{x\to a} f(x)$.

II. If r is a positive constant, then $\lim\limits_{x\to a}[f(x)]^r = \left[\lim\limits_{x\to a} f(x)\right]^r$.

III. $\lim\limits_{x\to a}[f(x) + g(x)] = \lim\limits_{x\to a} f(x) + \lim\limits_{x\to a} g(x)$.

IV. $\lim\limits_{x\to a}[f(x) - g(x)] = \lim\limits_{x\to a} f(x) - \lim\limits_{x\to a} g(x)$.

V. $\lim\limits_{x\to a}[f(x) \cdot g(x)] = \left[\lim\limits_{x\to a} f(x)\right]\left[\lim\limits_{x\to a} g(x)\right]$.

VI. If $\lim\limits_{x\to a} g(x) \neq 0$, then $\lim\limits_{x\to a} \dfrac{f(x)}{g(x)} = \dfrac{\lim\limits_{x\to a} f(x)}{\lim\limits_{x\to a} g(x)}$.

EXAMPLE 3   Use the limit theorems to compute the following limits:

(a) $\lim\limits_{x\to 2} x^3$ (b) $\lim\limits_{x\to 2} 5x^3$ (c) $\lim\limits_{x\to 2}[5x^3 - 15]$ (d) $\lim\limits_{x\to 2} \sqrt{5x^3 - 15}$

(e) $\lim\limits_{x\to 2}(\sqrt{5x^3 - 15}/x^5)$.

Solution   (a) Since $\lim\limits_{x\to 2} x = 2$, we have by Limit Theorem II that

$$\lim_{x\to 2} x^3 = \left(\lim_{x\to 2} x\right)^3 = 2^3 = 8.$$

(b) $\lim\limits_{x\to 2} 5x^3 = 5 \lim\limits_{x\to 2} x^3$ (Limit Theorem I with $k = 5$)

$\qquad\qquad = 5 \cdot 8$ [by part (a)]

$\qquad\qquad = 40.$

(c) $\lim\limits_{x\to 2}(5x^3 - 15) = \lim\limits_{x\to 2} 5x^3 - \lim\limits_{x\to 2} 15$ (Limit Theorem IV).

Note that $\lim\limits_{x\to 2} 15 = 15$. This is because the constant function $g(x) = 15$ always has the value 15, and so its limit as x approaches *any* number is 15. By part (b), $\lim\limits_{x\to 2} 5x^3 = 40$. Thus,

$$\lim_{x\to 2}(5x^3 - 15) = 40 - 15 = 25.$$

(d) $\lim\limits_{x\to 2} \sqrt{5x^3 - 15} = \lim\limits_{x\to 2}(5x^3 - 15)^{1/2}$

$\qquad\qquad = \left(\lim\limits_{x\to 2}(5x^3 - 15)\right)^{1/2}$ (Limit Theorem II with $r = \frac{1}{2}$, $f(x) = 5x^3 - 15$)

$\qquad\qquad = 25^{1/2}$ [by part (c)]

$\qquad\qquad = 5.$

(e) The limit of the denominator is $\lim\limits_{x\to 2} x^5$, which is $2^5 = 32$, a nonzero number. So by Limit Theorem VI we have

$$\lim_{x\to 2} \frac{\sqrt{5x^3 - 15}}{x^5} = \frac{\lim\limits_{x\to 2} \sqrt{5x^3 - 15}}{\lim\limits_{x\to 2} x^5}$$

$$= \frac{5}{32}\quad \text{[by part (d)].}$$

Many situations require algebraic simplifications before the limit theorems can be applied.

EXAMPLE 4   Compute the following limits.

(a) $\lim\limits_{x\to 3} \dfrac{x^2 - 9}{x - 3}$ (b) $\lim\limits_{x\to 0} \dfrac{\sqrt{x + 4} - 2}{x}.$

Solution   (a) The function $\dfrac{x^2 - 9}{x - 3}$ is not defined when $x = 3$, since $\dfrac{3^2 - 9}{3 - 3} = \dfrac{0}{0}$, which is undefined. That causes no difficulty, since the limit as x approaches 3 depends only on the values of x *near* 3 and excludes consideration of the value *at* $x = 3$ itself. To evaluate the limit, note that $x^2 - 9 = (x - 3)(x + 3)$. So for $x \neq 3$,

$$\frac{x^2 - 9}{x - 3} = \frac{(x - 3)(x + 3)}{x - 3} = x + 3.$$

As x approaches 3, $x + 3$ approaches 6. Therefore,

$$\lim_{x \to 3} \frac{x^2 - 9}{x - 3} = 6.$$

(b) Since the denominator approaches zero when taking the limit, we may not apply Limit Theorem VI directly. However, if we first apply an algebraic trick, the limit may be evaluated. Multiply numerator and denominator by $\sqrt{x + 4} + 2$.

$$\frac{(\sqrt{x + 4} - 2)}{x} \cdot \frac{(\sqrt{x + 4} + 2)}{(\sqrt{x + 4} + 2)} = \frac{(x + 4) - 4}{x(\sqrt{x + 4} + 2)}$$

$$= \frac{x}{x(\sqrt{x + 4} + 2)}$$

$$= \frac{1}{\sqrt{x + 4} + 2}.$$

Thus,

$$\lim_{x \to 0} \frac{\sqrt{x + 4} - 2}{x} = \lim_{x \to 0} \frac{1}{\sqrt{x + 4} + 2}$$

$$= \frac{\lim\limits_{x \to 0} 1}{\lim\limits_{x \to 0} (\sqrt{x + 4} + 2)} \qquad \text{(Limit Theorem VI)}$$

$$= \frac{1}{4}.$$

Now that we have the notion of a limit at our disposal, we may give a formal definition of the derivative of a function $f(x)$ at $x = a$. Recall that we defined the derivative $f'(x)$ as a "slope-formula" such that $f'(a)$ gives the slope of the tangent line at $x = a$. The secant line approximation of the derivative led to the result

$$f'(a) = \lim_{h \to 0} \frac{f(a + h) - f(a)}{h}. \tag{1}$$

This discussion depended on an intuitive geometric concept of the tangent line. However, the limit on the right may be considered independently of its geometric interpretation. In fact, we may use (1) to *define* $f'(a)$. We say that f is *differentiable at $x = a$* if the limit in (1) exists, and if so $f'(a)$ is defined to be the value of that limit. If the limit in (1) fails to exist, we say that f is *nondifferentiable at $x = a$*. The domain of the derivative $f'(x)$ consists of all values of x at which $f(x)$ is differentiable.

The limit definition of the derivative is the basis of all "theoretical" work in calculus. Indeed, in order to give rigorous proofs of the fundamental theorems about derivatives, geometrical intuition is rather unreliable and must be replaced by limit arguments. In the next example, we show how limits may be used to directly calculate several derivatives. Later in this chapter, limits will reappear when they are used to prove several of the basic rules for computing derivatives.

EXAMPLE 5 Use limits to compute the derivative $f'(5)$ for the following functions:

(a) $f(x) = 15 - x^2$ (b) $f(x) = \dfrac{1}{2x - 3}$.

Solution In each case, we must calculate $\lim\limits_{h \to 0} \dfrac{f(5 + h) - f(5)}{h}$.

(a)
$$\frac{f(5 + h) - f(5)}{h} = \frac{[15 - (5 + h)^2] - (15 - 5^2)}{h}$$

$$= \frac{15 - (25 + 10h + h^2) - (15 - 25)}{h}$$

$$= \frac{-10h - h^2}{h}$$

$$= -10 - h.$$

Therefore, $f'(5) = \lim\limits_{h \to 0} (-10 - h) = -10$.

(b)
$$\frac{f(5 + h) - f(5)}{h} = \frac{\dfrac{1}{2(5 + h) - 3} - \dfrac{1}{2(5) - 3}}{h}$$

$$= \frac{\dfrac{1}{7 + 2h} - \dfrac{1}{7}}{h}$$

$$= \frac{\dfrac{7 - (7 + 2h)}{(7 + 2h)7}}{h}$$

$$= \frac{-2h}{(7 + 2h)7 \cdot h}$$

$$= \frac{-2}{(7 + 2h)7}$$

$$= \frac{-2}{49 + 14h}.$$

$$f'(5) = \lim\limits_{h \to 0} \frac{-2}{49 + 14h} = -\frac{2}{49}.$$

Remark: When computing the limit in the example above, we considered only values of h near zero (and not $h = 0$ itself). Therefore, we were able to freely divide both numerator and denominator by h.

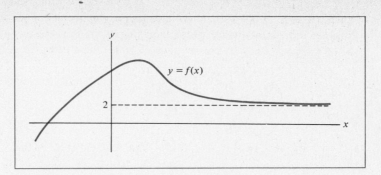

FIGURE 1

Infinity and Limits Consider the function $f(x)$ whose graph is sketched in Fig. 1. As x grows large the value of $f(x)$ approaches 2. In this circumstance, we say that 2 is *the limit of $f(x)$ as x approaches infinity*. Infinity is denoted by the symbol ∞. The above limit statement is expressed in the following notation:

$$\lim_{x \to \infty} f(x) = 2.$$

In a similar vein, consider the function whose graph is sketched in Fig. 2. As x grows large in the negative direction, the value of $f(x)$ approaches 0. In this circumstance, we say that 0 is *the limit of $f(x)$ as x approaches minus infinity*. In symbols,

$$\lim_{x \to -\infty} f(x) = 0.$$

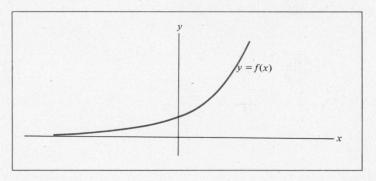

FIGURE 2

EXAMPLE 6 Calculate the following limits.

(a) $\displaystyle\lim_{x \to \infty} \frac{1}{x^2 + 1}$ (b) $\displaystyle\lim_{x \to \infty} \frac{x + 1}{x - 1}$.

Solution (a) As x increases without bound, so does $x^2 + 1$. Therefore, $1/(x^2 + 1)$ approaches zero as x approaches ∞.

(b) Both $x + 1$ and $x - 1$ increase without bound as x does. To determine the limit of their quotient, we employ an algebraic trick. Divide both numerator and denominator by x to obtain

$$\lim_{x \to \infty} \frac{x + 1}{x - 1} = \lim_{x \to \infty} \frac{1 + \dfrac{1}{x}}{1 - \dfrac{1}{x}}.$$

As x increases without bound, $1/x$ approaches zero, so that both $1 + (1/x)$ and $1 - (1/x)$ approach 1. Thus, the desired limit is $1/1 = 1$.

PRACTICE PROBLEMS 4

Determine which of the following limits exist. Compute the limits that exist.

1. $\displaystyle\lim_{x \to 6} \frac{x^2 - 4x - 12}{x - 6}$

2. $\displaystyle\lim_{x \to 6} \frac{4x + 12}{x - 6}$

EXERCISES 4

For each of the following functions, $g(x)$, determine whether or not $\displaystyle\lim_{x \to 3} g(x)$ exists. If so, give the limit.

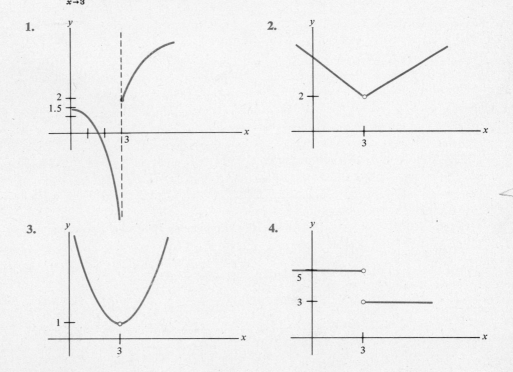

5. **6.**

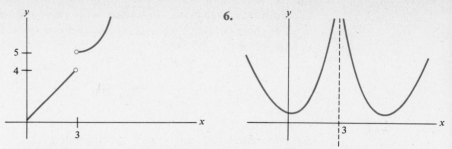

Determine which of the following limits exist. Compute the limits that exist.

7. $\lim_{x \to 1}(1 - 6x)$

8. $\lim_{x \to 2} \dfrac{x}{x - 2}$

9. $\lim_{x \to 3} \sqrt{x^2 + 16}$

10. $\lim_{x \to 4}(x^3 - 7)$

11. $\lim_{x \to 5} \dfrac{x^2 + 1}{5 - x}$

12. $\lim_{x \to 6}\left(\sqrt{6x} + 3x - \dfrac{1}{x}\right)(x^2 - 4)$

13. $\lim_{x \to 7}(x + \sqrt{x - 6})(x^2 - 2x + 1)$

14. $\lim_{x \to 8} \dfrac{\sqrt{5x - 4} - 1}{3x^2 + 2}$

15. $\lim_{x \to 9} \dfrac{\sqrt{x^2 - 5x - 36}}{8 - 3x}$

16. $\lim_{x \to 10}(2x^2 - 15x - 50)^{20}$

17. $\lim_{x \to 0} \dfrac{x^2 + 3x}{x}$

18. $\lim_{x \to 1} \dfrac{x^2 - 1}{x - 1}$

19. $\lim_{x \to 2} \dfrac{-2x^2 + 4x}{x - 2}$

20. $\lim_{x \to 3} \dfrac{x^2 - x - 6}{x - 3}$

21. $\lim_{x \to 4} \dfrac{x^2 - 16}{4 - x}$

22. $\lim_{x \to 5} \dfrac{2x - 10}{x^2 - 25}$

23. $\lim_{x \to 6} \dfrac{x^2 - 6x}{x^2 - 5x - 6}$

24. $\lim_{x \to 7} \dfrac{x^3 - 2x^2 + 3x}{x^2}$

25. $\lim_{x \to 8} \dfrac{x^2 + 64}{x - 8}$

26. $\lim_{x \to 9} \dfrac{1}{(x - 9)^2}$

27. $\lim_{x \to 0} \dfrac{-2}{\sqrt{x + 16} + 7}$

28. $\lim_{x \to 0} \dfrac{4x}{x(x^2 + 3x + 5)}$

Use limits to compute the following derivatives.

29. $f'(3)$ where $f(x) = x^2 + 1$

30. $f'(2)$ where $f(x) = x^3$

31. $f'(0)$ where $f(x) = x^3 + 3x + 1$

32. $f'(0)$ where $f(x) = x^2 + 2x + 2$

33. $f'(3)$ where $f(x) = \dfrac{1}{2x + 5}$

34. $f'(4)$ where $f(x) = \sqrt{2x - 1}$

35. $f'(2)$ where $f(x) = \sqrt{5 - x}$

36. $f'(3)$ where $f(x) = \dfrac{1}{7 - 2x}$

37. $f'(0)$ where $f(x) = \sqrt{1 - x^2}$ 38. $f'(2)$ where $f(x) = (5x - 4)^2$

39. $f'(0)$ where $f(x) = (x + 1)^3$ 40. $f'(0)$ where $f(x) = \sqrt{x^2 + x + 1}$

Compute the following limits.

41. $\lim\limits_{x \to \infty} \dfrac{1}{x^2}$ 42. $\lim\limits_{x \to -\infty} \dfrac{1}{x^2}$ 43. $\lim\limits_{x \to \infty} \dfrac{1}{x - 8}$

44. $\lim\limits_{x \to \infty} \dfrac{1}{3x + 5}$ 45. $\lim\limits_{x \to \infty} \dfrac{2x + 1}{x + 2}$ 46. $\lim\limits_{x \to \infty} \dfrac{x^2 + x}{x^2 - 1}$

SOLUTIONS TO PRACTICE PROBLEMS 4

1. The function under consideration is a quotient of two functions. Let us first consider the numerator. Clearly, $\lim\limits_{x \to 6} x^2 = 36$, $\lim\limits_{x \to 6} 4x = 24$, and $\lim\limits_{x \to 6} 12 = 12$. Therefore, by two applications of Limit Theorem IV,

$$\lim_{x \to 6}(x^2 - 4x - 12) = 36 - 24 - 12 = 0.$$

Now also, $\lim\limits_{x \to 6} (x - 6) = 0$. Since the function in the denominator has limit 0, we cannot apply Limit Theorem VI. However, since we are considering values of x different from 6, the quotient can be simplified by factoring and canceling.

$$\frac{x^2 - 4x - 12}{x - 6} = \frac{(x + 2)(x - 6)}{(x - 6)} = x + 2, \qquad \text{for } x \neq 6.$$

Now $\lim\limits_{x \to 6}(x + 2) = 8$. Therefore, $\lim\limits_{x \to 6} \dfrac{x^2 - 4x - 12}{x - 6} = 8$.

2. No limit exists. It is easily seen that $\lim\limits_{x \to 6}(4x + 12) = 36$ and $\lim\limits_{x \to 6}(x - 6) = 0$. As x approaches 6, the denominator gets very small and the numerator approaches 36. For example, if $x = 6.00001$, then the numerator is 36.00004 and the denominator is .00001. The quotient is 3,600,004. As x approaches 6 even more closely, the quotient gets arbitrarily large and cannot possibly approach a limit.

1.5. Differentiability and Continuity

In the preceding section, we defined differentiability of $f(x)$ at $x = a$ in terms of a limit. If this limit does not exist, then we say that $f(x)$ is *nondifferentiable* at $x = a$. Geometrically, the nondifferentiability of $f(x)$ at $x = a$ can manifest itself in several different ways. First of all, the graph of $f(x)$ could have no tangent line at $x = a$. Second, the graph could have a vertical tangent line at $x = a$. (Recall that slope is not defined for vertical lines.) Some of the various geometric possibilities are illustrated in Fig. 1.

The following example illustrates how nondifferentiable functions can arise in practice.

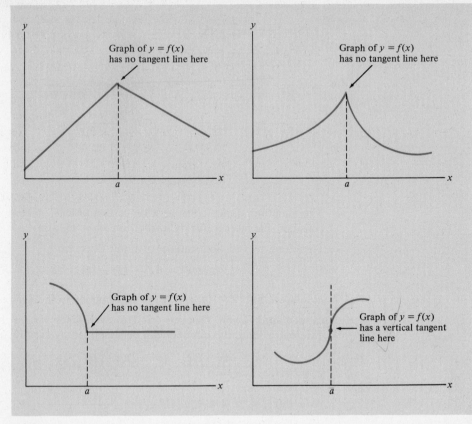

FIGURE 1

EXAMPLE 1 A railroad company charges $10 per mile to haul a boxcar up to 200 miles and $8 per mile for each mile exceeding 200. In addition, the railroad charges a $1000 handling charge per boxcar. Graph the cost of sending a boxcar x miles.

Solution If x is at most 200 miles, then the cost $C(x)$ is given by $C(x) = 1000 + 10x$ dollars. The cost for 200 miles is $C(200) = 1000 + 2000 = 3000$ dollars. If x exceeds 200 miles, then the total cost will be

$$C(x) = \underbrace{3000}_{\substack{\text{cost of first} \\ \text{200 miles}}} + \underbrace{8(x - 200)}_{\substack{\text{cost of miles in} \\ \text{excess of 200}}} = 1400 + 8x.$$

Thus

$$C(x) = \begin{cases} 1000 + 10x, & 0 < x \leq 200, \\ 1400 + 8x, & x > 200. \end{cases}$$

The graph of $C(x)$ is sketched in Fig. 2. Note that $C(x)$ is not differentiable at $x = 200$.

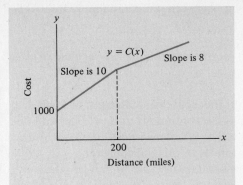

FIGURE 2

Closely related to the concept of differentiability is that of continuity. We say that a function $f(x)$ is *continuous* at $x = a$ provided that its graph has no breaks (or gaps) as it passes through the point $(a, f(a))$. That is, $f(x)$ is continuous at $x = a$ provided we can draw the graph through $(a, f(a))$ without lifting our pencil from the paper. The functions whose graphs are drawn in Figs. 1 and 2 are continuous for all values of x. By contrast, however, the function whose graph is drawn in Fig. 3(a) is not continuous (we say it is *discontinuous*) at $x = 1$ and $x = 2$, since the graph has breaks there. Similarly, the function whose graph is drawn in Fig. 3(b) is discontinuous at $x = 2$.

Discontinuous functions can occur in applications, as the following example shows.

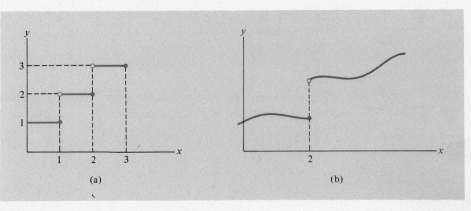

FIGURE 3

EXAMPLE 2 Suppose that a manufacturing plant is capable of producing 15,000 units in one shift of eight hours. For each shift worked, there is a fixed cost of $2000 (for light, heat, etc.). Suppose that the variable cost (the cost of labor and raw materials) is $2 per unit. Graph the cost $C(x)$ of manufacturing x units.

Solution If $x \le 15,000$, a single shift will suffice, so that

$$C(x) = 2000 + 2x, \qquad 0 \le x \le 15,000.$$

If x is between 15,000 and 30,000, one extra shift will be required, and

$$C(x) = 4000 + 2x, \qquad 15,000 < x \le 30,000.$$

If x is between 30,000 and 45,000, the plant will need to work three shifts, and

$$C(x) = 6000 + 2x, \qquad 30,000 < x \le 45,000.$$

The graph of $C(x)$ for $0 \le x \le 45,000$ is drawn in Fig. 4. Note that the graph has breaks at two points.

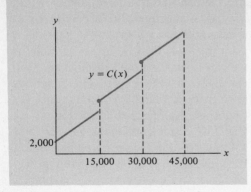

$y = C(x)$

2,000

15,000 30,000 45,000

FIGURE 4

The relationship between differentiability and continuity is this:

Theorem 1   If $f(x)$ is differentiable at $x = a$, then $f(x)$ is continuous at $x = a$.

Note, however, that the converse statement is definitely false: A function may be continuous at $x = a$ but still not be differentiable there. The functions whose graphs are drawn in Fig. 1 provide examples of this phenomenon.

Just as with differentiability, the notion of continuity can be phrased in terms of limits. In order for $f(x)$ to be continuous at $x = a$, the values of $f(x)$ for all x near a must be close to $f(a)$ (otherwise the graph would have a break at $x = a$.) In fact, the closer x is to a, the closer $f(x)$ must be to $f(a)$ (again, in order to avoid a break in the graph). In terms of limits, we must therefore have

$$\lim_{x \to a} f(x) = f(a).$$

Conversely, an intuitive argument shows that if the above limit relation holds, then the graph of $y = f(x)$ has no break at $x = a$. So we may state the following result.

> *Limit Criterion for Continuity:* **A function** $f(x)$ **is continuous at** $x = a$ **provided that the following limit relation holds:**
>
> $$\lim_{x \to a} f(x) = f(a).$$ (1)

In order for (1) to hold, two conditions must be fulfilled.

1. $\lim_{x \to a} f(x)$ must exist.

2. The limit $\lim_{x \to a} f(x)$ must have the value $f(a)$.

A function may fail to be continuous at $x = a$ because either condition 1 or condition 2 fails to hold. The various possibilities are illustrated in the next example.

EXAMPLE 3   Determine whether the functions whose graphs are drawn in Fig. 5 are continuous at $x = 3$. Use the limit criterion.

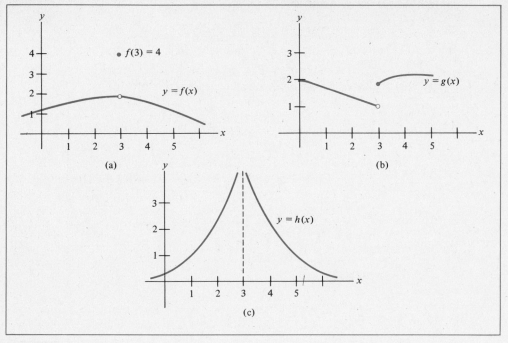

FIGURE 5

Solution   (a) Here $\lim\limits_{x \to 3} f(x) = 2$. However, $f(3) = 4$. So

$$\lim_{x \to 3} f(x) \neq f(3)$$

and $f(x)$ is not continuous at $x = 3$. (Geometrically, this is clear. The graph has a break at $x = 3$.)

(b) $\lim\limits_{x \to 3} g(x)$ does not exist, so $g(x)$ is not continuous at $x = 3$.

(c) $\lim\limits_{x \to 3} h(x)$ does not exist, so $h(x)$ is not continuous at $x = 3$.

In a more formal course in calculus, one proves many theorems which describe the general properties of continuous (and differentiable) functions. By using these theorems, it is possible to prove that the elementary functions of algebra are continuous for all x for which they are defined. Thus, for example, any polynomial

$$p(x) = a_0 + a_1x + \ldots + a_nx^n, \qquad a_0, \ldots, a_n \text{ constants,}$$

is continuous for all x. A rational function

$$\frac{p(x)}{q(x)} \qquad p(x), q(x) \text{ polynomials,}$$

is continuous for all x at which $q(x) \neq 0$. Similarly, $\sqrt{x}$ is continuous for all $x \geq 0$. For purposes of this book, we shall assume all these facts.

PRACTICE PROBLEMS 5

Let

$$f(x) = \begin{cases} \dfrac{x^2 - x - 6}{x - 3} & \text{for } x \neq 3 \\ 4 & \text{for } x = 3. \end{cases}$$

1. Is $f(x)$ continuous at $x = 3$?

2. Is $f(x)$ differentiable at $x = 3$?

EXERCISES 5

Is the function whose graph is drawn in Fig. 6 continuous at the following values of x?

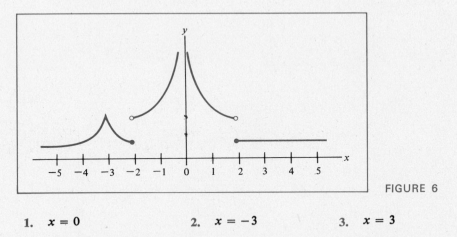

FIGURE 6

1. $x = 0$ 2. $x = -3$ 3. $x = 3$

4. $x = .001$ 5. $x = -2$ 6. $x = 2$

Is the function whose graph is drawn in Fig. 6 differentiable at the following values of x?

7. $x = 0$ 8. $x = -3$ 9. $x = 3$

10. $x = .001$ 11. $x = -2$ 12. $x = 2$

Determine whether each of the following functions is continuous and/or differentiable at $x = 1$.

13. $f(x) = \begin{cases} \dfrac{x^2 + 3x - 4}{x - 1} & \text{for } x \neq 1 \\ 5 & \text{for } x = 1. \end{cases}$

14. $f(x) = \begin{cases} \dfrac{x^2 + 4x - 5}{x - 1} & \text{for } x \neq 1 \\ 3 & \text{for } x = 1. \end{cases}$

15. $f(x) = \begin{cases} \dfrac{1}{x - 1} & \text{for } x \neq 1 \\ 10 & \text{for } x = 1. \end{cases}$

16. $f(x) = \begin{cases} \dfrac{1}{x} & \text{for } x \neq 0 \\ 0 & \text{for } x = 0. \end{cases}$

17. $f(x) = \begin{cases} 2 & \text{for } x \leq 1 \\ 4x - 2 & \text{for } x > 1. \end{cases}$

18. $f(x) = \begin{cases} 3 & \text{for } x \leq 1 \\ x^2 - 2x + 4 & \text{for } x > 1. \end{cases}$

19. $f(x) = \begin{cases} 1 & \text{for } x \leq 1 \\ -x^2 + 2x & \text{for } x > 1. \end{cases}$

20. $f(x) = \begin{cases} \sqrt{1 - x^2} & \text{for } -1 < x \leq 1 \\ \sqrt{1 - (x - 2)^2} & \text{for } 1 < x < 3. \end{cases}$

SOLUTIONS TO PRACTICE PROBLEMS 5

1. The function $f(x)$ is defined at $x = 3$, namely $f(3) = 4$. When computing $\lim_{x \to 3} f(x)$ we exclude consideration of $x = 3$; therefore, we can simplify the expression for $f(x)$ as follows:

$$f(x) = \frac{x^2 - x - 6}{x - 3} = \frac{(x - 3)(x + 2)}{x - 3} = x + 2.$$

Clearly,

$$\lim_{x \to 3} f(x) = \lim_{x \to 3} (x + 2) = 5.$$

Since $\lim_{x \to 3} f(x) = 5 \neq 4 = f(3)$, $f(x)$ is not continuous at $x = 3$.

2. There is no need to compute any limits in order to answer this question. By Theorem 1, since $f(x)$ is not continuous at $x = 3$, it cannot possibly be differentiable there.

1.6. Some Rules for Differentiation

Three additional rules of differentiation greatly extend the number of functions that we can differentiate.

> 1. **Constant-Multiple Rule**
>
> $$\frac{d}{dx}[k \cdot f(x)] = k \cdot \frac{d}{dx}[f(x)], \qquad k \text{ a constant.}$$
>
> 2. **Sum Rule**
>
> $$\frac{d}{dx}[f(x) + g(x)] = \frac{d}{dx}[f(x)] + \frac{d}{dx}[g(x)].$$
>
> 3. **General Power Rule**
>
> $$\frac{d}{dx}([g(x)]^r) = r \cdot [g(x)]^{r-1} \cdot \frac{d}{dx}[g(x)].$$

We shall discuss these rules and then prove the first two.

The Constant-Multiple Rule   Starting with a function $f(x)$, we can multiply it by a constant number k in order to obtain a new function $k \cdot f(x)$. For instance, if $f(x) = x^2 - 4x + 1$ and $k = 2$, then

$$2f(x) = 2(x^2 - 4x + 1) = 2x^2 - 8x + 2.$$

The constant-multiple rule says that the derivative of the new function $k \cdot f(x)$ is just k times the derivative of the original function.* In other words, when faced with the differentiation of a constant times a function, simply carry along the constant and differentiate the function.

EXAMPLE 1   Calculate:

(a) $\dfrac{d}{dx}(2x^5)$ (b) $\dfrac{d}{dx}\left(\dfrac{x^3}{4}\right)$ (c) $\dfrac{d}{dx}\left(-\dfrac{3}{x}\right)$ (d) $\dfrac{d}{dx}(5\sqrt{x}).$

Solution   (a) With $k = 2$ and $f(x) = x^5$, we have

$$\frac{d}{dx}(2 \cdot x^5) = 2 \cdot \frac{d}{dx}(x^5) = 2(5x^4) = 10x^4.$$

---

* More precisely, the constant multiple rule asserts that if $f(x)$ is differentiable for $x = a$ then so is the function $k \cdot f(x)$, and the derivative of $k \cdot f(x)$ at $x = a$ may be computed using the given formula.

(b) Write $\frac{x^3}{4}$ in the form $\frac{1}{4} \cdot x^3$. Then

$$\frac{d}{dx}\left(\frac{x^3}{4}\right) = \frac{1}{4} \cdot \frac{d}{dx}(x^3) = \frac{1}{4}(3x^2) = \frac{3}{4}x^2.$$

(c) Write $-\frac{3}{x}$ in the form $(-3) \cdot \frac{1}{x}$. Then

$$\frac{d}{dx}\left(-\frac{3}{x}\right) = (-3) \cdot \frac{d}{dx}\left(\frac{1}{x}\right) = (-3) \cdot \frac{-1}{x^2} = \frac{3}{x^2}.$$

(d) $\frac{d}{dx}(5\sqrt{x}) = 5\frac{d}{dx}(\sqrt{x}) = 5\frac{d}{dx}(x^{1/2}) = \frac{5}{2}x^{-1/2}.$

This answer may also be written in the form $\dfrac{5}{2\sqrt{x}}$.

The Sum Rule To differentiate a sum of functions, differentiate each function individually and add the derivatives together.* Another way of saying this is "the derivative of a sum of functions is the sum of the derivatives."

EXAMPLE 2 Find

(a) $\dfrac{d}{dx}(x^3 + 5x)$ (b) $\dfrac{d}{dx}\left(x^4 - \dfrac{3}{x^2}\right)$ (c) $\dfrac{d}{dx}(2x^7 - x^5 + 8)$.

Solution (a) Let $f(x) = x^3$ and $g(x) = 5x$. Then

$$\frac{d}{dx}(x^3 + 5x) = \frac{d}{dx}(x^3) + \frac{d}{dx}(5x) = 3x^2 + 5.$$

(b) The sum rule applies to differences as well as sums (see Exercise 36). Indeed, by the sum rule,

$$\frac{d}{dx}\left(x^4 - \frac{3}{x^2}\right) = \frac{d}{dx}(x^4) + \frac{d}{dx}\left(-\frac{3}{x^2}\right) \qquad \text{(sum rule)}$$

$$= \frac{d}{dx}(x^4) - 3\frac{d}{dx}(x^{-2}) \qquad \text{(constant-multiple rule)}$$

$$= 4x^3 - 3(-2x^{-3})$$

$$= 4x^3 + 6x^{-3}.$$

After some practice, one usually omits most or all of the intermediate steps and simply writes

$$\frac{d}{dx}\left(x^4 - \frac{3}{x^2}\right) = 4x^3 + 6x^{-3}.$$

---

* More precisely, the sum rule asserts that if both $f(x)$ and $g(x)$ are differentiable at $x = a$, then so is $f(x) + g(x)$, and the derivative (at $x = a$) of the sum is then the sum of the derivatives (at $x = a$).

(c) We apply the sum rule repeatedly and use the fact that the derivative of a constant function is 0:

$$\frac{d}{dx}(2x^7 - x^5 + 8) = \frac{d}{dx}(2x^7) - \frac{d}{dx}(x^5) + \frac{d}{dx}(8)$$

$$= 2(7x^6) - 5x^4 + 0$$

$$= 14x^6 - 5x^4.$$

The General Power Rule Frequently we will encounter expressions of the form $[g(x)]^r$—for instance, $(x^3 + 5)^2$, where $g(x) = x^3 + 5$ and $r = 2$. The general power rule says that, to differentiate $[g(x)]^r$, we must first treat $g(x)$ as if it were simply an x, form $r[g(x)]^{r-1}$, and then multiply it by a "correction factor" $g'(x)$.* Thus

$$\frac{d}{dx}(x^3 + 5)^2 = 2(x^3 + 5)^1 \cdot \frac{d}{dx}(x^3 + 5)$$

$$= 2(x^3 + 5) \cdot (3x^2)$$

$$= 6x^2(x^3 + 5).$$

In this special case it is easy to verify that the general power rule gives the correct answer. We first expand $(x^3 + 5)^2$ and then differentiate.

$$(x^3 + 5)^2 = (x^3 + 5)(x^3 + 5) = x^6 + 10x^3 + 25.$$

From the constant-multiple rule and the sum rule, we have

$$\frac{d}{dx}(x^3 + 5)^2 = \frac{d}{dx}(x^6 + 10x^3 + 25)$$

$$= 6x^5 + 30x^2 + 0$$

$$= 6x^2(x^3 + 5).$$

The two methods give the same answer.

Note that if we set $g(x) = x$ in the general power rule, we recover the power rule. So the general power rule contains the power rule as a special case.

EXAMPLE 3 Differentiate $\sqrt{1 - x^2}$.

Solution

$$\frac{d}{dx}(\sqrt{1 - x^2}) = \frac{d}{dx}((1 - x^2)^{1/2}) = \frac{1}{2}(1 - x^2)^{-1/2} \cdot \frac{d}{dx}(1 - x^2)$$

$$= \frac{1}{2}(1 - x^2)^{-1/2} \cdot (-2x)$$

$$= \frac{-x}{(1 - x^2)^{1/2}} = \frac{-x}{\sqrt{1 - x^2}}.$$

---

* More precisely, the general power rule asserts that if $g(x)$ is differentiable at $x = a$, and if $[g(x)]^r$ and $[g(x)]^{r-1}$ are both defined at $x = a$, then $[g(x)]^r$ is also differentiable at $x = a$ and its derivative is given by the formula stated.

EXAMPLE 4 Differentiate $y = \dfrac{1}{x^3 + 4x}$.

Solution

$$y = \frac{1}{x^3 + 4x} = (x^3 + 4x)^{-1}.$$

$$\frac{dy}{dx} = (-1)(x^3 + 4x)^{-2} \cdot \frac{d}{dx}(x^3 + 4x)$$

$$= \frac{-1}{(x^3 + 4x)^2}(3x^2 + 4)$$

$$= -\frac{3x^2 + 4}{(x^3 + 4x)^2}.$$

Proofs of the Constant-Multiple and Sum Rules

Let us verify both rules when x has the value a. Recall that if $f(x)$ is differentiable at $x = a$, then its derivative is the limit

$$\lim_{h \to 0} \frac{f(a + h) - f(a)}{h}.$$

Constant-Multiple Rule We assume that $f(x)$ is differentiable at $x = a$. We must prove that $k \cdot f(x)$ is differentiable at $x = a$ and that its derivative there is $k \cdot f'(a)$. This amounts to showing that the limit

$$\lim_{h \to 0} \frac{k \cdot f(a + h) - k \cdot f(a)}{h}$$

exists and has the value $k \cdot f'(a)$. However,

$$\lim_{h \to 0} \frac{k \cdot f(a + h) - k \cdot f(a)}{h} = \lim_{h \to 0} k \left[\frac{f(a + h) - f(a)}{h} \right]$$

$$= k \cdot \lim_{h \to 0} \frac{f(a + h) - f(a)}{h} \qquad \text{(by Limit Theorem I)}$$

$$= k \cdot f'(a) \qquad \text{(since } f(x) \text{ is differentiable at } x = a),$$

which is what we desired to show.

Sum Rule We assume that both $f(x)$ and $g(x)$ are differentiable at $x = a$. We must prove that $f(x) + g(x)$ is differentiable at $x = a$ and that its derivative is $f'(a) + g'(a)$. That is, we must show that the limit

$$\lim_{h \to 0} \frac{[f(a + h) + g(a + h)] - [f(a) + g(a)]}{h}$$

exists and equals $f'(a) + g'(a)$. Using Limit Theorem III and the fact that $f(x)$ and $g(x)$ are differentiable at $x = a$, we have

$$\lim_{h \to 0} \frac{[f(a + h) + g(a + h)] - [f(a) + g(a)]}{h}$$

$$= \lim_{h \to 0} \left[\frac{f(a + h) - f(a)}{h} + \frac{g(a + h) - g(a)}{h} \right]$$

$$= \lim_{h \to 0} \frac{f(a + h) - f(a)}{h} + \lim_{h \to 0} \frac{g(a + h) - g(a)}{h}$$

$$= f'(a) + g'(a).$$

The general power rule will be proven as a special case of the chain rule in Chapter 3.

PRACTICE PROBLEMS 6

1. Find $\dfrac{d}{dx}(x)$

2. Differentiate $y = \dfrac{x + (x^5 + 1)^{10}}{3}$

EXERCISES 6

Differentiate.

1. $y = x^3 + x^2$
2. $y = x^2 + (1/x)$
3. $y = x^2 + 3x - 1$
4. $y = x^3 + 2x + 5$
5. $f(x) = x^5 + (1/x)$
6. $f(x) = x^8 - x$
7. $f(x) = x^4 + x^3 + x$
8. $f(x) = x^5 + x^2 - x$
9. $y = 3x^2$
10. $y = 2x^3$
11. $y = x^3 + 7x^2$
12. $y = -2x$
13. $y = \dfrac{4}{x^2}$
14. $y = 2\sqrt{x}$
15. $y = 3x - (1/x)$
16. $y = -x^2 + 3x + 1$
17. $f(x) = \frac{1}{3}x^3 - \frac{1}{2}x^2 + x + 1$
18. $f(x) = 100x^{100}$
19. $f(x) = -1/5x^5$
20. $f(x) = x^2 - (1/x^2)$
21. $f(x) = 1 - \sqrt{x}$
22. $f(x) = -3x^2 + 7$
23. $f(x) = (3x + 1)^{10}$
24. $f(x) = 1/(x^2 + x + 1)$
25. $f(x) = \sqrt{3x^3 + x + 1}$
26. $y = 1/(x^2 - 7)^5$
27. $y = (2x^2 - x + 4)^6$
28. $y = \sqrt{-2x + 1}$

Find the slope of the graph of $y = f(x)$ at the designated point.

29. $f(x) = 3x^2 - 2x + 1$, $(1, 2)$

30. $f(x) = x^{10} + 1 + \sqrt{1 - x}$, $(0, 2)$

31. Find the slope of the tangent line to the curve $y = x^3 + 3x - 8$ at $(2, 6)$.

32. Write the equation of the tangent line to the curve $y = x^3 + 3x - 8$ at $(2, 6)$.

33. Find the slope of the tangent line to the curve $y = (x^2 - 15)^6$ at the point $(4, 1)$. Then write the equation of this tangent line.

34. Find the equation of the tangent line to the curve $y = \dfrac{8}{(x^2 + x + 2)}$ at $(2, 1)$.

35. Differentiate the function $f(x) = (3x^2 + x - 2)^2$ in two ways.

    (a) Use the general power rule.

    (b) Multiply $3x^2 + x - 2$ by itself and then differentiate the resulting polynomial.

36. Using the sum rule and the constant-multiple rule, show that for any functions $f(x)$ and $g(x)$

$$\frac{d}{dx}[f(x) - g(x)] = \frac{d}{dx}f(x) - \frac{d}{dx}g(x).$$

37. Find m such that the line $y = mx$ is tangent to the curve $y = (3x - 1)^2$. [Hint: The line and the curve must both pass through some point (x, y) and have the same slope at that point.]

38. There are two points on the curve $y = x^2 + 3$ where the tangent line to the curve passes through the point $(1, 0)$. Find these two points.

SOLUTIONS TO PRACTICE PROBLEMS 6

1. The problem asks for the derivative of the function $y = x$, a straight line of slope 1. Therefore, $\dfrac{d}{dx}(x) = 1$. The result can also be obtained from the power rule with $r = 1$. If $f(x) = x^1$, then $\dfrac{d}{dx}(f(x)) = 1 \cdot x^{1-1} = x^0 = 1$.

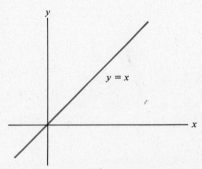

$y = x$

2. All three rules are required to differentiate this function.

$$\frac{dy}{dx} = \frac{d}{dx}\frac{1}{3} \cdot [x + (x^5 + 1)^{10}]$$

$$= \frac{1}{3}\frac{d}{dx}[x + (x^5 + 1)^{10}] \qquad \text{(constant-multiple rule)}$$

$$= \frac{1}{3}\left[\frac{d}{dx}(x) + \frac{d}{dx}(x^5 + 1)^{10}\right] \qquad \text{(sum rule)}$$

$$= \tfrac{1}{3}[1 + 10(x^5 + 1)^9 \cdot (5x^4)] \qquad \text{(general power rule)}$$

$$= \tfrac{1}{3}[1 + 50x^4(x^5 + 1)^9].$$

1.7. More About Derivatives

In many applications it is convenient to use variables other than x and y. One might, for instance, study the function $f(t) = t^2$ instead of writing $f(x) = x^2$. In this case, the notation for the derivative involves t rather than x, but the concept of the derivative as a slope formula is unaffected. (See Fig. 1.) When the independent variable is t instead of x, we write $\frac{d}{dt}$ in place of $\frac{d}{dx}$. For instance,

$$\frac{d}{dt}(t^3) = 3t^2, \qquad \frac{d}{dt}(2t^2 + 3t) = 4t + 3.$$

Recall that if y is a function of x, say $y = f(x)$, then we may write $\frac{dy}{dx}$ in place of $f'(x)$. We sometimes call $\frac{dy}{dx}$ "the derivative of y with respect to x." Similarly, if v is a function of t, then the derivative of v with respect to t is written as $\frac{dv}{dt}$. For example, if $v = 4t^2$, then $\frac{dv}{dt} = 8t$.

FIGURE 1

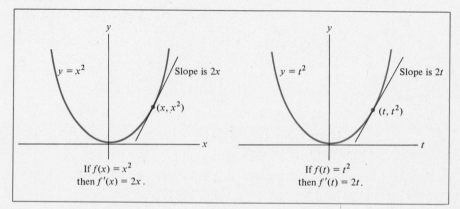

$y = x^2$ — Slope is $2x$ — (x, x^2) — If $f(x) = x^2$ then $f'(x) = 2x$.

$y = t^2$ — Slope is $2t$ — (t, t^2) — If $f(t) = t^2$ then $f'(t) = 2t$.

Of course, other letters can be used to denote variables. The formulas

$$\frac{d}{dP}(P^3) = 3P^2, \qquad \frac{d}{ds}(s^3) = 3s^2, \qquad \frac{d}{dz}(z^3) = 3z^2$$

are all expressing the same basic fact that the slope formula for the cubic curve $y = x^3$ is given by $3x^2$.

EXAMPLE 1 Compute

(a) $\dfrac{ds}{dp}$ if $s = 3(p^2 + 5p + 1)^{10}$.

(b) $\dfrac{d}{dt}(at^2 + St^{-1} + S^2)$.

Solution (a) $\dfrac{d}{dp} 3(p^2 + 5p + 1)^{10} = 30(p^2 + 5p + 1)^9 \cdot \dfrac{d}{dp}(p^2 + 5p + 1)$

$$= 30(p^2 + 5p + 1)^9(2p + 5).$$

(b) Although the expression $at^2 + St^{-1} + S^2$ contains several letters, the notation $\dfrac{d}{dt}$ indicates that all letters except t are to be considered as constants. Hence

$$\frac{d}{dt}(at^2 + St^{-1} + S^2) = \frac{d}{dt}(at^2) + \frac{d}{dt}(St^{-1}) + \frac{d}{dt}(S^2)$$

$$= a \cdot \frac{d}{dt}(t^2) + S \cdot \frac{d}{dt}(t^{-1}) + 0$$

$$= 2at - St^{-2}.$$

$$\left[\text{The derivative } \frac{d}{dt}(S^2) \text{ is zero because } S^2 \text{ is a constant.}\right]$$

The Second Derivative When we differentiate a function $f(x)$, we obtain a new function $f'(x)$ that is a formula for the slope of the curve $y = f(x)$. If we differentiate the function $f'(x)$, we obtain what is called the *second derivative* of $f(x)$, denoted by $f''(x)$. That is,

$$\frac{d}{dx} f'(x) = f''(x).$$

EXAMPLE 2 Find the second derivatives of the following functions.

(a) $f(x) = x^3 + (1/x)$.

(b) $f(x) = 2x + 1$.

(c) $f(t) = t^{1/2} + t^{-1/2}$.

Solution (a) $f(x) = x^3 + (1/x) = x^3 + x^{-1}$

$f'(x) = 3x^2 - x^{-2}$

$f''(x) = 6x + 2x^{-3}.$

(b) $f(x) = 2x + 1$

$f'(x) = 2$ (a constant function whose value is 2)

$f''(x) = 0.$ (The derivative of a constant function is zero.)

(c) $f(t) = t^{1/2} + t^{-1/2}$

$f'(t) = \frac{1}{2}t^{-1/2} - \frac{1}{2}t^{-3/2}$

$f''(t) = -\frac{1}{4}t^{-3/2} + \frac{3}{4}t^{-5/2}.$

The first derivative of a function $f(x)$ gives the slope of the graph of $f(x)$ at any point. The second derivative of $f(x)$ gives important additional information about the shape of the curve near any point. We shall examine this subject carefully in the next chapter.

Other Notation for Derivatives Unfortunately, the process of differentiation does not have a standardized notation. Consequently, it is important to become familiar with alternate terminology.

If y is a function of x, say $y = f(x)$, then we may denote the first and second derivatives of this function in several ways.

Prime notation	$\frac{d}{dx}$ *notation*
$f'(x)$	$\frac{d}{dx} f(x)$
y'	$\frac{dy}{dx}$
$f''(x)$	$\frac{d^2}{dx^2} f(x)$
y''	$\frac{d^2y}{dx^2}$

The notation $\frac{d^2}{dx^2}$ is purely symbolic. It reminds us that the second derivative is obtained by differentiating $\frac{d}{dx} f(x)$; that is,

$$f'(x) = \frac{d}{dx} f(x),$$

$$f''(x) = \frac{d}{dx}\left[\frac{d}{dx} f(x)\right].$$

If we evaluate the derivative $f'(x)$ at a specific value of x, say $x = a$, we get a number $f'(a)$ that gives the slope of the curve $y = f(x)$ at the point $(a, f(a))$. Another way of writing $f'(a)$ is

$$\frac{dy}{dx}\bigg|_{x=a}.$$

If we have a second derivative $f''(x)$, then its value when $x = a$ is written

$$f''(a) \quad \text{or} \quad \frac{d^2y}{dx^2}\bigg|_{x=a}.$$

EXAMPLE 3   If $y = x^4 - 5x^3 + 7$, find $\dfrac{d^2y}{dx^2}\bigg|_{x=3}$.

Solution

$$\frac{dy}{dx} = \frac{d}{dx}(x^4 - 5x^3 + 7) = 4x^3 - 15x^2$$

$$\frac{d^2y}{dx^2} = \frac{d}{dx}(4x^3 - 15x^2) = 12x^2 - 30x$$

$$\frac{d^2y}{dx^2}\bigg|_{x=3} = 12(3)^2 - 30(3) = 108 - 90 = 18.$$

EXAMPLE 4   If $s = t^3 - 2t^2 + 3t$, find

$$\frac{ds}{dt}\bigg|_{t=-2} \quad \text{and} \quad \frac{d^2s}{dt^2}\bigg|_{t=-2}.$$

Solution

$$\frac{ds}{dt} = \frac{d}{dt}(t^3 - 2t^2 + 3t) = 3t^2 - 4t + 3,$$

$$\frac{ds}{dt}\bigg|_{t=-2} = 3(-2)^2 - 4(-2) + 3 = 12 + 8 + 3 = 23$$

To find the value of the second derivative at $t = -2$, we must first differentiate $\dfrac{ds}{dt}$.

$$\frac{d^2s}{dt^2} = \frac{d}{dt}(3t^2 - 4t + 3) = 6t - 4$$

$$\frac{d^2s}{dt^2}\bigg|_{t=-2} = 6(-2) - 4 = -12 - 4 = -16.$$

PRACTICE PROBLEMS 7

1. Let $f(t) = t + (1/t)$. Find $f''(2)$.   2. Differentiate $g(r) = 2\pi rh$.

Find the first derivatives.

1. $f(t) = (t^2 + 1)^5$

2. $f(P) = P^4 - P^3 + 4P^2 - P$

3. $v = \sqrt{2t - 1}$

4. $g(z) = (z^3 - z + 1)^2$

5. $y = (T^3 + 5T)^{2/3}$

6. $s = \sqrt{t} + (1/\sqrt{t})$

7. Find $\dfrac{d}{dP}(3P^2 - \tfrac{1}{2}P + 1)$

8. Find $\dfrac{d}{dz}(\sqrt{z^2 - 1})$

9. Find $\dfrac{d}{dt}(a^2t^2 + b^2t + c^2)$

10. Find $\dfrac{d}{dx}(x^3 + t^3)$

Find the first and second derivatives.

11. $f(x) = \tfrac{1}{2}x^2 - 7x + 2$

12. $y = (1/x^2) + 1$

13. $y = \sqrt{x}$

14. $f(t) = t^{100} + t + 1$

15. $f(r) = \pi h r^2 + 2\pi r$

16. $v = t^{3/2} + t$

17. $g(x) = 2 - 5x$

18. $V(r) = \tfrac{4}{3}\pi r^3$

19. $f(P) = (3P + 1)^5$

20. $u = (t^6/30) - (t^4/12)$

Compute the following.

21. $\dfrac{d}{dx}(2x^2 - 3)\Big|_{x=5}$

22. $\dfrac{d}{dt}(1 - 2t - 3t^2)\Big|_{t=-1}$

23. $\dfrac{d}{dz}(z^2 - 4)^3\Big|_{z=1}$

24. $\dfrac{d}{dT}\left(\dfrac{1}{3T + 1}\right)\Big|_{T=2}$

25. $\dfrac{d^2}{dx^2}(3x^3 - x^2 + 7x - 1)\Big|_{x=2}$

26. $\dfrac{d}{dt}\left(\dfrac{dv}{dt}\right)$, where $v = 2t^{-3}$

27. $\dfrac{d}{dP}\left(\dfrac{dy}{dP}\right)$, where $y = \dfrac{k}{2P - 1}$

28. $\dfrac{d^2V}{dr^2}\Big|_{r=2}$, where $V = ar^3$

29. $f'(3)$ and $f''(3)$, when $f(x) = \sqrt{10 - 2x}$

30. $g'(2)$ and $g''(2)$, when $g(T) = (3T - 5)^{10}$

31. Suppose a company finds that the revenue R generated by spending x dollars on advertising is given by $R = 1000 + 80x - .02x^2$, for $0 \le x \le 2000$. Find $\dfrac{dR}{dx}\Big|_{x=1500}$.

32. A supermarket finds that its average daily volume of business V (in thousands of dollars) and the number of hours t the store is open for business each day are approximately related by the formula

$$V = 20\left(1 - \frac{100}{100 + t^2}\right), \qquad 0 \le t \le 24.$$

Find $\dfrac{dV}{dt}\bigg|_{t=10}$.

33. The *third derivative* of a function $f(x)$ is the derivative of the second derivative $f''(x)$ and is denoted $f'''(x)$. Compute $f'''(x)$ for the following functions.

(a) $f(x) = x^5 - x^4 + 3x$ (b) $f(x) = 4x^{5/2}$

34. Compute the third derivatives of the following functions.

(a) $f(t) = t^{10}$ (b) $f(z) = \dfrac{1}{z + 5}$

SOLUTIONS TO PRACTICE PROBLEMS 7

1. $f(t) = t + t^{-1}$.

$f'(t) = 1 + (-1)t^{(-1)-1} = 1 - t^{-2}$.

$f''(t) = -(-2)t^{(-2)-1} = 2t^{-3} = \dfrac{2}{t^3}$.

Therefore, $f''(2) = \dfrac{2}{2^3} = \dfrac{1}{4}$. [Note: It is essential to first compute the function $f''(t)$ and *then* evaluate the function at $t = 2$.]

2. The expression $2\pi rh$ contains two numbers, 2 and π, and two letters, r and h. The notation "$g(r)$" tells us that the expression $2\pi rh$ is to be regarded as a function in r. Therefore h and hence $2\pi h$ is to be treated like a constant, and differentiation is done with respect to the variable r. That is,

$$g(r) = (2\pi h)r$$
$$g'(r) = 2\pi h.$$

1.8. The Derivative as a Rate of Change

An important interpretation of the derivative of a function is as a rate of change. Consider the following situation, for example. The weight of an animal changes as time passes and we may think of the weight as a function of time, say $W(t)$. Suppose that we measure the weight at some time $t = a$ and at a later time $t = a + h$. Then the change in weight over this time interval is

$$W(a + h) - W(a).$$

The average *rate* of change of weight with respect to time during this time interval is found by dividing the change in weight by the length of the time interval:

$$\left.\begin{array}{l}\text{average rate of change}\\\text{of } W(t) \text{ with respect to}\\t, \text{ from } a \text{ to } a + h\end{array}\right\} = \frac{W(a + h) - W(a)}{h}$$

This average rate of change takes into account the change in weight as t increases from $t = a$ to $t = a + h$. In order to study the change in weight $W(t)$ only for t close to a, we must take h very small. But we know that as h becomes small, the quantity $[W(a + h) - W(a)]/h$ approaches the derivative $W'(a)$. Thus,

$$\left.\begin{array}{l}\text{average rate of change}\\\text{of } W(t) \text{ with respect to}\\t, \text{ from } a \text{ to } a + h\end{array}\right\} \approx W'(a),$$

for h very small.

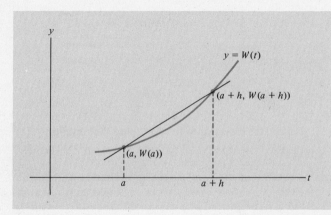

$y = W(t)$

$(a + h, W(a + h))$

$(a, W(a))$

a

$a + h$

FIGURE 1

In this sense, the derivative $W'(a)$ gives the rate of change of $W(t)$ with respect to t precisely at the point when $t = a$.

In geometric terms, the average rate of change of $W(t)$ from a to $a + h$ is the slope of the secant line in Fig. 1. The derivative $W'(a)$ is the slope of the tangent line through $(a, W(a))$ and gives the best measure of the rate of change of the weight at time a.

The derivative of any function $f(t)$ may be interpreted as a rate of change, just as we did for the weight function $W(t)$. We state it formally in the following definition.

> The derivative $f'(a)$ measures the rate of change of $f(t)$ with respect to t, at the point $t = a$.

The examples and exercises that follow show why it is often convenient to use a derivative to measure a rate of change. Other illustrations will be given later in the book.

EXAMPLE 1   A flu epidemic hits a midwestern town. Public Health officials estimate that the number of persons sick with the flu at time t (measured in days from the beginning of the epidemic) is approximated by $P(t) = 60t^2 - t^3$, provided that $0 \leq t \leq 40$. At what rate is the flu spreading at time $t = 30$?

The rate at which the flu spreads is given by the rate of change of $P(t)$, measured in people per day. Now

$$P'(t) = 120t - 3t^2$$
$$P'(30) = 120(30) - 3(30)^2 = 900.$$

Thus, 30 days after the beginning of the epidemic the flu is spreading at the rate of 900 people per day.

EXAMPLE 2 A common clinical procedure for studying a person's calcium metabolism (the rate at which the body assimilates and uses calcium) is to inject some chemically "labeled" calcium into the bloodstream and then measure how fast this calcium is removed from the blood by the person's bodily processes. Suppose that t days after an injection of calcium, the amount A of the labeled calcium remaining in the blood is $A = t^{-3/2}$ for $t \geq .5$, where A is measured in suitable units.* How fast (in "units of calcium per day") is the body removing calcium from the blood when $t = 1$ day?

Solution The rate of change (per day) of calcium in the blood is given by the derivative

$$\frac{dA}{dt} = -\tfrac{3}{2}t^{-5/2}$$

When $t = 1$, this rate equals

$$\left.\frac{dA}{dt}\right|_{t=1} = -\tfrac{3}{2}(1)^{-5/2} = -\tfrac{3}{2}.$$

The amount of calcium in the blood is changing at the rate of $-\tfrac{3}{2}$ units per day when $t = 1$. The negative sign indicates that the amount of calcium is decreasing rather than increasing.

The Marginal Concept in Economics Suppose a company determines that the cost of producing x units of its product is $C(x)$ dollars. Recall from Section 3 that $C'(x)$ is called the marginal cost function. The value of the derivative of $C(x)$ at $x = a$—that is, $C'(a)$—is called the *marginal cost at production level a*. Since the marginal cost is just a derivative, its value gives the rate at which costs are increasing with respect to the level of production, assuming that production is at level a. The use of marginal cost is illustrated in the next example.

EXAMPLE 3 Suppose that the cost function is $C(x) = .005x^3 - 3x$ and production is proceeding at 1000 units per day.

(a) What is the marginal cost at this production level?

(b) What is the cost of increasing production from 1000 to 1001 units per day?

---

* For a discussion of this mathematical model, see J. Defares, I. Sneddon, and M. Wise, *An Introduction to the Mathematics of Medicine and Biology*, (Chicago: Year Book Publishers, Inc., 1973), pp. 609–619.

Solution   (a) The marginal cost at production level 1000 is $C'(1000)$. Thus

$$C'(x) = .015x^2 - 3$$
$$C'(1000) = 14{,}997.$$

(b) The cost of increasing production by one unit, from 1000 to 1001, is $C(1001) - C(1000)$, which equals (rounded to the nearest dollar)

$$[.005(1001)^3 - 3(1001)] - [.005(1000)^3 - 3(1000)] = 5{,}012{,}012 - 4{,}997{,}000$$
$$= 15{,}012.$$

Note that the marginal cost approximately equals the cost of increasing production by one unit. This fact is true of marginal cost functions generally.

Velocity and Acceleration   An everyday illustration of rate of change is given by the velocity of a moving object. Suppose we are driving a car along a straight road and at each time t we let $s(t)$ be our position on the road, measured from some convenient reference point. See Fig. 2, where distances are positive to the right of the reference point and negative to the left. We call $s(t)$ the *position function* of the car. For the moment we shall assume that we are proceeding in only the positive direction along the road.

FIGURE 2

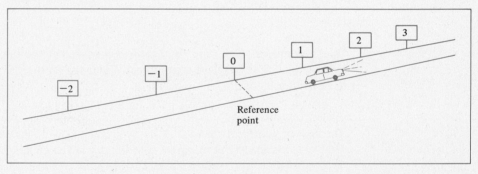

At any instant, the car's speedometer tells us how fast we are moving—that is, how fast our position $s(t)$ is changing. To show how the speedometer reading is related to our calculus concept of a derivative, let us examine what is happening at a specific time, say $t = 1$. Consider a short time interval of duration h from $t = 1$ to $t = 1 + h$. Our car will move from position $s(1)$ to position $s(1 + h)$, a distance of $s(1 + h) - s(1)$. Thus the *average velocity from $t = 1$ to $t = 1 + h$* is

$$\frac{[\text{distance traveled}]}{[\text{time elapsed}]} = \frac{s(1 + h) - s(1)}{h}. \tag{1}$$

If the car is traveling at a steady speed during this time period, then the speedometer reading will equal the average velocity in equation (1).

From our discussion in Section 3 the ratio (1) approaches the derivative $s'(1)$ as h approaches zero. For this reason we call $s'(1)$ *the* (instantaneous) *velocity at* $t = 1$. This number will agree with the speedometer reading at $t = 1$ because when h is very small, the car's speed will be nearly steady over the time interval from $t = 1$ to $t = 1 + h$, and so the average velocity over this time interval will be nearly the same as the speedometer reading at $t = 1$.

The reasoning used for $t = 1$ holds for an arbitrary t as well. Thus the following definition makes sense:

> If $s(t)$ denotes the position function of an object moving in a straight line, then the velocity $v(t)$ of the object at time t is given by
> $$v(t) = s'(t).$$

In our discussion we assumed that the car moved in the positive direction. If the car moves in the opposite direction, the ratio (1) and the limiting value $s'(1)$ will be negative. So we interpret negative velocity as movement in the negative direction along the road.

The derivative of the velocity function $v(t)$ is called the *acceleration* function and is often written as $a(t)$:

> $$a(t) = v'(t).$$

Since $v'(t)$ measures the rate of change of the velocity $v(t)$, this use of the word acceleration agrees with our common usage in connection with automobiles. Note that since $v(t) = s'(t)$, the acceleration is actually the second derivative of the position function $s(t)$:

$$a(t) = s''(t).$$

PRACTICE PROBLEMS 8

When a ball is thrown straight up into the air, it travels along a straight line and its motion can be described in the same manner as the motion of a car. Regard "up" as the positive direction and let $s(t)$ be the height of the ball in feet after t seconds. Suppose that $s(t) = -16t^2 + 128t + 5$.

1. What will be the velocity after 2 seconds?

2. What will be the acceleration after 2 seconds?

3. After how many seconds will the ball attain its greatest height?

1. If $f(t) = t^2 + 3t - 7$, what is the rate of change of $f(t)$ with respect to t when $t = 5$?

2. If $f(t) = 3t + 2 - \dfrac{5}{t}$, what is the rate of change of $f(t)$ with respect to t when $t = 2$?

3. An analysis of the daily output of a factory assembly line shows that about $60t + t^2 - \frac{1}{12}t^3$ units are produced after t hours of work, $0 \le t \le 8$. What is the rate of production (in units per hour) when $t = 2$?

4. Liquid is pouring into a large vat. After t hours, there are $5t - t^{1/2}$ gallons in the vat. At what rate is the liquid flowing into the vat (in gallons per hour) when $t = 4$?

5. Suppose that the weight in grams of a cancerous tumor at time t is $W(t) = .1t^2$, where t is measured in weeks. What is the rate of growth of the tumor (in grams per week) when $t = 5$?

6. After an advertising campaign, the sales of a product often increase and then decrease. Suppose that t days after the end of the advertising, the daily sales are $-3t^2 + 30t + 100$ units. At what rate (in units per day) are the sales changing when $t = 2$?

7. A sewage treatment plant accidentally discharged untreated sewage into a lake for a few days. This temporarily decreased the amount of dissolved oxygen in the lake. Let $f(t)$ be the amount of oxygen in the lake (measured in suitable units) t days after the sewage started flowing into the lake. Experimental data suggest that $f(t)$ is given approximately by

$$f(t) = 500\left[1 - \frac{10}{t + 10} + \frac{100}{(t + 10)^2}\right].$$

Find the rate of change (in units per day) of the oxygen content of the lake at $t = 5$ and at $t = 15$. Is the oxygen content increasing or decreasing when $t = 15$?

8. Suppose that t hours after being placed in a freezer, the temperature of a piece of meat is given by $T(t) = 70 - 12t + 4/(t + 1)$ degrees, where $0 \le t \le 5$. How fast is the temperature falling after one hour?

9. A manufacturer estimates that the hourly cost of producing x units of a product on an assembly line is $.1x^3 - 6x^2 + 136x + 200$ dollars.

   (a) Find the marginal cost when the production level is 20 units.

   (b) Show that this marginal cost is approximately the extra cost of raising the production from 20 to 21 units.

10. A market finds that if it prices a certain product so as to sell x units each week, then the revenue received will be approximately $2x - .001x^2$ dollars.

   (a) Find the marginal revenue at a sales level of 600 units.

   (b) Show that this marginal revenue is approximately the extra revenue produced by raising the sales from 600 to 601 units.

11. Suppose that the profit from producing x units of a product is given by $P(x) = .0003x^3 + .01x$.

   (a) Compute the marginal profit at a production level of 100 units.

   (b) Compute the additional profit gained from increasing sales from 100 to 101 units.

12. Suppose that the revenue from producing x units of a product is given by $R(x) = .01x^2 - 3x$.

   (a) Compute the marginal revenue at a production level of 20,000.

   (b) Compute the additional revenue gained by increasing production from 20,000 to 20,001 units.

13. An object moving in a straight line travels $s(t)$ kilometers in t hours, where $s(t) = \frac{1}{2}t^2 + 4t$.

   (a) What is the object's velocity when $t = 3$?

   (b) How far has the object traveled in three hours?

14. Suppose that the position of a car at time t is given by $s(t) = 50t - 7/(t + 1)$, where the position is measured in kilometers. Find the velocity and acceleration of the car at $t = 0$.

15. A toy rocket fired straight up into the air has height $s(t) = 160t - 16t^2$ feet after t seconds.

   (a) What is the rocket's initial velocity (when $t = 0$)?

   (b) What is the velocity after two seconds?

   (c) What is the acceleration when $t = 3$?

   (d) At what time will the rocket hit the ground?

   (e) At what velocity will the rocket be traveling just as it smashes into the ground?

16. A helicopter is rising straight up in the air. Its distance from the ground t seconds after takeoff is $s(t)$ feet, where $s(t) = t^2 + t$.

   (a) How long will it take for the helicopter to rise 20 feet?

   (b) Find the velocity and the acceleration of the helicopter when it is 20 feet above the ground.

17. During the epidemic in Example 1, is there any day when approximately 1200 people become sick with the flu?

$$t^2(60 - t) = 1200$$
$$60 - t = 1200$$
$$-t = 1140$$
$$t =$$

18. Suppose that the cost of producing x units of some product is $\frac{1}{3}x^3 - 5x^2 + 30x + 10$ dollars, while the revenue received from the sale of x units is $39x - x^2$ dollars. At what value(s) of x will the marginal revenue equal the marginal cost?

19. A subway train travels from Station A to Station B in 2 minutes. Its distance from Station A is $s(t) = t^2 - \frac{1}{3}t^3$ kilometers after t minutes. At what time(s) will the train be traveling at the rate of $\frac{1}{2}$ kilometer per minute?

20. Let $f(t)$ be the function in Exercise 7 that gives the amount of oxygen in the lake at time t. Find the time when the amount of oxygen stops decreasing and begins to increase.

SOLUTIONS TO PRACTICE PROBLEMS 8

1. Since velocity is rate of change of position, $v(t) = s'(t) = -32t + 128$. The velocity after two seconds is $v(2) = -32(2) + 128 = 64$. Therefore, after two seconds the ball will be traveling upward at a velocity of 64 feet per second.

2. The acceleration is the rate of change of velocity, so $a(t) = v'(t) = -32$. The acceleration is always -32. This constant acceleration is due to the downward (and therefore negative) force of gravity.

3. At the instant at which the ball attains its maximum height it will have zero velocity. Therefore, we can determine this time by setting $v(t) = 0$ and solving for t.

$$-32t + 128 = 0$$
$$t = 4.$$

So the ball attains its maximum height after four seconds.

Chapter 1 : CHECKLIST

□ Slope of a line
□ y-intercept
□ Slope Properties 1 to 5
□ Slope of a curve at a point
□ The derivative of a constant function is zero
□ Limit definition of the derivative
□ Power rule:

$$\frac{d}{dx}(x^r) = rx^{r-1}$$

□ General power rule:

$$\frac{d}{dx}([g(x)]^r) = r \cdot [g(x)]^{r-1} \cdot \frac{d}{dx}[g(x)]$$

□ Constant-multiple rule:

$$\frac{d}{dx}[k \cdot f(x)] = k \cdot \frac{d}{dx}[f(x)]$$

□ Sum rule:

$$\frac{d}{dx}[f(x) + g(x)] = \frac{d}{dx}[f(x)] + \frac{d}{dx}[g(x)]$$

□ Notation for first and second derivatives
□ The derivative $f'(a)$ measures the rate of change of $f(t)$ with respect to t at $t = a$
□ Marginal concept in economics
□ Position, velocity, and acceleration functions
□ Differentiable at $x = a$
□ Continuous at $x = a$
□ $\lim\limits_{x \to a} f(x)$
□ $\lim\limits_{x \to \infty} f(x)$, $\lim\limits_{x \to -\infty} f(x)$

Chapter 1 : SUPPLEMENTARY EXERCISES

Find the equation and sketch the graph of the following lines.

1. With slope -2, y-intercept $(0, 3)$.

2. With slope $\frac{3}{4}$, y-intercept $(0, -1)$.

3. Through $(2, 0)$, with slope 5.

4. Through $(1, 4)$, with slope $-\frac{1}{3}$.

5. Parallel to $y = -2x$, passing through $(3, 5)$.

6. Parallel to $-2x + 3y = 6$, passing through $(0, 1)$.

7. Through $(-1, 4)$, and $(3, 7)$.

8. Through $(2, 1)$ and $(5, 1)$.

Differentiate.

9. $y = x^7 + x^3$

10. $y = 5x^8$

11. $y = 6\sqrt{x}$

12. $y = x^7 + 3x^5 + 1$

13. $y = 3/x$

14. $y = x^4 - (4/x)$

15. $y = (3x^2 - 1)^8$

16. $y = \frac{3}{4}x^{4/3} + \frac{4}{3}x^{3/4}$

17. $y = \dfrac{1}{5x - 1}$

18. $y = (x^3 + x^2 + 1)^5$

19. $y = \sqrt{x^2 + 1}$

20. $y = \dfrac{5}{7x^2 + 1}$

21. $f(x) = x^{-4}$

22. $f(x) = (2x + 1)^3 - 5(2x + 1)^2$

23. $f(x) = 5$

24. $f(x) = \dfrac{5x}{2} - \dfrac{2}{5x}$

25. $f(x) = [x^5 - (x - 1)^5]^{10}$

26. $f(t) = t^{10} - 10t^9$

27. $g(t) = 3\sqrt{t} - \dfrac{3}{\sqrt{t}}$

28. $h(t) = 3\sqrt{2}$

29. $f(t) = \dfrac{2}{t - 3t^3}$

30. $g(P) = 4P^{.7}$

31. $h(x) = \frac{3}{2}x^{3/2} - 6x^{2/3}$

32. $f(x) = \sqrt{x + \sqrt{x}}$

33. If $f(t) = 3t^3 - 2t^2$, find $f'(2)$.

34. If $V(r) = 15\pi r^2$, find $V'(\frac{1}{3})$.

35. If $g(u) = 3u - 1$, find $g(5)$ and $g'(5)$.

36. If $h(x) = -\frac{1}{2}$, find $h(-2)$ and $h'(-2)$.

37. If $f(x) = x^{5/2}$, what is $f''(4)$?

38. If $g(t) = \frac{1}{4}(2t - 7)^4$, what is $g''(3)$?

39. Find the slope of the graph of $y = (3x - 1)^3 - 4(3x - 1)^2$ at $(0, -5)$.

40. Find the slope of the graph of $y = (4 - x)^5$ at the point where $x = 5$.

Compute.

41. $\dfrac{d}{dx}(x^4 - 2x^2)$

42. $\dfrac{d}{dt}(t^{5/2} + 2t^{3/2} - t^{1/2})$

43. $\dfrac{d}{dP}(\sqrt{1 - 3P})$

44. $\dfrac{d}{dn}(n^{-5})$

45. $\dfrac{d}{dz}(z^3 - 4z^2 + z - 3)\Big|_{z=-2}$

46. $\dfrac{d}{dx}(4x - 10)^5\Big|_{x=3}$

47. $\dfrac{d^2}{dx^2}(5x + 1)^4$

48. $\dfrac{d^2}{dt^2}(2\sqrt{t})$

49. $\dfrac{d^2}{dt^2}(t^3 + 2t^2 - t)\Big|_{t=-1}$

50. $\dfrac{d^2}{dP^2}(3P + 2)\Big|_{P=4}$

51. $\dfrac{d^2y}{dx^2}$ where $y = 4x^{3/2}$

52. $\dfrac{d}{dt}\left(\dfrac{dy}{dt}\right)$, where $y = \dfrac{1}{3t}$

53. What is the slope of the graph of $f(x) = x^3 - 4x^2 + 6$ at $x = 2$? Write the equation of the line tangent to the graph of $f(x)$ at $x = 2$.

54. What is the slope of the curve $y = 1/(3x - 5)$ at the point $(1, -\frac{1}{2})$? Write the equation of the line tangent to this curve at $(1, -\frac{1}{2})$.

55. Find the equation of the tangent line to the curve $y = x^2$ at the point $(\frac{3}{2}, \frac{9}{4})$. Sketch the graph of $y = x^2$ and sketch the tangent line at $(\frac{3}{2}, \frac{9}{4})$.

56. Find the equation of the tangent line to the curve $y = x^2$ at the point $(-2, 4)$. Sketch the graph of $y = x^2$ and sketch the tangent line at $(-2, 4)$.

57. Determine the equation of the tangent line to the curve $y = 3x^3 - 5x^2 + x + 3$ at $(1, 2)$.

58. Determine the equation of the tangent line to the curve $y = (2x^2 - 3x)^3$ at $(2, 8)$.

59. A helicopter is rising at a rate of 32 feet per second. At a height of 128 feet the pilot drops a pair of binoculars. After t seconds, the binoculars have height $s(t) = -16t^2 + 32t + 128$ feet from the ground. How fast will they be falling when they hit the ground?

60. The daily output of a coal mine after t hours of operation is approximately $40t + t^2 - \frac{1}{15}t^3$ tons, $0 \le t \le 12$. What is the rate of output (in tons of coal per hour) at $t = 5$ hours?

Determine whether the following limits exist. If so, compute the limit.

61. $\lim\limits_{x \to 2} \dfrac{x^2 - 4}{x - 2}$

62. $\lim\limits_{x \to 3} \dfrac{1}{x^2 - 4x + 3}$

63. $\lim\limits_{x \to 4} \dfrac{x - 4}{x^2 - 8x + 16}$

64. $\lim\limits_{x \to 5} \dfrac{x - 5}{x^2 - 7x + 2}$

Use limits to compute the following derivatives.

65. $f'(5)$, where $f(x) = 1/(2x)$.

66. $f'(3)$, where $f(x) = x^2 - 2x + 1$.

APPLICATIONS
OF THE
DERIVATIVE

Calculus techniques can be applied to a wide variety of problems in real life. We shall consider many examples in this chapter. In each case we shall construct a function as a "mathematical model" of some problem and then analyze the function and its derivatives in order to gain information about the original problem. Our principal method for analyzing a function will be to sketch its graph. For this reason we devote the first part of the chapter to curve sketching.

2.1. Describing Graphs of Functions

The graph of the function in Fig. 1 is either rising or falling, depending on whether we look at it from left to right or from right to left. To avoid confusion, we shall always follow the accepted practice of reading a graph from left to right. We say

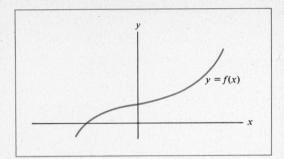

FIGURE 1

that a function $f(x)$ is *increasing* at $x = a$ if, in some small region near the point $(a, f(a))$, the graph is rising as we move from left to right. The function is *decreasing* at $x = a$ if, in some small region near the point $(a, f(a))$, the graph is falling. See Fig. 2.

An *extreme point* of a graph is a point at which the graph changes from increasing to decreasing or vice versa. We distinguish the two possibilities in an obvious way. A *maximum* point is a point at which the graph changes from increasing to decreasing; a *minimum* point is a point at which the graph changes from decreasing to increasing (Fig. 3).

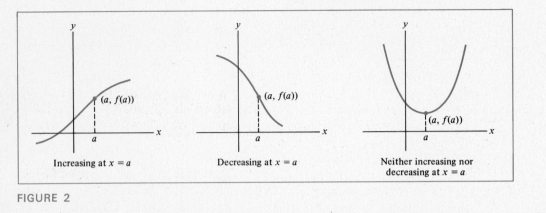

FIGURE 2

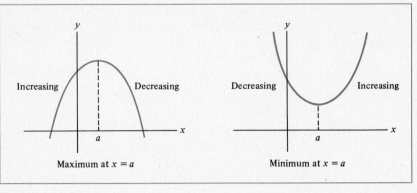

FIGURE 3

EXAMPLE 1   When a drug is injected intramuscularly (into a muscle), the concentration of the drug in the veins has the time-concentration curve shown in Fig. 4. Describe this graph, using the terms introduced above.

Solution   Initially (when $t = 0$), there is no drug in the veins. When the drug is injected into the muscle, it begins to diffuse into the bloodstream. The concentration in the

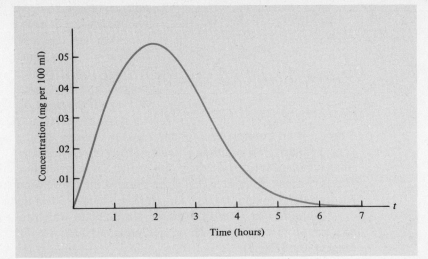

FIGURE 4

veins increases until it reaches a maximum at $t = 2$. After this time, the concentration begins to decrease, as the body's metabolic processes remove the drug from the blood. Eventually the drug concentration decreases to a level so small that, for all practical purposes, it is zero.

The graphs in Fig. 5 are both increasing, but there is a fundamental difference in the way they are increasing. Graph I, which describes the United States national debt per person, is steeper for 1970 than for 1960. That is, the *slope* of graph I is *increasing* as we move from left to right. On the other hand, the *slope* of graph II is *decreasing* as we move from left to right. Although the U.S. population is rising each year, the rate of increase was not as great in 1970 as it was in 1960.

The difference between the two graphs in Fig. 5 can also be described in geometric terms: graph I opens up and lies above its tangent line at each point,

FIGURE 5

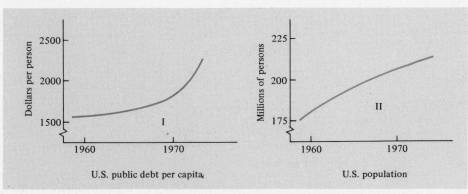

U.S. public debt per capita   U.S. population

FIGURE 6

whereas graph II opens down and lies below its tangent line at each point (Fig. 6).

We say that a function $f(x)$ is *concave up* at $x = a$ if, in some small region near the point $(a, f(a))$, the graph of $f(x)$ lies above its tangent line. Equivalently, $f(x)$ is concave up at $x = a$ if the slope of the graph increases as we move from left to right through $(a, f(a))$. Graph I is an example of a function that is concave up at each point.

Similarly, we say that a function $f(x)$ is *concave down* at $x = a$ if, in some small region near $(a, f(a))$, the graph of $f(x)$ lies below its tangent line. Equivalently, $f(x)$ is concave down at $x = a$ if the slope of the graph decreases as we move from left to right through $(a, f(a))$. Graph II is concave down at each point.

An *inflection point* of a graph is a point at which the graph changes from concave up to concave down or vice versa. At such a point, the graph crosses its tangent line (Fig. 7).

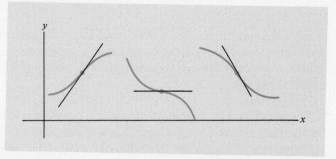

FIGURE 7

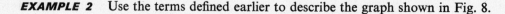

EXAMPLE 2   Use the terms defined earlier to describe the graph shown in Fig. 8.

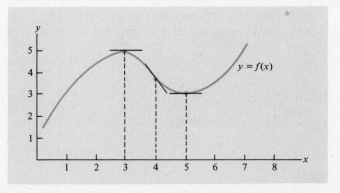

FIGURE 8

Solution   (a) For $x < 3$, $f(x)$ is increasing and concave down.

(b) Maximum at $x = 3$.

(c) For $3 < x < 4$, $f(x)$ is decreasing and concave down.

(d) Inflection point at $x = 4$.

(e) For $4 < x < 5$, $f(x)$ is decreasing and concave up.

(f) Minimum at $x = 5$.

(g) For $x > 5$, $f(x)$ is increasing and concave up.

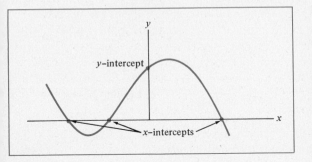

FIGURE 9

Intercepts, Undefined Points, Asymptotes
A point at which a graph crosses the y-axis is called a *y-intercept*, and a point at which it crosses the x-axis is called an *x-intercept*. The x-coordinate of an x-intercept is sometimes called a "zero" of the function, since the function has the value zero there. (See Fig. 9.)

Some functions are not defined for all values of x. For instance, $f(x) = 1/x$ is not defined for $x = 0$, and $f(x) = \sqrt{x}$ is not defined for $x < 0$. (See Fig. 10.) Many functions that arise in applications are defined only for $x \geq 0$. A properly drawn graph should leave no doubt as to the values of x for which the function is defined.

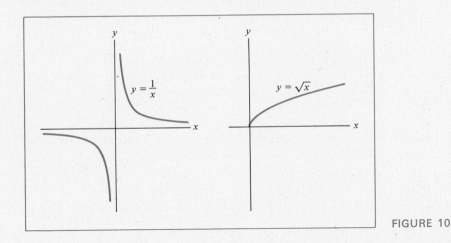

FIGURE 10

Graphs in applied problems sometimes straighten out and approach some straight line as x gets large (Fig. 11). Such a straight line is called an *asymptote* of the curve. The most common asymptotes are horizontal as in (a) and (b) of Fig. 11. In Example 1 the t-axis is an asymptote of the drug time-concentration curve.

The horizontal asymptotes of a graph may be determined by calculating the limits

$$\lim_{x \to \infty} f(x) \quad \text{and} \quad \lim_{x \to -\infty} f(x).$$

If either limit exists, then the value of the limit determines a horizontal asymptote.

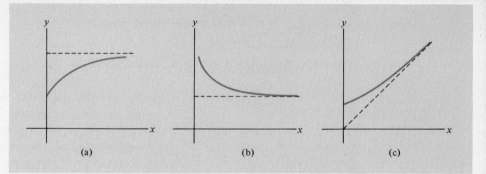

FIGURE 11   Graphs that approach asymptotes as x gets large.

Occasionally a graph will approach a vertical line as x approaches some fixed point, as in Fig. 12. Such a line is a *vertical asymptote*. Most often, we expect a vertical asymptote at a value x which would result in division by zero in the definition of $f(x)$. For example, $f(x) = 1/(x - 3)$ has a vertical asymptote $x = 3$.

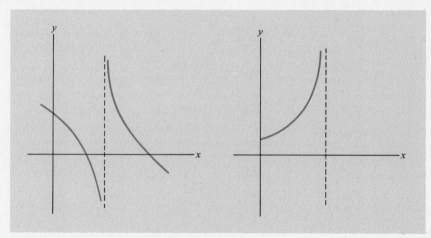

FIGURE 12   Examples of vertical asymptotes.

We now have five categories for describing the graph of a function:

1. Increasing, decreasing, maximum points, minimum points
2. Concave up, concave down, inflection points
3. x-intercept, y-intercept
4. Undefined points
5. Asymptotes

For us, the first two categories will be the most important. However, the last three categories should not be forgotten.

PRACTICE PROBLEMS 1

1. Does the slope of the curve in Fig. 13 increase or decrease as x increases?
2. At what value of x is the slope of the curve in Fig. 14 minimized?

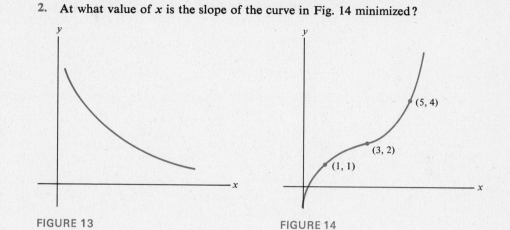

FIGURE 13   FIGURE 14

EXERCISES 1

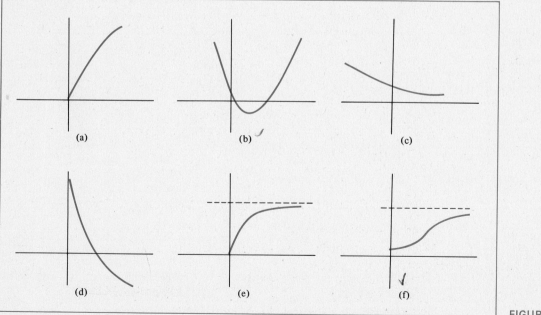

(a)   (b)   (c)

(d)   (e)   (f)

FIGURE 15

Exercises 1–4 refer to graphs (a)–(f) in Fig. 15.

1. Which functions are increasing for all x?

2. Which functions are decreasing for all x?

3. Which functions have the property that the slope always increases as x increases?

4. Which functions have the property that the slope always decreases as x increases?

Describe each graph below. Your description should include each of the five categories mentioned on page 108.

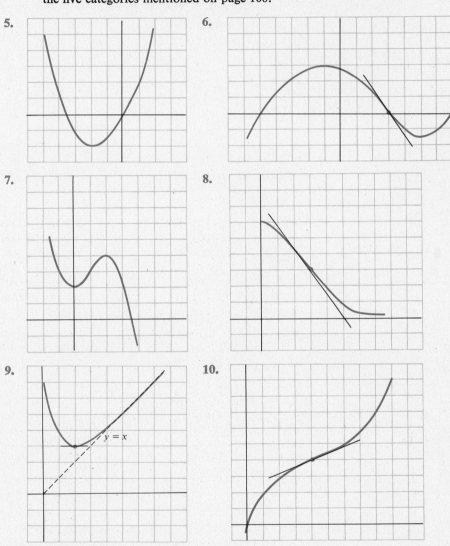

5.

6.

7.

8.

9.

$y = x$

10.

11. Describe the way the *slope* changes as you move along the graph (from left to right) in Exercise 5.

12. Describe the way the *slope* changes on the graph in Exercise 6.

13. Describe the way the *slope* changes on the graph in Exercise 8.

14. Describe the way the *slope* changes on the graph in Exercise 10.

15. Suppose that some organic waste products are dumped into a lake at time $t = 0$, and suppose that the oxygen content of the lake at time t is given by the graph in Fig. 16. Describe the graph in physical terms. Indicate the significance of the inflection point at $t = b$.

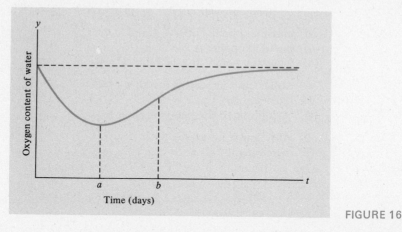

Time (days)

FIGURE 16

16. Let $C(x)$ denote the total cost of manufacturing x units of some product. Then $C(x)$ is an increasing function for all x. For small values of x, the rate of increase of $C(x)$ decreases. (This is because of the savings that are possible with "mass production.") Eventually, however, for large values of x, the cost $C(x)$ increases at an increasing rate. (This happens when production facilities are strained and become less efficient.) Sketch a graph that could represent $C(x)$.

17. The annual world consumption of oil rises each year. Furthermore, the amount of the annual *increase* in oil consumption is also rising each year. Sketch a graph that could represent the annual world consumption of oil.

18. In certain professions the average annual income has been rising at an increasing rate. Let $f(T)$ denote the average annual income at year T for persons in one of these professions and sketch a graph that could represent $f(T)$.

19. Let $s(t)$ be the distance (in feet) traveled by a parachutist after t seconds from the time of opening the chute and suppose that $s(t)$ has the line $y = -15t$ as an asymptote. What does this imply about the velocity of the parachutist? (Note: Distance traveled downwards is given a negative value.)

20. Let $P(t)$ be the population of a bacteria culture after t days and suppose that $P(t)$ has the line $y = 25,000,000$ as an asymptote. What does this imply about the size of the population?

21. Consider a smooth curve with no undefined points.
    (a) If it has two maximum points must it have a minimum point?
    (b) If it has two extreme points must it have an inflection point?

22. Can a point of a graph be both

    (a) a minimum point and an x-intercept?
    (b) an x-intercept and a y-intercept?
    (c) a maximum point and a minimum point?
    (d) an inflection point and an x-intercept?

In Exercises 23–26, sketch the graph of a function having the given properties.

23. (i) minimum point at $x = 0$
     (ii) inflection point at $x = 3$
     (iii) no maximum point.

24. (i) maximum points at $x = 1$ and $x = 5$
     (ii) minimum point at $x = 3$
     (iii) inflection points at $x = 2$ and $x = 4$.

25. (i) defined and increasing for all $x \geq 0$
     (ii) inflection point at $x = 5$
     (iii) asymptotic to the line $y = \frac{3}{4}x + 5$.

26. (i) inflection point at $x = 1$
     (ii) increasing for all x.

SOLUTIONS TO PRACTICE PROBLEMS 1

1. The curve is concave up and so the slope increases. Even though the curve itself is decreasing, the slope becomes less negative as we move from left to right.

2. At $x = 3$. We have drawn in tangent lines at various points. Note that as we move from left to right, the slopes decrease steadily until the point $(3, 2)$, at which time they start to increase. This is consistent with the fact that the graph is concave down (hence, slopes are decreasing) to the left of $(3, 2)$ and concave up (hence, slopes are increasing) to the right of $(3, 2)$. Extreme values of slopes always occur at inflection points.

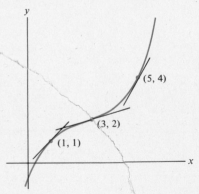

2.2. The First and Second Derivative Rules

We shall now show how properties of the graph of a function $f(x)$ are determined by properties of the derivatives, $f'(x)$ and $f''(x)$. These relationships will provide the key to the curve-sketching and optimization problems discussed in the rest of the chapter.*

We begin with a discussion of the first derivative of a function $f(x)$. Suppose that for some value of x, say $x = a$, the derivative $f'(a)$ is positive. Then the

---

*Throughout this chapter, we shall assume that we are dealing with functions which are not "too badly behaved." More precisely, it suffices to assume that all of our functions have continuous first and second derivatives in the interval(s) (in x) where we are considering their graphs.

tangent line at $(a, f(a))$ has positive slope and is a rising line (moving from left to right, of course). Since the graph of $f(x)$ near $(a, f(a))$ resembles its tangent line, the function must be increasing at $x = a$. Similarly, when $f'(a) < 0$, the function is decreasing at $x = a$. (See Fig. 1.)

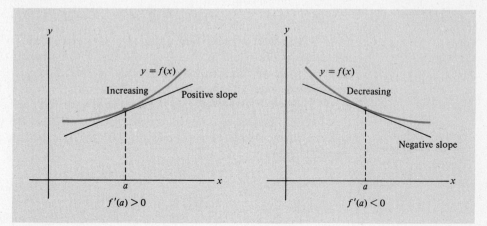

FIGURE 1

Thus we have the following useful result.

> *First Derivative Rule:* If $f'(a) > 0$, then $f(x)$ is increasing at $x = a$.
> If $f'(a) < 0$, then $f(x)$ is decreasing at $x = a$.

When $f'(a) = 0$, the first derivative rule gives no information. In this case, the function $f(x)$ might be increasing or decreasing, or have an extreme point, at $x = a$.

EXAMPLE 1   Sketch the graph of a function $f(x)$ that has all of the following properties:

(a) $f(3) = 4$,

(b) $f'(x) > 0$ for $x < 3$, $f'(3) = 0$ and $f'(x) < 0$ for $x > 3$.

Solution

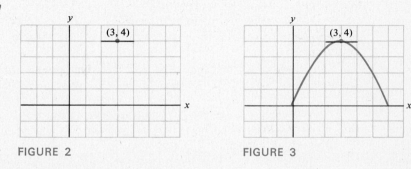

FIGURE 2   FIGURE 3

The only specific point on the graph is (3, 4) [condition (a)]. We plot this point and then use the fact that $f'(3) = 0$ to sketch the tangent line at $x = 3$ (Fig. 2).

From condition (b) and the first derivative rule, we know that $f(x)$ must be increasing for x less than 3 and decreasing for x greater than 3. A graph with these properties might look like the curve in Fig. 3.

The second derivative of a function $f(x)$ gives useful information about the concavity of the graph of $f(x)$. Suppose that $f''(a)$ is negative. Then since $f''(x)$ is the derivative of $f'(x)$, we conclude that $f'(x)$ has a negative derivative at $x = a$. In this case, $f'(x)$ must be a decreasing function at $x = a$; that is, the slope of the graph of $f(x)$ is decreasing as we move from left to right on the graph near $(a, f(a))$. (See Fig. 4.) This means that the graph of $f(x)$ is concave down at $x = a$. A similar analysis shows that if $f''(a)$ is positive, then $f(x)$ is concave up at $x = a$. Thus we have the following rule.

Second Derivative Rule: If $f''(a) > 0$, then $f(x)$ is concave up at $x = a$. If $f''(a) < 0$, then $f(x)$ is concave down at $x = a$.

When $f''(a) = 0$, the second derivative rule gives no information. In this case, the function might be concave up, concave down, or neither, at $x = a$.

The chart below shows how a graph may combine the properties of increasing, decreasing, concave up, and concave down.

Condition on the derivatives		Description of $f(x)$ at $x = a$	Graph of $y = f(x)$ near $x = a$
1. $f'(a)$ positive $f''(a)$ positive		$f(x)$ increasing $f(x)$ concave up	
2. $f'(a)$ positive $f''(a)$ negative		$f(x)$ increasing $f(x)$ concave down	
3. $f'(a)$ negative $f''(a)$ positive		$f(x)$ decreasing $f(x)$ concave up	
4. $f'(a)$ negative $f''(a)$ negative		$f(x)$ decreasing $f(x)$ concave down	

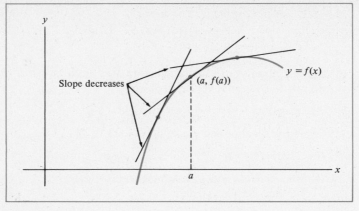

FIGURE 4

EXAMPLE 2   Sketch the graph of a function $f(x)$ with all of the following properties:

(a) $(2, 3)$, $(4, 5)$, and $(6, 7)$ are on the graph,

(b) $f'(6) = 0$ and $f'(2) = 0$,

(c) $f''(x) > 0$ for $x < 4$, $f''(4) = 0$ and $f''(x) < 0$ for $x > 4$.

Solution   First we plot the three points from condition (a) and then sketch two tangent lines, using the information from condition (b). (See Fig. 5.) From condition (c) and the second derivative rule, we know that $f(x)$ is concave up for $x < 4$. In particular, $f(x)$ is concave up at $(2, 3)$. Also, $f(x)$ is concave down for $x > 4$ and, in particular, at $(6, 7)$. Note that $f(x)$ must have an inflection point at $x = 4$ because the concavity changes there. We now sketch small portions of the curve near $(2, 3)$ and $(6, 7)$. (See Fig. 6.) We can now complete the sketch (Fig. 7), taking care to make the curve concave up for $x < 4$ and concave down for $x > 4$.

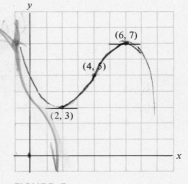

FIGURE 5

FIGURE 6

FIGURE 7

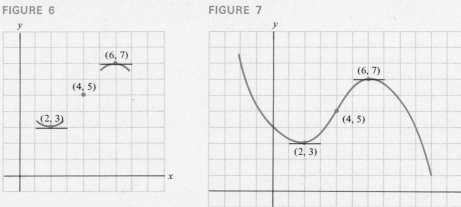

PRACTICE PROBLEMS 2

1. Make a good sketch of the function $f(x)$ near the point where $x = 2$, given that $f(2) = 5$, $f'(2) = 1$, $f''(2) = -3$.

2. The graph of $f(x) = x^3$ is on the right.

   (a) Is the function increasing at $x = 0$?

   (b) Compute $f'(0)$.

   (c) Reconcile your answers to (a) and (b) with the first derivative rule.

$y = x^3$

EXERCISES 2

Exercises 1–4 refer to the functions whose graphs are given in Fig. 8.

1. Which functions have a positive first derivative for all x?

2. Which functions have a negative first derivative for all x?

3. Which functions have a positive second derivative for all x?

4. Which functions have a negative second derivative for all x?

   In Exercises 5–10 sketch the graph of a function $f(x)$ that has the properties described.

5. $f(2) = 1$; $f'(2) = 0$; concave up for all x.

6. $f(-1) = 0$; $f'(x) < 0$ for $x < -1$, $f'(-1) = 0$ and $f'(x) > 0$ for $x > -1$.

7. $f(3) = 5$; $f'(x) > 0$ for $x < 3$, $f'(3) = 0$ and $f'(x) > 0$ for $x > 3$.

8. $(-2, -1)$ and $(2, 5)$ are on the graph; $f'(-2) = 0$ and $f'(2) = 0$; $f''(x) > 0$ for $x < 0$, $f''(0) = 0$, $f''(x) < 0$ for $x > 0$.

9. $(0, 6)$, $(2, 3)$, and $(4, 0)$ are on the graph; $f'(0) = 0$ and $f'(4) = 0$; $f''(x) < 0$ for $x < 2$; $f''(2) = 0$, $f''(x) > 0$ for $x > 2$.

10. $f(x)$ defined only for $x \geq 0$; $(0, 0)$ and $(5, 6)$ are on the graph; $f'(x) > 0$ for $x \geq 0$; $f''(x) < 0$ for $x < 5$, $f''(5) = 0$, $f''(x) > 0$ for $x > 5$.

11. By looking at the first derivative, decide which of the curves in Fig. 9 could *not* be the graph of $f(x) = (3x^2 + 1)^4$ for $x \geq 0$.

12. By looking at the first derivative, decide which of the curves in Fig. 9 could *not* be the graph of $f(x) = x^3 - 9x^2 + 24x + 1$ for $x \geq 0$. [Hint: Factor the formula for $f'(x)$.]

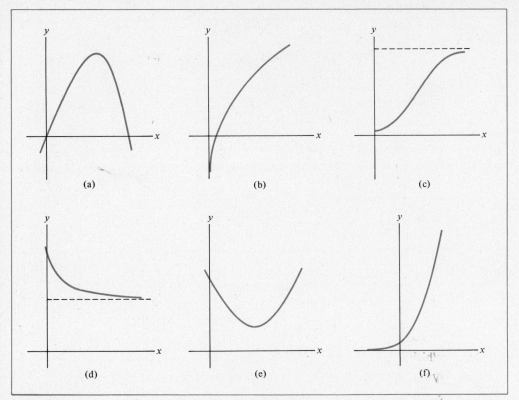

FIGURE 8

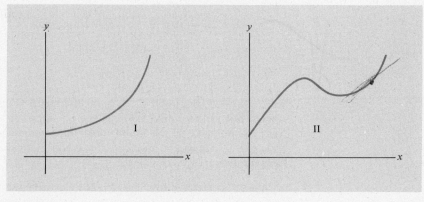

FIGURE 9

13. By looking at the second derivative, decide which of the curves in Fig. 10 could be the graph of $f(x) = \sqrt{x}$.

14. By looking at the second derivative, decide which of the curves in Fig. 10 could be the graph of $f(x) = x^{5/2}$.

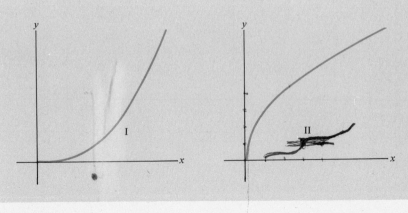

FIGURE 10

In Exercises 15–20, use the given information to make a good sketch of the function $f(x)$ near $x = 3$.

15. $f(3) = 4, f'(3) = -\frac{1}{2}, f''(3) = 5$

16. $f(3) = -2, f'(3) = 0, f''(3) = 1$

17. $f(3) = 1, f'(3) = 0, f''(3) = 0$, inflection point at $x = 3, f'(x) > 0$ for $x > 3$

18. $f(3) = 4, f'(3) = -\frac{3}{2}, f''(3) = -2$

19. $f(3) = -2, f'(3) = 2, f''(3) = 3$

20. $f(3) = 3, f'(3) = 1, f''(3) = 0$, inflection point at $x = 3, f''(x) < 0$ for $x > 3$

SOLUTIONS TO PRACTICE PROBLEMS 2

1. Since $f(2) = 5$, the point $(2, 5)$ is on the graph [Fig. 11(a)]. Since $f'(2) = 1$, the tangent line at the point $(2, 5)$ has slope 1. Draw in the tangent line. [Fig. 11(b)]. Near the point $(2, 5)$ the graph looks approximately like the tangent line. Since $f''(2) = -3$, a negative number, the graph is concave down at the point $(2, 5)$. Now we are ready to sketch the graph [Fig. 11(c)].

2. (a) Yes. The graph is steadily increasing as we pass through the point $(0, 0)$.

   (b) Since $f'(x) = 3x^2$, $f'(0) = 3 \cdot 0^2 = 0$.

   (c) There is no contradiction here. The first derivative rule says that if the derivative is positive, the function is increasing. However, it does not say that this is the only condition under which a function is increasing. As we have just seen, sometimes we can have the first derivative zero and the function still increasing.

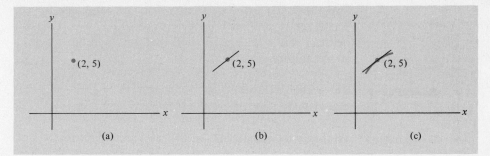

FIGURE 11

2.3. Curve Sketching (Introduction)

In this section and the next we develop our ability to sketch the graphs of functions. There are two important reasons for doing so. First, a geometric "picture" of a function is often easier to comprehend than its abstract formula. Second, the material in this section will provide a foundation for the applications in Sections 5 to 7.

A "sketch" of the graph of a function $f(x)$ should convey the general shape of the graph—it should show where $f(x)$ is increasing and decreasing and it should indicate, insofar as possible, where $f(x)$ is concave up and concave down. In addition, one or more key points should be accurately located on the graph. These points usually include extreme points, inflection points, and x- and y-intercepts. Other features of a graph may be important, too, but we shall discuss them as they arise in examples and applications.

Our general approach to curve sketching will involve four main steps:

1. Starting with $f(x)$, we compute $f'(x)$ and $f''(x)$.
2. Next, we locate all maximum and minimum points and make a partial sketch.
3. We study the concavity of $f(x)$ and locate all inflection points.
4. We consider other properties of the graph, such as the intercepts, and complete the sketch.

The first step was the main subject of the preceding chapter. We shall discuss the second and third steps in this section and then present several completely worked examples that combine all four steps in the next.

Locating Extreme Points The tangent line at a maximum or a minimum point of a function $f(x)$ has zero slope; that is, the derivative is zero there. Thus we may state the following useful rule.

> Look for possible extreme points of $f(x)$ by setting $f'(x) = 0$ and solving for x.   (1)

Suppose that $f'(a) = 0$. Then $x = a$ is a candidate for an extreme point of $f(x)$. There are several ways to determine if $f(x)$ has a maximum or a minimum (or neither) at $x = a$. The method that works in most cases is described in our first two examples.

EXAMPLE 1 The graph of the quadratic function $f(x) = \frac{1}{4}x^2 - x + 2$ is a parabola and so has one extreme point. Find it and sketch the graph.

Solution We begin by computing the first and second derivatives of $f(x)$.

$$f(x) = \tfrac{1}{4}x^2 - x + 2$$
$$f'(x) = \tfrac{1}{2}x - 1$$
$$f''(x) = \tfrac{1}{2}.$$

Setting $f'(x) = 0$, we have $\frac{1}{2}x - 1 = 0$, so that $x = 2$. Thus $f'(2) = 0$. Geometrically, this means that the graph of $f(x)$ will have a horizontal tangent line at the point where $x = 2$. To plot this point, we substitute the value 2 for x in the original expression for $f(x)$.

$$f(2) = \tfrac{1}{4}(2)^2 - (2) + 2 = 1.$$

Figure 1 shows the point $(2, 1)$ together with the horizontal tangent line. Is $(2, 1)$ an extreme point? In order to decide, we look at $f''(2)$. Since $f''(2) = \frac{1}{2}$, which is positive, the graph of $f(x)$ is concave up at $x = 2$. So a partial sketch of the graph near $(2, 1)$ should look something like Fig. 2.

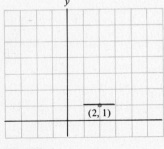

(2, 1)

FIGURE 1

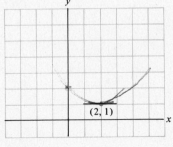

(2, 1)

FIGURE 2

We see that $(2, 1)$ is a minimum point. In fact, it is the only extreme point, for there is no other place where the tangent line is horizontal. Since the graph has no other "turning points," it must be decreasing before it gets to $(2, 1)$ and then increasing to the right of $(2, 1)$. Note that since $f''(x)$ is positive (and equal to $\frac{1}{2}$) for all x, the graph is concave upward at each point. A completed sketch is given in Fig. 3.

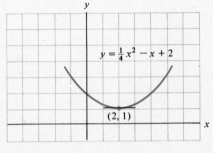

$$y = \tfrac{1}{4}x^2 - x + 2$$

(2, 1)

FIGURE 3

EXAMPLE 2   Locate all possible extreme points on the graph of the function $f(x) = x^3 - 3x^2 + 5$. Check the concavity at these points and use this information to sketch the graph of $f(x)$.

Solution   We have

$$f(x) = x^3 - 3x^2 + 5$$

$$f'(x) = 3x^2 - 6x$$

$$f''(x) = 6x - 6.$$

The easiest way to find those values of x for which $f'(x)$ is zero is to factor the expression for $f'(x)$:

$$3x^2 - 6x = 3x(x - 2).$$

From this factorization it is clear that $f'(x)$ will be zero if and only if $x = 0$ or $x = 2$. In other words, the graph will have horizontal tangent lines when $x = 0$ and $x = 2$, and nowhere else.

To plot the points on the graph where $x = 0$ and $x = 2$, we substitute these values back into the original expression for $f(x)$. That is, we compute

$$f(0) = (0)^3 - 3(0)^2 + 5 = 5$$

$$f(2) = (2)^3 - 3(2)^2 + 5 = 1.$$

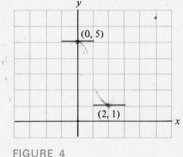

FIGURE 4

Figure 4 shows the points $(0, 5)$ and $(2, 1)$, along with the corresponding tangent lines.

Next, we check the concavity of the graph at these points by evaluating $f''(x)$ at $x = 0$ and $x = 2$:

$$f''(0) = 6(0) - 6 = -6$$

$$f''(2) = 6(2) - 6 = +6.$$

Since $f''(0)$ is negative, the graph is concave down at $x = 0$; since $f''(2)$ is positive, the graph is concave up at $x = 2$. A partial sketch of the graph is given in Fig. 5.

It is clear from Fig. 5 that $(0, 5)$ is a maximum point and $(2, 1)$ is a minimum point. Since they are the only turning points, the graph must be increasing before it gets to $(0, 5)$, decreasing from $(0, 5)$ to $(2, 1)$, and then increasing again to the right of $(2, 1)$. A sketch incorporating these properties appears in Fig. 6.

The facts that we used to sketch Fig. 6 could equally well be used to produce the graph in Fig. 7. Which graph really corresponds to $f(x) = x^3 - 3x^2 + 5$? The answer will be clear when we find the inflection points on the graph of $f(x)$.

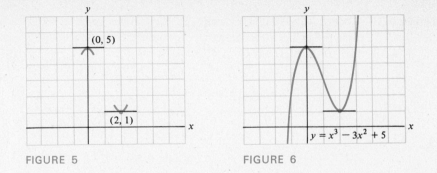

FIGURE 5

FIGURE 6

Locating Inflection Points An inflection point of a function $f(x)$ can occur only at a value of x for which $f''(x)$ is zero, because the curve is concave up where $f''(x)$ is positive and concave down where $f''(x)$ is negative. Thus we have the following test.

Look for possible inflection points by setting $f''(x) = 0$ and solving for x. $\qquad$ (2)

Once we have a value of x where the second derivative is zero, say at $x = b$, we must check the concavity of $f(x)$ at nearby points to see if the concavity really changes at $x = b$.

EXAMPLE 3 Find the inflection points of the function $f(x) = x^3 - 3x^2 + 5$ and explain why the graph in Fig. 6 has the correct shape.

Solution From Example 2 we have $f''(x) = 6x - 6 = 6(x - 1)$. Clearly, $f''(x) = 0$ if and only if $x = 1$. We will want to plot the corresponding point on the graph, so we compute

$$f(1) = (1)^3 - 3(1)^2 + 5 = 3.$$

Therefore the only possible inflection point is (1, 3).

Now look back at Fig. 5, where we indicated the concavity of the graph at the extreme points. Since $f(x)$ is concave down at (0, 5) and concave up at (2, 1), the concavity must reverse somewhere between these points. Hence (1, 3) must be an inflection point. Furthermore, since the concavity of $f(x)$ reverses nowhere else, the concavity at all points to the left of (1, 3) must be the same (i.e., concave down). Similarly, the concavity at all points to the right of (1, 3) must be the same (that is, concave up). Thus the graph in Fig. 6 has the correct shape. The graph in Fig. 7 has too many "wiggles," caused by frequent changes in concavity; that is, there are too many inflection points. A correct sketch showing the one inflection point at (1, 3) is given in Fig. 8.

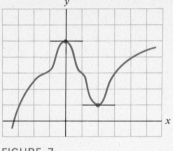

FIGURE 7

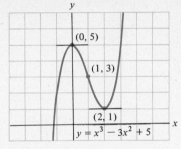

FIGURE 8

EXAMPLE 4 Sketch the graph of $y = -\frac{1}{3}x^3 + 3x^2 - 5x$.

Solution Let

$$f(x) = -\tfrac{1}{3}x^3 + 3x^2 - 5x.$$

Then

$$f'(x) = -x^2 + 6x - 5$$
$$f''(x) = -2x + 6.$$

We set $f'(x) = 0$ and solve for x:

$$-(x^2 - 6x + 5) = 0$$
$$-(x - 1)(x - 5) = 0$$
$$x = 1 \quad \text{or} \quad x = 5.$$

Substituting these values of x back into $f(x)$, we find that

$$f(1) = -\tfrac{1}{3}(1)^3 + 3(1)^2 - 5(1) = -\tfrac{7}{3}$$
$$f(5) = -\tfrac{1}{3}(5)^3 + 3(5)^2 - 5(5) = \tfrac{25}{3}.$$

The information we have so far is given in Fig. 9(a). The sketch in Fig. 9(b) is obtained by computing

$$f''(1) = -2(1) + 6 = 4$$
$$f''(5) = -2(5) + 6 = -4.$$

The curve is concave up at $x = 1$ because $f''(1)$ is positive, and the curve is concave down at $x = 5$ because $f''(5)$ is negative.

Since the concavity reverses somewhere between $x = 0$ and $x = 5$, there must be at least one inflection point. However, if we set $f''(x) = 0$, we find that

$$-2x + 6 = 0$$
$$x = 3.$$

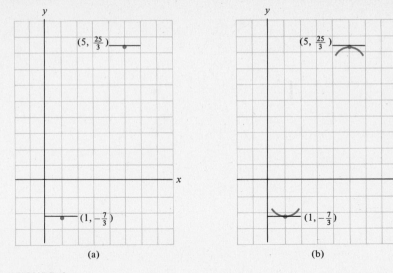

(a)   (b)

FIGURE 9

And so the inflection point must occur at $x = 3$. In order to plot the inflection point, we compute

$$f(3) = -\tfrac{1}{3}(3)^3 + 3(3)^2 - 5(3) = 3.$$

The final sketch of the graph is given in Fig. 10.

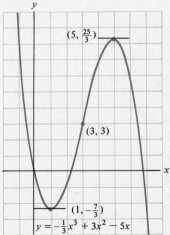

FIGURE 10

The argument in Example 4 about the existence of an inflection point is valid whenever $f(x)$ is a polynomial. However, it does not always apply to a function whose graph has a break in it. For example, the function $f(x) = 1/x$ is concave down at $x = -1$ and concave up at $x = 1$, but there is no inflection point in between.

A summary of curve-sketching techniques appears at the end of the next section. You may find steps 1, 2, and 3 helpful when working the exercises.

PRACTICE PROBLEMS 3

1. Which of the curves below could possibly be the graph of a function of the form $f(x) = ax^2 + bx + c$, where $a \neq 0$?

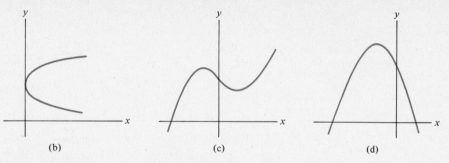

(a)　　　　　　　(b)　　　　　　　(c)　　　　　　　(d)

2. Which of the curves below could possibly be the graph of a function of the form $f(x) = ax^3 + bx^2 + cx + d$, where $a \neq 0$?

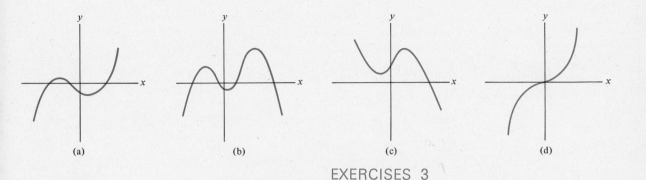

(a)　　　　　　　(b)　　　　　　　(c)　　　　　　　(d)

EXERCISES 3

Each of the graphs of the functions in Exercises 1–8 has one extreme point. Plot this point and check the concavity there. Using only this information, sketch the graph. [As you work the problems, observe that if $f(x) = ax^2 + bx + c$, then $f(x)$ has a minimum when $a > 0$ and a maximum when $a < 0$.]

1. $f(x) = 2x^2 - 8$

2. $f(x) = 3x^2 + 6x - 5$

3. $f(x) = \frac{1}{2}x^2 + x - 4$

4. $f(x) = -\frac{1}{2}x^2 + x - 4$

5. $f(x) = 1 + 6x - x^2$

6. $f(x) = 1 + x + x^2$

7. $f(x) = -x^2 - 8x - 10$

8. $f(x) = -3x^2 + 18x - 20$

Each of the graphs of the functions in Exercises 9–16 has one maximum and one minimum point. Plot these two points and check the concavity there. Using only this information, sketch the graph.

9. $f(x) = x^3 + 6x^2 + 9x$

10. $f(x) = \frac{1}{9}x^3 - x^2$

11. $f(x) = x^3 - 12x$

12. $f(x) = -\frac{1}{3}x^3 + 9x - 2$

13. $f(x) = -\frac{1}{9}x^3 + x^2 + 9x$

14. $f(x) = 2x^3 - 15x^2 + 36x - 24$

15. $f(x) = -\frac{1}{3}x^3 + 2x^2 - 12$

16. $f(x) = \frac{1}{3}x^3 + 2x^2 - 5x + \frac{8}{3}$

Sketch the following curves, indicating all extreme points and inflection points.

17. $y = x^3 - 3x + 2$

18. $y = x^3 - 6x^2 + 9x + 3$

19. $y = 1 + 3x^2 - x^3$

20. $y = -x^3 + 12x - 4$

21. $y = \frac{1}{3}x^3 - x^2 - 3x + 5$

22. $y = x^3 + \frac{3}{2}x^2 - 6x + 4$

23. $y = 2x^3 - 3x^2 - 36x + 20$

24. $y = 11 + 9x - 3x^2 - x^3$

25. Let a, b, c be fixed numbers with $a \neq 0$ and let $f(x) = ax^2 + bx + c$. Is it possible for the graph of $f(x)$ to have an inflection point? Explain your answer.

26. Let a, b, c, d be fixed numbers with $a \neq 0$ and let $f(x) = ax^3 + bx^2 + cx + d$. Is it possible for the graph of $f(x)$ to have more than one inflection point? Explain your answer.

The graph of each function below has one extreme point. Find it (giving both x- and y-coordinates) and determine if it is a maximum or a minimum. Do not include a sketch of the graph of the function.

27. $f(x) = \frac{1}{4}x^2 - 2x + 7$

28. $f(x) = 5 - 12x - 2x^2$

29. $g(x) = 3 + 4x - 2x^2$

30. $g(x) = x^2 + 10x + 10$

31. $f(x) = 5x^2 + x - 3$

32. $f(x) = 30x^2 - 1800x + 29{,}000$

SOLUTIONS TO PRACTICE PROBLEMS 3

1. Answer: (a) and (d). Curve (b) has the shape of a parabola, but it is not the graph of any function, since vertical lines cross it twice. Curve (c) has two extreme points, but the derivative of $f(x)$ is a linear function which could not be zero for two different values of x.

2. Answer: (a), (c), (d). Curve (b) has three extreme points, but the derivative of $f(x)$ is a quadratic function which could not be zero for three different values of x.

2.4. Curve Sketching (Conclusion)

In Section 3 we discussed the main techniques for curve sketching. Here we add a few finishing touches and examine some slightly more complicated curves.

The more points we plot on a graph, the more accurate the graph becomes. This statement is true even for the simple quadratic and cubic curves in Section 3. Of course, the most important points on a curve are the extreme points and the inflection points. In addition, the x- and y-intercepts often have some intrinsic interest in an applied problem. The y-intercept is $(0, f(0))$. To find the x-intercepts on the graph of $f(x)$, we must find those values of x for which $f(x) = 0$. Since this can be a difficult (or impossible) problem, we shall find x-intercepts only when they are easy to find or when a problem specifically requires us to find them.

When $f(x)$ is a quadratic function, as in Example 1, we can easily compute the x-intercepts (if they exist) either by factoring the expression for $f(x)$ or by using the quadratic formula.

EXAMPLE 1   Sketch the graph of $y = \frac{1}{2}x^2 - 4x + 7$.

Solution   Let

$$f(x) = \tfrac{1}{2}x^2 - 4x + 7.$$

Then

$$f'(x) = x - 4$$

$$f''(x) = 1.$$

Since $f'(x) = 0$ only when $x = 4$ and since $f''(4)$ is positive, $f(x)$ must have a minimum at $x = 4$. The minimum point is $(4, f(4)) = (4, -1)$.

The y-intercept is $(0, f(0)) = (0, 7)$. To find the x-intercepts, we set $f(x) = 0$ and solve for x:

$$\tfrac{1}{2}x^2 - 4x + 7 = 0.$$

The expression for $f(x)$ is not easily factored, so we use the quadratic formula to solve the equation.

$$x = \frac{-(-4) \pm \sqrt{(-4)^2 - 4(\tfrac{1}{2})(7)}}{2(\tfrac{1}{2})} = 4 \pm \sqrt{2}.$$

The x-intercepts are $(4 - \sqrt{2}, 0)$ and $(4 + \sqrt{2}, 0)$. To plot these points we use the approximation: $\sqrt{2} \approx 1.4$. (See Fig. 1.)

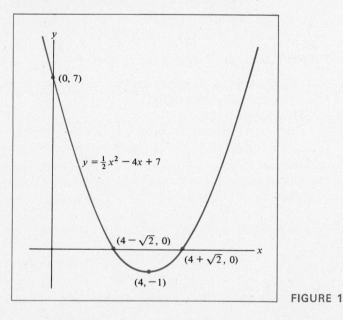

FIGURE 1

EXAMPLE 2 Sketch the graph of $f(x) = \frac{1}{6}x^3 - \frac{3}{2}x^2 + 5x + 1$.

Solution

$$f(x) = \frac{1}{6}x^3 - \frac{3}{2}x^2 + 5x + 1$$
$$f'(x) = \frac{1}{2}x^2 - 3x + 5$$
$$f''(x) = x - 3.$$

Let us set $f'(x) = 0$ and try to solve for x:

$$\frac{1}{2}x^2 - 3x + 5 = 0. \tag{1}$$

If we apply the quadratic formula with $a = \frac{1}{2}$, $b = -3$, and $c = 5$, we see that $b^2 - 4ac$ is negative, and so there is no solution to (1). In other words, $f'(x)$ is never zero. Thus the graph cannot have extreme points. If we evaluate $f'(x)$ at some x, say $x = 0$, we see that the first derivative is positive, and so $f(x)$ is increasing there. Since the graph of $f(x)$ is a smooth curve with no extreme points and no breaks, $f(x)$ must be increasing for all x. [For if $f(x)$ is increasing at $x = a$ and decreasing at $x = b$, then $f(x)$ has an extreme point between a and b.]

Now let us check the concavity.

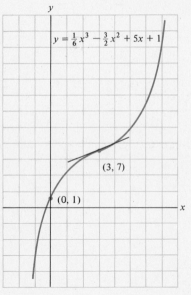

$y = \frac{1}{6}x^3 - \frac{3}{2}x^2 + 5x + 1$

$(3, 7)$

$(0, 1)$

$f''(x) = x - 3$		*Graph of $f(x)$*
$x < 3$	negative	concave down
$x = 3$	zero	concavity reverses
$x > 3$	positive	concave up

The inflection point is $(3, f(3)) = (3, 7)$. The y-intercept is $(0, f(0)) = (0, 1)$. We omit the x-intercept because it is difficult to solve the cubic equation $\frac{1}{6}x^3 - \frac{3}{2}x^2 + 5x + 1 = 0$.

The quality of our sketch of the curve will be improved if we first sketch the tangent line at the inflection point. To do this, we need to know the slope of the graph at $(3, 7)$:

$$f'(3) = \frac{1}{2}(3)^2 - 3(3) + 5 = \frac{1}{2}.$$

We draw a line through $(3, 7)$ with slope $\frac{1}{2}$ and then complete the sketch as shown in Fig. 2.

FIGURE 2

The methods used so far will work for most of the functions we shall study. Occasionally, however, $f'(x)$ and $f''(x)$ are both zero at some value of x, say $x = a$, and we cannot tell from the second derivative if the function has a minimum or a maximum at $x = a$. We show how to handle this unusual situation in the next example.

EXAMPLE 3 Sketch the graph of $f(x) = (x - 2)^4 - 1$.

Solution

$$f(x) = (x - 2)^4 - 1$$
$$f'(x) = 4(x - 2)^3$$
$$f''(x) = 12(x - 2)^2.$$

Clearly, $f'(x) = 0$ only if $x = 2$. So the curve has a horizontal tangent at $(2, f(2)) = (2, -1)$. Since $f''(2) = 0$, the second derivative rule tells us nothing about the concavity at $x = 2$. Let us look at $f''(x)$ for x near 2. Since $f''(x) = 12(x - 2)^2 > 0$ for $x \neq 2$, the graph is concave up at all points close to $(2, -1)$. In other words, the graph lies above its tangent line at each point near $(2, -1)$. Consequently, the graph cannot drop below its tangent line at $(2, -1)$, and it is concave up at $(2, -1)$.

The y-intercept is $(0, f(0)) = (0, 15)$. To find the x-intercepts, we set $f(x) = 0$ and solve for x:

$$(x - 2)^4 - 1 = 0$$

$$(x - 2)^4 = 1$$

$$x - 2 = 1 \quad \text{or} \quad x - 2 = -1$$

$$x = 3 \quad \text{or} \quad x = 1.$$

(See Fig. 3.)

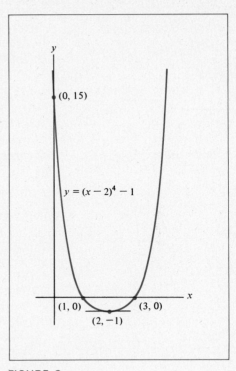

FIGURE 3

A Graph with Asymptotes Graphs similar to the one in the next example will arise in several applications later in this chapter.

EXAMPLE 4 Sketch the graph of $f(x) = x + (1/x)$ $(x > 0)$.

Solution
$$f(x) = x + \frac{1}{x}$$

$$f'(x) = 1 - \frac{1}{x^2}$$

$$f''(x) = \frac{2}{x^3}.$$

We set $f'(x) = 0$ and solve for x:

$$1 - \frac{1}{x^2} = 0$$

$$1 = \frac{1}{x^2}$$

$$x^2 = 1$$

$$x = 1.$$

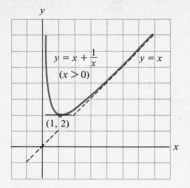

FIGURE 4

(We exclude the case $x = -1$ because we are only considering positive values of x.) The graph has a horizontal tangent at $(1, f(1)) = (1, 2)$. Now, $f''(1) = 2 > 0$, and so the graph is concave up at $x = 1$ and $(1, 2)$ is a minimum point. In fact, $f''(x) = (2/x^3) > 0$ for all positive x, and therefore the graph is concave up at all points.

Before sketching the graph, notice that as x approaches zero (a point at which $f(x)$ is not defined), the term $1/x$ in the formula for $f(x)$ becomes arbitrarily large. Thus $f(x)$ has the y-axis as an asymptote. For large values of x, $f(x) = x + (1/x)$ is only slightly larger than x; that is, the graph of $f(x)$ is slightly above the graph of $y = x$. As x increases, the graph of $f(x)$ has the line $y = x$ as an asymptote. (See Fig. 4.)

Summary of Curve-Sketching Techniques

1. Compute $f'(x)$ and $f''(x)$.

2. Find all maximum and minimum points.
   (a) Set $f'(x) = 0$ and solve for x. Suppose that $f'(a) = 0$.
      (i) If $f''(a) > 0$, the curve is concave up at $x = a$, and so $f(x)$ has a minimum at $x = a$. The minimum point is $(a, f(a))$.
      (ii) If $f''(a) < 0$, the curve is concave down at $x = a$, and so $f(x)$ has a maximum at $x = a$. The maximum point is $(a, f(a))$.
      (iii) If $f''(a) = 0$, examine $f'(x)$ to the left and right of $x = a$ in order to determine if the function changes from increasing to decreasing or vice versa.
   (b) Make a partial sketch of the graph near each point where $f(x)$ has a horizontal tangent line.

3. Determine the concavity of $f(x)$.
   (a) Set $f''(x) = 0$ and solve for x. Suppose that $f''(b) = 0$. Next, test the concavity for x near to b. If the concavity changes at $x = b$, then $(b, f(b))$ is an inflection point. Otherwise the concavity at $x = b$ is the same as at other nearby points.

4. Consider other properties of the function and complete the sketch.
   (a) If $f(x)$ is defined at $x = 0$, the y-intercept is $(0, f(0))$.
   (b) Does the partial sketch suggest that there are x-intercepts? If so, they are found by setting $f(x) = 0$ and solving for x. (Solve only in easy cases or when a problem essentially requires you to calculate the x-intercepts.)
   (c) Observe where $f(x)$ is defined. Sometimes the function is given only for restricted values of x. Sometimes the formula for $f(x)$ is meaningless for certain values of x.
   (d) Look for possible asymptotes.
      (i) Examine the formula for $f(x)$. If some terms become insignificant as x gets large and if the rest of the formula gives the equation of a straight line, then that straight line is an asymptote.
      (ii) Suppose that there is some point a such that $f(x)$ is defined for x near a but not at a (for example, $1/x$ at $x = 0$). If $f(x)$ gets arbitrarily large (in the positive or negative sense) as x approaches a, then the vertical line $x = a$ is an asymptote for the graph.
   (e) Complete the sketch.

PRACTICE PROBLEMS 4

1. Determine whether each of the following functions has an asymptote as x gets large. If so, give the equation of the straight line which is the asymptote.

   (a) $f(x) = (3/x) - 2x + 1$

   (b) $f(x) = \sqrt{x} + x$

   (c) $f(x) = 7 + 1/(2x + 6)$.

2. Sketch the graph of $f(x) = 1/(2 - x) + 5$, for $x > 2$.

EXERCISES 4

Find the x-intercepts of the following curves.

1. $y = x^2 - 3x + 1$ 2. $y = x^2 + 5x + 5$

3. $y = 2x^2 + 5x + 2$ 4. $y = 4 - 2x - x^2$

5. $y = 4x - 4x^2 - 1$ 6. $y = 3x^2 + 7x + 2$

7. Show that the function $f(x) = \frac{1}{3}x^3 - 2x^2 + 5x$ has no extreme points.

8. Show that the function $f(x) = 5 - 11x + 6x^2 - \frac{2}{3}x^3$ is always decreasing.

Sketch the graphs of the following functions.

9. $f(x) = x^3$ 10. $f(x) = -x^3$

11. $f(x) = x^3 + 3x + 1$ 12. $f(x) = 4 - x - x^3$

13. $f(x) = 5 - 13x + 6x^2 - x^3$ 14. $f(x) = 2x^3 + x - 2$

15. $f(x) = \frac{4}{3}x^3 - 2x^2 + x$ 16. $f(x) = 1 - 4x - 3x^2 - x^3$

17. $f(x) = 1 - 3x + 3x^2 - x^3$ 18. $f(x) = x^3 - 6x^2 + 12x - 5$

19. $f(x) = x^4 - 6x^2$ 20. $f(x) = 1 + 6x^2 - 3x^4$

21. $f(x) = (x - 3)^4$ 22. $f(x) = (x + 2)^4 - 1$

Sketch the graphs of the following functions.

23. $y = (1/x) + 3$ 24. $y = 1/(x - 1)$

25. $y = (1/x) + \frac{1}{4}x$ 26. $y = (1/x) + 9x$

27. $y = (9/x) + x + 1 \ (x > 0)$ 28. $y = (12/x) + 3x + 1 \ (x > 0)$

29. $y = (360/x) + 10x \ (x > 0)$

30. $y = (40{,}000/x) + 2500x + 320{,}000 \ (x > 0)$

31. $y = \dfrac{1}{x - 2} + \dfrac{x}{4} \ (x > 2)$ 32. $y = \dfrac{4}{2x - 5} + 2x \ (x > \frac{5}{2})$

33. $y = 6\sqrt{x} - x \ (x > 0)$ 34. $y = \dfrac{1}{x} + \dfrac{1}{4 - x} \ (0 < x < 4)$

SOLUTIONS TO PRACTICE PROBLEMS 4

1. Functions with asymptotes as x gets large have the form $f(x) = g(x) + mx + b$ where $g(x)$ approaches zero as x gets large. The function $g(x)$ often looks like c/x or $c/(ax + b)$. The asymptote will be the straight line $y = mx + b$.

(a) Here $g(x)$ is $3/x$ and the asymptote is $y = -2x + 1$.

(b) This function has no asymptote as x gets large. Of course, it can be written as $g(x) + mx + b$, where $mx + b$ is x. However $g(x) = \sqrt{x}$ does not approach zero as x gets large.

(c) Here $g(x)$ is $1/(2x + 6)$ and the asymptote is the line $y = 7$.

2. $f(x) = \dfrac{1}{2 - x} + 5 = (2 - x)^{-1} + 5$

$f'(x) = (2 - x)^{-2} = \dfrac{1}{(2 - x)^2}$

$f''(x) = 2(2 - x)^{-3} = \dfrac{2}{(2 - x)^3}.$

The first derivative, $1/(2 - x)^2$ is never zero and so there are no extreme points. The second derivative, $2/(2 - x)^3$, is also never zero and so there are no inflection points.

   Just to have a point of reference, let us find some point on the graph. For instance, let $x = 3$; then $f(3) = 4$. So the point $(3, 4)$ is on the graph. [See Fig. 5(a).] The first derivative is always positive. Thus the graph is always increasing. The second derivative is always negative, since $x > 2$ implies $2 - x$, and hence $(2 - x)^3$, is negative. Therefore, the graph is concave down. Now for the asymptotes. As x gets large, the term $1/(2 - x)$ approaches zero, and so the graph of $f(x)$ has the straight line $y = 5$ as an asymptote [See Fig. 5(b).]

   The function also has a vertical asymptote at $x = 2$. For values of x just to the right of 2, $1/(2 - x)$ is negative and very large in magnitude. [See Fig. 5(c).]

FIGURE 5

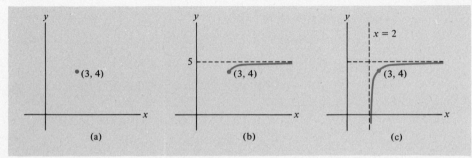

(a)   (b)   (c)

2.5. Optimization Problems

One of the most important applications of the derivative concept is to "optimization" problems, in which some quantity must be maximized or minimized. Examples of such problems abound in many areas of life. An airline must decide how many daily flights to schedule between two cities in order to maximize its profits. A doctor wants to find the minimum amount of a drug that will produce a desired response in one of her patients. A manufacturer needs to determine how often to replace certain equipment in order to minimize maintenance and replacement costs.

Our purpose in this section is to illustrate how calculus can be used to solve optimization problems. In each example we will find or construct a function that provides a "mathematical model" for the problem. Then, by sketching the graph of this function, we will be able to determine the answer to the original optimization problem.

The first two examples below are quite simple because the functions to be studied are explicitly given.

EXAMPLE 1 Find the minimum value of the function $f(x) = 2x^3 - 15x^2 + 24x + 19$ for $x \geq 0$.

Solution Using the curve-sketching techniques from Section 3, we obtain the graph in Fig. 1.

FIGURE 1

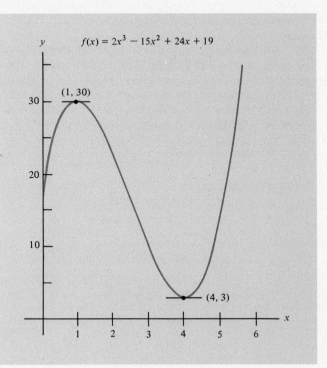

$f(x) = 2x^3 - 15x^2 + 24x + 19$

(1, 30)

(4, 3)

The lowest point on the graph is (4, 3). The minimum *value* of the function $f(x)$ is the *y*-coordinate of this point—namely, 3.

EXAMPLE 2 Suppose that a ball is thrown straight up into the air and its height after *t* seconds is $4 + 48t - 16t^2$ feet. Determine how long it will take for the ball to reach its maximum height and determine the maximum height.

Solution Consider the function $f(t) = 4 + 48t - 16t^2$. For each value of t, $f(t)$ is the height of the ball at time t. We want to find the value of t for which $f(t)$ is the greatest. Using the techniques of Section 3, we sketch the graph of $f(t)$. (See Fig. 2.) Note that we may neglect the portions of the graph corresponding to points for which either $t < 0$ or $f(t) < 0$. [A negative value of $f(t)$ would correspond to the ball being underneath the ground.] We see that $f(t)$ is greatest when $t = \frac{3}{2}$. At this value

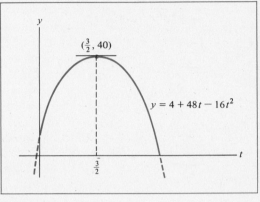

FIGURE 2

of t, the ball attains a height of 40 feet. [Note that the curve in Fig. 2 is the graph of $f(t)$, *not* a picture of the physical path of the ball.]

Answer The ball reaches its maximum height of 40 feet in 1.5 seconds.

EXAMPLE 3 A man wants to plant a rectangular garden along one side of his house, with a picket fence on the other three sides of the garden. Find the dimensions of the largest garden that can be enclosed using 40 feet of fencing.

Solution The first step is to make a simple diagram and assign letters to the quantities that may vary. Let us denote the dimensions of the rectangular garden by w and x (Fig. 3). The fencing on three sides must total 40 running feet; that is,

$$2x + w = 40. \tag{1}$$

The phrase "largest garden" above indicates that we must maximize the area, A, of the garden. In terms of the variables w and x,

$$A = wx. \tag{2}$$

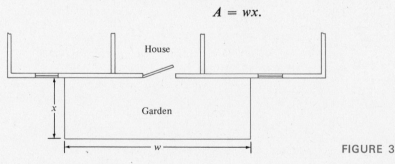

FIGURE 3

We now solve equation (1) for w in terms of x:

$$w = 40 - 2x. \tag{3}$$

Substituting this expression for w into equation (2), we have

$$A = (40 - 2x)x = 40x - 2x^2. \tag{4}$$

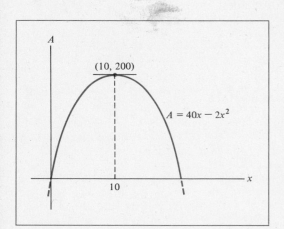

FIGURE 4

We now have a formula for the area A that depends on just one variable, and so we may graph A as a function of x.

Using curve-sketching techniques, we obtain the graph in Fig. 4. We see from the graph that the area is maximized when $x = 10$. (The maximum area is 200 square feet, but this fact is not needed for the problem.) From equation (3) we find that when $x = 10$,

$$w = 40 - 2(10) = 20.$$

Answer $w = 20$ feet, $x = 10$ feet.

Equation (1) in Example 3 is called a *constraint equation* because it places a limit or constraint on the way x and w may vary. Equation (2) is called the *objective equation*. It expresses the quantity to be maximized (the area of the garden) in terms of the variables w and x. The key step in the solution was to use the constraint equation to simplify the objective equation—that is, to write down the objective equation in a way that involves only one variable, as in (4).

EXAMPLE 4   The manager of a department store wants to build a 600-square-foot rectangular enclosure on the store's parking lot in order to display some equipment. Three sides of the enclosure will be built of redwood fencing, at a cost of $7 per running foot. The fourth side will be built of cement blocks, at a cost of $14 per running foot. Find the dimensions of the enclosure that minimizes the total cost of the building materials.

Solution   Let x be the length of the side built out of cement blocks and let y be the length of an adjacent side, as shown in Fig. 5. Since the area of the enclosure must be 600 square feet, the constraint equation is

$$xy = 600. \tag{5}$$

The phrase "minimizes the total cost..." tells us that the objective equation should

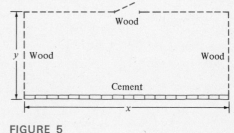

FIGURE 5

be a formula giving the total cost of the building materials.

$$[\text{cost of redwood}] = [\text{length of redwood fencing}] \cdot [\text{cost per foot}]$$

$$= (x + 2y) \ 7 = 7x + 14y.$$

$$[\text{cost of cement blocks}] = [\text{length of cement wall}] \cdot [\text{cost per foot}]$$

$$= x \cdot 14.$$

If C denotes the total cost of the materials, then

$$C = (7x + 14y) + 14x$$

$$C = 21x + 14y \quad \text{(objective equation)}. \tag{6}$$

We simplify this objective equation by solving (5) for one of the variables, say y, and substituting into (6):

$$C = 21x + 14\left(\frac{600}{x}\right) = 21x + \frac{8400}{x}.$$

We may now graph C as a function of the one variable x, where $x > 0$ (Fig. 6). (A similar curve was sketched in Example 4 of Section 4.) The minimum cost of $840 occurs where $x = 20$. From equation (5) we find that the corresponding value of y is $\frac{600}{20} = 30$.

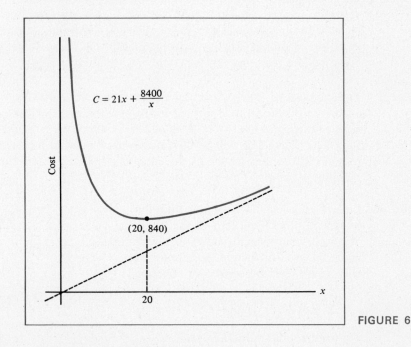

FIGURE 6

Answer $x = 20$ feet, $y = 30$ feet.

EXAMPLE 5 U.S. Parcel Post regulations state that packages must have length plus girth of no more than 84 inches. Find the dimensions of the cylindrical package of greatest volume that is mailable by parcel post.

Solution Let l be the length of the package and let r be the radius of the circular end. (See Fig. 7.) The girth equals the circumference of the end—that is, $2\pi r$. Since we want the package to be as large as possible, we must use the entire 84 inches allowable:

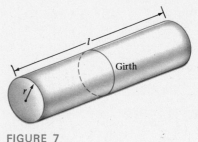

$$\text{length} + \text{girth} = 84$$

$$l + 2\pi r = 84 \qquad \text{(constraint equation).} \qquad (7)$$

FIGURE 7

The phrase "greatest volume . . ." tells us that the objective equation should express the volume of the package in terms of the dimensions l and r. Let V denote the volume. Then

$$V = [\text{area of base}] \cdot [\text{length}]$$

$$V = \pi r^2 l \qquad \text{(objective equation).} \qquad (8)$$

We now solve equation (7) for one of the variables, say $l = 84 - 2\pi r$. Substituting this expression into (8), we obtain

$$V = \pi r^2(84 - 2\pi r) = 84\pi r^2 - 2\pi^2 r^3. \qquad (9)$$

Let $f(r) = 84\pi r^2 - 2\pi^2 r^3$. Then, for each value of r, $f(r)$ is the volume of the parcel with end radius r that meets the postal regulations. We want to find that value of r for which $f(r)$ is as large as possible.

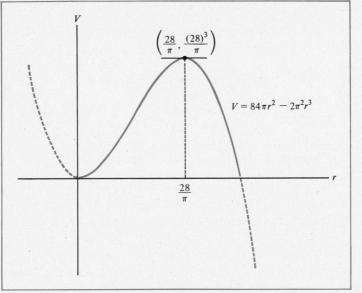

FIGURE 8

Using curve-sketching techniques, we obtain the graph in Fig. 8. [The graph is shown with a dotted curve where r is negative and where $f(r)$ is negative, for these portions of the graph have no physical significance.] We see that the volume is greatest when $r = 28/\pi$. From (7) we find that the corresponding value of l is

$$l = 84 - 2\pi r = 84 - 2\pi\left(\frac{28}{\pi}\right)$$

$$= 84 - 56 = 28.$$

The girth when $r = \dfrac{28}{\pi}$ is

$$2\pi r = 2\pi\left(\frac{28}{\pi}\right) = 56.$$

Answer $l = 28$ inches, $r = 28/\pi$ inches, girth $= 56$ inches.

Suggestions for Solving an Optimization Problem

1. Draw a picture, if possible.
2. Assign letters to quantities that may vary.
3. Find the "constraint equation" that relates the variables to each other and to any constants that are given in the problem.
4. The main objective of the problem is to maximize or minimize some quantity Q. Determine the "objective equation" that expresses Q as a function of the variables assigned in step 2.
5. Use the constraint equation to simplify the objective equation in such a way that Q becomes a function of only one variable.
6. Sketch the graph of the function obtained in step 5 and use this graph to solve the optimization problem.

PRACTICE PROBLEMS 5

1. A canvas wind shelter for the beach has a back, two square sides, and a top. Suppose that 96 square feet of canvas are to be used. Find the dimensions of the shelter for which the space inside the shelter (that is, the volume) will be maximized.

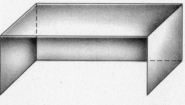

FIGURE 9

2. In Practice Problem 1, what are the objective equation and the constraint equation?

1. Find the maximum value of the function $f(x) = 12x - x^2$.

2. Find the minimum value of the function $g(x) = \frac{1}{2}x^2 - 8x + 35$.

3. Find the minimum value of the function $f(t) = t^3 - 6t^2 + 40$, $t \geq 0$, and give the value of t where this minimum occurs.

4. Find the maximum value of the function $f(s) = 9s^2 - s^3$, $s \geq 0$, and give the value of s where this maximum occurs.

5. For what x does the function $g(x) = 10 + 40x - x^2$ have its maximum?

6. For what t does the function $f(t) = t^2 - 24t$ have its minimum?

7. A ball thrown straight up into the air has a height of $-16t^2 + 96t + 6$ feet after t seconds. How high will the ball go?

8. The height of a bullet t seconds after it is fired from a rifle is $-16t^2 + 50t$ meters. How soon will the bullet reach its maximum height?

9. A manufacturer estimates that his profit from the daily production and sale of x power saws will be approximately $-.1x^2 + 50x - 1000$ dollars. How many saws should be produced each day in order to maximize the profit?

10. A manufacturer of costume jewelry estimates that the profit from producing x hundred necklaces per week is $300x - x^3 - 500$ dollars. What is the largest possible weekly profit?

11. Find two positive numbers whose sum is 100 and whose product is as large as possible. (Note: Denote the two numbers by x and h; then $x + h = 100$ and the product is $x \cdot h$.)

12. Find two positive numbers whose product is 100 and whose sum is as small as possible.

13. Figure 10(a) shows an open rectangular box with a square base. Find x and h such that the volume is 32 cubic inches and the amount of material needed to construct the box is minimal. (Note: The amount of material needed is $x^2 + 4xh$ square inches.)

FIGURE 10

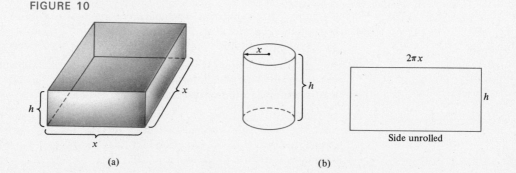

(a)

$2\pi x$

h

x

h

Side unrolled

(b)

14. A large soup can is to be designed so that the can will hold 16π cubic inches (about 28 fluid ounces) of soup. [See Fig. 10(b).] Find the values of x and h such that the amount of metal needed is as small as possible. (Note: The side requires $2\pi xh$ square inches of metal and each end requires πx^2 square inches.)

15. Figure 11(a) shows a "Norman window," which consists of a rectangle capped by a semicircular region. Find the value of x such that the perimeter of the window will be 14 feet and the area of the window will be as large as possible. (Note: The perimeter of the window is $2x + 2h + \pi x$.)

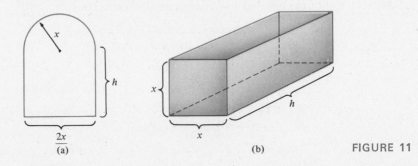

2x

(a)

x

(b)

FIGURE 11

16. Postal requirements specify that parcels must have length plus girth of at most 84 inches. Find the dimensions of the square-ended rectangular package [Fig. 11(b)] of greatest volume that is mailable. (Note: The length plus girth is $h + 4x$.)

17. A rectangular garden of area 75 square feet is to be surrounded on three sides by a brick wall costing $10 per foot and on one side by a fence costing $5 per foot. Find the dimensions of the garden such that the cost of the materials is minimized. [Note: Letting x be the length of the fence and h be the length of the other dimension, the cost of the materials is $5x + 10(2h + x)$.]

18. A closed rectangular box with square base and a volume of 12 cubic feet is to be constructed using two different types of materials. The top is made of a metal costing $2 per square foot and the remainder of wood costing $1 per square foot. Find the dimensions of the box for which the cost of materials is minimized.

19. Find the dimensions of the closed rectangular box with square base and volume 8000 cubic centimeters that can be constructed using the least amount of material.

20. Find the dimensions of the rectangular garden of greatest area that can be fenced off (all four sides) with 300 meters of fencing.

21. A storage shed is to be built in the shape of a box with a square base. It is to have a volume of 150 cubic feet. The concrete for the base costs $4 per square foot, the material for the roof costs $2 per square foot, and the material for the sides costs $2.50 per square foot. Find the dimensions of the most economical shed.

22. A canvas wind shelter for the beach has a back, two square sides, and a top. Find the dimensions for which the volume will be 250 cubic feet and which requires the least possible amount of canvas.

23. In Example 3 one can solve the constraint equation (1) for x instead of w to get $x = 20 - \frac{1}{2}w$. Substituting this for x in (2), one has $A = xw = (20 - \frac{1}{2}w)w$. Sketch the graph of the equation $A = 20w - \frac{1}{2}w^2$ and show that the maximum area occurs when $w = 20$ and $x = 10$.

24. In Example 4 one can solve the constraint equation (5) for x instead of y to get $x = 600/y$. Substituting this for x in (6), one has $C = 21x + 14y = (12,600/y) + 14y$. Sketch the graph of C as a function of y and show that the minimum cost occurs when $y = 30$ feet.

SOLUTIONS TO PRACTICE PROBLEMS 5

1. Since the sides of the wind shelter are square, we may let x represent the length of each side of the square. The remaining dimension of the wind shelter can be denoted by the letter h. (See Fig. 12.) The volume of the shelter is x^2h, and this is to be maximized. Since we have learned to maximize only functions of a single variable, we must express h in terms of x. We must use the information that 96 feet of canvas are used—that is, $2x^2 + 2xh = 96$. (Note: the roof and the back each have area xh, and each end has area x^2.) We now solve this equation for h.

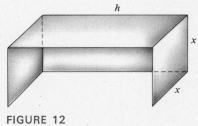

FIGURE 12

$$2x^2 + 2xh = 96$$

$$2xh = 96 - 2x^2$$

$$h = \frac{96}{2x} - \frac{2x^2}{2x} = \frac{48}{x} - x.$$

The volume, V, is

$$x^2h = x^2\left(\frac{48}{x} - x\right) = 48x - x^3.$$

By sketching the graph of $V = 48x - x^3$, we see that V has a maximum when $x = 4$. Then, $h = \frac{48}{4} - 4 = 12 - 4 = 8$. So each end of the shelter should be a 4-foot-by-4-foot square and the top should be 8 feet long.

2. The objective equation is: $V = x^2h$, since it expresses the volume (the quantity to be maximized) in terms of the variables. The constraint equation is $2x^2 + 2xh = 96$, for it relates the variables to each other; that is, it can be used to express one of the variables in terms of the other.

2.6. Further Optimization Problems

In this section we apply the optimization techniques developed in the previous section to some practical situations.

EXAMPLE 1   Suppose that, on a certain route, an airline carries 8000 passengers per month, each paying \$50. The airline wants to increase the fare. However, the market research department estimates that for each $1 increase in fare, the airline will lose 100 passengers. Determine the price which maximizes the airline's revenue.

Solution   Since the problem calls for setting an optimum price, let x be the price per ticket. The other variable is the number of passengers, which we can denote by n. The goal is to maximize revenue.

$$[\text{revenue}] = [\text{number of passengers}][\text{price per ticket}]$$
$$= n \cdot x.$$

If R denotes revenue, then the objective equation is

$$R = nx.$$

Now the constraint equation is given by

$$[\text{number of passengers}] = [\text{original number}] - [\text{loss due to fare increase}]$$
$$n \qquad = \qquad 8000 \qquad - \qquad (x - 50) \cdot 100$$
$$= 13{,}000 - 100x.$$

Therefore,

$$R = nx = (13{,}000 - 100x)x = 13{,}000x - 100x^2.$$

Figure 1 shows that revenue is maximized when the price per ticket is $65.

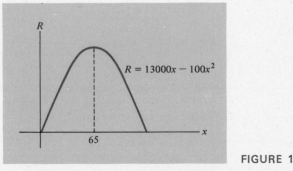

FIGURE 1

EXAMPLE 2   A supermarket manager wants to establish an optimal inventory policy for frozen orange juice. It is estimated that a total of 1200 cases will be sold at a steady rate during the next quarter-year. Since the orange juice requires costly freezer space for storage, she does not want to order all 1200 cases at once. On the other hand, since each delivery incurs additional handling costs, she cannot place too many separate orders. She decides to place several orders of the same size, say each consisting of x cases, equally spaced throughout the quarter-year. Use the data below to determine the optimal reorder quantity x that minimizes total purchasing and inventory costs.

1. The purchasing cost for each delivery is $75.

2. Inventory cost per quarter is $4x$ dollars. [It costs \$8 to store one case for one quarter-year. Since the stock begins at each delivery with x cases and is depleted steadily, there will be an average of $x/2$ cases in stock at any time. Therefore, inventory cost for the quarter will be $(x/2) \cdot 8$ or $4x$ dollars.]

Solution Let r be the number of orders placed during the quarter. Since each order calls for x cases, the constraint equation is

$$r \cdot x = 1200.$$

Now

$$[\text{total cost}] = [\text{purchasing cost}] + [\text{inventory cost}]$$

$$= 75r + 4x.$$

If C denotes the total cost, then the objective equation is

$$C = 75r + 4x.$$

The constraint equation says that $r = 1200/x$. Substitution into the objective equation yields

$$C = \frac{90{,}000}{x} + 4x.$$

Figure 2 is the graph of C as a function of x, for $x > 0$. The total cost is at a minimum when $x = 150$. Therefore, the optimum inventory policy is to order 150 cases at a time and to place $1200/150 = 8$ orders during the quarter-year.

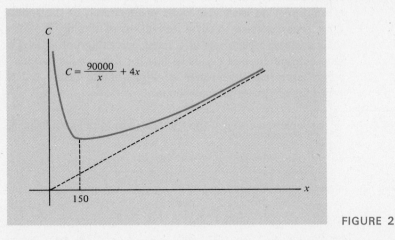

FIGURE 2

EXAMPLE 3 Rework Example 2, assuming that sales of frozen orange juice increase fourfold (that is, 4800 cases are sold each quarter).

Solution The only change in our previous solution is in the constraint equation, which now becomes

$$r \cdot x = 4800.$$

The objective equation is, as before,

$$C = 75r + 4x.$$

Since $r = 4800/x$,

$$C = 75 \cdot \frac{4800}{x} + 4x$$

$$= \frac{360{,}000}{x} + 4x.$$

Now

$$C' = -\frac{360{,}000}{x^2} + 4.$$

Setting $C' = 0$ yields

$$\frac{360{,}000}{x^2} = 4$$

$$90{,}000 = x^2$$

$$x = 300.$$

Therefore, the optimal reorder quantity is 300 cases.

Notice that although the sales increased by a factor of four, the optimal reorder quantity only increased by a factor of two $(=\sqrt{4})$. In general, a store's inventory of an item should be proportional to the square root of the expected sales. (See Exercise 16 for a derivation of this result.) Many stores tend to keep their average inventories at a fixed percentage of sales. For example, each order may contain enough goods to last for four or five weeks. This policy is likely to create excessive inventories of high-volume items and uncomfortably low inventories of slower-moving items.

EXAMPLE 4   When a person coughs, the trachea (windpipe) contracts. (See Fig. 3.)

FIGURE 3   Cross-section of trachea.

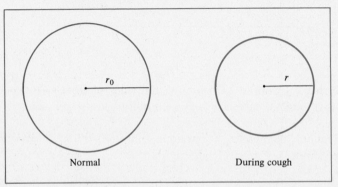

Normal   During cough

Let

r_0 = normal radius of the trachea,

r = radius during a cough,

P = increase in air pressure in the trachea during cough,

v = velocity of air through trachea during cough.

Use the principles of fluid flow given below to determine how much the trachea should contract in order to create the greatest air velocity—that is, the most effective condition for clearing the lungs and the trachea.

1. $r_0 - r = aP$, for some positive constant a. (Experiment has shown that, during coughing, the decrease in the radius of the trachea is nearly proportional to the increase in the air pressure.)

2. $v = b \cdot P \cdot \pi r^2$, for some positive constant b. (The theory of fluid flow requires that the velocity of the air forced through the trachea is proportional to the product of the increase in the air pressure and the area of a cross section of the trachea.)

Solution   In this problem the constraint equation **1.** and the objective equation **2.** are given directly. Solving equation **1.** for P and substituting this result into equation **2.**, we have

$$v = b\left(\frac{r_0 - r}{a}\right)\pi r^2 = k(r_0 - r)r^2,$$

where $k = b\pi/a$. To find the radius at which the velocity v is a maximum, we first compute the derivatives:

$$v = k(r_0 r^2 - r^3)$$

$$\frac{dv}{dr} = k(2r_0 r - 3r^2) = kr(2r_0 - 3r)$$

$$\frac{d^2v}{dr^2} = k(2r_0 - 6r).$$

FIGURE 4

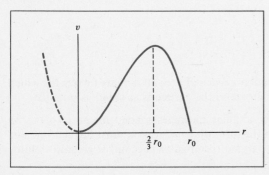

We see that $\frac{dv}{dr} = 0$ when $r = 0$ or when $2r_0 - 3r = 0$; that is, when $r = \frac{2}{3}r_0$. It is easy to see that $\frac{d^2v}{dr^2}$ is positive at $r = 0$ and is negative at $r = \frac{2}{3}r_0$. The graph of v as a function of r is drawn in Fig. 4. The air velocity is maximized at $r = \frac{2}{3}r_0$.

When solving optimization problems, we have always made a rough sketch of the graph of the function being optimized before drawing any conclusions. This procedure is necessary as the next example shows.

EXAMPLE 5 A rancher has 204 meters of fencing from which to build two corrals; one square and the other rectangular with length that is twice the width. Find the dimensions which result in the greatest combined area.

Solution Let x be the width of the rectangular corral and h the length of each side of the square corral. See Fig. 5. Then

$$204 = [\text{perimeter of square}] + [\text{perimeter of rectangle}] = 4h + 6x.$$

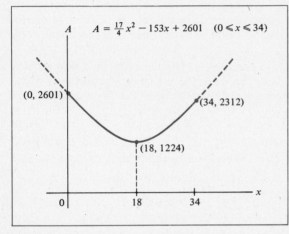

FIGURE 5

Let A be the combined area. Then

$$A = [\text{area of square}] + [\text{area of rectangle}]$$
$$= h^2 + 2x^2.$$

Since the perimeter of the rectangle cannot exceed 204, we must have $0 \le 6x \le 204$ or $0 \le x \le 34$. Solving the constraint equation for h and substituting into the objective equation leads to the function graphed in Fig. 6. The graph reveals that the area is minimum when $x = 18$. However, the problem asks for the *maximum* possible area. From Fig. 6 we see that this occurs at the endpoint where $x = 0$. Therefore, the rancher should build only the square corral, with $h = 204/4 = 51$ meters.

$A \qquad A = \frac{17}{4}x^2 - 153x + 2601 \quad (0 \le x \le 34)$

(0, 2601)

(34, 2312)

(18, 1224)

FIGURE 6

PRACTICE PROBLEMS 6

1. An apple orchard produces a profit of $40 a tree when planted with 1000 trees. Owing to overcrowding, the profit per tree (for each tree in the orchard) is reduced by 2 cents for each additional tree planted. How many trees should be planted in order to maximize the total profit from the orchard?

2. In the inventory problem of Example 2, suppose that the sales of frozen orange juice increase ninefold; that is, 10,800 cases are sold each quarter. What is now the optimal reorder quantity?

EXERCISES 6

1. Find the dimensions of the rectangular garden of area 100 square meters for which the amount of fencing needed to surround the garden is as small as possible.

2. A certain airline requires that rectangular packages carried on an airplane by passengers be such that the sum of the three dimensions is at most 120 centimeters. Find the dimensions of the square-ended rectangular package of greatest volume that meets this requirement.

3. An athletic field [Fig. 7(a)] consists of a rectangular region with a semicircular region at each end. The perimeter will be used for a 440-yard track. Find the value of x for which the area of the rectangular region is as large as possible.

4. An open rectangular box is to be constructed by cutting square corners out of a 16 × 16-inch piece of cardboard and folding up the flaps. [See Fig. 7(b).] Find the value for x for which the volume of the box will be as large as possible.

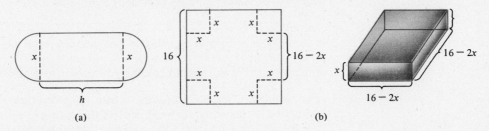

(a)   (b)

FIGURE 7

5. A long rectangular sheet of metal 30 inches wide is to be made into a gutter by turning up strips vertically along the two sides. How many inches should be turned up on each side so as to maximize the amount of water that the gutter can carry?

6. Design an open rectangular box with square ends, having volume 36 cubic inches, that minimizes the amount of material required for construction.

7. A supermarket is to be designed in the form of a rectangle whose area is 12,000 square feet. The front of the building will be mostly glass and will cost $70 per running foot for materials. The other three sides will be constructed of brick and cement block, at a cost of $50 per running foot. Ignore all other costs (labor, cost of foundation and roof, etc.) and find the dimensions of the building that will minimize the cost of the materials for the four sides of the building.

8. A manufacturer can produce x units of a certain commodity at a cost of $C(x) = .01x^3 - .04x^2 + 2x + 100$ dollars and receives a revenue of $R(x) = 50x - .04x^2$ dollars. How many units should be produced in order to maximize the profit? [*Note*: The profit from producing x units is $P(x) = R(x) - C(x)$.]

9. A rectangular corral of 54 square meters is to be fenced off and then divided by a fence into two sections. (See Fig. 8.) Find the dimensions of the corral so that the amount of fencing required is minimized.

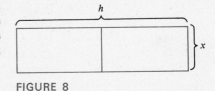

FIGURE 8

10. Referring to Exercise 9, suppose that the cost of the fencing for the boundary is $5 per meter and the dividing fence costs $2 per meter. Find the dimensions of the corral that minimize the cost of the fencing.

11. A swimming club offers memberships at the rate of $200, provided that a minimum of 100 people join. For each member in excess of 100, the membership fee will be

reduced by $1 per person (for every member). At most 160 memberships will be sold. How many memberships should the club try to sell in order to maximize its revenue?

12. In the planning of a sidewalk cafe, it is estimated that if there are 12 tables, the daily profit will be $10 per table. Owing to overcrowding, for each additional table the profit per table (for every table in the cafe) will be reduced by $.50. How many tables should be provided in order to maximize the profit from the cafe?

13. A certain toll road averages 36,000 cars per day when charging $1 per car. A survey concludes that increasing the toll will result in 300 fewer cars for each cent of increase. What toll should be charged in order to maximize the revenue?

14. A small orchard yields 25 bushels of fruit per tree when planted with 40 trees. Owing to overcrowding, the yield per tree (for each tree in the orchard) is reduced by one-half bushel for each additional tree that is planted. How many trees should be planted in order to maximize the total yield of the orchard?

15. A publishing company sells 400,000 copies of a certain book each year. Ordering the entire amount printed at the beginning of the year ties up valuable storage space. However, running off the copies in several partial runs throughout the year results in added costs for setting up each printing run. Using the data below, find the production run size, x, that minimizes the total setting up and storage costs.

   (i) Setting up each production run costs $1000.

   (ii) The storage cost is $x/4$, where x is the size of each production run. (There will be an average of $x/2$ books in storage during the year and storage costs are 50 cents per book per year.)

16. A store manager wants to establish an optimal inventory policy for an item. Sales are expected to be at a steady rate and should total Q items sold during the year. Each time an order is placed, a handling cost of h dollars is incurred. If the size of each order is x, then storage costs for the year will be $(x/2) \cdot s$ dollars. (That is, there will be an average of $x/2$ items in storage during the year, and the cost of storage is s dollars per item per year.) Show that total costs for handling and storage are minimized when each order calls for $\sqrt{2hQ/s}$ items.

17. Foggy Optics, Inc. makes a laboratory microscope at a cost of $200 for materials, labor, and similar items. Each production run costs $2500. The expense of holding one microscope in inventory for one year is $50. Suppose that the company expects to sell 1600 of these microscopes at a fairly uniform rate throughout the year. Determine the production run size that will minimize the company's expenses.

18. The Great American Tire Co. expects to sell 600,000 tires of a particular size and grade during the next year. Sales tend to be roughly the same from month to month. Each tire costs $20 for materials, labor, and so on, and each production run costs the company an additional $15,000. The inventory expense for one tire is $5 per year. What size production runs should be scheduled for the coming year in order to minimize the overall cost of producing these tires?

19. Let $f(t)$ be the amount of oxygen in a lake t days after sewage is dumped into the lake, and suppose that $f(t)$ is given approximately by

$$f(t) = 500\left(1 - \frac{10}{t + 10} + \frac{10}{(t + 10)^2}\right).$$

At what time is the oxygen content increasing the fastest?

20. The daily output of a coal mine after t hours of operation is approximately $40t + t^2 - \frac{1}{15}t^3$ tons, $0 \leq t \leq 12$. Find the maximum rate of output (in tons of coal per hour).

21. Consider a parabolic arch whose shape may be represented by the graph of $y = 9 - x^2$, where the base of the arch lies on the x-axis from $x = -3$ to $x = 3$. Find the dimensions of the rectangular window of maximum area that can be constructed inside the arch.

22. Advertising for a certain product is terminated, and t weeks later the weekly sales are $f(t)$ cases, where $f(t) = 1000(t + 8)^{-1} - 4000(t + 8)^{-2}$. At what time is the weekly sales amount falling the fastest?

SOLUTIONS TO PRACTICE PROBLEMS 6

1. Since the question asks for the optimum "number of trees," let x be the number of trees to be planted. The other quantity that varies is "profit per tree." So let p be the profit per tree. The objective is to maximize total profit, call it T. Then

$$[\text{total profit}] = [\text{profit per tree}] \cdot [\text{number of trees}]$$

$$T = p \cdot x.$$

Now,

$$[\text{profit per tree}] = [\text{original profit per tree}] - [\text{loss per tree due to increase}]$$

$$p = 40 - (x - 1000)(.02)$$

$$p = 60 - .02x.$$

Therefore,

$$T = p \cdot x = (60 - .02x)x = 60x - .02x^2.$$

By computing first and second derivatives and sketching the graph, we easily find that total profit is maximum when $x = 1500$. Therefore, 1500 trees should be planted.

2. This problem can be solved in the same manner that Example 3 was solved. However, the comment made at the end of Example 3 indicates that the optimal reorder quantity should increase by a factor of three, since $3 = \sqrt{9}$. Therefore, the optimal reorder quantity is $3 \cdot 150 = 450$ cases.

2.7. Applications of Calculus to Business and Economics

In recent years economic decision making has become more and more mathematically oriented. Faced with huge masses of statistical data, depending on hundreds or even thousands of different variables, business analysts and economists have increasingly turned to mathematical methods to help them describe what is happening, predict the effects of various policy alternatives, and choose reasonable courses of action from the myriad possibilities. Among the mathematical methods employed is calculus. In this section we will illustrate just a few of the many applications of calculus to business and economics. All our applications will center around what economists call *the theory of the firm*. In other words, we will study the activity of a business (or possibly a whole industry) and will restrict our analysis to a time period during which background conditions (such as supplies of raw materials, wage rates, taxes) are fairly constant. We will then show how calculus can help the management of such a firm make vital production decisions.

Management, whether or not it knows calculus, utilizes many functions of the sort we have been considering. Examples of such functions are

$C(x) =$ the cost of producing x units of the product,

$R(x) =$ the revenue generated by producing x units of the product,

$P(x) = R(x) - C(x) =$ the profit (or loss) generated by producing x units of the product.

Note that the functions $C(x)$, $R(x)$, $P(x)$ are defined only for nonnegative integers—that is, for $x = 0, 1, 2, 3, \ldots$. The reason is that it does not make sense to speak about the cost of producing -1 cars or the revenue generated by selling 3.62 refrigerators. Thus each of these functions gives rise to a set of discrete points on

FIGURE 1

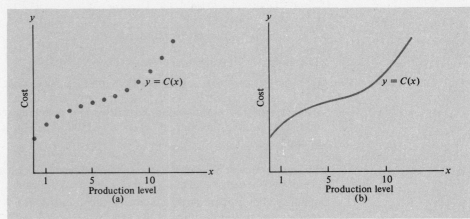

a graph, as in Fig. 1(a). In studying these functions, however, economists usually draw a smooth curve through the points and assume that $C(x)$ is actually defined for all positive x. Of course, we must often interpret answers to problems in light of the fact that x is, in most cases, a nonnegative integer.

Cost Functions If we assume that a cost function $C(x)$ has a smooth graph as in Fig. 1(b), we can use the tools of calculus to study it. A typical cost function is analyzed in Example 1.

EXAMPLE 1 Suppose that the cost function for a manufacturer is given by $C(x) = (10^{-6})x^3 - .003x^2 + 5x + 1000$ dollars.

(a) Describe the behavior of the marginal cost.

(b) Sketch the graph of $C(x)$.

Solution The first two derivatives of $C(x)$ are given by

$$C'(x) = (3 \cdot 10^{-6})x^2 - .006x + 5$$
$$C''(x) = (6 \cdot 10^{-6})x - .006.$$

Let us sketch the marginal cost $C'(x)$ first. From the behavior of $C'(x)$, we will be able to graph $C(x)$. The marginal cost function $y = (3 \cdot 10^{-6})x^2 - .006x + 5$ has as its graph a parabola that opens upward. Since $y' = C''(x) = .000006(x - 1000)$. we see that the parabola has a horizontal tangent at $x = 1000$. So the minimum value of y occurs at $x = 1000$. The corresponding y-coordinate is

$$(3 \cdot 10^{-6})(1000)^2 - .006 \cdot (1000) + 5 = 3 - 6 + 5 = 2.$$

The graph of $y = C'(x)$ is shown in Fig. 2. Consequently, at first the marginal cost decreases. It reaches a minimum of 2 at production level 1000 and increases thereafter. This answers part (a). Let us now graph $C(x)$. Since the graph shown in Fig. 2 is the graph of the derivative of $C(x)$, we see that $C'(x)$ is never zero, so that there are no extreme points. Since $C'(x)$ is always positive, $C(x)$ is always increasing (as any cost curve should be). Moreover, since $C'(x)$ decreases for x less than 1000 and increases for x greater than 1000, we see that $C(x)$ is concave down for x less

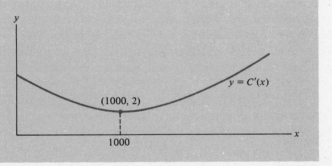

FIGURE 2

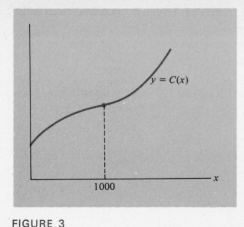

FIGURE 3

than 1000, concave up for x greater than 1000, and has an inflection point at $x = 1000$. The graph of $C(x)$ is drawn in Fig. 3. Note that the inflection point of $C(x)$ occurs at the value of x for which marginal cost is a minimum.

Actually, most marginal cost functions have the same general shape as the marginal cost curve of Example 1. For when x is small, production of additional units is subject to economies of production, which lower unit costs. Thus, for x small, marginal cost decreases. However, increased production eventually leads to overtime, use of less efficient, older plants, and competition for scarce raw materials. As a result, the cost of additional units will increase for very large x. And so we see that $C'(x)$ initially decreases and then increases.

Revenue Functions   In general, a business is concerned not only with its costs, but also with its revenues. Recall that if $R(x)$ is the revenue received from the sale of x units of some commodity, then the derivative $R'(x)$ is called the *marginal revenue*. Economists use this to measure the rate of increase in revenue per unit increase in sales.

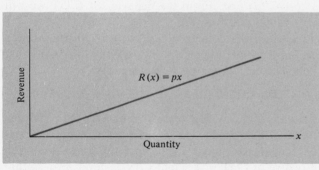

FIGURE 4

If x units of a product are sold at a price p per unit, then the total revenue $R(x)$ is given by

$$R(x) = x \cdot p.$$

If a firm is small and is in competition with many other companies, its sales have little effect on the market price. Then since the price is constant as far as the one firm is concerned, the marginal revenue $R'(x)$ equals the price p [that is, $R'(x)$ is the amount that the firm receives from the sale of one additional unit]. In this case, the revenue function will have a graph as in Fig. 4.

An interesting problem arises when a single firm is the only supplier of a certain product or service—that is, when the firm has a "monopoly." Consumers will buy large amounts of the commodity if the price per unit is low and less if the price is raised. For each quantity x, let $f(x)$ be the highest price per unit that can be set in order to sell all x units to consumers. Since selling greater quantities requires a lowering of the price, $f(x)$ will be a decreasing function. Figure 5 shows a typical "demand curve" that relates the quantity demanded, x, to the price $p = f(x)$.

The *demand equation* $p = f(x)$ determines the total revenue function. If the firm wants to sell x units, the highest price it can set is $f(x)$ dollars per unit, and so the total revenue from the sale of x units is

$$R(x) = x \cdot p = x \cdot f(x).$$

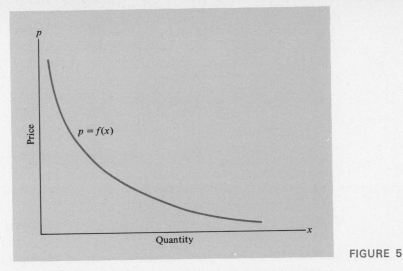

Price

$p = f(x)$

Quantity

FIGURE 5

The concept of a demand curve applies to an entire industry (with many producers) as well as to a single monopolistic firm. In this case, many producers offer the same product for sale. If x denotes the total output of the industry, then $f(x)$ is the market price per unit of output and $x \cdot f(x)$ is the total revenue earned from the sale of the x units.

EXAMPLE 2 The demand equation for a certain product is $p = 6 - \frac{1}{2}x$. Find the level of production that results in maximum revenue.

Solution The revenue function $R(x)$, in this case, is

$$R(x) = x \cdot p = x(6 - \frac{1}{2}x) = 6x - \frac{1}{2}x^2.$$

The marginal revenue is given by

$$R'(x) = 6 - x.$$

The graph of $R(x)$ is a parabola that opens downward (Fig. 6). It has a horizontal tangent precisely at those x for which $R'(x) = 0$—that is, for those x at which marginal revenue is 0. The only such x is $x = 6$. The corresponding value of revenue is

$$R(6) = 6 \cdot 6 - \frac{1}{2}(6)^2 = 18.$$

Thus the rate of production resulting in maximum revenue is $x = 6$, which results in total revenue of 18.

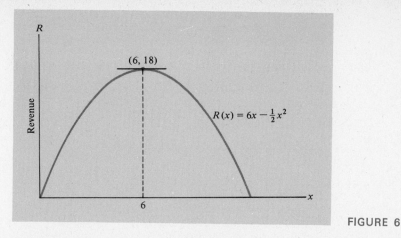

FIGURE 6

EXAMPLE 3 The WMA Bus Lines offers sightseeing tours of Washington, D.C. One of the tours, priced at $7 per person, had an average demand of about 1000 customers per week. When the price was lowered to $6, the weekly demand jumped to about 1200 customers. Assuming that the demand equation is linear, find the tour price that should be charged per person in order to maximize the total revenue each week.

Solution First we must find the demand equation. Let x be the number of customers per week and let p be the price of a tour ticket. Then $(x, p) = (1000, 7)$ and $(x, p) = (1200, 6)$ are on the demand curve (Fig. 7). Using the point-slope formula for the line through these two points, we have

$$p - 7 = \frac{7 - 6}{1000 - 1200} \cdot (x - 1000)$$

$$= -\frac{1}{200}(x - 1000)$$

$$= -\frac{1}{200}x + 5$$

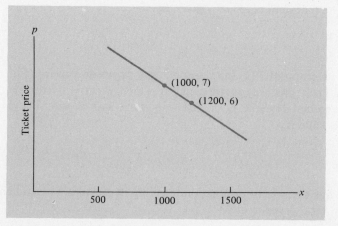

FIGURE 7

and so

$$p = 12 - \frac{1}{200}x. \tag{1}$$

From equation (1) we obtain the revenue function

$$R(x) = x(12 - \tfrac{1}{200}x) = 12x - \tfrac{1}{200}x^2.$$

The marginal revenue is

$$R'(x) = 12 - \tfrac{1}{100}x = -\tfrac{1}{100}(x - 1200).$$

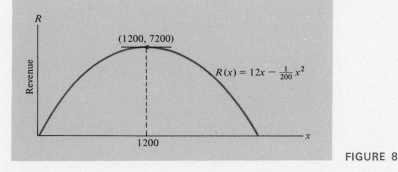

FIGURE 8

Using $R(x)$ and $R'(x)$, we can sketch the graph of $R(x)$ (Fig. 8). The maximum revenue occurs when the marginal revenue is zero—that is, when $x = 1200$. The price corresponding to this number of customers is found from the demand equation (1),

$$p = 12 - \tfrac{1}{200}(1200) = 6.$$

Thus the price of \$6 is most likely to bring in the greatest revenue per week.

Profit Functions   Once we know the cost function $C(x)$ and the revenue function $R(x)$, we can compute the profit function $P(x)$ from

$$P(x) = R(x) - C(x).$$

EXAMPLE 4   Suppose that the demand equation for a monopolist is $p = 100 - .01x$ and the cost function is $C(x) = 50x + 10,000$. Find the value of x that maximizes the profit and determine the corresponding price and total profit for this level of production. (See Fig. 9.)

Solution   The total revenue function is

$$R(x) = x \cdot p = x(100 - .01x) = 100x - .01x^2.$$

Hence the profit function is

$$\begin{aligned}
P(x) &= R(x) - C(x) \\
&= 100x - .01x^2 - (50x + 10,000) \\
&= -.01x^2 + 50x - 10,000.
\end{aligned}$$

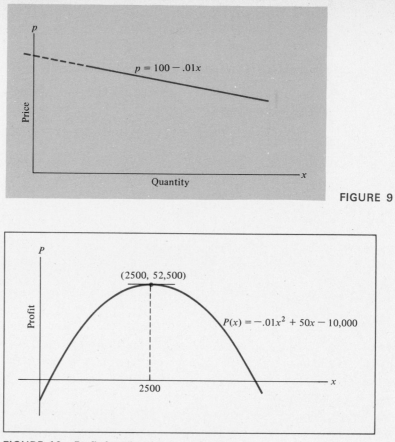

FIGURE 9

FIGURE 10   Profit function for a monopolist.

The graph of this function is a parabola that opens downward. (See Fig. 10.) Its highest point will be where the curve has zero slope—that is, where the marginal profit $P'(x)$ is zero. Now

$$P'(x) = -.02x + 50 = -.02(x - 2500).$$

So $P'(x) = 0$ when $x = 2500$. The profit for this level of production is

$$P(2500) = -.01(2500)^2 + 50(2500) - 10{,}000$$
$$= 52{,}500.$$

Finally, we return to the demand equation to find the highest price that can be charged per unit in order to sell all 2500 units:

$$p = 100 - .01(2500)$$
$$= 100 - 25 = 75.$$

Answer   Produce 2500 units and sell them at $75 per unit. The profit will be $52,500.

EXAMPLE 5   Rework Example 4 under the condition that the government imposes an excise tax of $10 per unit.

Solution   For each unit sold, the manufacturer will have to pay $10 to the government. In other words, $10x$ dollars is added to the cost of producing and selling x units. The cost function now will be

$$C(x) = (50x + 10,000) + 10x = 60x + 10,000.$$

The demand equation is unchanged by this tax, and so the revenue function is still

$$R(x) = 100x - .01x^2.$$

Proceeding as before, we have

$$P(x) = R(x) - C(x)$$
$$= 100x - .01x^2 - (60x + 10,000)$$
$$= -.01x^2 + 40x - 10,000$$
$$P'(x) = -.02x + 40 = -.02(x - 2000).$$

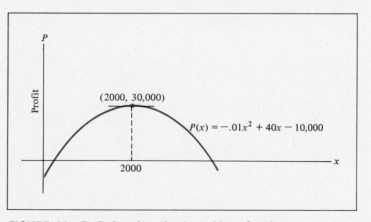

FIGURE 11   Profit function after imposition of excise tax

The graph of $P(x)$ is still a parabola that opens downward, and the highest point is where $P'(x) = 0$—that is, where $x = 2000$. (See Fig. 11.) The corresponding profit is

$$P(2000) = -.01(2000)^2 + 40(2000) - 10,000$$
$$= 30,000.$$

From the demand equation, $p = 100 - .01x$, we find the price that corresponds to $x = 2000$:

$$p = 100 - .01(2000) = 80.$$

Answer   Produce 2000 units and sell them at $80 per unit. The profit will be $30,000.

Notice in Example 5 that the optimal price is raised from $75 to $80. If the monopolist wishes to maximize profits, he should pass only half the $10 tax on to the consumer. The monopolist cannot avoid the fact that profits will be substantially lowered by the imposition of the tax. This is one reason why industries lobby against taxation.

Setting Production Levels   Suppose that a firm has cost function $C(x)$ and revenue function $R(x)$. In a free enterprise economy the firm will set production x in such a way as to maximize its profit function

$$P(x) = R(x) - C(x).$$

We have seen that if $P(x)$ has a maximum at $x = a$, then $P'(a) = 0$. In other words, since $P'(x) = R'(x) - C'(x)$, we see that

$$R'(a) - C'(a) = 0$$
$$R'(a) = C'(a).$$

Thus, profit is maximized at a production level for which marginal revenue equals marginal cost. (See Fig. 12.)

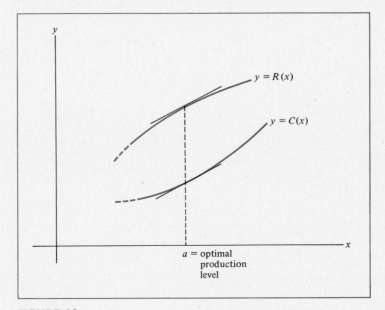

FIGURE 12

PRACTICE PROBLEMS 7

1. Rework Example 4 by finding the production level at which marginal revenue equals marginal cost.

2. Rework Example 4 under the condition that the fixed cost is increased from $10,000 to $15,000.

EXERCISES 7

1. Given the cost function $C(x) = x^3 - 6x^2 + 13x + 15$, find the minimum marginal cost.

2. Suppose that a total cost function is $C(x) = .0001x^3 - .06x^2 + 12x + 100$. Is the marginal cost increasing, decreasing, or not changing at $x = 100$? Find the minimum marginal cost.

3. The revenue function for a one-product firm is

$$R(x) = 200 - \frac{1600}{x + 8} - x.$$

Find the value of x that results in maximum revenue.

4. The revenue function for a particular product is $R(x) = x(4 - .0001x)$. Find the largest possible revenue.

5. A one-product firm estimates that its daily total cost function (in suitable units) is $C(x) = x^3 - 6x^2 + 13x + 15$ and its total revenue function is $R(x) = 28x$. Find the value of x that maximizes the daily profit.

6. A small tie shop sells men's ties for $3.50 each. The daily cost function is estimated to be $C(x)$ dollars, where x is the number of ties sold on a typical day and $C(x) = .0006x^3 - .03x^2 + 2x + 80$. Find the value of x that will maximize the store's daily profit.

7. The demand equation for a certain commodity is $p = \frac{1}{12}x^2 - 10x + 300$, $0 \le x \le 60$. Find the value of x and the corresponding price p that maximize the revenue.

8. The demand equation for a product is $p = 2 - .001x$. Find the value of x and the corresponding price p that maximize the revenue.

9. Some years ago it was estimated that the demand for steel approximately satisfied the equation $p = 256 - 50x$, and the total cost of producing x units of steel was $C(x) = 182 + 56x$. (The quantity x was measured in millions of tons and the price and total cost were measured in millions of dollars.) Determine the level of production and the corresponding price that maximize the profits.

10. Consider a rectangle in the x-y plane, with corners at $(0, 0)$, $(a, 0)$, $(0, b)$, and (a, b). Suppose that (a, b) lies on the graph of the equation $y = 30 - x$. Find a and b such that the area of the rectangle is maximized. What economic interpretation can be given to your answer if the equation $y = 30 - x$ represents a demand curve and y is the price corresponding to the demand x?

11. Until recently hamburgers at the city sports arena cost $1 each. The food concessionaire sold an average of 10,000 hamburgers on a game night. When the price was raised to $1.20, hamburger sales dropped off to an average of 8000 per night.

   (a) Assuming a linear demand curve, find the price of a hamburger that will maximize the nightly hamburger revenue.

(b) Suppose that the concessionaire has fixed costs of $1000 per night and the variable cost is $.30 per hamburger. Find the price of a hamburger that will maximize the nightly hamburger profit.

12. The average ticket price for a concert at the opera house was $16. The 4000-seat auditorium was filled for nearly every performance. When the ticket price was raised to $17, attendance declined to an average of 3800 persons per performance. Salaries, electricity, and maintenance expenses total $60,000 per performance. Costs that vary with the size of the audience are negligible. What should the average ticket price be in order to maximize the profit for the opera house? (Assume a linear demand curve.)

13. The monthly demand equation for an electric utility company is estimated to be·

$$p = 60 - (10^{-5})x,$$

where p is measured in dollars and x is measured in thousands of kilowatt hours. The utility has fixed costs of $7,000,000 per month and variable costs of $30 per thousand kilowatt hours of electricity generated, so that the cost function is

$$C(x) = 7 \cdot 10^6 + 30x.$$

(a) Find the value of x and the corresponding price for a thousand kilowatt hours that maximize the utility's profit.

(b) Suppose that rising fuel costs increase the utility's variable costs from $30 to $40, so that its new cost function is

$$C_1(x) = 7 \cdot 10^6 + 40x.$$

Should the utility pass all this increase of $10 per thousand kilowatt hours on to consumers? Explain your answer.

14. The demand equation for a monopolist is $p = 200 - 3x$, and the cost function is $C(x) = 75 + 80x - x^2, 0 \leq x \leq 40$.

(a) Determine the value of x and the corresponding price that maximizes the profit.

(b) Suppose that the government imposes a tax on the monopolist of $4 per unit quantity produced. Determine the new price that maximizes the profit.

(c) Suppose that the government imposes a tax of T dollars per unit quantity produced, so that the new cost function is

$$C(x) = 75 + (80 + T)x - x^2, \qquad 0 \leq x \leq 40.$$

Determine the new value of x that maximizes the monopolist's profit as a function of T. Assuming that the monopolist cuts back production to this level, express the tax revenues received by the government as a function of T. Finally, determine the value of T that will maximize the tax revenue received by the government.

SOLUTIONS TO PRACTICE PROBLEMS 7

1. The revenue function is $R(x) = 100x - .01x^2$. So the marginal revenue function is $R'(x) = 100 - .02x$. The cost function is $C(x) = 50x + 10,000$ and so the marginal cost function is $C'(x) = 50$. Let us now equate the two marginal functions and solve for x.

$$R'(x) = C'(x)$$
$$100 - .02x = 50$$
$$-.02x = -50$$
$$x = \frac{-50}{-.02} = \frac{5000}{2} = 2500.$$

Of course, we obtain the same level of production as before.

2. If the fixed cost is increased from $10,000 to $15,000, the new cost function will be $C(x) = 50x + 15,000$ but the marginal cost function will still be $C'(x) = 50$. Therefore, the solution will be the same: 2500 units should be produced and sold at $75 per unit. [Increases in fixed costs should not necessarily be passed on to the consumer if the objective is to maximize the profit.]

Chapter 2: CHECKLIST

- ☐ Increasing, decreasing
- ☐ Extreme point, maximum point, minimum point
- ☐ Concave up, concave down
- ☐ Inflection point
- ☐ x-intercept, y-intercept
- ☐ Asymptote
- ☐ First derivative rule
- ☐ Second derivative rule
- ☐ How to look for possible extreme points
- ☐ How to decide whether an extreme point is a maximum or minimum point
- ☐ How to look for possible inflection points
- ☐ Summary of curve-sketching techniques
- ☐ Objective equation
- ☐ Constraint equation
- ☐ Suggestions for solving an optimization problem

Chapter 2: SUPPLEMENTARY EXERCISES

Properties of various functions are described below. In each case draw some conclusion about the graph of the function.

1. $f(1) = 2, f'(1) > 0$

2. $g(1) = 5, g'(1) = -1$

3. $h'(3) = 4, h''(3) = 1$

4. $F'(2) = -1, F''(2) < 0$

5. $G(10) = 2, G'(10) = 0, G''(10) > 0$

6. $f(4) = -2, f'(4) > 0, f''(4) = -1$

7. $g(5) = -1, g'(5) = -2, g''(5) = 0$

8. $H(0) = 0, H'(0) = 0, H''(0) = 1$

9. $F(-2) = 0, F'(-2) = 0, F''(-2) = -1$

10. $h(-3) = 4, h'(-3) = 1, h''(-3) = 0$

Sketch the following parabolas. Include their x- and y-intercepts.

11. $y = 3 - x^2$

12. $y = 7 + 6x - x^2$

13. $y = x^2 + 3x - 10$

14. $y = 4 + 3x - x^2$

15. $y = -2x^2 + 10x - 10$

16. $y = x^2 - 9x + 19$

17. $y = x^2 + 3x + 2$

18. $y = -x^2 + 8x - 13$

19. $y = -x^2 + 20x - 90$

20. $y = 2x^2 + x - 1$

Sketch the following curves.

21. $y = 2x^3 + 3x^2 + 1$

22. $y = x^3 - \frac{3}{2}x^2 - 6x$

23. $y = x^3 - 3x^2 + 3x - 2$

24. $y = 100 + 36x - 6x^2 - x^3$

25. $y = \frac{11}{3} + 3x - x^2 - \frac{1}{3}x^3$

26. $y = x^3 - 3x^2 - 9x + 7$

27. $y = -\frac{1}{3}x^3 - 2x^2 - 5x$

28. $y = x^3 - 6x^2 - 15x + 50$

29. $y = x^4 - 2x^2$

30. $y = x^4 - 4x^3$

31. Let $f(x) = (x^2 + 2)^{3/2}$. Show that the graph of $f(x)$ has a possible extreme point at $x = 0$.

32. Show that the function $f(x) = (2x^2 + 3)^{3/2}$ is decreasing for $x < 0$ and increasing for $x > 0$.

33. Let $f(x)$ be a function whose *derivative* is

$$f'(x) = \frac{1}{1 + x^2}.$$

Note that $f'(x)$ is always positive. Show that the graph of $f(x)$ definitely has an inflection point at $x = 0$.

34. Let $f(x)$ be a function whose *derivative* is

$$f'(x) = \sqrt{5x^2 + 1}.$$

Show that the graph of $f(x)$ definitely has an inflection point at $x = 0$.

35. A closed rectangular box is to be constructed such that the base has a length twice as long as its width. Suppose that the volume must be 9 cubic feet. Find the dimensions of the box that will minimize the total surface area and show that this minimum surface area is 27 square feet.

36. A closed rectangular box is to be constructed such that the base has a length twice as long as its width. Suppose that the total surface area must be 27 square feet. Find the dimensions of the box that will maximize the volume and show that this maximum volume is 9 cubic feet.

37. The strength S of a rectangular beam is proportional to its width x and the square of its depth y, so that $S = kxy^2$, where k is a positive constant (Fig. 13). Find the dimensions of the strongest beam that can be cut from a circular log of diameter 75 centimeters.

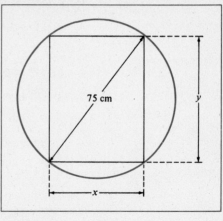

FIGURE 13

38. A fabric store purchases scissors for $5 a pair and resells them for $9. It costs $10 to place an order with the manufacturer, and the yearly inventory cost per pair of scissors is $.80. If the store sells about 900 pairs of scissors each year at a relatively constant rate, how many scissors should be ordered at one time?

39. A California distributor of sporting equipment expects to receive orders during the coming year for 100,000 cans of tennis balls. Yearly inventory costs per can are about $.50, and the cost of placing an order with the manufacturer is $10. Assuming a fairly constant demand for the tennis balls, determine the optimal reorder quantity for the distributor.

40. Suppose the demand equation for a monopolist is $p = 150 - .02x$ and the cost function is $C(x) = 10x + 300$. Find the value of x that maximizes the profit.

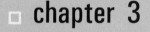

chapter 3

TECHNIQUES OF DIFFERENTIATION

We have seen that the derivative is useful in many applications. However, our ability to differentiate functions is somewhat limited. For example, we cannot yet readily differentiate

$$(x^{3/2} + 1)\sqrt{1 - x^2}, \qquad \frac{x}{x + 1}, \qquad \frac{x^3}{3x^2 - 3x + 1}.$$

In this chapter we develop differentiation techniques that apply to functions like those given above. Two new rules are the *product rule* and the *quotient rule*. In Section 2 we extend the general power rule into a powerful formula called the *chain rule*.

3.1. The Product and Quotient Rules

We observed in our discussion of the sum rule for derivatives that the derivative of the sum of two differentiable functions is the sum of the derivatives. Unfortunately, however, the derivative of the product $f(x)g(x)$ is *not* the product of the derivatives. Rather, the derivative of a product is determined from the following rule.

> *Product Rule*
>
> $$\frac{d}{dx}[f(x)g(x)] = f(x)g'(x) + g(x)f'(x).$$

The derivative of the product of two functions is the first function times the derivative of the second plus the second function times the derivative of the first. At the end of the section we shall show why this statement is true.

EXAMPLE 1 Show that the product rule works for the case $f(x) = x^2$, $g(x) = x^3$.

Solution Since $x^2 \cdot x^3 = x^5$, we know that

$$\frac{d}{dx}[x^2 \cdot x^3] = \frac{d}{dx}[x^5] = 5x^4.$$

On the other hand, using the product rule,

$$\frac{d}{dx}(x^2 \cdot x^3) = x^2 \frac{d}{dx}(x^3) + x^3 \frac{d}{dx}(x^2)$$

$$= x^2(3x^2) + x^3(2x)$$

$$= 3x^4 + 2x^4 = 5x^4.$$

Thus the product rule gives the correct answer.

EXAMPLE 2 Differentiate the product $(2x^3 - 5x)(3x + 1)$.

Solution Let $f(x) = 2x^3 - 5x$ and $g(x) = 3x + 1$. Then

$$\frac{d}{dx}[(2x^3 - 5x)(3x + 1)] = (2x^3 - 5x) \cdot \frac{d}{dx}(3x + 1) + (3x + 1) \cdot \frac{d}{dx}(2x^3 - 5x)$$

$$= (2x^3 - 5x)(3) + (3x + 1)(6x^2 - 5)$$

$$= 6x^3 - 15x + 18x^3 - 15x + 6x^2 - 5$$

$$= 24x^3 + 6x^2 - 30x - 5.$$

EXAMPLE 3   Apply the product rule to $y = g(x) \cdot g(x)$.

Solution
$$\frac{d}{dx}[g(x) \cdot g(x)] = g(x) \cdot g'(x) + g(x) \cdot g'(x)$$
$$= 2g(x)g'(x).$$

This answer is the same as that given by the general power rule:
$$\frac{d}{dx}[g(x) \cdot g(x)] = \frac{d}{dx}[g(x)]^2 = 2g(x)g'(x).$$

EXAMPLE 4   Differentiate the function $(x^3 - 1)(x^2 + 1)^5$.

Solution   Let $f(x) = x^3 - 1$ and $g(x) = (x^2 + 1)^5$. When we compute $g'(x)$ we will need the general power rule. Using the product rule first, we find that

$$\frac{d}{dx}[(x^3 - 1)(x^2 + 1)^5] = (x^3 - 1)\frac{d}{dx}(x^2 + 1)^5 + (x^2 + 1)^5\frac{d}{dx}(x^3 - 1)$$

$$= (x^3 - 1) \cdot 5(x^2 + 1)^4(2x) + (x^2 + 1)^5 \cdot (3x^2)$$

$$= 10x(x^3 - 1)(x^2 + 1)^4 + 3x^2(x^2 + 1)^5.$$

The Quotient Rule   Another useful formula for differentiating functions is the quotient rule.

Quotient Rule

$$\frac{d}{dx}\left[\frac{f(x)}{g(x)}\right] = \frac{g(x)f'(x) - f(x)g'(x)}{[g(x)]^2}.$$

One must be careful to remember the order of the terms in this formula because of the minus sign in the numerator.

EXAMPLE 5   Differentiate $\dfrac{x^2 + 1}{x^3 + 5}$.

Solution   Let $f(x) = x^2 + 1$ and $g(x) = x^3 + 5$.

$$\frac{d}{dx}\left(\frac{x^2 + 1}{x^3 + 5}\right) = \frac{(x^3 + 5)\dfrac{d}{dx}(x^2 + 1) - (x^2 + 1)\dfrac{d}{dx}(x^3 + 5)}{(x^3 + 5)^2}$$

$$= \frac{(x^3 + 5)(2x) - (x^2 + 1)(3x^2)}{(x^3 + 5)^2}$$

$$= \frac{-x^4 - 3x^2 + 10x}{(x^3 + 5)^2}.$$

EXAMPLE 6 Suppose that the total cost of manufacturing x units of a certain product is given by the function $C(x)$. Then the *average cost per unit*, AC, is defined by

$$AC = \frac{C(x)}{x}.$$

Recall that the *marginal cost*, MC, is defined by

$$MC = C'(x).$$

Show that at the level of production where the average cost is at a minimum, the average cost equals the marginal cost.

Solution In practice, the marginal cost and average cost curves will have the general shapes shown in Fig. 1. So the minimum point on the average cost curve will occur when $\frac{d}{dx}(AC) = 0$. To compute the derivative, we need the quotient rule,

$$\frac{d}{dx}(AC) = \frac{d}{dx}\left[\frac{C(x)}{x}\right] = \frac{x \cdot C'(x) - C(x)}{x^2}.$$

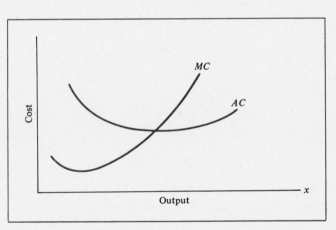

FIGURE 1   Marginal cost and average cost functions.

Setting the derivative equal to zero and multiplying by x^2, we obtain

$$0 = x \cdot C'(x) - C(x)$$

$$C(x) = x \cdot C'(x)$$

$$\frac{C(x)}{x} = C'(x)$$

$$AC = MC.$$

Thus when the output x is chosen so that the average cost is minimized, the average cost equals the marginal cost.

Verification of the Product and Quotient Rules

Verification of the Product Rule

From our discussion of limits we compute the derivative of $f(x)g(x)$ at $x = a$ as the limit

$$\frac{d}{dx}[f(x)g(x)]\bigg|_{x=a} = \lim_{h \to 0} \frac{f(a+h)g(a+h) - f(a)g(a)}{h}.$$

Let us add and subtract the quantity $f(a)g(a + h)$ in the numerator. After factoring and applying Limit Theorem III, we obtain

$$\lim_{h \to 0} \frac{[f(a+h)g(a+h) - f(a)g(a+h)] + [f(a)g(a+h) - f(a)g(a)]}{h}$$

$$= \lim_{h \to 0} g(a+h) \cdot \frac{f(a+h) - f(a)}{h} + \lim_{h \to 0} f(a) \cdot \frac{g(a+h) - g(a)}{h}.$$

And this expression may be rewritten by Limit Theorem V as

$$\lim_{h \to 0} g(a+h) \cdot \lim_{h \to 0} \frac{f(a+h) - f(a)}{h} + \lim_{h \to 0} f(a) \cdot \lim_{h \to 0} \frac{g(a+h) - g(a)}{h}$$

Note, however, that since $g(x)$ is differentiable at $x = a$, it is continuous there, so that $\lim_{h \to 0} g(a + h) = g(a)$. Therefore, the above expression equals

$$g(a)f'(a) + f(a)g'(a).$$

That is, we have proved that

$$\frac{d}{dx}[f(x)g(x)]\bigg|_{x=a} = g(a)f'(a) + f(a)g'(a),$$

which is the product rule.

Verification of the Quotient Rule

From the general power rule, we know that

$$\frac{d}{dx}\left[\frac{1}{g(x)}\right] = \frac{d}{dx}[g(x)]^{-1} = (-1)[g(x)]^{-2} \cdot g'(x).$$

We can now derive the quotient rule from the product rule.

$$\frac{d}{dx}\left[\frac{f(x)}{g(x)}\right] = \frac{d}{dx}\left[\frac{1}{g(x)} \cdot f(x)\right]$$

$$= \frac{1}{g(x)} \cdot f'(x) + f(x) \cdot \frac{d}{dx}\left[\frac{1}{g(x)}\right]$$

$$= \frac{g(x)f'(x)}{[g(x)]^2} + f(x) \cdot (-1)[g(x)]^{-2} \cdot g'(x)$$

$$= \frac{g(x)f'(x) - f(x)g'(x)}{[g(x)]^2}.$$

1. Consider the function $y = (\sqrt{x} + 1)x$.

    (a) Differentiate y by the product rule.

    (b) Multiply out first and then differentiate.

2. Differentiate $y = x/(x^4 - x^3 + 1)$.

EXERCISES 1

Differentiate the following functions.

1. $(x + 1)(x^3 + 5x + 2)$

2. $(2x - 1)(x^2 - 3)$

3. $(3x^2 - x + 2)(2x^2 - 1)$

4. $(x^3 + x^2 + x)(5x - 4)$

5. $(x^4 + 1)(3x + 5)$

6. $(10x^2 - 2x + 1)(x^2 + 7x - 8)$

7. $(x^2 - 3x + 1)\left(x + \dfrac{2}{x}\right)$

8. $\left(4 - \dfrac{1}{x^2}\right)(3x^3 - 5x + 8)$

9. $(2x - 7)(x - 1)^5$

10. $(x^2 + 4)(x + 3)^6$

11. $(x^2 + 3)(x^2 - 3)^{10}$

12. $(x + 1)^2(x - 3)^3$

13. $(2x + 9)^{-2}(5 - 3x)$

14. $(1 - x^2)(x^2 - 4x - 1)^{-4}$

15. $\sqrt{x}(3x^2 - 1)^3$

16. $(5x - 1)^4\left(\sqrt{x} + \dfrac{1}{\sqrt{x}}\right)$

17. $(1 - 4x)\sqrt{x^2 - 1}$

18. $(x^{3/2} + 2x^{1/2})(4x^{5/2} - x^{3/2})$

19. $\dfrac{x^2 - 1}{x^2 + 1}$

20. $\dfrac{x^2 + x + 1}{5 - x}$

21. $\dfrac{1}{5x^2 + 2x + 5}$

22. $\dfrac{2x - 1}{x}$

23. $\dfrac{\sqrt{x}}{x + 1}$

24. $\dfrac{3x^2 + 5x + 2}{x^2 - 3}$

25. $\dfrac{x}{(x^2 + 1)^2}$

26. $\dfrac{4 + x - x^2}{1 + \sqrt{x}}$

27. $\sqrt{\dfrac{x - 2}{x + 2}}$

28. $\dfrac{x - 1}{(2x + 1)^2}$

29. $\dfrac{3 - x}{\sqrt{4x + x^2}}$

30. $\dfrac{1 + \sqrt{2x}}{1 - \sqrt{2x}}$

31. $\dfrac{(x^5 + 1)^3}{(3x^2 + 1)^2}$

32. $\left(\dfrac{4x - 1}{3x + 1}\right)^3$

33. (A Voting Model.) Let x be the proportion of the total popular vote a Democratic candidate for president receives in a U.S. national election. Political scientists have observed that a good estimate of the proportion of seats in the House of Representatives going to Democratic candidates is given by

$$f(x) = \frac{x^3}{3x^2 - 3x + 1} \quad (0 \leq x \leq 1).$$

This formula is referred to as the "cube law" by some political scientists, primarily because $f(x)$ may also be written as

$$f(x) = \frac{x^3}{x^3 + (1 - x)^3}.$$

(Note that $1 - x$ is approximately the proportion of votes received by the Republicans.)

(a) Are $f(0)$, $f(\frac{1}{2})$, and $f(1)$ what you would expect them to be?

(b) Show that $f'(\frac{1}{2}) = 3$.

(c) It can be shown (by the quotient rule) that

$$f''(x) = \frac{6x(x - 1)(2x - 1)}{(3x^2 - 3x + 1)^3}.$$

What does this tell you about $f(x)$?

(d) Sketch the graph of $y = f(x)$ for $0 \leq x \leq 1$.

34. Let $f(x) = 1/x$ and $g(x) = x^3$.

(a) Show that the product rule yields the correct derivative of $(1/x) \cdot x^3 = x^2$.

(b) Compute the product $f'(x)g'(x)$ and note that it is *not* the derivative of $f(x)g(x)$.

35. The derivative of $(x^3 - 4x)/x$ is obviously $2x$ for $x \neq 0$, because $(x^3 - 4x)/x = x^2 - 4$ for $x \neq 0$. Verify that the quotient rule gives the same derivative.

36. Use the fact that $\frac{d}{dx}(x^3) = 3x^2$ to show that $\frac{d}{dx}(x^4) = 4x^3$. (Write $x^4 = x \cdot x^3$.)

37. Use the fact that $\frac{d}{dx}(x^4) = 4x^3$ to show that $\frac{d}{dx}(x^5) = 5x^4$.

38. Suppose you know that $\frac{d}{dx}x^k = kx^{k-1}$ for some positive integer k. Use this fact to show that $\frac{d}{dx}(x^{k+1}) = (k + 1)x^k$. (This exercise proves by mathematical induction that $\frac{d}{dx}(x^n) = nx^{n-1}$ for all positive integers n.)

39. Use the fact that $\frac{d}{dx}(x^2) = 2x$ to show that $\frac{d}{dx}(x^{-2}) = -2x^{-3}$. (Write $x^{-2} = \frac{1}{x^2}$ and use the quotient rule.)

40. Use the fact that $\dfrac{d}{dx}(x^3) = 3x^2$ to show that $\dfrac{d}{dx}(x^{-3}) = -3x^{-4}$. (Write $x^{-3} = \dfrac{1}{x^3}$ and use the quotient rule.)

41. Let n be a positive integer. Use the fact that $\dfrac{d}{dx}(x^n) = nx^{n-1}$ to show that $\dfrac{d}{dx}(x^{-n}) = -nx^{-n-1}$.

42. Let $f(x)$, $g(x)$, $h(x)$ be differentiable functions. Find a formula for the derivative of $f(x)g(x)h(x)$.

43. Suppose $f(x)$ is a function whose derivative is $f'(x) = \dfrac{1}{x}$. Find the derivative of $xf(x) - x$.

44. Suppose $f(x)$ is a function whose derivative is $f'(x) = \dfrac{1}{1+x^2}$. Find the derivative of $\dfrac{f(x)}{1+x^2}$.

SOLUTIONS TO PRACTICE PROBLEMS 1

1. (a) Apply the product rule to $y = (\sqrt{x} + 1)x$ with
$$f(x) = \sqrt{x} + 1 = x^{1/2} + 1$$
$$g(x) = x$$
$$\frac{dy}{dx} = (x^{1/2} + 1) \cdot 1 + x \cdot \tfrac{1}{2}x^{-1/2}$$
$$= x^{1/2} + 1 + \tfrac{1}{2}x^{1/2}$$
$$= \tfrac{3}{2}\sqrt{x} + 1.$$

(b) $y = (\sqrt{x} + 1)x = (x^{1/2} + 1)x = x^{3/2} + x.$
$$\frac{dy}{dx} = \tfrac{3}{2}x^{1/2} + 1 = \tfrac{3}{2}\sqrt{x} + 1.$$

Comparing (a) and (b), we note that sometimes it is helpful to simplify the function before differentiating.

2. Apply the quotient rule to $y = x/(x^4 - x^3 + 1).$
$$\frac{dy}{dx} = \frac{(x^4 - x^3 + 1) \cdot 1 - x(4x^3 - 3x^2)}{(x^4 - x^3 + 1)^2}$$
$$= \frac{x^4 - x^3 + 1 - 4x^4 + 3x^3}{(x^4 - x^3 + 1)^2}$$
$$= \frac{-3x^4 + 2x^3 + 1}{(x^4 - x^3 + 1)^2}.$$

3.2. The Chain Rule

A useful way of combining functions $f(x)$ and $g(x)$ is to replace every occurrence of the variable x in $f(x)$ by the function $g(x)$. The resulting function is called the *composition* (or *composite*) of $f(x)$ and $g(x)$ and is denoted $f(g(x))$.

EXAMPLE 1 Let $f(x) = x^2$, $g(x) = x^3 - 3x + 1$. What is $f(g(x))$?

Solution Replace each occurrence of x in $f(x)$ by $g(x)$ to obtain

$$f(g(x)) = g(x)^2 = (x^3 - 3x + 1)^2.$$

EXAMPLE 2 Let $f(x) = \dfrac{x-1}{x+1}$, $g(x) = x^3$. What is $f(g(x))$?

Solution Replace each occurrence of x in $f(x)$ by $g(x)$ to obtain

$$f(g(x)) = \frac{g(x) - 1}{g(x) + 1}$$

$$= \frac{x^3 - 1}{x^3 + 1}.$$

EXAMPLE 3 Let $f(x) = \dfrac{1}{x}$, $g(x) = x^2 + 3x + 1$. What are $f(g(x))$ and $g(f(x))$?

Solution
$$f(g(x)) = \frac{1}{g(x)} = \frac{1}{x^2 + 3x + 1}.$$

$$g(f(x)) = (f(x))^2 + 3(f(x)) + 1$$

$$= \left(\frac{1}{x}\right)^2 + 3 \cdot \left(\frac{1}{x}\right) + 1$$

$$= \frac{1}{x^2} + \frac{3}{x} + 1.$$

Not only is it important to be able to compute $f(g(x))$ when $f(x)$ and $g(x)$ are given, but it is also important (see below) to be able to recognize a given function as the composite of two simpler functions.

EXAMPLE 4 Let $h(x) = (x^5 + 9x + 7)^{18}$, $k(x) = \sqrt{3x^2 + 1}$. Write $h(x)$ and $k(x)$ as composites of simpler functions.

Solution (a) Let $f(x) = x^{18}$, $g(x) = x^5 + 9x + 7$. Then $f(g(x)) = (x^5 + 9x + 7)^{18} = h(x)$.

(b) Let $f(x) = \sqrt{x}$, $g(x) = 3x^2 + 1$. Then $f(g(x)) = \sqrt{3x^2 + 1} = k(x)$.

A function of the form $[g(x)]^r$ is obviously a composite $f(g(x))$, where $f(x) = x^r$. Moreover, we have already given a rule for differentiating this function—namely,

$$\frac{d}{dx} [g(x)]^r = r[g(x)]^{r-1} \cdot g'(x).$$ (1)

This rule is a special case of the *chain rule*, which describes how to differentiate a composite $f(g(x))$. This rule enables us to calculate the derivative of $f(g(x))$ in terms of the derivatives of $f(x)$ and $g(x)$.

> *The Chain Rule:* To differentiate $f(g(x))$, first differentiate $f(x)$ and substitute $g(x)$ for x in the result. Then multiply by the derivative of $g(x)$. Symbolically,
>
> $$\frac{d}{dx} [f(g(x))] = f'(g(x)) \cdot g'(x).$$

EXAMPLE 5   Use the chain rule to compute the derivative of $h(x) = (x^5 + 9x + 7)^{18}$.

Solution   We saw in Example 4(a) that $h(x)$ can be represented as the composite $f(g(x))$, where $f(x) = x^{18}$, $g(x) = x^5 + 9x + 7$. Therefore, we apply the chain rule to compute $h'(x)$.

$$f'(x) = 18x^{17}, \qquad g'(x) = 5x^4 + 9$$

$$f'(g(x)) = 18(x^5 + 9x + 7)^{17}$$

$$f'(g(x)) \cdot g'(x) = 18(x^5 + 9x + 7)^{17} \cdot (5x^4 + 9).$$

Therefore,

$$h'(x) = 18(x^5 + 9x + 7)^{17} \cdot (5x^4 + 9).$$

Note that this is the same answer as derived using the general power rule.

EXAMPLE 6   Use the chain rule to compute the derivative of

$$h(x) = \frac{(x^3 + 1)^5 - 1}{(x^3 + 1)^5 + 1}.$$

Solution   Note that $h(x) = f(g(x))$, where

$$f(x) = \frac{x^5 - 1}{x^5 + 1}, \qquad g(x) = x^3 + 1.$$

By the quotient rule,

$$f'(x) = \frac{(x^5 + 1) \cdot 5x^4 - (x^5 - 1) \cdot 5x^4}{(x^5 + 1)^2} = \frac{10x^4}{(x^5 + 1)^2}.$$

Therefore,

$$f'(g(x)) = \frac{10(x^3 + 1)^4}{[(x^3 + 1)^5 + 1]^2}$$

$$g'(x) = 3x^2$$

$$h'(x) = f'(g(x)) \cdot g'(x)$$

$$= \frac{10(x^3 + 1)^4}{[(x^3 + 1)^5 + 1]^2} \cdot 3x^2$$

$$= \frac{30x^2(x^3 + 1)^4}{[(x^3 + 1)^5 + 1]^2}.$$

Note that this answer can also be derived using the quotient rule together with the general power rule.

At this stage, the chain rule does not allow us to differentiate any functions which we could not already differentiate via the general power rule. However, as we shall see in our discussion of the exponential and logarithmic functions, the chain rule does indeed expand the repertoire of functions we can differentiate.

When the power rule and the general power rule were introduced in Chapter 1, we promised to justify these rules later in the text. The general power rule can be deduced from the chain rule and the ordinary power rule. The chain rule is justified below. As for the ordinary power rule, we have outlined its justification in the case of integer exponents in Exercises 38 and 41 of Section 1. The case of the ordinary power rule for fractional exponents is treated in Exercises 39 and 40 below.

Verification of the Chain Rule

Suppose $f(x)$ and $g(x)$ are differentiable, and let $x = a$ be a number in the domain of $f(g(x))$. Since every differentiable function is continuous, we have

$$\lim_{h \to 0} g(a + h) = g(a),$$

which implies that

$$\lim_{h \to 0} [g(a + h) - g(a)] = 0. \tag{2}$$

Now $g(a)$ is a number in the domain of f, and the limit definition of the derivative gives us

$$f'(g(a)) = \lim_{k \to 0} \frac{f(g(a) + k) - f(g(a))}{k}. \tag{3}$$

Let $k = g(a + h) - g(a)$. By equation (2), k approaches zero as h approaches zero. Also, $g(a + h) = g(a) + k$. Therefore, (3) may be rewritten in the form

$$f'(g(a)) = \lim_{h \to 0} \frac{f(g(a + h)) - f(g(a))}{g(a + h) - g(a)}. \tag{4}$$

(Strictly speaking, we must assume that the denominator in (4) is never zero. This assumption may be avoided by a somewhat different and more technical argument which we omit.) Finally, we show that the function $f(g(x))$ has a derivative at $x = a$. We use the limit definition of the derivative, Limit Theorem V, and (4) above.

$$\frac{d}{dx}[f(g(x))]\Big|_{x=a} = \lim_{h\to 0}\frac{f(g(a+h)) - f(g(a))}{h}$$

$$= \lim_{h\to 0}\left[\frac{f(g(a+h)) - f(g(a))}{g(a+h) - g(a)} \cdot \frac{g(a+h) - g(a)}{h}\right]$$

$$= \lim_{h\to 0}\frac{f(g(a+h)) - f(g(a))}{g(a+h) - g(a)} \cdot \lim_{h\to 0}\frac{g(a+h) - g(a)}{h}$$

$$= f'(g(a)) \cdot g'(a).$$

PRACTICE PROBLEMS 2

Consider the function $h(x) = (2x^3 - 5)^5 + (2x^3 - 5)^4$.

1. Write $h(x)$ as a composite function, $f(g(x))$.

2. Compute $f'(x)$ and $f'(g(x))$.

3. Use the chain rule to differentiate $h(x)$.

EXERCISES 2

Compute $f(g(x))$, where $f(x)$ and $g(x)$ are the following:

1. $f(x) = \dfrac{1}{x + 1}$, $g(x) = x^3$
   2. $f(x) = \dfrac{1}{x^2 - 1}$, $g(x) = \dfrac{1}{x}$

3. $f(x) = \sqrt{x + 1}$, $g(x) = \dfrac{1}{x^2 - 1}$
   4. $f(x) = \sqrt{x}$, $g(x) = \dfrac{x - 1}{x + 1}$

5. $f(x) = x^2$, $g(x) = x^2$
   6. $f(x) = \sqrt{x}$, $g(x) = \sqrt[3]{x}$

Write the following functions as composites of two functions:

7. $\sqrt{x^2 + 3x - 1}$
   8. $(2x - 3)^3 + (2x - 3)^2 + \dfrac{1}{2x - 3}$

9. $\sqrt{\dfrac{x - 1}{x + 1}}$
   10. $\dfrac{\sqrt{x} - 1}{\sqrt{x} + 1}$

11. $(x^3 + 9x - 2)^5$
    12. $(9x^2 + 2x - 5)^7$

13. $\dfrac{x^2 - 1}{x^2 + 1}$
    14. $\dfrac{x^3 + 3x^6}{1 + x^9}$

15. $\sqrt{x^{1/3} + 1}$
    16. $(\sqrt{x})^{1/3}$

17. $(x^3 + 1)^2 + 1$
    18. $(x^2 + 1)^3 + 1$

19. $x^2(x^4 + 1)^2$
    20. $(x^2 + 1)(x^4 + 1)^3$

Differentiate the following functions using the product rule, quotient rule, and chain rule.

21. $(x^2 + 1)^{15}$
22. $(x^4 + x^2 + 1)^{10}$

23. $\sqrt{1 + x^4}$
24. $(2x^3 + 3x - 1)^{-7}$

25. $\left(\dfrac{2}{x} + 5(3x - 1)^4\right)^3$
26. $[(2x^2 + 1)^3 + 5]^4$

27. $(x^3 - 1)(x^2 + 1)^4$
28. $(2x^2 - 4)^5(5x + 1)^3$

29. $(4 - x)^3(4 + x)^3$
30. $(2x - 1)^{1/4}(2x + 1)^{3/4}$

31. $\sqrt{2 + \sqrt{1 + x^2}}$
32. $\left(\dfrac{x - 1}{x + 1}\right)^3$

33. $\dfrac{3 - x}{\sqrt{4x + x^2}}$
34. $x^2\sqrt{x^2 - 9}$

35. $\dfrac{1 + \sqrt{2x}}{1 - \sqrt{2x}}$
36. $\dfrac{(x^5 + 1)^3}{(3x^2 + 1)^4}$

37. Sketch the graph of $y = 4x/(x + 1)^2$, $(x > -1)$.

38. Sketch the graph of $y = 2/(1 + x^2)$.

39. The chain rule along with the results of Exercises 38 and 41 of Section 1 imply that

$$\frac{d}{dx}[g(x)]^n = n \cdot [g(x)]^{n-1} \cdot g'(x)$$

for all integers n. Use this formula to show that

$$\frac{d}{dx}(x^{1/n}) = \frac{1}{n} \cdot x^{(1/n)-1}$$

for any nonzero integer n. [If $g(x) = x^{1/n}$, then $[g(x)]^n = x$. Differentiate both sides and solve for $g'(x)$.]

40. Use the formula given in Exercise 39 to show that

$$\frac{d}{dx}(x^r) = r \cdot x^{r-1}.$$

for any rational number r. (Write $r = m/n$, where m and n are integers and $n \neq 0$. Then $x^r = (x^m)^{1/n}$.)

41. Suppose $f(x)$ and $g(x)$ are differentiable functions. Find $g(x)$ if you know that

$$\frac{d}{dx}f(g(x)) = 3x^2 \cdot f'(x^3 + 1).$$

42. Suppose $f(x)$ and $g(x)$ are differentiable functions. Find $g(x)$ if you know that $f'(x) = \dfrac{1}{x}$ and

$$\frac{d}{dx}f(g(x)) = \frac{2x + 5}{x^2 + 5x - 4}.$$

1. Let $f(x) = x^5 + x^4$ and $g(x) = 2x^3 - 5$.

2. $f'(x) = 5x^4 + 4x^3$, $f'(g(x)) = 5(2x^3 - 5)^4 + 4(2x^3 - 5)^3$.

3. We have $g'(x) = 6x^2$. Then, from the chain rule and the result of problem 2, we have

$$h'(x) = f'(g(x))g'(x) = [5(2x^3 - 5)^4 + 4(2x^3 - 5)^3](6x^2).$$

Chapter 3: CHECKLIST

☐ $\dfrac{d}{dx}[f(x) \cdot g(x)] = f(x)g'(x) + g(x)f'(x)$

☐ $\dfrac{d}{dx}\left[\dfrac{f(x)}{g(x)}\right] = \dfrac{g(x)f'(x) - f(x)g'(x)}{[g(x)]^2}$

☐ $\dfrac{d}{dx}[f(g(x))] = f'(g(x)) \cdot g'(x)$

Chapter 3: SUPPLEMENTARY EXERCISES

Differentiate the following functions.

1. $\sqrt{x}(x^3 + 1)$

2. $\sqrt{x + 2}(x^2 + 5x + 1)$

3. $\dfrac{\sqrt{x}}{\sqrt{x} + 1}$

4. $(x^2 - 1)^5(x^3 + 5)^4$

5. $(x^2 + 5x + 9)(x^3 - 2)^{10}$

6. $(\sqrt{x} - 1)^3(\sqrt[3]{x} - 2)^2$

7. $\dfrac{(x + 2)^2 - (x - 2)^2}{x^2 + 2}$

8. $\dfrac{x - \sqrt{x}}{x + \sqrt{x}}$

9. $(x + 2 + (x + 2)^2)^2$

10. $(\sqrt{x} + (\sqrt{x} + 1))^2$

11. $[(3x + 1)^3 + (3x + 1)^2 + 1](\sqrt{x} - 2)$

12. $\dfrac{(5x^2 + 2)^2 - 1}{x + 2}$

13. $\sqrt{x + \sqrt{x + \sqrt{x}}}$

14. $\sqrt{x + \sqrt[3]{x + \sqrt{x}}}$

15. $\sqrt[3]{x + \sqrt{x + \sqrt{x}}}$

16. $\sqrt[3]{x + \sqrt[3]{x + \sqrt[3]{x}}}$

17. Suppose that $y = f(x)$ is a function with the property that the slope of the tangent line at each point (x, y) on its graph is $\frac{1}{y}$. In addition, suppose that $(1, 2)$ is a point on the graph. What is the slope of the tangent line to the curve $y = f(x^3)$ at $(1, 2)$?

18. Rework Exercise 17 with $\frac{1}{y}$ replaced by $\frac{1}{x}$.

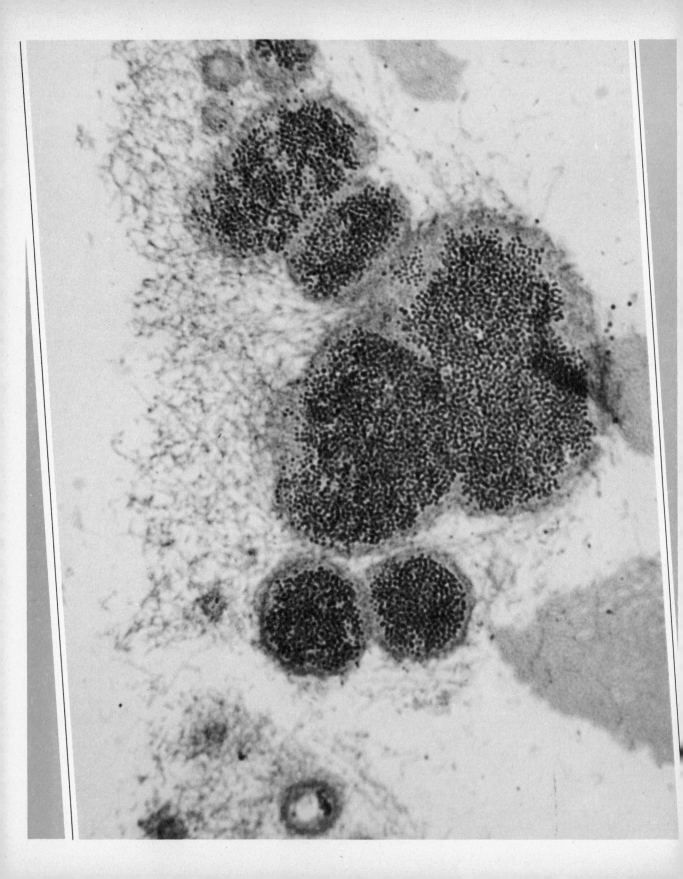

THE EXPONENTIAL FUNCTION

When a bacteria culture is grown in a laboratory dish, the rate of growth of the culture at any given moment is proportional to the total number of bacteria in the dish at that moment. This is an example of what is called *exponential growth*. A pile of radioactive uranium U^{235} constantly decays. The rate of decay of the uranium at any given moment is proportional to the amount of U^{235} remaining at that moment. The decay of uranium (or of radioactive elements in general) is an example of what is called *exponential decay*. Both exponential growth and exponential decay can be described and studied in terms of *exponential functions*, whose properties are investigated in this chapter. Subsequently, we will use exponential functions to study exponential growth, exponential decay, advertising, and the spread of epidemics, to mention only a few of the applications.

4.1. Properties of Exponents

Let us precede our discussion of exponential functions with a review of exponents. That is, we will discuss the operation of raising a number b to some power.

Consider the powers of 2 as an example. As the reader may recall from elementary algebra, if n is a positive integer (e.g., 1, 2, 3, 4, ...), then 2^n just means $2 \cdot 2 \cdot 2 \cdot \ldots \cdot 2$ (n times). Further, recall that $2^0 = 1$. Thus, for example,

$$2^0 = 1, \qquad 2^1 = 2, \qquad 2^2 = 2 \cdot 2 = 4, \qquad 2^3 = 2 \cdot 2 \cdot 2 = 8,$$

$$2^4 = 2 \cdot 2 \cdot 2 \cdot 2 = 16, \qquad \ldots$$

Moreover, $2^{-n} = \dfrac{1}{2^n}$, so that we may raise 2 to negative powers:

$$2^{-1} = \frac{1}{2^1} = \frac{1}{2}, \qquad 2^{-2} = \frac{1}{2^2} = \frac{1}{4}, \qquad 2^{-3} = \frac{1}{2^3} = \frac{1}{8}, \qquad \ldots$$

We may also raise 2 to fractional powers. Recall that

$2^{1/2}$ is the positive number whose square is 2: $2^{1/2} = \sqrt{2}$;

$2^{1/3}$ is the positive number whose cube is 2: $2^{1/3} = \sqrt[3]{2}$;

$2^{1/4}$ is the positive number whose fourth power is 2: $2^{1/4} = \sqrt[4]{2}$;

and similarly for other powers of 2 of the form $2^{1/n}$. These powers may be numerically computed by using tables or a pocket calculator with a y^x key. For example, from a table of square roots we find that $2^{1/2} = \sqrt{2} \approx 1.4142\ldots$.

We compute the power $2^{3/2}$ as $(2^{1/2})^3$. That is, to compute $2^{3/2}$, we first calculate $2^{1/2}$ and then raise it to the third power. Thus

$$2^{3/2} = (2^{1/2})^3 \approx (1.4142)^3 \approx 2.8283.$$

More generally, we can compute $2^{m/n}$ by the formula

$$2^{m/n} = (2^{1/n})^m.$$

This shows how to raise 2 to any fractional exponent. We can handle decimal exponents by converting them to fractions. For example, $2^{1.4} = 2^{14/10}$ and $2^{.05} = 2^{5/100}$.

Laws of Exponents The following laws of exponents summarize the properties of exponents that are most useful in dealing with calculations. They are usually covered in a course in algebra.

Let a, b be positive numbers and let x, y be any numbers (positive, negative, or zero).

EI $b^x \cdot b^y = b^{x+y}$.

EII $\dfrac{b^x}{b^y} = b^{x-y}$.

EIII $(b^x)^y = b^{xy}$.

EIV $a^x b^x = (ab)^x$.

EXAMPLE 1   Use the laws of exponents EI to EIV to calculate

(a) $2^3 \cdot 2^4$ (b) $(2^3)^4$ (c) $2^3/2^{1/2}$ (d) $2^{1/2} \cdot 10^{1/2}$.

Solution   (a) By EI, $2^3 \cdot 2^4 = 2^{3+4} = 2^7 = 128$.

(b) By EIII, $(2^3)^4 = 2^{3 \cdot 4} = 2^{12} = 4096$.

(c) By EII, $2^3/2^{1/2} = 2^{3-1/2} = 2^{5/2} = (2^{1/2})^5 = (\sqrt{2})^5$.

(d) By EIV, $2^{1/2} \cdot 10^{1/2} = (2 \cdot 10)^{1/2} = 20^{1/2} = \sqrt{20}$.

The law of exponents EIV is often used in algebraic calculations. For example, EIV implies that

$$(2t)^5 = 2^5 t^5 = 32 t^5.$$

Two very useful consequences of the laws of exponents are worthwhile remembering. If we set $x = 0$ in EI, then we have that $b^0 \cdot b^y = b^{0+y} = b^y$. Thus, on dividing both sides by b^y, we see that

$$b^0 = 1. \tag{1}$$

If we set $x = 0$ in EII, then we see that

$$\frac{b^0}{b^y} = b^{0-y} = b^{-y}.$$

But we just showed that $b^0 = 1$. Therefore we have that

$$\frac{1}{b^y} = b^{-y}. \tag{2}$$

We wrote down properties (1) and (2) in our discussion of powers of 2 above. Now we see that they are general properties of exponents that follow easily from the four fundamental laws of exponents EI to EIV.

PRACTICE PROBLEMS 1

1. Simplify $(2^x)^3 \cdot 2^x$

2. Solve for x: $7 \cdot 2^{6-3x} = 28$

EXERCISES 1

Calculate the following.

1. $8^{2/3}$, $9^{3/2}$, $16^{-3/4}$

2. $9^{-1/2}$, $8^{4/3}$, $27^{-2/3}$

3. $(\frac{1}{4})^2$, $(\frac{1}{8})^{-3}$, $(\frac{4}{9})^{1/2}$

4. $(\frac{1}{9})^2$, $(\frac{1}{27})^{1/3}$, $(\frac{9}{16})^{-1/2}$

5. $16^3 \cdot (16)^{-5/2}$, $9^2(\frac{1}{9})^3$

6. $8^{8/3} \cdot (\frac{1}{8})^2$, $25(25)^{-1/2}$

7. $9^{1/3} \cdot (3)^{4/3}$, $8^{3/2} \cdot (\frac{1}{2})^{3/2}$

8. $16^{-2/3} \cdot (2)^{2/3}$, $5^{1/3} \cdot (25)^{1/3}$

9. $(8^{1/2} \cdot 8^{1/3})^2$, $(25^{1/2} \cdot 25^4)^{1/9}$

10. $(9^{-1} \cdot 9^{1/3})^3$, $(16^{1/4} \cdot 16^{-3/4})^3$

From a table we have $2^{1/2} \approx 1.414$, $2^{1/10} \approx 1.072$, $(2.7)^{1/2} \approx 1.643$, and $(2.7)^{1/10} \approx 1.104$. (These figures are accurate to three decimal places.) Compute the following numbers, rounding off your answer to two decimal places, if necessary.

11. (a) 2^x for $x = 3$ (b) 2^x for $x = -3$

(c) 2^x for $x = \frac{5}{2}$ [Hint: $\frac{5}{2} = 2 + \frac{1}{2}$]   (d) 2^x for $x = 4.1$

(e) 2^x for $x = .2$ (f) 2^x for $x = .9$ [Hint: $.9 = 1 - .1$]

(g) 2^x for $x = -2.5$
[Hint: $-2.5 = -3 + .5$]   (h) 2^x for $x = -3.9$

12. (a) $(2.7)^x$ for $x = 0.2$ (b) $(2.7)^x$ for $x = 1.5$ (c) $(2.7)^x$ for $x = 0$

(d) $(2.7)^x$ for $x = -1$ (e) $(2.7)^x$ for $x = 1.1$ (f) $(2.7)^x$ for $x = .6$

Solve the following equations for x.

13. $3^{2x} = 3^2$

14. $10^{-x} = 10^2$

15. $(2.5)^{2x+1} = (2.5)^5$

16. $(3.2)^{x-3} = (3.2)^5$

17. $10^{1-x} = 100$

18. $2^{4-x} = 8$

19. $3(2.7)^{5x} = 8.1$

20. $4(2.7)^{2x-1} = 10.8$

21. $(2^{x+1} \cdot 2^{-3})^2 = 2$

22. $(3^{2x} \cdot 3^2)^4 = 3$

Simplify the following.

23. $(2^{3x})^5$

24. $\dfrac{3^{5x}}{3^{6x}}$

25. $\dfrac{(5^{7x/2} - 5^{x/2})}{\sqrt{5^x}}$

26. $(e^x)^{-x}$

27. $\dfrac{3^{10x} - 1}{3^{5x} - 1}$

28. $\dfrac{2^{6x} + 3 \cdot 2^{3x} + 2}{2^{3x} + 1}$

SOLUTIONS TO PRACTICE PROBLEMS 1

1. By EIII, $(2^x)^3 = 2^{3x}$ and so $(2^x)^3 \cdot 2^x = 2^{3x} \cdot 2^x$. By EI, $2^{3x} \cdot 2^x = 2^{3x+x} = 2^{4x}$.

2. Divide both sides of the equation by 7. We then obtain

$$2^{6-3x} = 4.$$

Now 4 can be written as 2^2. So we have

$$2^{6-3x} = 2^2.$$

Equate exponents.

$$6 - 3x = 2$$

$$4 = 3x$$

$$x = \tfrac{4}{3}.$$

The general idea for solving equations in which the variable appears as an exponent is to try to put the equation into the form: $b^{[\text{expression in } x]} = [\text{some number}]$. Then express the number on the right as a power of b. In order to solve for x, we equate the exponents on both sides of the equation.

4.2. Graphs of Exponential Functions

Let b be a fixed positive number. The function

$$y = b^x$$

is called an *exponential function* because the variable x is the exponent. In this section we will study the graphs of exponential functions for various values of b. Let us begin with the special case $b = 2$ and sketch the graph of $y = 2^x$.

We have tabulated the values of 2^x for $x = 0, \pm 1, \pm 2, \pm 3$ and plotted these values in Fig. 1. Other intermediate values of 2^x for $x = \pm .1 \pm .2, \pm .3, \ldots$, may be obtained from tables or from a calculator with a y^x key. [See Fig. 2(a).] By passing a smooth curve through these points, we obtain the graph of $y = 2^x$, in Fig. 2(b).

In the same manner, we have sketched the graph of $y = 3^x$ (Fig. 3). The graphs of $y = 2^x$ and $y = 3^x$ have the same basic shape. Also note that they both pass through the point $(0, 1)$ (because $2^0 = 1$, $3^0 = 1$).

In Fig. 4 we have sketched the graphs of several more exponential functions. Notice that the graph of $y = 5^x$ has a large slope at $x = 0$, since the graph at $x = 0$ is quite steep;

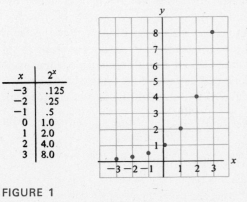

x	2^x
-3	.125
-2	.25
-1	.5
0	1.0
1	2.0
2	4.0
3	8.0

FIGURE 1

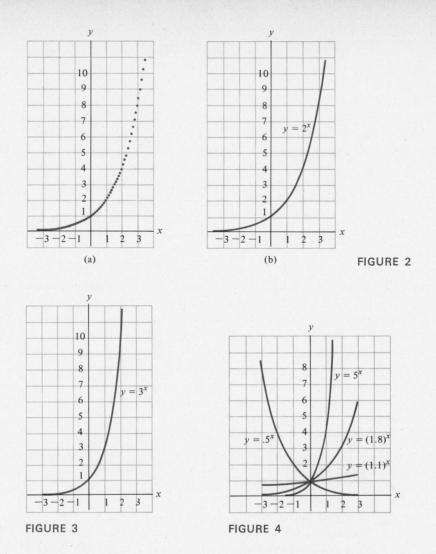

(a)

(b)

FIGURE 2

FIGURE 3

FIGURE 4

however, the graph of $y = (1.1)^x$ is nearly horizontal at $x = 0$, and hence the slope is close to zero.

Our next task is to discover formulas for the slopes of curves of the form $y = b^x$. We begin with the special case $y = 2^x$. Before computing the slope at an arbitrary x, let us determine the slope at $x = 0$. We will denote this slope by m and use the secant-line approximation of the derivative to approximate m. We proceed by constructing a secant line in Fig. 5. The slope of the secant line through $(0, 1)$ and $(h, 2^h)$ is $\dfrac{2^h - 1}{h}$. As h approaches zero, the slope of the secant line approaches the slope of $y = 2^x$ at $x = 0$. That is,

$$\lim_{h \to 0} \frac{2^h - 1}{h} = m. \tag{1}$$

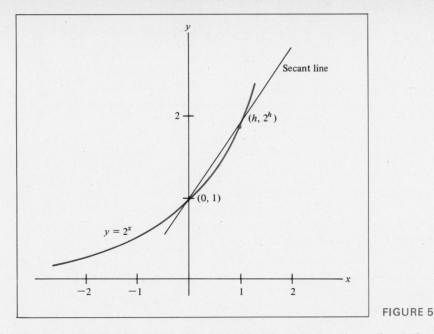

FIGURE 5

We can estimate the value of m by taking h smaller and smaller. When $h = .1$, we have $2^h \approx 1.072$ (from a table of values of 2^x), and

$$\frac{2^h - 1}{h} \approx \frac{.072}{.1} = .72.$$

When $h = .01$, we have $2^h \approx 1.00696$, and

$$\frac{2^h - 1}{h} \approx \frac{.00696}{.01} \approx .696.$$

When $h = .001$, $2^h \approx 1.0006934$, so that

$$\frac{2^h - 1}{h} \approx \frac{.0006934}{.001} \approx .693.$$

Thus it is reasonable to conclude that $m \approx .69$.* Since m equals the slope of $y = 2^x$ at $x = 0$, we have

$$m = \frac{d}{dx}(2^x)\Big|_{x=0} \approx .69. \tag{2}$$

Now that we have estimated the slope of $y = 2^x$ at $x = 0$, let us compute the slope for an arbitrary value of x. We construct a secant line through $(x, 2^x)$ and a nearby point $(x + h, 2^{x+h})$ on the graph. The slope of the secant line is

$$\frac{2^{x+h} - 2^x}{h}. \tag{3}$$

_____

* To ten decimal places, m is .6931471806.

By the law of exponents EI, we have $2^{x+h} - 2^x = 2^x(2^h - 1)$ so that, by (1), we see that

$$\lim_{h \to 0} \frac{2^{x+h} - 2^x}{h} = \lim_{h \to 0} 2^x \frac{2^h - 1}{h} = 2^x \lim_{h \to 0} \frac{2^h - 1}{h} = m2^x. \tag{4}$$

However, the slope of the secant (3) approaches the derivative of 2^x as h approaches zero. Consequently, we have that

$$\frac{d}{dx}(2^x) = m2^x, \qquad \text{where } m = \frac{d}{dx}(2^x)\Big|_{x=0}. \tag{5}$$

EXAMPLE 1   Calculate $\dfrac{d}{dx}(2^x)\Big|_{x=3}, \ \dfrac{d}{dx}(2^x)\Big|_{x=-1}$.

Solution   (a) $\dfrac{d}{dx}(2^x)\Big|_{x=3} = m \cdot 2^3 = 8m \approx 8(.69) = 5.52$.

(b) $\dfrac{d}{dx}(2^x)\Big|_{x=-1} = m \cdot 2^{-1} = .5m \approx .5(.69) = .345$.

The calculations just carried out for $y = 2^x$ can be carried out for $y = b^x$, where b is any positive number. Equation (5) will read exactly the same except that 2 will be replaced by b.

$$\frac{d}{dx}(b^x) = mb^x, \qquad \text{where } m = \frac{d}{dx}(b^x)\Big|_{x=0}. \tag{6}$$

Our calculations showed that if $b = 2$, then $m \approx .69$. If $b = 3$, then it turns out that $m \approx 1.1$. (See Exercise 1 below.) The function of the type $y = b^x$ for which the formula (6) is simplest is the one for which $m = 1$—that is, the one whose slope is 1 at $x = 0$. We wish to find the value of b for which this situation occurs. From our preceding calculations it seems that the appropriate value of b lies between 2 and 3. It can be shown (using tables or a calculator) that

$$\frac{d}{dx}(2.7)^x\Big|_{x=0} \approx .993251773$$

$$\frac{d}{dx}(2.71)^x\Big|_{x=0} \approx .996948635$$

$$\frac{d}{dx}(2.718)^x\Big|_{x=0} \approx .999896316$$

$$\frac{d}{dx}(2.719)^x\Big|_{x=0} \approx 1.000264166.$$

Thus the value of b for which $m = 1$ lies somewhere between 2.718 and 2.719. The *exact* value of b for which

$$\frac{d}{dx}(b^x)\Big|_{x=0} = 1$$

is denoted by the letter e. Thus the number e has the property that

$$\frac{d}{dx}(e^x)\Big|_{x=0} = 1. \tag{7}$$

That is, the function $y = e^x$ has slope 1 at $x = 0$. The number e is an important constant of nature that has been calculated to thousands of decimal places. To ten figures, $e = 2.718281828$. For our purposes, it is usually sufficient to think of e as "approximately 2.7." From equations (6) and (7) we see that

$$\frac{d}{dx}(e^x) = 1 \cdot e^x = e^x. \tag{8}$$

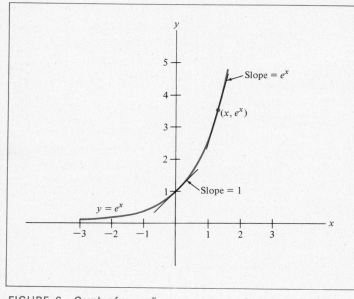

FIGURE 6   Graph of $y = e^x$.

The graphical interpretation of (7) is that the curve $y = e^x$ has slope 1 at $x = 0$. The graphical interpretation of (8) is that the slope of the curve $y = e^x$ at an arbitrary value of x is exactly equal to the value of the function e^x at that point. (See Fig. 6.)

The function e^x is the same type of function as 2^x and 3^x except that taking derivatives of e^x is much easier. For this reason, the function e^x is used in almost all applications that require an exponential-type function to describe a physical phenomenon. As a result, e^x is usually called *the* exponential function. Extensive tables are available from which to determine the value of e^x for a wide range of values of x. Also, scientific calculators provide for calculating e^x at the push of a button.

PRACTICE PROBLEMS 2

In the following problems, use the number 20 as the (approximate) value of e^3.

1. Find the equation of the tangent line to the graph of $y = e^x$ at $x = 3$.
2. Solve for x: $4e^{6x} = 80$.

1. Use the table below to show that

$$\frac{d}{dx}(3^x)\Big|_{x=0} \approx 1.1.$$

That is, calculate the slope

$$\frac{3^h - 1}{h}$$

of the secant line passing through the point $(0, 1)$ and $(h, 3^h)$. Take $h = .1, .01, .001$.

x	3^x
0	1.00000
.001	1.00110
.010	1.01105
.100	1.11612

2. Use the table below to show that

$$\frac{d}{dx}(2.7)^x\Big|_{x=0} \approx 0.99.$$

That is, calculate

$$\frac{(2.7)^h - 1}{h}$$

for $h = .1, .01,$ and $.001$.

x	$(2.7)^x$
0	1.00000
.001	1.00099
.010	1.00998
.100	1.10443

3. Consider the secant line on the graph of e^x passing through $(0, 1)$ and (h, e^h). Its slope is

$$\frac{e^h - 1}{h}.$$

Compute this quantity for $h = .01, .005,$ and $.001$.

x	e^x
0	1.00000
.001	1.00100
.005	1.00501
.010	1.01005

4. Use (8) and a familiar rule for differentiation to find

$$\frac{d}{dx}(5e^x).$$

5. Use (8) and a familiar rule for differentiation to find

$$\frac{d}{dx}(e^x)^{10}.$$

6. Use the fact that $e^{2+x} = e^2 \cdot e^x$ to find

$$\frac{d}{dx}(e^{2+x}).$$

[Remember that e^2 is just a constant—approximately $(2.7)^2$.]

7. Use the fact that $e^{4x} = (e^x)^4$ to find

$$\frac{d}{dx}(e^{4x}).$$

8. Find $\frac{d}{dx}(e^x + x^2)$.

Simplify:

9. $e^{2x}(1 + e^{3x})$ 10. $(e^x)^2$ 11. $e^{1-x} \cdot e^{2x}$

12. $5e^{3x}/e^x$ 13. $1/e^{-2x}$ 14. $e^3 \cdot e^{x+1}$

Use Table 1 of the Appendix to determine the following numbers:

15. e^2 16. $e^{-1.30}$ 17. $e^{-.5}$ 18. $e^{3/2}$

Solve the following equations for x:

19. $e^{5x} = e^{20}$ 20. $e^{1-x} = e^2$

21. $e^{x^2-2x} = e^8$ 22. $e^{-x} = 1$

Differentiate the following functions:

23. xe^x 24. e^x/x 25. $e^x/(1 + e^x)$

26. $(1 + x^2)e^x$ 27. $(1 + 5e^x)^4$ 28. $(xe^x - 1)^{-3}$

SOLUTIONS TO PRACTICE PROBLEMS 2

1. When $x = 3$, $y = e^3 = 20$. So the point $(3, 20)$ is on the tangent line. Since $\frac{d}{dx}(e^x) = e^x$, the slope of the tangent line is e^3 or 20. Therefore, the equation of the tangent line in point-slope form is $y - 20 = 20(x - 3)$.

2. This problem is similar to the second practice problem of Section 1. First **divide** both sides of the equation by 4.

$$e^{6x} = 20.$$

The idea is to express 20 as a power of e and then equate exponents.

$$e^{6x} = e^3$$

$$6x = 3$$

$$x = \tfrac{1}{2}.$$

4.3. Differentiation of Exponential Functions

We have shown that $\dfrac{d}{dx}(e^x) = e^x$. Using this fact and the chain rule, we may differentiate functions of the form $e^{g(x)}$, where $g(x)$ is any differentiable function. This is because $e^{g(x)}$ is the composite of two functions. Indeed, if $f(x) = e^x$, then

$$e^{g(x)} = f(g(x)).$$

Thus, by the chain rule, we have

$$\frac{d}{dx}(e^{g(x)}) = f'(g(x))g'(x)$$

$$= f(g(x))g'(x) \qquad [\text{since } f'(x) = f(x)]$$

$$= e^{g(x)}g'(x).$$

So we have the following result:

> *Chain Rule for Exponential Functions:* Let $g(x)$ be any differentiable function. Then
>
> $$\frac{d}{dx}(e^{g(x)}) = e^{g(x)}g'(x).$$

(1)

EXAMPLE 1 Differentiate e^{x^2+1}.

Solution Here $g(x) = x^2 + 1$, $g'(x) = 2x$, so

$$\frac{d}{dx}(e^{x^2+1}) = e^{x^2+1} \cdot 2x$$

$$= 2xe^{x^2+1}.$$

EXAMPLE 2   Differentiate $e^{3x^2 - (1/x)}$.

Solution
$$\frac{d}{dx}\left(e^{3x^2 - (1/x)}\right) = e^{3x^2 - (1/x)} \cdot \frac{d}{dx}\left(3x^2 - \frac{1}{x}\right)$$

$$= e^{3x^2 - (1/x)}\left(6x + \frac{1}{x^2}\right).$$

EXAMPLE 3   Differentiate e^{5x}.

Solution
$$\frac{d}{dx}\left(e^{5x}\right) = e^{5x} \cdot \frac{d}{dx}(5x) = e^{5x} \cdot 5 = 5e^{5x}.$$

Using a computation similar to that used in Example 3, we may differentiate e^{kx} for any constant k. (In Example 3 we have $k = 5$.) The result is the following useful formula.

$$\frac{d}{dx}\left(e^{kx}\right) = ke^{kx}. \qquad (2)$$

Many applications involve exponential functions of the form $y = Ce^{kx}$, where C and k are constants. In the next example we differentiate such functions.

EXAMPLE 4   Differentiate the following exponential functions.

(a) $3e^{5x}$.

(b) $3e^{kx}$, where k is a constant.

(c) Ce^{kx}, where C and k are constants.

Solution   (a) $\dfrac{d}{dx}(3e^{5x}) = 3\dfrac{d}{dx}(e^{5x}) = 3 \cdot 5e^{5x} = 15e^{5x}.$

(b) $\dfrac{d}{dx}(3e^{kx}) = 3\dfrac{d}{dx}(e^{kx})$

$\qquad = 3 \cdot ke^{kx} \qquad$ (by (2))

$\qquad = 3ke^{kx}.$

(c) $\dfrac{d}{dx}(Ce^{kx}) = C\dfrac{d}{dx}(e^{kx}) = Cke^{kx}.$

The result of part (c) may be summarized in an extremely useful fashion as follows: Suppose that we let $y = Ce^{kx}$. By part (c) we have

$$y' = Cke^{kx}$$
$$= k \cdot (Ce^{kx})$$
$$= ky.$$

In other words, the derivative of the function Ce^{kx} is k times the function itself. Let us record this fact.

Let C, k be any constants and let $y = Ce^{kx}$. Then y satisfies the equation

$$y' = ky.$$

The equation $y' = ky$ expresses a relationship between the function y and its derivative y'. Any equation expressing a relationship between a function y and one or more of its derivatives is called a *differential equation*.

Very often an applied problem will involve a function $y = f(x)$ which satisfies the differential equation $y' = ky$. It can be shown that y must then necessarily be an exponential function of the form Ce^{kx}. That is, we have the following result.

Suppose that $y = f(x)$ satisfies the differential equation

$$y' = ky.$$

Then y is an exponential function of the form

$$y = Ce^{kx}, \qquad C \text{ a constant.}$$

(3)

We shall verify this result in Exercise 47.

EXAMPLE 5   Determine all functions $y = f(x)$ such that $y' = -.2y$.

Solution   The equation $y' = -.2y$ has the form $y' = ky$ with $k = -.2$. Therefore, any solution of the equation has the form

$$y = Ce^{-.2x},$$

where C is a constant.

EXAMPLE 6   Determine all functions $y = f(x)$ such that $y' = y/2$ and $f(0) = 4$.

Solution   The equation $y' = y/2$ has the form $y' = ky$ with $k = \frac{1}{2}$. Therefore,

$$f(x) = Ce^{(1/2)x}$$

for some constant C. We also require that $f(0) = 4$. That is,

$$4 = f(0) = Ce^{(1/2) \cdot 0} = Ce^0 = C.$$

So $C = 4$ and

$$f(x) = 4e^{(1/2)x}.$$

The Functions e^{kx} Exponential functions of the form e^{kx} occur in many applications. Figure 1 shows the graphs of several functions of this type when k is a positive number. These curves $y = e^{kx}$, k positive, have several properties in common:

1. $(0, 1)$ is on the graph.
2. The graph lies strictly above the x-axis (e^{kx} is never zero).
3. The x-axis is an asymptote as x becomes large negatively.
4. The graph is always increasing and concave up.

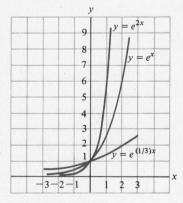

FIGURE 1

When k is negative, the graph of $y = e^{kx}$ is decreasing. (See Fig. 2.) Note the following properties of the curves $y = e^{kx}$, k negative:

1. $(0, 1)$ is on the graph.
2. The graph lies strictly above the x-axis.
3. The x-axis is an asymptote as x becomes large positively.
4. The graph is always decreasing and concave up.

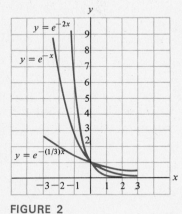

FIGURE 2

The Functions b^x If b is a positive number, then the function b^x may be written in the form e^{kx} for some k. For example, take $b = 2$. From Fig. 6 of the preceding section it is clear that there is some value of x such that $e^x = 2$. Call this value k, so that $e^k = 2$. Then

$$2^x = (e^k)^x = e^{kx}$$

for all x. In general, if b is any positive number, there is a value of x, say $x = k$, such that $e^k = b$. In this case, $b^x = (e^k)^x = e^{kx}$. Thus all the curves $y = b^x$ discussed in Section 2 can be written in the form $y = e^{kx}$. This is one reason why we have focused on exponential functions with base e instead of studying 2^x, 3^x, and so on.

1. Differentiate $[e^{-3x}(1 + e^{6x})]^{12}$.

2. Determine all functions $y = f(x)$ such that $y' = -\dfrac{y}{20}$, $f(0) = 2$.

EXERCISES 3

Differentiate the following.

1. $y = e^{-x}$

2. $f(x) = e^{10x}$

3. $f(x) = 5e^x$

4. $y = \dfrac{e^x + e^{-x}}{2}$

5. $f(t) = e^{t^2}$

6. $f(t) = e^{-2t}$

7. $f(x) = \dfrac{e^x - e^{-x}}{2}$

8. $f(x) = 2e^{1-x}$

9. $y = e^{-2x} - 2x$

10. $f(x) = \frac{1}{10}e^{-x^2/2}$

11. $g(x) = (e^x + e^{-x})^3$

12. $y = (e^{-x})^2$

13. $y = \frac{1}{3}e^{3-2x}$

14. $g(x) = e^{1/x}$

15. $f(t) = e^t(e^{2t} - e^{-t})$

16. $f(t) = \dfrac{e^t + e^{-t}}{e^t}$

[Hint: In Exercises 15 and 16, simplify $f(t)$ before differentiating.]

17. $y = e^{x^3 + x - (1/x)}$

18. $y = (e^{x^2} + x^2)^5$

19. $f(x) = (2x + 1 - e^{2x+1})^4$

20. $f(x) = e^{1/(3x-7)}$

21. $x^3 e^{x^2}$

22. $x^2 e^{-3x}$

23. e^{x^3}/x

24. $e^{(x-1)/x}$

25. $(x + 1)e^{-x+2}$

26. $x\sqrt{2 + e^x}$

27. $[(1/x) + 3]e^x$

28. $e^{-3x}/(1 - 3x)$

29. $(e^x - 1)/(e^x + 1)$

30. $(xe^x - 3)/(x + 1)$

In Exercises 31–40 find all values of x such that the function has a possible maximum or minimum. Use the second derivative test to determine the nature of the function at these points. (Recall that e^x is positive for all x.)

31. $f(x) = (1 + x)e^{-x/2}$

32. $f(x) = (1 - x)e^{-x/2}$

33. $f(x) = \dfrac{3 - 2x}{e^{x/4}}$

34. $f(x) = \dfrac{4x - 3}{e^{x/2}}$

35. $f(x) = (8 - 2x)e^{x+5}$

36. $f(x) = (4x - 1)e^{3x-2}$

37. $f(x) = \dfrac{(x - 1)^2}{e^x}$

38. $f(x) = (x + 3)^2 e^x$

39. $f(x) = (x + 5)^2 e^{2x-1}$

40. $f(x) = \dfrac{(x - 3)^2}{e^{2x}}$

41. Let a and b be positive numbers: A curve whose equation is $y = e^{-ae^{-bx}}$ is called a *Gompertz growth curve*. These curves are used in biology to describe certain types of population growth. Compute the derivative of $y = e^{-2e^{-.01x}}$.

42. Find $\dfrac{dy}{dx}$ if $y = e^{-(1/10)e^{-x/2}}$.

43. Determine all solutions of the differential equation

$$y' = -4y.$$

44. Determine all solutions of the differential equation

$$y' = \tfrac{1}{3}y.$$

45. Determine all functions $y = f(x)$ such that

$$y' = -.5y \quad \text{and} \quad f(0) = 1.$$

46. Determine all functions $y = f(x)$ such that

$$y' = 3y \quad \text{and} \quad f(0) = \tfrac{1}{2}.$$

47. Verify the result (3). [Hint: Let $g(x) = f(x)e^{-kx}$. Show that $g'(x) = 0$.] You may assume that only a constant function has a zero derivative.

48. Let $f(x)$ be a function with the property that $f'(x) = \dfrac{1}{x}$. Let $g(x) = f(e^x)$, and compute $g'(x)$.

SOLUTIONS TO PRACTICE PROBLEMS 3

1. We must use the general power rule. However, this is most easily done if we first use the laws of exponents to simplify the function inside the brackets.

$$e^{-3x}(1 + e^{6x}) = e^{-3x} + e^{-3x} \cdot e^{6x}$$

$$= e^{-3x} + e^{3x}.$$

Now

$$\frac{d}{dx}[e^{-3x} + e^{3x}]^{12} = 12 \cdot [e^{-3x} + e^{3x}]^{11} \cdot (-3e^{-3x} + 3e^{3x})$$

$$= 36 \cdot [e^{-3x} + e^{3x}]^{11} \cdot (-e^{-3x} + e^{3x}).$$

2. The differential equation $y' = -\dfrac{y}{20}$ is of the type $y' = ky$, where $k = -\tfrac{1}{20}$. Therefore, any solution has the form $f(x) = Ce^{-(1/20)x}$. Now $f(0) = Ce^{-(1/20)\cdot 0} = Ce^0 = C$, so that $f(0) = 2$ when $C = 2$. Therefore the desired function is $f(x) = 2e^{-(1/20)x}$.

Chapter 4: CHECKLIST

- $b^x \cdot b^y = b^{x+y}$
- $b^x / b^y = b^{x-y}$
- $(b^x)^y = b^{xy}$
- $a^x b^x = (ab)^x$
- $b^0 = 1$
- $b^{-y} = 1/b^y$
- e
- e^x
- $\dfrac{d}{dx}(e^{kx}) = ke^{kx}$
- $\dfrac{d}{dx}(e^{g(x)}) = e^{g(x)}g'(x)$
- Graph of e^{kx}, k positive
- Graph of e^{kx}, k negative
- If $y = f(x)$ satisfies $y' = ky$, then $y = Ce^{kx}$ for some constant C.

Chapter 4: SUPPLEMENTARY EXERCISES

Calculate the following.

1. $27^{4/3}$
2. $4^{1.5}$
3. 5^{-2}
4. $16^{-.25}$
5. $(2^{5/7})^{14/5}$
6. $8^{1/2} \cdot 2^{1/2}$
7. $9^{5/2}/9^{3/2}$
8. $4^{.2} \cdot 4^{.3}$

Simplify the following.

9. $(e^{x^2})^3$
10. $e^{5x} \cdot e^{2x}$
11. e^{3x}/e^x
12. $2^x \cdot 3^x$
13. $(e^{8x} + 7e^{-2x})e^{3x}$
14. $\dfrac{e^{5x/2} - e^{3x}}{\sqrt{e^x}}$

Solve the following equations for x.

15. $e^{-3x} = e^{-12}$
16. $e^{x^2-x} = e^2$
17. $(e^x \cdot e^2)^3 = e^{-9}$
18. $e^{-5x} \cdot e^4 = e$

Differentiate the following functions.

19. $10e^{7x}$
20. $e^{\sqrt{x}}$
21. xe^{x^2}
22. $\dfrac{(e^x + 1)}{(x - 1)}$

23. $e^{(e^x)}$ 24. $(\sqrt{x} + 1)e^{-2x}$ 25. $\dfrac{x^2 - x + 5}{e^{3x} + 3}$ 26. x^e

27. Determine all solutions of the differential equation $y' = -y$.

28. Determine all functions $y = f(x)$ such that $y' = -1.5y$ and $f(0) = 2000$.

29. Determine all solutions of the differential equation $y' = 1.5y$ and $f(0) = 2$.

30. Determine all solutions of the differential equation $y' = \frac{1}{3}y$.

   Graph the following functions.

31. $e^{-x} + x$ 32. $e^x - x$ 33. $e^{-(1/2)x^2}$ 34. xe^{-x} for $x \geq 0$.

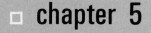

chapter 5

THE NATURAL LOGARITHM FUNCTION

In this chapter we study the natural logarithm function. This function is closely related to the exponential function and will appear in a number of applications throughout the book.

5.1. The Natural Logarithm Function

As a preparation for the definition of the natural logarithm, we shall make a geometrical digression. In Fig. 1 we have plotted several pairs of points. Observe how they are related to the line $y = x$.

The points $(5, 7)$ and $(7, 5)$, for example, are the same distance from the line $y = x$. If we were to plot the point $(5, 7)$ with wet ink and then fold the page

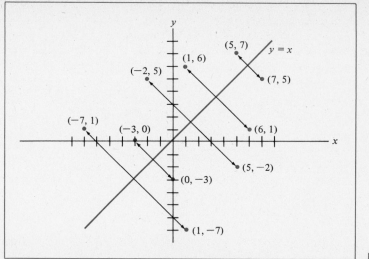

FIGURE 1

along the line $y = x$, the ink blot would produce a second blot at the point (7, 5). If we think of the line $y = x$ as a mirror, then (7, 5) is the mirror image of (5, 7). We say that (7, 5) is the *reflection* of (5, 7) through the line $y = x$. Similarly, (5, 7) is the reflection of (7, 5) through the line $y = x$.

Now let us consider all points lying on the graph of the exponential function $y = e^x$ (see Fig. 2(a)). If we reflect each such point through the line $y = x$, we obtain a new graph (see Fig. 2(b)). For each positive x, there is exactly one value of y such that (x, y) is on the new graph. We call this value of y the *natural logarithm of x*, denoted $\ln x$. Thus the reflection of the graph of $y = e^x$ through the line $y = x$ is the graph of the natural logarithm function $y = \ln x$.

FIGURE 2

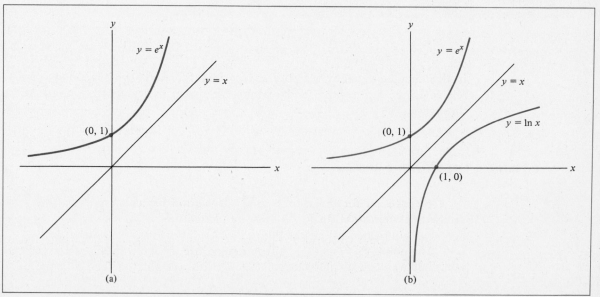

We may deduce some properties of the natural logarithm function from an inspection of its graph.

1. The point $(1, 0)$ is on the graph of $y = \ln x$ (because $(0, 1)$ is on the graph of $y = e^x$). In other words,

$$\ln 1 = 0. \tag{1}$$

2. $\ln x$ is defined only for positive values of x.
3. $\ln x$ is negative for x between 0 and 1.
4. $\ln x$ is positive for x greater than 1.
5. $\ln x$ is an increasing function.

Let us study the relationship between the natural logarithm and exponential functions more closely. From the way in which the graph of $\ln x$ was obtained we know that (a, b) is on the graph of $\ln x$ if and only if (b, a) is on the graph of e^x. However, a typical point on the graph of $\ln x$ is of the form $(a, \ln a)$, $a > 0$. So for any positive value of a, the point $(\ln a, a)$ is on the graph of e^x. That is,

$$e^{\ln a} = a.$$

Since a was an arbitrary positive number, we have the following important relationship between the natural logarithm and exponential functions.

$$e^{\ln x} = x \quad \text{for} \quad x > 0. \tag{2}$$

Equation (2) can be put into verbal form.

For each positive number x, $\ln x$ is that exponent to which we must raise e in order to get x.

If b is any number, then e^b is positive and hence $\ln(e^b)$ makes sense. What is $\ln(e^b)$? Since (b, e^b) is on the graph of e^x, we know that (e^b, b) must be on the graph of $\ln x$. That is, $\ln(e^b) = b$. Thus we have shown that

$$\ln(e^x) = x \quad \text{for any } x. \tag{3}$$

The identities (2) and (3) express the fact that the natural logarithm is the *inverse* of the exponential function. For instance, if we take a number x and compute e^x, then, by (3), we can undo the effect of the exponentiation by taking the natural logarithm; that is, the logarithm of e^x equals the original number x. Similarly, if we take a positive number x and compute ln x, then, by (2), we can undo the effect of the logarithm by raising e to the ln x power; that is, $e^{\ln x}$ equals the original number x.

Scientific calculators have an "ln x" key that will compute the natural logarithm of a number to as many as ten significant figures. For instance, entering the number 2 into the calculator and pressing the ln x key, one obtains ln 2 = .6931471806 (to ten significant figures). If a scientific calculator is unavailable, one may use a table of logarithms, such as the table in the Appendix of this text, which gives the values of ln x to five significant figures. For instance, this table shows that ln .8 = −.22314 (to five significant figures).

The relationships (2) and (3) between e^x and ln x may be used to solve equations, as the next examples show.

EXAMPLE 1   Solve the equation $5e^{x-3} = 4$ for x.

Solution   First divide each side by 5,

$$e^{x-3} = .8.$$

Taking the logarithm of each side and using (3), we have

$$\ln(e^{x-3}) = \ln .8$$

$$x - 3 = \ln .8$$

$$x = 3 + \ln .8.$$

[If desired, the numerical value of x can be obtained by using a scientific calculator or a natural logarithm table, namely, $x = 3 - .22314 = 2.77686$ (to five decimal places).]

EXAMPLE 2   Solve the equation $2 \ln x + 7 = 0$ for x.

Solution

$$2 \ln x = -7$$

$$\ln x = -3.5$$

$$e^{\ln x} = e^{-3.5}$$

$$x = e^{-3.5} \qquad \text{[by (2)]}.$$

Other Exponential and Logarithm Functions   In our discussion of the exponential function, we mentioned that all exponential functions of the form a^x, where a is a fixed positive number, can be expressed in terms of *the* exponential function e^x. Now we can be quite explicit. For since $a = e^{\ln a}$, we see that

$$a^x = (e^{\ln a})^x = e^{(\ln a) \cdot x}.$$

Hence we have shown that

$$a^x = e^{kx}, \quad \text{where } k = \ln a.$$

The natural logarithm function is sometimes called the *logarithm to the base e*, for it is the inverse of the exponential function e^x. If we reflect the graph of the function $y = 2^x$ through the line $y = x$, we obtain the graph of a function called the *logarithm to the base 2*, denoted by $\log_2 x$. Similarly, if we reflect the graph of $y = 10^x$ through the line $y = x$, we obtain the graph of a function called the *logarithm to the base 10*, denoted by $\log_{10} x$. (See Fig. 3.)

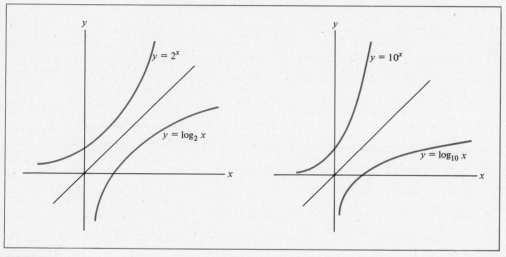

FIGURE 3

Logarithms to the base 10 are sometimes called *common* logarithms. Common logarithms are usually introduced into algebra courses for the purpose of simplifying certain arithmetic calculations. However, with the advent of the modern digital computer and the widespread availability of pocket electronic calculators, the need for common logarithms has diminished considerably. It can be shown that

$$\log_{10} x = \frac{1}{\ln 10} \cdot \ln x,$$

so that $\log_{10} x$ is simply a constant multiple of $\ln x$. However, we shall not need this fact.

The natural logarithm function is used in calculus because differentiation and integration formulas are simpler than for $\log_{10} x$ or $\log_2 x$, and so on. (Recall that we prefer the function e^x over the functions 10^x and 2^x for the same reason.)

Also, ln x arises "naturally" in the process of solving certain differential equations that describe various growth processes.

PRACTICE PROBLEMS 1

1. Find ln e.

2. Solve $e^{-3x} = 2$ using the natural logarithm function.

EXERCISES 1

1. Find $\ln(1/e)$.

2. Find $\ln(\sqrt{e})$.

3. If $e^{-x} = 1.7$, write x in terms of the natural logarithm.

4. If $e^x = 3.5$, write x in terms of the natural logarithm.

5. If $\ln x = 2.2$, write x using the exponential function.

6. If $\ln x = -5.7$, write x using the exponential function.

Simplify the following expressions.

7. $\ln e^2$

8. $e^{\ln 1.37}$

9. $e^{e \ln 1}$

10. $\ln(e^{.73 \ln e})$

11. $e^{5 \ln 1}$

12. $\ln(\ln e)$

Solve the following equations for x.

13. $e^{2x} = 5$

14. $e^{3x-1} = 4$

15. $\ln(4 - x) = \frac{1}{2}$

16. $\ln 3x = 2$

17. $\ln x^2 = 6$

18. $e^{x^2} = 7$

19. $6e^{-.00012x} = 3$

20. $2 - \ln x = 0$

21. $\ln 5x = \ln 3$

22. $\ln(x^2 - 3) = 0$

23. $\ln(\ln 2x) = 0$

24. $3 \ln x = 8$

25. $2e^{x/3} - 9 = 0$

26. $4 - 3e^{x+6} = 0$

27. $300e^{.2x} = 1800$

28. $750e^{-.4x} = 375$

29. $e^{5x} \cdot e^{\ln 5} = 2$

30. $e^{x^2-5x+6} = 1$

31. $4e^x \cdot e^{-2x} = 6$

32. $(e^x)^2 \cdot e^{2-3x} = 4$

In Exercises 33–36, find the coordinates of each extreme point of the given function and determine if the point is a maximum point or a minimum point.

33. $f(x) = e^{-x} + 3x$

34. $f(x) = 5x - 2e^x$

35. $f(x) = \frac{1}{3}e^{2x} - x + \frac{1}{2} \ln \frac{3}{2}$

36. $f(x) = 5 - \frac{1}{2}x - e^{-3x}$

37. When a drug or vitamin is administered intramuscularly (into a muscle), the concentration in the blood at time t after injection can be approximated by a function of the form $f(t) = c(e^{-k_1 t} - e^{-k_2 t})$. Sketch the graph of the function $f(t) = 5(e^{-.01t} - e^{-.51t})$ for $t \geq 0$ and discuss why this graph could be a reasonable model for the concentration of a drug injected intramuscularly.

38. Under certain geographic conditions, the wind velocity v at a height x cm above the ground is given by $v = K \ln(x/x_0)$, where K is a positive constant (depending on the air density, average wind velocity, etc.), and x_0 is a roughness parameter (depending on the roughness of the vegetation on the ground).* Suppose $x_0 = .7$ cm (a value that applies to lawn grass 3 cm high) and $K = 300$ cm/sec.

   (a) At what height above the ground is the wind velocity zero?

   (b) At what height is the wind velocity 1200 cm/sec?

39. Use the tables of values of e^x and $\ln x$ to estimate $(1.6)^{10}$.

SOLUTIONS TO PRACTICE PROBLEMS 1

1. Answer: 1. The number $\ln e$ is that exponent to which e must be raised in order to obtain e.

2. Take the logarithm of each side and use (3) to simplify the left side:

$$\ln e^{-3x} = \ln 2$$

$$-3x = \ln 2$$

$$x = -\frac{\ln 2}{3}.$$

5.2. The Derivative of ln x

Let us now compute the derivative of $\ln x$ for $x > 0$. Since $e^{\ln x} = x$, we have

$$\frac{d}{dx}(e^{\ln x}) = \frac{d}{dx}(x) = 1. \tag{1}$$

On the other hand, if we differentiate $e^{\ln x}$ by the chain rule, we find that

$$\frac{d}{dx}(e^{\ln x}) = e^{\ln x} \cdot \frac{d}{dx}(\ln x) = x \cdot \frac{d}{dx}(\ln x), \tag{2}$$

where the last equality used the fact that $e^{\ln x} = x$. By combining equations (1) and (2) we obtain

$$x \cdot \frac{d}{dx}(\ln x) = 1.$$

---

* G. Cox, B. Collier, A. Johnson, and P. Miller, *Dynamic Ecology* (Englewood Cliffs, N.J.: Prentice-Hall, Inc., 1973), pp. 113–115.

In other words,

$$\frac{d}{dx}(\ln x) = \frac{1}{x}, \quad x > 0. \tag{3}$$

By combining this differentiation formula with the chain rule, product rule, and quotient rule, we may differentiate many functions involving $\ln x$.

EXAMPLE 1 Differentiate

(a) $(\ln x)^5$ (b) $x \ln x$ (c) $\ln(x^3 + 5x^2 + 8)$.

Solution (a) By the general power rule,

$$\frac{d}{dx}(\ln x)^5 = 5(\ln x)^4 \cdot \frac{d}{dx}(\ln x)$$

$$= 5(\ln x)^4 \cdot \frac{1}{x}$$

$$= \frac{5(\ln x)^4}{x}.$$

(b) By the product rule,

$$\frac{d}{dx}(x \ln x) = x \cdot \frac{d}{dx}(\ln x) + (\ln x) \cdot 1$$

$$= x \cdot \frac{1}{x} + \ln x$$

$$= 1 + \ln x.$$

(c) By the chain rule,

$$\frac{d}{dx}\ln(x^3 + 5x^2 + 8) = \frac{1}{x^3 + 5x^2 + 8} \cdot \frac{d}{dx}(x^3 + 5x^2 + 8)$$

$$= \frac{3x^2 + 10x}{x^3 + 5x^2 + 8}.$$

Let $g(x)$ be any differentiable function. For any value of x for which $g(x)$ is positive, the function $\ln(g(x))$ is defined. For such a value of x, the derivative is given by the chain rule as

$$\frac{d}{dx}[\ln g(x)] = \frac{1}{g(x)} \cdot \frac{d}{dx}g(x)$$

$$= \frac{g'(x)}{g(x)}.$$

Example 1(c) illustrates a special case of this formula.

EXAMPLE 2 The function $f(x) = (\ln x)/x$ has an extreme point for some $x > 0$. Find the point and determine whether it is a maximum or a minimum point.

Solution By the quotient rule,

$$f'(x) = \frac{x \cdot \dfrac{1}{x} - \ln x \cdot 1}{x^2} = \frac{1 - \ln x}{x^2}$$

$$f''(x) = \frac{x^2 \cdot \left(-\dfrac{1}{x}\right) - (1 - \ln x)(2x)}{x^4} = \frac{2 \ln x - 3}{x^3}.$$

If we set $f'(x) = 0$, then

$$1 - \ln x = 0$$

$$\ln x = 1$$

$$e^{\ln x} = e^1 = e$$

$$x = e.$$

Therefore, the only possible extreme point is at $x = e$. When $x = e$, $f(e) = (\ln e)/e = 1/e$. Furthermore,

$$f''(e) = \frac{2 \ln e - 3}{e^3} = -\frac{1}{e^3} < 0,$$

which implies that the graph of $f(x)$ is concave down at $x = e$. Therefore, $(e, 1/e)$ is a maximum point of the graph of $f(x)$.

The next example introduces a function that will be needed later when we study integration.

EXAMPLE 3 The function $\ln|x|$ is defined for all nonzero values of x. Its graph is sketched in Fig. 1. Compute the derivative of $\ln|x|$.

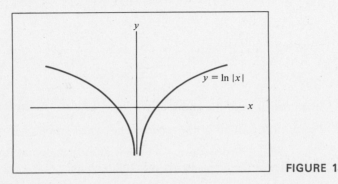

FIGURE 1

Solution   If x is positive, then $|x| = x$, and so

$$\frac{d}{dx} \ln|x| = \frac{d}{dx} \ln x = \frac{1}{x}.$$

If x is negative, then $|x| = -x$; and, by the chain rule,

$$\frac{d}{dx} \ln|x| = \frac{d}{dx} \ln(-x) = \frac{1}{-x} \cdot \frac{d}{dx}(-x) = \frac{1}{-x} \cdot (-1) = \frac{1}{x}.$$

Therefore we have established the following useful fact.

$$\frac{d}{dx} \ln|x| = \frac{1}{x}, \quad x \neq 0.$$

PRACTICE PROBLEMS 2

Differentiate:

1. $f(x) = \dfrac{1}{\ln(x^4 + 5)}$

2. $f(x) = \ln(\ln x)$.

EXERCISES 2

Differentiate the following functions.

1. $\ln 2x$
2. $\ln x^2$
3. $\ln(x + 5)$

4. $x^2 \ln x$
5. $\dfrac{1}{x} \ln(x + 1)$
6. $\sqrt{\ln x}$

7. $e^{\ln x + x}$
8. $\ln[x/(x - 3)]$
9. $4 + \ln(x/2)$

10. $\ln \sqrt{x}$
11. $(\ln x)^2 + \ln x$
12. $\ln(x^3 + 2x + 1)$

13. $\ln(kx),\ k$ constant
14. $x/(\ln x)$
15. $x/(\ln x)^2$

16. $(\ln x)e^{-x}$
17. $e^{2x} \ln x$
18. $(\ln x + 1)^3$

19. $\ln(e^{5x} + 1)$
20. $\ln(e^{e^x})$

Find

21. $\dfrac{d}{dt}(t^2 \ln 4)$
22. $\dfrac{d^2}{dx^2} \ln(1 + x^2)$
23. $\dfrac{d^2}{dt^2}(\ln t)^3$

24. Find the slope of the graph of $y = \ln|x|$ at $x = 3$ and $x = -3$.

25. Write the equation of the tangent line to the graph of $y = \ln(x^2 + e)$ at $x = 0$.

26. The function $f(x) = (\ln x + 1)/x$ has an extreme point for $x > 0$. Find the coordinates of the point. Is it a maximum point?

27. The function $f(x) = (\ln x)/\sqrt{x}$ has an extreme point for $x > 0$. Find the coordinates of the point. Is it a maximum point?

28. The function $f(x) = x/(\ln x + x)$ has an extreme point for $x > 1$. Find the coordinates of the point. Is it a minimum point?

29. Sketch the graph of the function $y = \ln x + (1/x) - \frac{1}{2}$.

30. Sketch the graph of $y = 1 + \ln(x^2 - 6x + 10)$.

31. If a cost function is $C(x) = (\ln x)/(40 - 3x)$, find the marginal cost when $x = 10$.

32. Suppose that the demand equation for a certain commodity is $p = 45/(\ln x)$. Determine the marginal revenue function for this commodity, and compute the marginal revenue when $x = 20$.

33. Suppose that the total revenue function for a manufacturer is $R(x) = 300 \ln(x + 1)$, so that the sale of x units of a product brings in about $R(x)$ dollars. Suppose also that the total cost of producing x units is $C(x)$ dollars, where $C(x) = 2x$. Find the value of x at which the profit function $R(x) - C(x)$ will be maximized. Show that the profit function has a maximum and not a minimum at this value of x.

34. Evaluate $\lim\limits_{h \to 0} \dfrac{\ln(7 + h) - \ln 7}{h}$.

35. Find the maximum area of a rectangle in the first quadrant with one corner at the origin, two sides on the coordinate axes, and one corner on the graph of $y = -\ln x$.

SOLUTIONS TO PRACTICE PROBLEMS 2

1. Here $f(x) = [\ln(x^4 + 5)]^{-1}$. By the chain rule,

$$f'(x) = -[\ln(x^4 + 5)]^{-2} \cdot \frac{d}{dx} \ln(x^4 + 5)$$

$$= -[\ln(x^4 + 5)]^{-2} \cdot \frac{4x^3}{x^4 + 5}.$$

2. $f'(x) = \dfrac{d}{dx} \ln(\ln x) = \dfrac{1}{\ln x} \cdot \dfrac{d}{dx} \ln x$

$$= \frac{1}{\ln x} \cdot \frac{1}{x} = \frac{1}{x \ln x}.$$

5.3. Properties of the Natural Logarithm Function

The natural logarithm function $\ln x$ possesses many of the familiar properties of logarithms to base 10 (or common logarithms) that are encountered in algebra.

> Let x and y be positive numbers, b any number.
>
> LI $\ln(xy) = \ln x + \ln y.$
>
> LII $\ln\left(\dfrac{1}{x}\right) = -\ln x.$
>
> LIII $\ln\left(\dfrac{x}{y}\right) = \ln x - \ln y.$
>
> LIV $\ln(x^b) = b \ln x.$

Verification of LI   By equation (2) of Section 1 we have $e^{\ln(xy)} = xy$, $e^{\ln x} = x$, and $e^{\ln y} = y$. Therefore

$$e^{\ln(xy)} = xy = e^{\ln x} \cdot e^{\ln y} = e^{\ln x + \ln y}.$$

By equating exponents, we get LI.

Verification of LII   Since $e^{\ln(1/x)} = 1/x$, we have

$$e^{\ln(1/x)} = \frac{1}{x} = \frac{1}{e^{\ln x}} = e^{-\ln x}.$$

By equating exponents, we get LII.

Verification of LIII   By LI and LII, we have

$$\ln\left(\frac{x}{y}\right) = \ln\left(x \cdot \frac{1}{y}\right)$$

$$= \ln x + \ln\left(\frac{1}{y}\right)$$

$$= \ln x - \ln y.$$

Verification of LIV   Since $e^{\ln(x^b)} = x^b$, we have

$$e^{\ln(x^b)} = x^b = (e^{\ln x})^b = e^{b \ln x}.$$

Equating exponents, we get LIV.

These properties of the natural logarithm should be learned thoroughly. You will find them useful in many calculations involving $\ln x$ and the exponential function.

EXAMPLE 1   Simplify $\ln 5 + 2 \ln 3$.

Solution

$$\ln 5 + 2 \ln 3 = \ln 5 + \ln 3^2 \qquad \text{(LIV)}$$

$$= \ln 5 + \ln 9$$

$$= \ln 45. \qquad \text{(LI)}$$

EXAMPLE 2   Simplify $\frac{1}{2}\ln(4t) - \ln(t^2 + 1)$.

Solution

$$\frac{1}{2}\ln(4t) - \ln(t^2 + 1) = \ln[(4t)^{1/2}] - \ln(t^2 + 1) \qquad \text{(LIV)}$$

$$= \ln(2\sqrt{t}) - \ln(t^2 + 1)$$

$$= \ln\left(\frac{2\sqrt{t}}{t^2 + 1}\right). \qquad \text{(LIII)}$$

EXAMPLE 3   Simplify $\ln x + \ln 3 + \ln y - \ln 5$.

Solution   Use (LI) twice and (LIII) once.

$$(\ln x + \ln 3) + \ln y - \ln 5 = \ln 3x + \ln y - \ln 5$$

$$= \ln 3xy - \ln 5$$

$$= \ln\left(\frac{3xy}{5}\right).$$

EXAMPLE 4   Differentiate $f(x) = \ln[x(x + 1)(x + 2)]$.

Solution   First rewrite $f(x)$, using (LI).

$$f(x) = \ln[x(x + 1)(x + 2)]$$

$$= \ln x + \ln(x + 1) + \ln(x + 2).$$

Then $f'(x)$ is easily calculated:

$$f'(x) = \frac{1}{x} + \frac{1}{x + 1} + \frac{1}{x + 2}.$$

The natural logarithm function can be used to simplify the task of differentiating products. Suppose, for example, that we wish to differentiate the function

$$g(x) = x(x + 1)(x + 2).$$

As we showed in Example 4,

$$\frac{d}{dx}\ln g(x) = \frac{1}{x} + \frac{1}{x + 1} + \frac{1}{x + 2}.$$

However,

$$\frac{d}{dx}\ln g(x) = \frac{g'(x)}{g(x)}.$$

Therefore, equating the two expressions for $\frac{d}{dx}\ln g(x)$ we have

$$\frac{g'(x)}{g(x)} = \frac{1}{x} + \frac{1}{x + 1} + \frac{1}{x + 2}.$$

Finally, we solve for $g'(x)$:

$$g'(x) = g(x) \cdot \left(\frac{1}{x} + \frac{1}{x+1} + \frac{1}{x+2}\right)$$

$$= x(x+1)(x+2)\left(\frac{1}{x} + \frac{1}{x+1} + \frac{1}{x+2}\right).$$

In a similar way, we can differentiate the product of any number of terms by first taking natural logarithms, then differentiating, and finally solving for the desired derivative. This procedure is called *logarithmic differentiation*.

EXAMPLE 5 Differentiate the function $g(x) = (x^2 + 1)(x^3 - 3)(2x + 5)$ using logarithmic differentiation.

Solution Begin by taking the natural logarithm of both sides of the given equation:

$$\ln g(x) = \ln[(x^2 + 1)(x^3 - 3)(2x + 5)]$$

$$= \ln(x^2 + 1) + \ln(x^3 - 3) + \ln(2x + 5).$$

Now differentiate and solve for $g'(x)$:

$$\frac{g'(x)}{g(x)} = \frac{2x}{x^2 + 1} + \frac{3x^2}{x^3 - 3} + \frac{2}{2x + 5}$$

$$g'(x) = g(x)\left(\frac{2x}{x^2 + 1} + \frac{3x^2}{x^3 - 3} + \frac{2}{2x + 5}\right)$$

$$= (x^2 + 1)(x^3 - 3)(2x + 5)\left(\frac{2x}{x^2 + 1} + \frac{3x^2}{x^3 - 3} + \frac{2}{2x + 5}\right).$$

EXAMPLE 6 Use logarithmic differentiation to show that

$$\frac{d}{dx}(x^r) = rx^{r-1}.$$

Solution Let $f(x) = x^r$. Then

$$\ln f(x) = \ln x^r = r \ln x.$$

Differentiation of this equation yields

$$\frac{f'(x)}{f(x)} = r \cdot \frac{1}{x}$$

$$f'(x) = r \cdot \frac{1}{x} \cdot f(x) = r \cdot \frac{1}{x} \cdot x^r = rx^{r-1}.$$

PRACTICE PROBLEMS 3

1. Differentiate $f(x) = \ln\left[\dfrac{e^x \sqrt{x}}{(x + 1)^6}\right]$.

2. Use logarithmic differentiation to differentiate $f(x) = (x + 1)^7(x + 2)^8(x + 3)^9$.

EXERCISES 3

Simplify the following expressions.

1. $\ln 5 + \ln x$
2. $\ln x^5 - \ln x^3$
3. $\frac{1}{2} \ln 9$
4. $3 \ln \frac{1}{2} + \ln 16$
5. $\ln 4 + \ln 6 - \ln 12$
6. $\ln 2 - \ln x + \ln 3$
7. $e^{2 \ln x}$
8. $\frac{3}{2} \ln 4 - 5 \ln 2$
9. $5 \ln x - \frac{1}{2} \ln y + 3 \ln z$
10. $e^{\ln x^2 + 3 \ln y}$
11. $\ln x - \ln x^2 + \ln x^4$
12. $\frac{1}{2} \ln xy + \frac{3}{2} \ln \dfrac{x}{y}$

13. Which is larger, $2 \ln 5$ or $3 \ln 3$?

14. Which is larger, $\frac{1}{2} \ln 16$ or $\frac{1}{3} \ln 27$?

Differentiate.

15. $\ln[(x + 5)(2x - 1)(4 - x)]$
16. $\ln[x^3(x + 1)^4]$
17. $\ln\left[\dfrac{(x + 1)(3x - 2)}{x + 2}\right]$
18. $\ln\left[\dfrac{x^2}{(3 - x)^3}\right]$
19. $\ln\left[\dfrac{\sqrt{x}}{x^2 + 1}\right]$
20. $\ln[e^{x^2}(x^4 + x^2 + 1)]$

Use logarithmic differentiation to differentiate the following functions.

21. $f(x) = (x + 1)^3(4x - 1)^2$
22. $f(x) = e^x(x - 4)^8$
23. $f(x) = (x - 2)^3(x - 3)^5(x + 2)^{-7}$
24. $f(x) = (x + 1)(2x + 1)(3x + 1)(4x + 1)$
25. $f(x) = x^x$
26. $f(x) = x^{1/x}$
✳27. $f(x) = e^x\sqrt{x^2 - 1}$
28. $f(x) = 2^x$
29. $f(x) = x^{\ln x}$
30. $f(x) = (2x - 1)^{2x}$
31. $f(x) = \dfrac{\sqrt{x - 1}(x - 2)}{x^2 - 3}$
32. $f(x) = \dfrac{xe^x}{\sqrt{3x^2 + 1}}$

33. There is substantial empirical data to show that if x and y measure the sizes of two organs of a particular animal, then x and y are related by an *allometric equation* of the form

$$\ln y - k \ln x = \ln c,$$

where k and c are positive constants that depend only on the type of parts or organs that are measured, and are constant among animals belonging to the same species.* Solve this equation for y in terms of x, k, and c.

34. In the study of epidemics, one finds the equation

$$\ln(1 - y) - \ln y = C - rt,$$

where y is the fraction of the population that has a specific disease at time t. Solve the equation for y in terms of t and the constants C and r.

35. Determine the values of h and k for which the graph of $y = he^{kx}$ passes through the points $(1, 6)$ and $(4, 48)$.

36. Find values of k and r for which the graph of $y = kx^r$ passes through the points $(2, 3)$ and $(4, 15)$.

SOLUTIONS TO PRACTICE PROBLEMS 3

1. Use the properties of the natural logarithm to express $f(x)$ as a sum of simple functions before differentiating.

$$f(x) = \ln\left[\frac{e^x\sqrt{x}}{(x + 1)^6}\right]$$

$$= \ln e^x + \ln \sqrt{x} - \ln(x + 1)^6$$

$$= x + \tfrac{1}{2} \ln x - 6 \ln(x + 1).$$

$$f'(x) = 1 + \frac{1}{2x} - \frac{6}{x + 1}.$$

2.
$$f(x) = (x + 1)^7(x + 2)^8(x + 3)^9$$

$$\ln f(x) = 7 \ln(x + 1) + 8 \ln(x + 2) + 9 \ln(x + 3).$$

Now we differentiate both sides of the equation.

$$\frac{f'(x)}{f(x)} = \frac{7}{x + 1} + \frac{8}{x + 2} + \frac{9}{x + 3}$$

$$f'(x) = f(x)\left(\frac{7}{x + 1} + \frac{8}{x + 2} + \frac{9}{x + 3}\right)$$

$$= (x + 1)^7(x + 2)^8(x + 3)^9\left(\frac{7}{x + 1} + \frac{8}{x + 2} + \frac{9}{x + 3}\right).$$

Chapter 5: CHECKLIST

☐ Reflection in the line $y = x$
☐ Definition of $\ln x$

---

* E. Batschelet, *Introduction to Mathematics for Life Scientists* (New York: Springer-Verlag, 1971), pp. 305–307.

- □ $\ln 1 = 0$
- □ $\ln e = 1$
- □ $e^{\ln x} = x,\ x > 0$
- □ $\ln e^x = x$
- □ $\dfrac{d}{dx}(\ln x) = \dfrac{1}{x}$
- □ $\ln(xy) = \ln x + \ln y$
- □ $\ln\left(\dfrac{x}{y}\right) = \ln x - \ln y$
- □ $\ln\left(\dfrac{1}{x}\right) = -\ln x$
- □ $\ln(x^b) = b \ln x$

Chapter 5: SUPPLEMENTARY EXERCISES

Simplify the following expressions.

1. $\dfrac{\ln x^2}{\ln x^3}$ 2. $e^{2\ln 2}$ 3. $e^{-5\ln 1}$

4. $[e^{\ln x}]^2$ 5. $e^{(\ln 5)/2}$ 6. $e^{\ln(x^2)}$

Solve the following equations for t.

7. $3e^{2t} = 15$ 8. $3e^{t/2} - 12 = 0$ 9. $2\ln t = 5$

10. $2e^{-.3t} = 1$ 11. $t^{\ln t} = e$ 12. $\ln(\ln 3t) = 0$

Differentiate the following functions.

13. $\ln(5x - 7)$ 14. $\ln(9x)$ 15. $(\ln x)^2$

16. $(x \ln x)^3$ 17. $\ln(x^6 + 3x^4 + 1)$ 18. $\dfrac{x}{\ln x}$

19. $\ln\left(\dfrac{xe^x}{\sqrt{1 + x}}\right)$ 20. $\ln[(x^2 + 3)^5(x^3 + 1)^{-4}]$ 21. $\ln(\ln\sqrt{x})$

22. $1/\ln x$ 23. $x \ln x - x$ 24. $e^{2\ln(x+1)}$

25. $e^x \ln x$ 26. $\ln(x^2 + e^x)$

Use logarithmic differentiation to differentiate the following functions.

27. $f(x) = (x^2 + 5)^6(x^3 + 7)^8(x^4 + 9)^{10}$ 28. $f(x) = x^{1+x}$

29. $f(x) = 10^x$ 30. $f(x) = \sqrt{x^2 + 5}\,e^{x^2}$

Sketch the following curves.

31. $y = x - \ln x$ 32. $y = \ln(x^2 + 1)$

33. $y = (\ln x)^2$ 34. $y = (\ln x)^3$

APPLICATIONS OF THE EXPONENTIAL AND NATURAL LOGARITHM FUNCTIONS

Earlier, we introduced the exponential function e^x and the natural logarithm function $\ln x$ and studied their most important properties. From the way we introduced these functions, it is by no means clear that they have any substantial connection with the physical world. However, as this chapter will demonstrate, the exponential and natural logarithm functions intrude into the study of many physical problems, often in a very curious and unexpected way.

Here the most significant fact that we require is that the exponential function is uniquely characterized by its differential equation. In other words, we will constantly make use of the following fact, which was stated previously.

> The function* $y = Ce^{kt}$ satisfies the differential equation
> $$y' = ky.$$
> Conversely, if $y = f(t)$ satisfies the differential equation
> then $y = Ce^{kt}$ for some constant C.

If $f(t) = Ce^{kt}$, then, by setting $t = 0$, we have

$$f(0) = Ce^0 = C.$$

Therefore C is the value of $f(t)$ at $t = 0$.

6.1. Exponential Growth and Decay

In biology, chemistry, and economics it is often necessary to study the behavior of a quantity that is increasing as time passes. If, at every instant, the rate of increase of the quantity is proportional to the quantity at that instant, then we say that the quantity is *growing exponentially* or is *exhibiting exponential growth*. A simple example of exponential growth is exhibited by the growth of bacteria in a culture. Under ideal laboratory conditions a bacteria culture grows at a rate proportional to the number of bacteria present. It does so because the growth of the culture is accounted for by the division of the bacteria. The more bacteria there are at a given instant, the greater the possibilities for division and hence the more rapid is the rate of growth.

Let us study the growth of a bacteria culture as a typical example of exponential growth. Suppose that $P(t)$ denotes the number of bacteria in a certain culture at time t. The rate of growth of the culture at time t is $P'(t)$. We assume that this rate of growth is proportional to the size of the culture at time t, so that

$$P'(t) = kP(t), \tag{1}$$

where k is a positive constant of proportionality. If we let $y = P(t)$, then (1) can be written as

$$y' = ky.$$

Therefore from our discussion at the beginning of this chapter we see that

$$y = P(t) = P_0 e^{kt}, \tag{2}$$

where P_0 is the number of bacteria in the culture at time $t = 0$. The number k is called the *growth constant*.

---

* Note that we use the variable t instead of x. The reason is that, in most applications, the variable of our exponential function is time. The variable t will be used throughout this chapter.

EXAMPLE 1   Suppose that a certain bacteria culture grows at a rate proportional to its size. At time $t = 0$, approximately 20,000 bacteria are present. In five hours there are 400,000 bacteria. Determine a function that expresses the size of the culture as a function of time, measured in hours.

Solution   Let $P(t)$ be the number of bacteria present at time t. By assumption, $P(t)$ satisfies a differential equation of the form $y' = ky$, and so $P(t)$ has the form
$$P(t) = P_0 e^{kt},$$
where the constants P_0 and k must be determined. The values of P_0 and k can be obtained from the data that give the population size at two different times. We are told that
$$P(0) = 20,000, \qquad P(5) = 400,000. \qquad (3)$$
The first condition immediately implies that $P_0 = 20,000$, and so
$$P(t) = 20,000 e^{kt}.$$
Using the second condition in (3), we have
$$20,000 e^{k \cdot 5} = P(5) = 400,000$$
$$e^{5k} = 20 \qquad (4)$$
$$5k = \ln 20$$
$$k = \frac{\ln 20}{5} \approx .60*$$

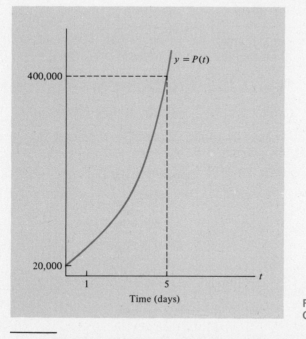

$y = P(t)$

400,000

20,000

1   5

Time (days)

t

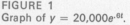

FIGURE 1
Graph of $y = 20,000 e^{.6t}$.

---

* Here and elsewhere in this chapter, we have used the ln table from the appendix and have carried out calculations to two significant figures.

So we may take

$$P(t) = 20{,}000e^{.6t}.$$

This function is a mathematical model of the growth of the bacteria culture. (See Fig. 1.)

EXAMPLE 2 Suppose that a colony of fruit flies is growing according to the exponential law $P(t) = P_0e^{kt}$ and suppose that the size of the colony doubles in 12 days. Determine the growth constant k.

Solution We do not know the initial size of the population at $t = 0$. However, we are told that $P(12) = 2P(0)$; that is,

$$P_0e^{k \cdot 12} = 2P_0$$

$$e^{12k} = 2$$

$$12k = \ln 2$$

$$k = \tfrac{1}{12} \ln 2 \approx .058.$$

Notice that the initial size P_0 of the population was not given in Example 2. We were able to determine the growth constant because we were told the amount of time required for the colony to double in size. Thus the growth constant does not depend on the initial size of the population. This property is characteristic of exponential growth.

EXAMPLE 3 Suppose that the initial size of the colony in Example 2 was 300. At what time will the colony contain 1800 fruit flies?

Solution From Example 2 we have $P(t) = P_0e^{.058t}$. Since $P(0) = 300$, we conclude that

$$P(t) = 300e^{.058t}.$$

Now that we have the explicit formula for the size of the colony, we can set $P(t) = 1800$ and solve for t:

$$300e^{.058t} = 1800$$

$$e^{.058t} = 6$$

$$.058t = \ln 6$$

$$t = \frac{\ln 6}{.058} \approx 31 \text{ days.}$$

The table below shows the growth of the colony in Example 3. Notice that 1800 is exactly halfway between 1200 ($t = 24$) and 2400 ($t = 36$). It is incorrect to guess that the population will equal 1800 when t is halfway between $t = 24$ and $t = 36$—that is, when $t = 30$. We saw in Example 3 that it takes approximately 31 days for the colony to reach 1800 fruit flies.

Population Size	Day
300	0
600	12
1200	24
2400	36
4800	48
9600	60
19200	72
⋮	⋮

Exponential Decay   An example of negative exponential growth, or *exponential decay*, is given by the disintegration of a radioactive element such as uranium 235. It is known that, at any instant, the rate at which a radioactive substance is decaying is proportional to the amount of the substance that has not yet disintegrated. If $P(t)$ is the quantity present at time t, then $P'(t)$ is the rate of decay. Of course, $P'(t)$ must be negative, since $P(t)$ is decreasing. Thus we may write $P'(t) = kP(t)$ for some negative constant k. To emphasize the fact that the constant is negative, k is often replaced by $-\lambda$,* where λ is a positive constant. Then $P(t)$ satisfies the differential equation

$$P'(t) = -\lambda P(t). \tag{5}$$

The general solution of (5) has the form

$$P(t) = P_0 e^{-\lambda t}$$

for some positive number P_0. We call such a function an *exponential decay function*. The constant λ is called the *decay constant*.

EXAMPLE 4   The decay constant for strontium 90 is $\lambda = .0244$, where the time is measured in years. How long will it take for a quantity P_0 of strontium 90 to decay to one-half its original size?

Solution   We have

$$P(t) = P_0 e^{-.0244t}.$$

Next, set $P(t)$ equal to $\frac{1}{2}P_0$ and solve for t:

$$P_0 e^{-.0244t} = \tfrac{1}{2}P_0$$

$$e^{-.0244t} = \tfrac{1}{2} = .5$$

$$-.0244t = \ln .5$$

$$t = \frac{\ln .5}{-.0244} \approx 28 \text{ years.}$$

The *half-life* of a radioactive element is the length of time required for a given quantity of that element to decay to one-half its original size. Thus strontium 90

---

* λ is the Greek letter lambda.

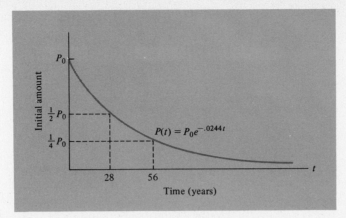

FIGURE 2 Half-life of radioactive strontium 90.

has a half-life of about 28 years. (See Fig. 2.) Notice from Example 4 that the half-life does not depend on the initial amount P_0.

EXAMPLE 5 Radioactive carbon 14 has a half-life of about 5730 years. Find its decay constant.

Solution If P_0 denotes the initial amount of carbon 14, then the amount after t years will be

$$P(t) = P_0 e^{-\lambda t}.$$

After 5730 years, $P(t)$ will equal $\frac{1}{2}P_0$. That is,

$$P_0 e^{-\lambda(5730)} = P(5730) = \tfrac{1}{2}P_0 = .5P_0.$$

Solving for λ gives

$$e^{-5730\lambda} = .5$$

$$-5730\lambda = \ln .5$$

$$\lambda = \frac{\ln .5}{-5730} \approx .00012.$$

One of the problems connected with above-ground nuclear explosions is the radioactive debris that falls on plants and grass, thereby contaminating the food supply of animals. Strontium 90 is one of the most dangerous components of "fallout" because it has a relatively long half-life and because it is chemically similar to calcium and is absorbed into the bone structure of animals (and humans) who eat contaminated food. Iodine 131 is also produced by nuclear explosions, but it presents less of a hazard because it has a half-life of eight days.

EXAMPLE 6 If dairy cows eat hay containing too much iodine 131, their milk will be unfit to drink. Suppose that some hay contains ten times the maximum allowable level of iodine 131. How many days should the hay be stored before it is fed to dairy cows?

Solution Let P_0 be the amount of iodine 131 present in the hay. Then the amount at time t is $P(t) = P_0 e^{-\lambda t}$ (t in days). The half-life of iodine 131 is eight days, so

$$P_0 e^{-8\lambda} = .5 P_0$$
$$e^{-8\lambda} = .5$$
$$-8\lambda = \ln .5$$
$$\lambda = \frac{\ln .5}{-8} \approx .087,$$

and

$$P(t) = P_0 e^{-.087t}.$$

Now that we have the formula for $P(t)$, we want to find t such that $P(t) = \frac{1}{10}P_0$. We have

$$P_0 e^{-.087t} = .1 P_0$$

and so

$$e^{-.087t} = .1$$
$$-.087t = \ln .1$$
$$t = \frac{\ln .1}{-.087} \approx 26 \text{ days.}$$

Radiocarbon Dating Knowledge about radioactive decay is valuable to social scientists who want to estimate the age of objects belonging to ancient civilizations. Several different substances are useful for radioactive-dating techniques; the most common is radiocarbon 14, C^{14}. Carbon 14 is produced in the upper atmosphere when cosmic rays react with atmospheric nitrogen. Because the C^{14} eventually decays, the concentration of C^{14} cannot rise above certain levels. An equilibrium is reached where C^{14} is produced at the same rate as it decays. Scientists usually assume that the total amount of C^{14} in the biosphere has remained constant over the past 50,000 years. Consequently, it is assumed that the *ratio* of C^{14} to ordinary nonradioactive carbon 12, C^{12}, has been constant during this same period. (The ratio is about one part C^{14} to 10^{12} parts of C^{12}.) Both C^{14} and C^{12} are in the atmosphere in the form of carbon dioxide. All living vegetation and most forms of animal life contain C^{14} and C^{12} in the same proportion as the atmosphere. The reason is that plants absorb carbon dioxide through photosynthesis. The C^{14} and C^{12} in plants is distributed through the various food chains to almost all animal life.

When an organism dies, it stops replacing its carbon, and therefore the amount of C^{14} begins to decrease through radioactive decay. (The C^{12} in the dead organism remains constant.) At a later date, the ratio of C^{14} to C^{12} can be measured in order to determine when the organism died. (See Fig. 3.)

FIGURE 3 C^{14}–C^{12} ratio compared to the ratio in living plants.

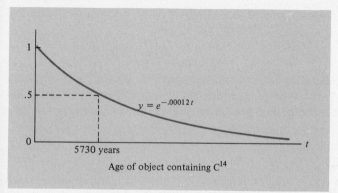

$$y = e^{-.00012t}$$

5730 years

Age of object containing C^{14}

EXAMPLE 7 A parchment fragment was discovered that had a C^{14}–C^{12} ratio of about .8 times the ratio found today in living matter. Estimate the age of the parchment.

Solution We assume that the original C^{14}–C^{12} ratio in the parchment was the same as the ratio in living organisms today. Consequently, about eight-tenths of the original C^{14} remains. From Example 5 we obtain the formula for the amount of C^{14} present t years after the parchment was made from an animal skin:

$$P(t) = P_0 e^{-.00012t},$$

where P_0 = initial amount. We want to find t such that $P(t) = .8P_0$.

$$P_0 e^{-.00012t} = .8P_0$$

$$e^{-.00012t} = .8$$

$$-.00012t = \ln .8$$

$$t = \frac{\ln .8}{-.00012} \approx 1900 \text{ years old.}$$

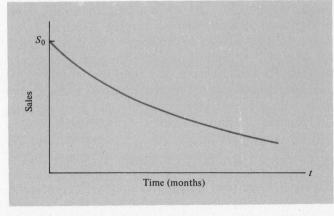

FIGURE 4   Exponential decay of sales.

A Sales Decay Curve Marketing studies* have demonstrated that if advertising and other promotion of a particular product are stopped and if other market conditions remain fairly constant, then, at any time t, the sales of that product will be declining at a rate proportional to the amount of current sales at t. (See Fig. 4.) If S_0 is the number of sales in the last month during which advertising occurred and if $S(t)$ is the number of sales in the tth month following the cessation of promotional effort, then a good mathematical model for $S(t)$ is

$$S(t) = S_0 e^{-\lambda t},$$

where λ is a positive number called the *sales decay constant*. The value of λ depends on many factors, such as the type of product, the number of years of prior advertising, the number of competing products, and other characteristics of the market.

* M. Vidale and H. Wolfe, "An operations-research study of sales response to advertising," *Operations Research* **5** (1957), 370–381. Reprinted in F. Bass et al., *Mathematical Models and Methods in Marketing* (Homewood, Ill.: Richard D. Irwin, Inc., 1961).

PRACTICE PROBLEMS 1

1. (a) Solve the differential equation: $P'(t) = -.6P(t)$, $P(0) = 50$.

   (b) Solve the differential equation $P'(t) = kP(t)$, $P(0) = 4000$, where k is some constant.

   (c) Find the value of k in (b) for which $P(2) = 100P(0)$.

2. Under ideal conditions a colony of *E. coli* bacteria can grow by a factor of 100 every two hours. If initially 4000 bacteria are present, how long will it take before there are 1,000,000 bacteria?

EXERCISES 1

1. Solve the following differential equations.

   (a) $P'(t) = .03P(t)$, $P(0) = 5000$

   (b) $f'(x) = 2f(x)$, $f(0) = 30$

   (c) $P'(t) = -.2P(t)$, $P(0) = 10^4$

   (d) $f'(x) = -\frac{1}{2}f(x)$, $f(0) = 5.3$

   (e) $S'(t) = -.05S(t)$, $S(0) = 50,000$

   (f) $f'(x) = \frac{2}{3}f(x)$, $f(0) = 4$

2. Find the equation $y = f(x)$ of a curve that passes through $(0, 3)$ and has the property that

$$f'(x) = -\tfrac{1}{2}f(x).$$

3. Sketch the graph of a function $f(x)$ whose slope at $(x, f(x))$ is twice the value of $f(x)$ and that passes through the point $(0, 1)$.

4. The growth rate of a certain cell culture is proportional to its size. Initially, 2×10^5 cells were present. In 10 hours there were approximately 8×10^5 cells. Construct a mathematical model for the growth of this culture.

5. The decay constant for cobalt 60 is $\lambda = .13$ when time is measured in years. Find the half-life of cobalt 60.

6. Radioactive potassium is also used for dating fossils. It has a half-life of 1.3 billion years. Determine its decay constant.

7. A large bakery decided to stop advertising its bread and then noticed that its sales of bread declined by about 10% in the first year after the advertising ended. Construct an exponential model for the annual sales $S(t)$ of the bread.

8. The size of a certain insect population is given by

$$P(t) = 300e^{.01t},$$

where t is measured in days. At what time will the population equal 600? 1200?

9. Sandals woven from strands of tree bark were found in Fort Rock Cave in Oregon. The bark has a C^{14}–C^{12} ratio of .34 times the ratio found in living bark. Estimate the age of the sandals.

10. A 4500-year-old wooden chest was found in the tomb of the twenty-fifth century B.C. Chaldean king Meskalamdug of Ur. What C^{14}–C^{12} ratio would you expect to find in the wooden chest?

11. An island in the Pacific Ocean is contaminated by fallout from a nuclear explosion. If the strontium 90 is 100 times the level that scientists believe is "safe," how many years will it take for the island to once again be "safe" for human habitation? The half-life of strontium 90 is 28 years.

12. If a bacteria culture doubles in size every 20 minutes, how long will it take for a population of 10^4 to grow to 10^8 bacteria?

13. A certain cell culture grows at a rate proportional to the size of the culture. During a 10-hour experiment the culture doubled in size every three hours. At the end of the experiment approximately 10^5 cells were present. How many cells were present at the beginning of the experiment?

14. A common infection of the urinary tract in humans is caused by the bacterium *E. coli*. The infection is generally noticed when the bacteria colony reaches a population of about 10^8. The colony doubles in size about every 20 minutes. When a full bladder is emptied, about 90% of the bacteria are eliminated. Suppose that at the beginning of a certain time period, a person's bladder and urinary tract contain 10^8 *E. coli* bacteria. During an interval of T minutes the person drinks enough liquid to fill the bladder. Find the value of T such that if the bladder is emptied after T minutes, about 10^8 bacteria will still remain. (Note: The average bladder holds about one liter of urine. It is seldom possible to eliminate an *E. coli* infection by diuresis without drugs—such as by drinking large amounts of water.)

15. By 1974 the United States had an estimated 80 million gallons of radioactive products from nuclear power plants and other nuclear reactors. These waste products were stored in various sorts of containers (made of such materials as stainless steel and cement), and the containers were buried in the ground and the ocean. Scientists feel that the waste products must be prevented from contaminating the rest of the earth until more than 99.99% of the radioactivity is gone (that is, until the level is less than 0.0001 times the original level). If a storage cylinder contains waste products whose half-life is 1500 years, how many years must the container survive without leaking? (Note: Some of the containers are already leaking.)

16. In 1950 the world's population required 1×10^9 hectares* of arable land for food growth. It is estimated that in 1980 2×10^9 hectares will be required. Current population trends indicate that if $A(t)$ denotes the amount of land needed t years after 1950, then

$$\frac{dA}{dt} = kA$$

for some constant k.

---

* A hectare equals 2.471 acres.

(a) Derive a formula for $A(t)$.

(b) The total amount of arable land on the earth's surface is estimated at 3.2 × 10^9 hectares. In what year will the earth exhaust its supply of land for growing food?

[Data based on the Club of Rome's report, *The Limits to Growth* by D. H. and D. L. Meadows, J. Randers, W. Behrens III (New York: Universe Books, 1972).]

17. A drought in the African veldt causes the death of much of the animal population. A typical herd of wildebeests suffers a death rate proportional to its size. The herd numbers 500 at the onset of the drought and only 200 remain four months later.

(a) Find a formula for the herd's population at time t months.

(b) How long will it take for the herd to diminish to one-tenth its original size?

SOLUTIONS TO PRACTICE PROBLEMS 1

1. (a) Answer: $P(t) = 50e^{-.6t}$. Differential equations of the type $y' = ky$ have as their solution $P(t) = Ce^{kt}$, where C is $P(0)$.

(b) Answer: $P(t) = 4000e^{kt}$. This problem is like the previous one except that the constant is not specified. Additional information is needed if one wants to determine a specific value for k.

(c) Answer: $P(t) = 4000e^{2.3t}$. From the solution to (b) we know that $P(t) = 4000e^{kt}$. We are given that $P(2) = 100P(0) = 100(4000) = 400,000$. So,

$$P(2) = 4000e^{k(2)} = 400,000$$
$$e^{2k} = 100$$
$$2k = \ln 100$$
$$k = \frac{\ln 100}{2} \approx 2.3.$$

2. Let $P(t)$ be the number of bacteria present after t hours. We must first find an expression for $P(t)$ and then determine the value of t for which $P(t) = 1,000,000$.

From the discussion at the beginning of the section we know that $P'(t) = k \cdot P(t)$. Also, we are given that $P(2)$ (the population after two hours) is $100P(0)$ (100 times the initial population). From 1(c) above we have an expression for $P(t)$:

$$P(t) = 4000e^{2.3t}.$$

Now we must solve $P(t) = 1,000,000$ for t.

$$4000e^{2.3t} = 1,000,000$$
$$e^{2.3t} = 250$$
$$2.3t = \ln 250$$
$$t = \frac{\ln 250}{2.3} \approx 2.4.$$

Therefore, after 2.4 hours there will be 1,000,000 bacteria.

6.2. Compound Interest

When money is deposited in a savings account, interest is paid at stated intervals. If this interest is added to the account and thereafter earns interest itself, then the interest is called *compound interest*. The original amount deposited is called the *principal amount*. The principal amount plus the compound interest is called the *compound amount*. The interval between interest payments is referred to as the *interest period*. In formulas for compound interest, the interest rate is expressed as a decimal rather than a percent. Thus 6% is written as .06.

If $1000 is deposited at 6% annual interest, compounded annually, the compound amount at the end of the first year will be

$$A_1 = \underset{\text{principal}}{1000} + \underset{\text{interest}}{1000(.06)} = 1000(1 + .06).$$

At the end of the second year the compound amount will be

$$A_2 = \underset{\substack{\text{compound} \\ \text{amount}}}{A_1} + \underset{\text{interest}}{A_1(.06)} = A_1(1 + .06)$$

$$= [1000(1 + .06)](1 + .06) = 1000(1 + .06)^2.$$

At the end of three years

$$A_3 = A_2 + A_2(.06) = A_2(1 + .06)$$

$$= [1000(1 + .06)^2](1 + .06) = 1000(1 + .06)^3.$$

After n years the compound amount will be

$$A = 1000(1 + .06)^n.$$

In this example the interest period was one year. The important point to note, however, is that at the end of each interest period the amount on deposit grew by a factor of $(1 + .06)$. In general, if the interest rate is i instead of .06, the compound amount will grow by a factor of $(1 + i)$ at the end of each interest period.

Suppose that a principal amount P is invested at a compound interest rate i per interest period, for a total of n interest periods. Then the compound amount A at the end of the nth period will be

$$\boxed{A = P(1 + i)^n.} \tag{1}$$

EXAMPLE 1 Suppose that $5000 is invested at 8% per year, with interest compounded annually. What is the compound amount after three years?

Solution Substituting $P = 5000$, $i = .08$, and $n = 3$ into formula (1), we have

$$A = 5000(1 + .08)^3 = 5000(1.08)^3$$

$$= 5000(1.259712) = 6298.56 \text{ dollars.}$$

It is a common practice to state the interest rate as a percent per year ("per annum"), even though each interest period is often shorter than one year. If the annual rate is r and if interest is paid and compounded m times per year, then the interest rate i for each period is given by

$$\text{(rate per period)} \quad i = \frac{r}{m} \quad \left(\frac{\text{annual interest rate}}{\text{periods per year}}\right).$$

Many banks pay interest quarterly. If the stated annual rate is 5%, then $i = .05/4 = .0125$.

If interest is compounded for t years, with m interest periods each year, there will be a total of mt interest periods. If in formula (1) we replace n by mt and replace i by r/m, we obtain the following formula for the compound amount:

$$A = P\left(1 + \frac{r}{m}\right)^{mt},$$

where P = principal amount,

r = interest rate per annum,

m = number of interest periods per year,

t = number of years.

(2)

EXAMPLE 2   Suppose that $1000 is deposited in a savings account that pays 6% per annum, compounded quarterly. If no additional deposits or withdrawals are made, how much will be in the account at the end of one year?

Solution   We use (2) with $P = 1000$, $r = .06$, $m = 4$, and $t = 1$.

$$A = 1000\left(1 + \frac{.06}{4}\right)^4 = 1000(1.015)^4$$

$$= 1000(1.06136355) \approx 1061.36 \text{ dollars.}$$

Note that the $1000 in Example 2 earned a total of $61.36 in (compound) interest. This is 6.136% of $1000. Savings institutions sometimes advertise this rate as the *effective* annual interest rate. That is, the savings institutions mean that *if* they paid interest only once a year, they would have to pay a rate of 6.136% in order to produce the same earnings as their 6% rate compounded quarterly. The stated annual rate of 6% is often called the *nominal rate*.

The effective annual rate can be increased by compounding the interest more often. Some savings institutions compound interest monthly or even daily.

EXAMPLE 3   Suppose that the interest in Example 2 were compounded monthly. How much would be in the account at the end of one year? What about the case when 6% annual interest is compounded daily?

Solution For monthly compounding, $m = 12$. From (2) we have

$$A = 1000\left(1 + \frac{.06}{12}\right)^{12} = 1000(1.005)^{12}$$

$$\approx 1000(1.06167781) \approx 1061.68 \text{ dollars.}$$

The effective rate in this case is 6.168%.

A "bank year" usually consists of 360 days (in order to simplify calculations). So, for daily compounding, we take $m = 360$. Then

$$A = 1000\left(1 + \frac{.06}{360}\right)^{360} \approx 1000(1.00016667)^{360}$$

$$\approx 1000(1.06183133) \approx 1061.83 \text{ dollars.}$$

With daily compounding, the effective rate is 6.183%.

What would happen if the interest in Example 3 were compounded more often than once a day? Would the total interest be much more than $61.83 if the interest were compounded every hour? Every minute? To answer these questions, we shall digress for a moment and look at the graph of the exponential function.

Recall that we *defined* e to be the number such that the graph of $y = e^x$ has a slope of 1 at $x = 0$. Consider a secant line that approximates the tangent line at $x = 0$. Its slope is $(e^h - 1)/h$ (Fig. 1). Thus

$$[\text{slope of secant}] = \frac{e^h - 1}{h} \approx 1 = [\text{slope of tangent}] \tag{3}$$

when h is very close to zero. Then

$$e^h - 1 \approx h$$

and

$$e^h \approx 1 + h.$$

If we let $h = r/m$, then

$$e^{r/m} \approx 1 + \frac{r}{m}. \tag{4}$$

When m gets very large, r/m gets small—that is, h gets small. We know that the approximation in (3) improves as h approaches zero, so the approximation in (4) improves as m gets very large. Let t be a fixed number and raise both sides of (4) to the power mt:

$$(e^{r/m})^{mt} \approx \left(1 + \frac{r}{m}\right)^{mt}$$

$$e^{rt} \approx \left(1 + \frac{r}{m}\right)^{mt}.$$

Finally, multiplying both sides by P, we have

$$\boxed{Pe^{rt} \approx P\left(1 + \frac{r}{m}\right)^{mt}.} \tag{5}$$

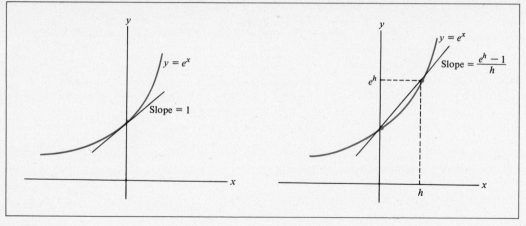

FIGURE 1

Thus Pe^{rt} is a good approximation for the formula in (2) when m is a large number. We conclude that *if P dollars are invested at an annual interest rate r, then the compound amount at the end of t years will be approximately Pe^{rt} dollars, provided that the interest is compounded frequently.* The compound amount calculated from (2) gets closer and closer to Pe^{rt} as the number of interest periods per year is increased. When the formula Pe^{rt} is used to calculate the compound interest, we say that the interest is *compounded continuously.*

When interest is compounded continuously, the compound amount $A(t)$ is an exponential function of the number of years t that interest is earned, $A(t) = Pe^{rt}$. Hence $A(t)$ satisfies the differential equation

$$\frac{dA}{dt} = rA.$$

The rate of growth of the compound amount is proportional to the amount of money present.

We can now answer the question posed following Example 3. Suppose that $1000 is deposited for one year in an account paying 6% per annum. Then $P = 1000$, $r = .06$, and $t = 1$. If the interest is compounded continuously, the compound amount at the end of the year will be

$$1000e^{.06} \approx 1061.84 \text{ dollars.}$$

Recall from Example 3 that daily compounding of the 6% interest produced only $1061.83. Consequently, more frequent compounding (such as every hour or every second) will produce at most one cent more.

In some parts of the country, savings institutions offer interest compounded continuously. Furthermore, continuously compounded interest is an important theoretical tool in business and economics. Problems involving depreciation of equipment, for example, may be analyzed by calculus techniques when the cost of money is computed by using continuously compounded interest.

If P dollars are invested today, the formula $A = Pe^{rt}$ gives the value of this investment after t years (assuming continuously compounded interest). We say that P is the *present value* of the amount A to be received in t years. If we solve for P in terms of A, we obtain

$$P = Ae^{-rt}. \qquad (6)$$

EXAMPLE 4   Find the present value of \$5000 to be received in two years if money can be invested at 6% compounded continuously.

Solution   Use (6) with $A = 5000$, $r = .06$, and $t = 2$.

$$P = 5000e^{-(.06)(2)} = 5000e^{-.12}$$

$$\approx 5000(0.88692) = 4434.60 \text{ dollars.}$$

PRACTICE PROBLEMS 2

1. One thousand dollars is to be invested in a bank for four years. Would 8% interest compounded semiannually be better than $7\frac{3}{4}$% interest compounded continuously?

2. A building was bought for \$150,000 and sold 10 years later for \$400,000. What interest rate (compounded continuously) was earned on the investment?

EXERCISES 2

1. Ten thousand dollars is invested at 8% interest (per annum), compounded quarterly. Give the formula that describes the value of the investment after three years.

2. One thousand dollars is invested at 5% interest, compounded continuously. Compute the value of the investment at the end of six years.

3. A savings account pays 6% interest, compounded continuously. How much should be deposited now in order to have \$1000 in the account at the end of four years?

4. One hundred dollars is deposited in a savings account paying 5% interest, compounded continuously. What is the effective annual rate of interest?

5. Five hundred dollars is deposited in a savings account paying 5% interest, compounded daily. Estimate the balance in the account at the end of three years (assuming that there are no additional deposits or withdrawals).

6. How long will it take for an investment to double itself if interest is paid at 10% per annum, compounded continuously?

7. What annual rate of interest will make an investment double in five years if the interest is compounded continuously?

8. Find the present value of $1000 payable at the end of three years if money may be invested at 8%, with interest compounded continuously.

9. A parcel of land bought in 1975 for $10,000 was worth $16,000 in 1980. If it continues to appreciate at this rate, in what year will it be worth $45,000?

10. Investment A is currently worth $70,200 and is growing at the rate of 8% per year compounded continuously. Investment B is currently worth $60,000 and is growing at the rate of 9% per year compounded continuously. After how many years will the two investments have the same value?

11. An annual interest rate of 16% compounded continuously is equivalent to what interest rate compounded quarterly?

SOLUTIONS TO PRACTICE PROBLEMS 2

1. Let us compute the balance after four years for each type of interest.
   8% compounded semiannually: Use formula (2). Here $P = 1000$, $r = .08$, $m = 2$ (semiannually means that there are two interest periods per year), and $t = 4$. Therefore

$$A = 1000\left(1 + \frac{.08}{2}\right)^{2\cdot 4}$$

$$= 1000(1.04)^8 = 1368.57.$$

   7¾% compounded continuously: Use the formula $A = Pe^{rt}$ where $P = 1000$, $r = .0775$, and $t = 4$. Then

$$A = 1000e^{(.0775)\cdot 4}$$

$$= 1000e^{.31} = 1363.43.$$

   Therefore, 8% compounded semiannually is best.

2. If the $150,000 had been compounded continuously for 10 years at interest rate r, the balance would be $150,000e^{r\cdot 10}$. The question asks, "For what value of r will the balance be 400,000?" We need just solve an equation for r.

$$150,000e^{r\cdot 10} = 400,000$$

$$e^{r\cdot 10} \approx 2.67$$

$$r\cdot 10 = \ln 2.67$$

$$r = \frac{\ln 2.67}{10} \approx .098.$$

   Therefore, the investment earned 9.8% interest per year.

6.3. Further Exponential Models

A skydiver, on jumping out of an airplane, falls at an increasing rate. However, the wind rushing past the skydiver's body creates an upward force that begins to counterbalance the downward force of gravity. This air friction finally becomes so great that the skydiver's velocity reaches a limiting speed called the *terminal velocity*. If we let $v(t)$ be the downward velocity of the skydiver after t seconds of free fall, then a good mathematical model for $v(t)$ is given by

$$v(t) = M(1 - e^{-kt}), \qquad (1)$$

where M is the terminal velocity and k is some positive constant (Fig. 1). When t is close to zero, e^{-kt} is close to one and the velocity is small. As t increases, e^{-kt} becomes small and so $v(t)$ approaches M.

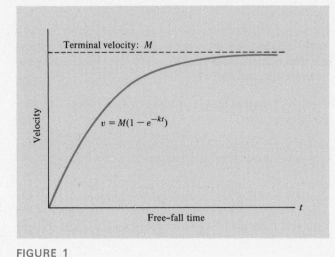

FIGURE 1

EXAMPLE 1 Show that the velocity given in (1) satisfies the differential equation

$$\frac{dv}{dt} = k[M - v(t)], \qquad v(0) = 0. \qquad (2)$$

Solution From (1) we have $v(t) = M - Me^{-kt}$. Then

$$\frac{dv}{dt} = Mke^{-kt}.$$

However,

$$k[M - v(t)] = k[M - (M - Me^{-kt})] = kMe^{-kt},$$

so that the differential equation $\frac{dv}{dt} = k[M - v(t)]$ holds. Also,

$$v(0) = M - Me^0 = M - M = 0.$$

The differential equation (2) says that the rate of change in v is proportional to the difference between the terminal velocity M and the actual velocity v. It is not difficult to show that the only solution of (2) is given by the formula in (1).

The two equations (1) and (2) arise as mathematical models in a variety of situations. Some of these applications are described below.

The Learning Curve Psychologists have found that in many learning situations a person's rate of learning is rapid at first and then slows down. Finally, as the task is mastered, the person's level of performance reaches a level above which it is almost physically impossible to rise. For example, within reasonable limits, each person seems to have a certain maximum capacity for memorizing a list of nonsense syllables. Suppose that a subject can memorize M syllables in a row if given sufficient time, say an hour, to study the list but cannot memorize $M + 1$ syllables in a row even if allowed several hours of study. By giving the subject different lists of syllables and varying lengths of time to study the lists, the psychologist can determine an empirical relationship between the number of nonsense syllables memorized accurately and the number of minutes of study time. It turns out that a good model for this situation is

$$y = M(1 - e^{-kt}) \qquad (3)$$

for some appropriate positive constant k. (See Fig. 2.)

The *slope* of this learning curve at time t is approximately the number of additional syllables that can be memorized if the subject is given one more minute of study time. Thus the slope is a measure of the *rate of learning*. The differential equation satisfied by the function in (3) is

$$y' = k(M - y), \qquad f(0) = 0.$$

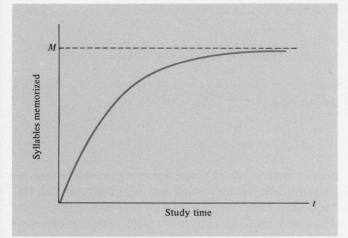

This equation says that if the subject is given a list of M nonsense syllables, then the rate of memorization is proportional to the number of syllables remaining to be memorized.

FIGURE 2   A learning curve, $y = M(1 - e^{-kt})$.

Diffusion of Information by Mass Media Sociologists have found that the differential equation (2) provides a good model for the way information is spread (or "diffused") through a population when the information is being propagated constantly by mass media, such as television or magazines.* Given a fixed population P, let $f(t)$ be the number of people who have already heard a certain piece of information by time t. Then $P - f(t)$ is the number who have not yet heard the information. Also, $f'(t)$ is the rate of increase of the number of people who have heard the news (the "rate of diffusion" of the information). If the information is being publicized often by some mass media, then it is likely that the number of *newly informed* people per unit time is proportional to the number of people who have not yet heard the news. Therefore

$$f'(t) = k[P - f(t)].$$

---

* J. Coleman, *Introduction to Mathematical Sociology* (New York: The Free Press, 1964), p. 43.

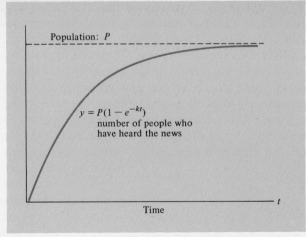

Population: P

$y = P(1 - e^{-kt})$
number of people who
have heard the news

Time t

FIGURE 3 Diffusion of information by mass media.

Assume that $f(0) = 0$ (that is, there was a time $t = 0$ when nobody had heard the news). Then the remark following Example 1 shows that

$$f(t) = P(1 - e^{-kt}).\tag{4}$$

(See Fig. 3.)

EXAMPLE 2 Suppose that a certain piece of news (such as the resignation of a public official) is broadcast frequently by radio and television stations. Also suppose that one-half of the residents of a city have heard the news within four hours of its initial release. Use the exponential model (4) to estimate when 90% of the residents will have heard the news.

Solution We must find the value of k in (4). If P is the number of residents, then the number who have heard the news in four hours is given by (4) with $t = 4$. By assumption, this number is half the population. So

$$\tfrac{1}{2}P = P(1 - e^{-k \cdot 4})$$
$$.5 = 1 - e^{-4k}$$
$$e^{-4k} = 1 - .5 = .5.$$

Solving for k, we find that $k \approx .17$. So the model for this particular situation is

$$f(t) = P(1 - e^{-.17t}).$$

Now we want to find t such that $f(t) = .90P$. We solve for t:

$$.90P = P(1 - e^{-.17t})$$
$$.90 = 1 - e^{-.17t}$$
$$e^{-.17t} = 1 - .90 = .10$$
$$-.17t = \ln .10$$
$$t = \frac{\ln .10}{-.17} \approx 14.$$

Therefore 90% of the residents will hear the news within 14 hours of its initial release.

Intravenous Infusion of Glucose The human body both manufactures and uses glucose ("blood sugar"). Usually there is a balance in these two processes, so that the bloodstream has a certain "equilibrium level" of glucose. Suppose that a patient is given a single intravenous injection of glucose and let $A(t)$ be the amount of glucose (in milligrams) above the equilibrium level. Then the body will start using up the excess glucose at a rate proportional to the amount of excess glucose; that is,

$$A'(t) = -\lambda A(t), \tag{5}$$

where λ is a positive constant called the *velocity constant of elimination*. This constant depends on how fast the patient's metabolic processes eliminate the excess glucose from the blood. Equation (5) describes a simple exponential decay process.

Now suppose that, instead of a single shot, the patient receives a continuous intravenous infusion of glucose. A bottle of glucose solution is suspended above the patient, and a small tube carries the glucose down to a needle that runs into a vein. In this case, there are two influences on the amount of excess glucose in the blood: the glucose being added steadily from the bottle and the glucose being removed from the blood by metabolic processes. Let r be the rate of infusion of glucose (often from 10 to 100 milligrams per minute). If the body did not remove any glucose, the excess glucose would increase at a constant rate of r milligrams per minute; that is,

$$A'(t) = r. \tag{6}$$

Taking into account the two influences on $A'(t)$ described by (5) and (6), we can write

$$A'(t) = r - \lambda A(t). \tag{7}$$

If we let $M = r/\lambda$, then

$$A'(t) = \lambda(M - A(t)).$$

It can be shown that a solution of this differential equation is given by

$$A(t) = M(1 - e^{-\lambda t}) = \frac{r}{\lambda}(1 - e^{-\lambda t}). \tag{8}$$

Reasoning as in Example 1, we conclude that the amount of excess glucose rises until it reaches a stable level. (See Fig. 4.)

The Logistic Growth Curve The model for simple exponential growth discussed in Section 1 is adequate for describing the growth of many types of populations, but obviously a population cannot increase exponentially forever. The simple exponential growth model becomes inapplicable when the environment begins to inhibit the growth of the population. The logistic growth curve is an important exponential model that takes into account some of the effects of the

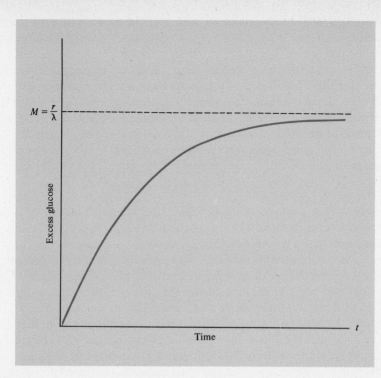

FIGURE 4 Continuous infusion of glucose.

$$y = \frac{r}{\lambda}(1 - e^{-\lambda t}).$$

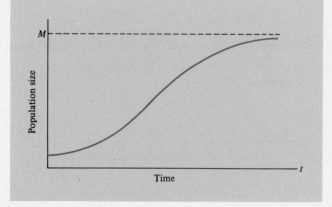

FIGURE 5 Logistic growth.

environment on a population (Fig. 5). For small values of t, the curve has the same basic shape as an exponential growth curve. Then when the population begins to suffer from overcrowding or lack of food, the growth rate (the slope of the population curve) begins to slow down. Eventually the growth rate tapers off to zero as the population reaches the maximum size that the environment will support. This latter part of the curve resembles the growth curves studied earlier in this section.

The equation for logistic growth has the general form

$$y = \frac{M}{1 + Be^{-Mkt}}, \qquad (9)$$

where B, M, and k are positive constants. It can be shown that y satisfies the differential equation

$$y' = ky(M - y). \qquad (10)$$

The factor y reflects the fact that the growth rate (y') depends in part on the size y of the population. The factor $M - y$ reflects the fact that the growth rate also depends on how close y is to the maximum level M.

The logistic curve is often used to fit experimental data that lie along an "S-shaped" curve. Examples are given by the growth of a fish population in a lake and the growth of a fruit fly population in a laboratory container. Also, certain enzyme reactions in animals follow a logistic law. One of the earliest applications of the logistic curve occurred in about 1840 when the Belgian sociologist P. Verhulst fit a logistic curve to six U.S. census figures, 1790 to 1840, and predicted the U.S. population for 1940. His prediction missed by less than one million persons (an error of about 1%).

EXAMPLE 3   Suppose that a lake is stocked with 100 fish. After three months there are 250 fish. A study of the ecology of the lake predicts that the lake can support 1000 fish. Find a formula for the number $P(t)$ of fish in the lake t months after it has been stocked.

Solution   The limiting population M is 1000. Therefore we have

$$P(t) = \frac{1000}{1 + Be^{-1000kt}}.$$

At $t = 0$ there are 100 fish, so that

$$100 = P(0) = \frac{1000}{1 + Be^0} = \frac{1000}{1 + B}.$$

Thus $1 + B = 10$, or $B = 9$. Finally, since $P(3) = 250$, we have

$$250 = \frac{1000}{1 + 9e^{-3000k}}$$

$$1 + 9e^{-3000k} = 4$$

$$e^{-3000k} = \tfrac{1}{3}.$$

$$-3000k = \ln \tfrac{1}{3}$$

$$k \approx .00037.$$

Therefore

$$P(t) = \frac{1000}{1 + 9e^{-.37t}}.$$

Several theoretical justifications can be given for using (9) and (10) in situations where the environment prevents a population from exceeding a certain size. A discussion of this topic may be found in *Mathematical Models and Applications* by Maki and Thompson (Prentice-Hall, 1973), pp. 312–317.

An Epidemic Model   To conclude this section we shall illustrate how one can actually "build" a mathematical model. Our example concerns the spread of a highly contagious disease. We begin by making several simplifying assumptions:

1.   The population is a fixed number P and each member of the population is susceptible to the disease.

2. The duration of the disease is long, so that no cures occur during the time period under study.
3. All infected individuals are contagious and circulate freely among the population.
4. During each unit time period (such as one day or one week) each infected person makes c contacts, and each contact with an uninfected person results in transmission of the disease.

Consider a short period of time from t to $t + h$. Each infected person makes $c \cdot h$ contacts. How many of these contacts are with uninfected persons? If $f(t)$ is the number of infected persons at time t, then $P - f(t)$ is the number of uninfected persons, and $[P - f(t)]/P$ is the fraction of the population that is uninfected. Thus, of the $c \cdot h$ contacts made,

$$\left[\frac{P - f(t)}{P}\right] \cdot c \cdot h$$

will be with uninfected persons. This is the number of new infections produced by one infected person during the time period of length h. The total number of *new* infections during this period is

$$f(t)\left[\frac{P - f(t)}{P}\right]ch.$$

But this number must equal $f(t + h) - f(t)$, where $f(t + h)$ is the total number of infected persons at time $t + h$. So

$$f(t + h) - f(t) = f(t)\left[\frac{P - f(t)}{P}\right]ch.$$

Dividing by h, the length of the time period, we obtain the average number of new infections per unit time (during the small time period):

$$\frac{f(t + h) - f(t)}{h} = \frac{c}{P} f(t)[P - f(t)].$$

If we let h approach zero and let y stand for $f(t)$, the left-hand side approaches the rate of change in the number of infected persons and we derive the following equation:

$$\frac{dy}{dt} = \frac{c}{P} y(P - y).$$

This is the same type of equation as that used in (10) for logistic growth, although the two situations leading to this model appear to be quite dissimilar.

Comparing (11) with (10), we see that the number of infected individuals at time t is described by a logistic curve with $M = P$ and $k = c/P$. Therefore, by (9), we can write

$$f(t) = \frac{P}{1 + Be^{-ct}}.$$

B and c can be determined from the characteristics of the epidemic. (See Example 4 below.)

The logistic curve has an inflection point at that value of t for which $f(t) = P/2$. The position of this inflection point has great significance for applications of the logistic curve. From inspecting a graph of the logistic curve, we see that the inflection point is the point at which the curve has greatest slope. In other words, the inflection point corresponds to the instant of fastest growth of the logistic curve. This means, for example, that in the foregoing epidemic model the disease is spreading with the greatest rapidity precisely when half the population is infected. Any attempt at disease control (through immunization, for example) must strive to reduce the incidence of the disease to as low a point as possible, but in any case at least below the inflection point at $P/2$, at which point the epidemic is spreading fastest.

EXAMPLE 4 The Public Health Service monitors the spread of an epidemic of a particularly long-lasting strain of flu in a city of 500,000 people. At the beginning of the first week of monitoring, 200 cases have been reported; during the first week 300 new cases are reported. Estimate the number of infected individuals after six weeks.

Solution Here $P = 500,000$. If $f(t)$ denotes the number of cases at the end of t weeks, then

$$f(t) = \frac{P}{1 + Be^{-ct}}$$

$$= \frac{500,000}{1 + Be^{-ct}}.$$

Moreover, $f(0) = 200$, so that

$$200 = \frac{500,000}{1 + Be^0} = \frac{500,000}{1 + B},$$

and $B = 2499$. Consequently, since $f(1) = 300 + 200 = 500$, we have

$$500 = f(1) = \frac{500,000}{1 + 2499e^{-c}},$$

so that $e^{-c} \approx .4$ and $c \approx .92$. Finally,

$$f(t) = \frac{500,000}{1 + 2499e^{-.92t}},$$

and

$$f(6) = \frac{500,000}{1 + 2499e^{-.92(6)}} \approx 45,000.$$

After six weeks, about 45,000 individuals are infected.

This epidemic model is used by sociologists (where it is still called an epidemic model) to describe the spread of a rumor. In economics the model is used to describe the diffusion of knowledge about a product. An "infected person" represents an individual who possesses knowledge of the product. In both cases, it is assumed that the members of the population are themselves primarily responsible for the spread of the rumor or knowledge of the product. This situation is in contrast to the model described earlier where information was spread through a population by external sources, such as radio and television.

There are several limitations to this epidemic model. Each of the four simplifying assumptions made at the outset is unrealistic in varying degrees. More complicated models can be constructed that rectify one or more of these defects, but they require more advanced mathematical tools.

PRACTICE PROBLEMS 3

1. A sociological study* was made to examine the process by which doctors decide to adopt a new drug. The doctors were divided into two groups. The doctors in group A had little interaction with other doctors and so received most of their information via mass media. The doctors in group B had extensive interaction with other doctors and so received most of their information via word of mouth. For each group, let $f(t)$ be the number who have learned about a new drug after t days. Examine the appropriate differential equations to explain why the two graphs were of the types shown below.

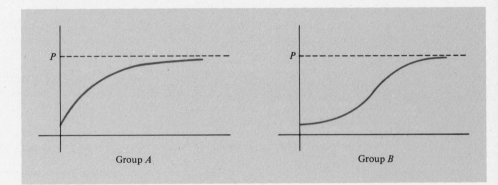

Group A   Group B

EXERCISES 3

1. Consider the function $f(x) = 5(1 - e^{-2x})$, $x \geq 0$.

(a) Show that $f(x)$ is increasing and concave down for all $x \geq 0$.

(b) Explain why $f(x)$ approaches 5 as x gets large.

(c) Sketch the graph of $f(x)$, $x \geq 0$.

---

* James S. Coleman, Elihu Katz, and Herbert Menzel: "The Diffusion of an Innovation Among Physicians," *Sociometry*, 20 (1957), 253–270.

2. Consider the function $g(x) = 10 - 10e^{-.1x}$, $x \geq 0$.

   (a) Show that $g(x)$ is increasing and concave down for $x \geq 0$.

   (b) Explain why $g(x)$ approaches 10 as x gets large.

   (c) Sketch the graph of $g(x)$, $x \geq 0$.

3. Suppose that $y = 2(1 - e^{-x})$. Compute y' and show that $y' = 2 - y$.

4. Suppose that $y = 5(1 - e^{-2x})$. Compute y' and show that $y' = 10 - 2y$.

5. Suppose that $f(x) = 3(1 - e^{-10x})$. Show that $y = f(x)$ satisfies the differential equation

$$y' = 10(3 - y), \qquad f(0) = 0.$$

6. (Ebbinghaus model for forgetting.) Suppose that a student learns a certain amount of material for some class. Let $f(t)$ denote the percentage of the material that the student can recall t weeks later. The psychologist Ebbinghaus found that this percent retention can be modeled by a function of the form

$$f(t) = (100 - a)e^{-\lambda t} + a,$$

where λ and a are positive constants and $0 < a < 100$. Sketch the graph of the function $f(t) = 85e^{-.5t} + 15$, $t \geq 0$.

7. When a grand jury indicted the mayor of a certain town for accepting bribes, the newspaper, radio, and television immediately began to publicize the news. Within an hour one-quarter of the citizens heard about the indictment. Estimate when three-quarters of the town heard the news.

8. Examine formula (8) for the amount $A(t)$ of excess glucose in the bloodstream of a patient at time t. Describe what would happen if the rate r of infusion of glucose were doubled.

9. Describe an experiment that a doctor could perform in order to determine the velocity constant of elimination of glucose for a particular patient.

10. Physiologists usually describe the continuous intravenous infusion of glucose in terms of the excess *concentration* of glucose, $C(t) = A(t)/V$, where V is the total volume of blood in the patient. In this case, the rate of increase in the concentration of glucose due to the continuous injection is r/V. Find a differential equation that gives a model for the rate of change of the excess concentration of glucose.

SOLUTIONS TO PRACTICE PROBLEMS 3

1. The difference between transmission of information via mass media and via word of mouth is that in the second case the rate of transmission depends not only on the number of people who have not yet received the information, but also on the number of people who know the information and therefore are capable of spreading it. Therefore, for group A, $f'(t) = k[P - f(t)]$, and for group B, $f'(t) = kf(t)[P - f(t)]$. Note that the spread of information by word of mouth follows the same pattern as the spread of an epidemic.

Chapter 6: CHECKLIST

- □ $y' = ky$
- □ Exponential growth
- □ Exponential decay
- □ Half-life of a radioactive element
- □ Continuous compounding of interest
- □ $y = M(1 - e^{-kt})$
- □ $y = M/(1 + Be^{-Mkt})$ (logistic growth)

Chapter 6: SUPPLEMENTARY EXERCISES

Find all functions $y = f(x)$ such that

1. $y' = 3y$, $f(0) = \frac{5}{2}$
2. $y' = -y$, $f(0) = 7$
3. $y' = -.1y$, $f(0) = -1$
4. $y' = 7y$, $f(0) = -3.1$
5. $y' = y$, $f(2) = 8$
6. $y' = 3y$, $f(-1) = \frac{3}{2}$
7. $y' = -6y$, $f(3) = 1$
8. $y' = \frac{1}{4}y$, $f(2) = -4$

Find a function of the form $f(t) = Ae^{kt}$ such that

9. $f(1) = 1$, $f(2) = 2$
10. $f(1) = e$, $f(2) = e^3$

11. In many cases the concentration of a drug in the body is given by a function $f(t) = c_0 e^{-t/T}$, where c_0 is the initial concentration when the time $t = 0$ and T is a positive constant called the *elimination time*. Suppose that c_0 is .1 milligram per cubic centimeter of blood and that T is 10 hours. Find the concentration when $t = 10$ hours.

12. A person is administered a dose of 50 milligrams of a drug intravenously. Suppose that the volume of blood is 5000 cubic centimeters and the elimination time is 30 hours. (See Exercise 11.)

    (a) What is the initial concentration?

    (b) How many hours are required for the concentration to drop to .001 milligram per cubic centimeter?

13. A person is administered a dose of 100 milligrams of a drug intravenously. Suppose that the volume of blood is 5000 cubic centimeters and the elimination time is 100 hours. Determine the size of a second dose to be administered after 72 hours, such that the concentration of the drug is returned to its initial concentration. (Assume that the volume of blood is not appreciably altered by administration of the drug.) (See Exercise 11.)

14. A physician wants to maintain a minimum concentration of a drug in a patient's bloodstream of .01 milligram per cubic centimeter. Suppose that the volume of blood is 6000 cubic centimeters and the elimination time is 120 hours. Determine

the amount of the drug to be administered intravenously if one dose is to be given every seven days. (See Exercise 11.)

15. Two different bacteria colonies are growing near a pool of stagnant water. Suppose that the first colony initially has 100 bacteria and doubles after one hour. The second colony initially has 5000 bacteria and doubles after two hours. How much time will elapse before the first colony becomes as large as the second?

16. A certain type of cell culture doubles in size after three hours. How long before its size is multiplied by 10?

17. How long does it take for money to triple when compounded continuously at 6% interest?

18. Suppose that a businessman initially invests $10,000 in a speculative venture. Suppose that the investment pays 10% interest compounded continuously for five years and 3% interest compounded continuously for five years thereafter.

(a) How much does the $10,000 grow to after 10 years?

(b) Suppose that the businessman has the alternative of an investment paying 7% interest compounded continuously. Which investment is superior over a 10-year period and by how much?

INTEGRATION

The derivative is one of the two fundamental concepts of calculus. The other is the *integral*, which we introduce in this chapter.

7.1. Antidifferentiation

We have developed several techniques for calculating the derivative $F'(x)$ of a function $F(x)$. In many applications, however, it is necessary to proceed in reverse. We are given the derivative $F'(x)$ and must determine the function $F(x)$. The process of determining $F(x)$ from $F'(x)$ is called *antidifferentiation*. The next example gives a typical application involving antidifferentiation.

EXAMPLE 1 For a number of years, the annual worldwide rate of oil consumption has been growing exponentially with a growth constant of about .07. At the beginning of 1970, the rate was about 16.1 billion barrels of oil per year. Let $C(t)$ denote the rate of oil consumption at time t, where t is the number of years since the beginning of 1970. Then a reasonable model for $C(t)$ is given by

$$C(t) = 16.1e^{.07t}. \tag{1}$$

Use this formula for $C(t)$ to estimate the total amount of oil that will be consumed from 1970 to 1990.

Solution Let $T(t)$ be the total amount of oil consumed from time 0 (1970) until time t. We wish to calculate $T(20)$, the amount of oil consumed from 1970 to 1990. We do this by first determining a formula for $T(t)$. Since $T(t)$ is the total oil consumed, the derivative $T'(t)$ is the *rate* of oil consumption, namely $C(t)$. Thus, although we do not yet have a formula for $T(t)$, we do know that

$$T'(t) = C(t).$$

Thus, the problem of determining a formula for $T(t)$ has been reduced to a problem of antidifferentiation: Find a function whose derivative is $C(t)$. We shall solve this particular problem after developing some techniques for solving antidifferentiation problems in general.

Suppose $f(x)$ is a given function and $F(x)$ is a function having $f(x)$ as its derivative—that is, $F'(x) = f(x)$. We call $F(x)$ an *antiderivative* of $f(x)$.

EXAMPLE 2 Find an antiderivative of $f(x) = x^2$.

Solution One such function is $F(x) = \frac{1}{3}x^3$, since

$$F'(x) = \frac{1}{3} \cdot 3x^2 = x^2.$$

Another antiderivative is $F(x) = \frac{1}{3}x^3 + 2$, since

$$\frac{d}{dx}\left(\frac{1}{3}x^3 + 2\right) = \frac{1}{3} \cdot 3x^2 + 0 = x^2.$$

In fact, if C is any constant, the function $F(x) = \frac{1}{3}x^3 + C$ is also an antiderivative of x^2, since

$$\frac{d}{dx}\left(\frac{1}{3}x^3 + C\right) = \frac{1}{3} \cdot 3x^2 + 0 = x^2.$$

(The derivative of a constant function is zero.)

EXAMPLE 3 Find an antiderivative of the function $f(x) = 2x - (1/x^2)$.

Solution Since

$$\frac{d}{dx}(x^2) = 2x \quad \text{and} \quad \frac{d}{dx}\left(\frac{1}{x}\right) = -\frac{1}{x^2},$$

we see that one antiderivative of $f(x)$ is given by

$$F(x) = x^2 + \frac{1}{x}.$$

However, any function of the form $x^2 + (1/x) + C$, C a constant, will do, since

$$\frac{d}{dx}\left(x^2 + \frac{1}{x} + C\right) = 2x - \frac{1}{x^2} + 0 = 2x - \frac{1}{x^2}.$$

Using the same reasoning as in Examples 2 and 3, we see that if $F(x)$ is an antiderivative of $f(x)$, then so is $F(x) + C$, where C is any constant. Thus if we know one antiderivative $F(x)$ of a function $f(x)$, we can write down an infinite number by adding all possible constants C to $F(x)$. It turns out that in this way we obtain all antiderivatives of $f(x)$. That is, we have the following fundamental result.

> Theorem I If $F_1(x)$ and $F_2(x)$ are two antiderivatives of the same function $f(x)$, then $F_1(x)$ and $F_2(x)$ differ by a constant. In other words, there is a constant C such that
>
> $$F_2(x) = F_1(x) + C.$$

Our verification of this theorem will be based on the following fact, which is important in its own right.

> Theorem II   If $F'(x) = 0$ for all x, then $F(x) = C$ for some constant C.

Verification of Theorem II   If $F'(x) = 0$ for all x, then the curve $y = F(x)$ has slope equal to zero at every point. Thus the tangent line to $y = F(x)$ at any point is horizontal, which implies that the graph of $y = F(x)$ is a horizontal line. (Try to draw the graph of a function with horizontal tangent everywhere. There is no choice but to keep your pencil moving on a constant, horizontal line!) If the horizontal line is $y = C$, then $F(x) = C$ for all x.

Verification of Theorem I   If $F_1(x)$ and $F_2(x)$ are two antiderivatives of $f(x)$, then the function $F(x) = F_2(x) - F_1(x)$ has derivative

$$F'(x) = F_2'(x) - F_1'(x)$$

$$= f(x) - f(x)$$

$$= 0.$$

So, by Theorem II, we know that $F(x) = C$ for some constant C. In other words, $F_2(x) - F_1(x) = C$, so that

$$F_2(x) = F_1(x) + C,$$

which is Theorem I.

Using Theorem I, we can find *all* antiderivatives of a given function once we know one antiderivative. For instance, since one antiderivative of x^2 is $\frac{1}{3}x^3$ (Example 2), all antiderivatives of x^2 have the form $\frac{1}{3}x^3 + C$, where C is a constant.

Suppose that $f(x)$ is a function whose antiderivatives are $F(x) + C$. The standard way to express this fact is to write

$$\int f(x)\,dx = F(x) + C.$$

The symbol $\int$ is called an *integral sign*. The entire notation $\int f(x)\,dx$ is called an *indefinite integral* and stands for antidifferentiation of the function $f(x)$. We always record the variable of interest by writing it, prefaced by the letter d. For example, if the variable of interest is t rather than x, then we write $\int f(t)\,dt$ for the antiderivative of $f(t)$.

EXAMPLE 4 Determine

(a) $\displaystyle\int x^r\,dx, \quad r \neq -1$ (b) $\displaystyle\int e^{kx}\,dx, \quad k \neq 0.$

Solution (a) By the constant multiple and power rules,

$$\frac{d}{dx}\left(\frac{1}{r+1}\,x^{r+1}\right) = \frac{1}{r+1} \cdot \frac{d}{dx}\,x^{r+1} = \frac{1}{r+1} \cdot (r+1)x^r = x^r.$$

Thus $x^{r+1}/(r+1)$ is an antiderivative of x^r. Letting C represent any constant, we have

$$\int x^r\,dx = \frac{1}{r+1}\,x^{r+1} + C, \qquad r \neq -1. \tag{2}$$

(b) An antiderivative of e^{kx} is e^{kx}/k, since

$$\frac{d}{dx}\left(\frac{1}{k}\,e^{kx}\right) = \frac{1}{k} \cdot \frac{d}{dx}\,e^{kx} = \frac{1}{k}(ke^{kx}) = e^{kx}.$$

Hence

$$\int e^{kx}\,dx = \frac{1}{k}\,e^{kx} + C, \quad k \neq 0. \tag{3}$$

Formula (2) does not give an antiderivative of x^{-1} because $1/(r+1)$ is undefined for $r = -1$. However, we know that for $x \neq 0$, the derivative of $\ln|x|$ is $1/x$. Hence $\ln|x|$ is an antiderivative of $1/x$, and we have

$$\int \frac{1}{x}\,dx = \ln|x| + C, \quad x \neq 0. \tag{4}$$

Formulas (2), (3), and (4) each followed by "reversing" a familiar differentiation rule. In a similar fashion, one may use the sum rule and constant-multiple rule for derivatives to obtain corresponding rules for antiderivatives:

$$\int [f(x) + g(x)]\,dx = \int f(x)\,dx + \int g(x)\,dx \tag{5}$$

$$\int kf(x)\,dx = k \int f(x)\,dx, \quad k \text{ a constant.} \tag{6}$$

In words, (5) says that a sum of functions may be antidifferentiated term by term, and (6) says that a constant multiple may be moved through the integral sign.

EXAMPLE 5 Compute

$$\int \left(x^{-3} + 7e^{5x} + \frac{4}{x} \right) dx.$$

Solution Using the rules given above, we have

$$\int \left(x^{-3} + 7e^{5x} + \frac{4}{x} \right) dx = \int x^{-3}\,dx + \int 7e^{5x}\,dx + \int \frac{4}{x}\,dx$$

$$= \int x^{-3}\,dx + 7\int e^{5x}\,dx + 4\int \frac{1}{x}\,dx$$

$$= \frac{1}{-2} x^{-2} + 7\left(\frac{1}{5} e^{5x} \right) + 4\ln|x| + C$$

$$= -\frac{1}{2} x^{-2} + \frac{7}{5} e^{5x} + 4\ln|x| + C.$$

After some practice, most of the steps shown in the solution of Example 5 can be omitted.

A function $f(x)$ has infinitely many different antiderivatives, corresponding to the various choices of the constant C. In applications, it is often necessary to

EXAMPLE 6   Find the antiderivative $F(x)$ of $3x^2 - 5$ for which $F(0) = 2$.

Solution   One antiderivative of $3x^2 - 5$ is $x^3 - 5x$. Therefore, by Theorem I, the anti-derivatives are precisely the functions

$$F(x) = x^3 - 5x + C, \quad C \text{ a constant.}$$

If $F(0) = 2$, then

$$2 = F(0) = 0^3 - 5 \cdot 0 + C = C,$$

so that $C = 2$, and the unique antiderivative satisfying the given condition is

$$F(x) = x^3 - 5x + 2.$$

Having introduced the basics of antidifferentiation, let us now solve the oil-consumption problem.

Solution
of Example 1
(Continued)   The rate of oil consumption at time t is $C(t) = 16.1e^{.07t}$ billion barrels per year. Moreover, we observed that the total consumption $T(t)$, from time 0 to time t, is an antiderivative of $C(t)$. Using (3) and (6) we have

$$T(t) = \int 16.1e^{.07t}\, dt = \frac{16.1}{.07} e^{.07t} + C = 230e^{.07t} + C,$$

where C is a constant. However, in our particular example, $T(0) = 0$, since $T(0)$ is the amount of oil used from time 0 to time 0. Therefore, the constant C must satisfy

$$0 = T(0) = 230e^{.07(0)} + C = 230 + C,$$

$$C = -230.$$

Therefore,

$$T(t) = 230e^{.07t} - 230 = 230(e^{.07t} - 1).$$

The total oil that will be consumed from 1970 to 1990 is

$$T(20) = 230(e^{.07(20)} - 1) \approx 703 \text{ billion barrels.}$$

Antidifferentiation can be used to solve a variety of applied problems, of which the next two examples are typical.

EXAMPLE 7   A rocket is fired vertically into the air. Its velocity at t seconds after liftoff is $v(t) = 20t + 50$ meters per second. How far will the rocket travel during the first 100 seconds?

Solution   If $s(t)$ denotes the height (in meters) of the rocket at time t seconds after liftoff, then $s'(t)$ is the rate (in meters per second) at which the height is changing at time t, which is just $v(t)$. In other words, $s(t)$ is an antiderivative of $v(t)$. Thus,

$$s(t) = \int v(t)\,dt = \int (20t + 50)\,dt$$

$$= 10t^2 + 50t + C,$$

where C is a constant. At time 0 the height of the rocket is 0, so $s(0) = 0$ and

$$0 = s(0) = 10(0)^2 + 50(0) + C = C$$

$$C = 0$$

$$s(t) = 10t^2 + 50t.$$

At $t = 100$, the height of the rocket is

$$s(100) = 10(100)^2 + 50(100) = 105{,}000 \text{ meters.}$$

EXAMPLE 8 A factory's marginal cost function is $\frac{1}{100}x^2 - 2x + 120$, where x denotes the number of units produced per day. The factory has fixed costs of \$1000 per day. Find the cost of producing x units per day.

Solution Let $C(x)$ be the cost of producing x units per day. The derivative $C'(x)$ is just the marginal cost function. In other words, $C(x)$ is an antiderivative of the marginal cost function. Thus,

$$C(x) = \int (\tfrac{1}{100}x^2 - 2x + 120)\,dx$$

$$= \tfrac{1}{300}x^3 - x^2 + 120x + C$$

for some constant C. However, the fixed costs are just the costs experienced when producing 0 units. That is, the fixed costs equal $C(0)$. So the given data imply that $C(0) = 1000$. Thus,

$$1000 = C(0) = \tfrac{1}{300}(0)^3 - (0)^2 - 120(0) + C$$

$$C = 1000.$$

Therefore, the cost function $C(x)$ is given by

$$C(x) = \tfrac{1}{300}x^3 - x^2 + 120x + 1000.$$

PRACTICE PROBLEMS 1

1. Determine

    (a) $\int (x^3 + 4x)\,dx$ (b) $\int t^{7/2}\,dt$

2. Let $R(x)$ be the revenue received from the sale of x units of a product. The marginal revenue function is $5 - 3x$. Find $R(x)$. [Note that the revenue is 0 if no units are sold. That is, $R(0) = 0$.]

EXERCISES 1

Find all antiderivatives of each of the following functions.

1. $f(x) = x$ 2. $f(x) = 9x^8$ 3. $f(x) = e^{3x}$

4. $f(x) = e^{-3x}$ 5. $f(x) = 3$ 6. $f(x) = -4x$

In Exercises 7–22 find the value of k that makes the antidifferentiation formula true. (Note: You can check your answer without looking in the answer section. How?)

7. $\int x^{-5} dx = kx^{-4} + C$ 8. $\int x^{1/3} dx = kx^{4/3} + C$

9. $\int \sqrt{x}\, dx = kx^{3/2} + C$ 10. $\int \frac{6}{x^3} dx = \frac{k}{x^2} + C$

11. $\int \frac{10}{t^6} dt = kt^{-5} + C$ 12. $\int \frac{3}{\sqrt{t}} dt = k\sqrt{t} + C$

13. $\int 5e^{-2t} dt = ke^{-2t} + C$ 14. $\int 3e^{t/10} dt = ke^{t/10} + C$

15. $\int 2e^{4x-1} dx = ke^{4x-1} + C$ 16. $\int \frac{4}{e^{3x+1}} dx = \frac{k}{e^{3x+1}} + C$

17. $\int (x-7)^{-2} dx = k(x-7)^{-1} + C$ 18. $\int \sqrt{x+1}\, dx = k(x+1)^{3/2} + C$

19. $\int (x+4)^{-1} dx = k \ln|x+4| + C$ 20. $\int \frac{5}{(x-8)^4} dx = \frac{k}{(x-8)^3} + C$

21. $\int (3x+2)^4 dx = k(3x+2)^5 + C$ 22. $\int (2x-1)^3 dx = k(2x-1)^4 + C$

Determine the following:

23. $\int (x^2 - x - 1)\, dx$ 24. $\int (x^3 + 6x^2 - x)\, dx$

25. $\int \left(\frac{2}{\sqrt{x}} - 3\sqrt{x} \right) dx$ 26. $\int \left[\frac{\sqrt{t}}{4} - 4(t-3)^{-2} \right] dt$

27. $\int \left(4 - 5e^{-5t} + \frac{e^{2t}}{3} \right) dt$ 28. $\int (e^2 + 3t^2 - 2e^{3t})\, dt$

Find all functions $f(t)$ with the following property.

29. $f'(t) = t^{3/2}$ 30. $f'(t) = \dfrac{4}{6+t}$

31. $f'(t) = 0$ 32. $f'(t) = t^2 - 5t - 7$

Find all functions $f(x)$ with the following properties:

33. $f'(x) = x, \quad f(0) = 3$ 34. $f'(x) = 8x^{1/3}, \quad f(1) = 4$

35. $f'(x) = \sqrt{x} + 1$, $f(4) = 0$ 36. $f'(x) = x^2 + \sqrt{x}$, $f(1) = 3$

37. $f'(x) = 2/x$, $f(1) = 2$ 38. $f'(x) = 3$, $f(2) = 7$

39. A ball is thrown upward from a height of 256 feet above the ground, with an initial velocity of 96 feet per second. From physics it is known that the velocity at time t is $96 - 32t$ feet per second.

    (a) Find $s(t)$, the function giving the height of the ball at time t.

    (b) How long will it take for the ball to reach the ground?

    (c) How high will the ball go?

40. A rock is dropped from the top of a 400-foot cliff. Its velocity at time t seconds is $v(t) = -32t$ feet per second.

    (a) Find $s(t)$, the height of the rock above the ground at time t.

    (b) How long will it take to reach the ground?

    (c) What will be its velocity when it hits the ground?

41. Let $P(t)$ be the total output of a factory assembly line after t hours of work. Suppose that the rate of production at time t is $60 + 2t - \frac{1}{4}t^2$ units per hour. Find the formula for $P(t)$. [Hint: The rate of production is $P'(t)$, and $P(0) = 0$.]

42. After t hours of operation a coal mine is producing coal at the rate of $40 + 2t - \frac{1}{5}t^2$ tons of coal per hour. Find a formula for the total output of the coal mine after t hours of operation.

43. A package of frozen strawberries is taken from a freezer at $-5°C$ into a room at $20°C$. At time t the average temperature of the strawberries is increasing at the rate of $10e^{-.4t}$ degrees Celsius per hour. Find the temperature of the strawberries at time t.

44. A flu epidemic hits a town. Let $P(t)$ be the number of persons sick with the flu at time t, where time is measured in days from the beginning of the epidemic and $P(0) = 100$. Suppose that after t days the flu is spreading at the rate of $120t - 3t^2$ people per day. Find the formula for $P(t)$.

45. A small tie shop finds that at a sales level of x ties per day its marginal profit is $MP(x)$ dollars per tie, where $MP(x) = 1.30 + .06x - .0018x^2$. Also, the shop will lose \$95 per day at a sales level of $x = 0$. Find the profit from operating the shop at a sales level of x ties per day.

46. A soap manufacturer estimates that its marginal cost of producing soap powder is $.2x + 1$ hundred dollars per ton at a production level of x tons per day. Fixed costs are \$200 per day. Find the cost of producing x tons of soap powder per day.

47. The United States has been consuming iron ore at the rate of $C(t)$ million metric tons per year at time t, where $t = 0$ corresponds to 1980 and $C(t) = 94e^{.016t}$. Find a formula for the total U.S. consumption of iron ore from 1980 until time t.

48. The rate of production of natural and manufactured gas in the U.S. has been $P(t)$ quadrillion Btu per year at time t, with $t = 0$ corresponding to 1976 and $P(t) = 20e^{.02t}$. Find a formula for the total U.S. production of natural and manufactured gas from 1976 until time t.

1. (a) $\int (x^3 + 4x)\, dx = \frac{1}{4}x^4 + 2x^2 + C$

   (b) $\int t^{7/2}\, dt = \dfrac{1}{\frac{9}{2}}\, t^{9/2} + C = \frac{2}{9}t^{9/2} + C$

2. $R(x)$ is an antiderivative of the marginal revenue function. So

$$R(x) = \int (5 - 3x)\, dx = 5x - \tfrac{3}{2}x^2 + C$$

for some constant C. Since $R(0) = 0$,

$$0 = R(0) = 5(0) - \tfrac{3}{2}(0)^2 + C = C$$
$$C = 0.$$

Thus, $R(x) = 5x - \tfrac{3}{2}x^2$.

7.2. Definite Integrals

In order to motivate the definite integral, let us return for a few moments to Example 1 of Section 1. Recall that we were given a formula $C(t)$ for the annual rate of oil consumption at time t. From this formula we determined the function $T(t)$, which gives the total oil consumed from time 0 to time t. Recall that $t = 0$ corresponds to 1970 and that $T(t)$ is an antiderivative of $C(t)$.

Consider the following related question: How much oil was consumed between 1975 and 1980? This is easy to determine using the function $T(t)$. The year 1975 corresponds to $t = 5$, and 1980 corresponds to $t = 10$. So the oil consumed between 1975 and 1980 is

$$T(10) - T(5). \tag{1}$$

[Indeed, $T(10)$ gives the oil consumed between 1970 and 1980 and $T(5)$ gives the oil consumed during the first five of those years.] The quantity $T(10) - T(5)$ is called the *net change* of $T(t)$ over the interval $t = 5$ to $t = 10$. More generally, if $F(x)$ is any function and a, b are numbers, then the *net change of $F(x)$ over the interval $x = a$ to $x = b$* is the number

$$F(b) - F(a).$$

The quantity $F(b) - F(a)$ is often abbreviated by the symbol $F(x)\big|_a^b$. Thus, for example, the net change (1) can be written in this notation as $T(t)\big|_5^{10}$.

Since $T(t)$ is an antiderivative of $C(t)$, the quantity (1) is the net change of an antiderivative of $C(t)$ over a certain interval. This suggests the following definition.

> *The Definite Integral* Let $f(x)$ be a function and a, b be numbers. Then the definite integral of $f(x)$ over the interval from $x = a$ to $x = b$, denoted $\int_a^b f(x)\,dx$, is the net change in an antiderivative of $f(x)$ over that interval. Thus, if $F(x)$ is an antiderivative of $f(x)$, we have
>
> $$\int_a^b f(x)\,dx = F(x)\Big|_a^b = F(b) - F(a).$$

For example, the quantity (1) can be expressed as the definite integral

$$\int_5^{10} C(t)\,dt.$$

EXAMPLE 1   Evaluate

(a) $\displaystyle\int_2^5 3x^2\,dx$ (b) $\displaystyle\int_{-1}^2 e^{3x}\,dx.$

Solution   (a) An antiderivative of $3x^2$ is x^3. So

$$\int_2^5 3x^2\,dx = x^3\Big|_2^5 = 5^3 - 2^3 = 117.$$

(b) An antiderivative of e^{3x} is $\frac{1}{3}e^{3x}$. Therefore,

$$\int_{-1}^2 e^{3x}\,dx = \tfrac{1}{3}e^{3x}\Big|_{-1}^2 = \tfrac{1}{3}e^{3\cdot 2} - \tfrac{1}{3}e^{3\cdot(-1)} = \tfrac{1}{3}e^6 - \tfrac{1}{3}e^{-3}.$$

In computing the definite integral we may use *any* antiderivative of $f(x)$. The value of the definite integral does not depend on the choice of antiderivative. For instance, consider the definite integral $\int_2^5 3x^2\,dx$. The antiderivatives of $3x^2$ are all of the form $x^3 + C$, where C is any constant. Moreover

$$(x^3 + C)\Big|_2^5 = (5^3 + C) - (2^3 + C) = 5^3 - 2^3 = 117.$$

So the value of the definite integral does not depend on the choice of C. The same argument works for a general function $f(x)$ in place of $3x^2$.

EXAMPLE 2   Refer to the oil-consumption data of Example 1, Section 1. How much oil will be consumed from 1978 to 1985?

Solution   As we saw above, the total consumption during a given time interval can be represented as a definite integral over that interval. Since $t = 8$ corresponds to 1978 and $t = 15$ to 1985, the desired quantity equals

$$\int_8^{15} C(t)\,dt = \int_8^{15} 16.1e^{.07t}\,dt.$$

Since an antiderivative of $16.1e^{.07t}$ is $230e^{.07t}$, the above definite integral equals

$$230e^{.07t}\Big|_8^{15} = 230e^{.07(15)} - 230e^{.07(8)}$$

$$\approx 255 \text{ billion barrels.}$$

EXAMPLE 3 Suppose that the velocity $v(t)$ of a rocket t seconds after liftoff is $v(t) = 20t + 50$ meters per second. Determine the distance it travels during the time interval from $t = 60$ to $t = 70$.

Solution Let $s(t)$ denote the height of the rocket at time t. The desired distance is the net change $s(70) - s(60)$. Since $s(t)$ is an antiderivative of $v(t)$, this net change is a definite integral:

$$s(70) - s(60) = \int_{60}^{70} v(t)\, dt$$

$$= \int_{60}^{70} (20t + 50)\, dt$$

$$= (10t^2 + 50t)\Big|_{60}^{70}$$

$$= 13,500 \text{ meters.}$$

Thus, the rocket travels 13,500 meters during the time interval from $t = 60$ to $t = 70$.

EXAMPLE 4 The marginal cost function for a certain factory is $\frac{1}{100}x^2 - 2x + 120$ dollars per unit, where x is the total daily production. Find the net increase in cost if production is raised from 100 to 105 units per day.

Solution Let $C(x)$ denote the cost of producing x units per day. Then $C'(x) = \frac{1}{100}x^2 - 2x + 120$. Hence,

$$C(105) - C(100) = \int_{100}^{105} C'(x)\, dx$$

$$= \int_{100}^{105} \left(\frac{1}{100} x^2 - 2x + 120\right) dx$$

$$= (\tfrac{1}{300}x^3 - x^2 + 120x)\Big|_{100}^{105}$$

$$= 5433.75 - 5333.33 = 100.42.$$

Thus, if the factory is currently producing 100 units per day, it will cost an additional $100.42 to produce five additional units per day.

One of the most important applications of the definite integral is to the calculation of areas under curves.

PROBLEM Let $f(x)$ be a function that is continuous and nonnegative for all x between a and b, where $a < b$. Determine the area of the region lying beneath the curve $y = f(x)$ and above the x-axis, from $x = a$ to $x = b$.

The region referred to is the shaded region in Fig. 1. We see that the region is bounded on the left by the vertical line $x = a$ and on the right by the vertical line $x = b$. We shall refer to the area of this region simply as the *area under the curve $y = f(x)$ from $x = a$ to $x = b$*.

This problem has a rather simple (and surprising) solution.

FIGURE 1

The Fundamental Theorem of Calculus Let $f(x)$ be continuous, and nonnegative for x between a and b. Then the area under the curve $y = f(x)$ from $x = a$ to $x = b$ is given by the definite integral

$$\int_a^b f(x)\, dx.$$

We shall explain why this theorem is true after we examine two examples of its application.

EXAMPLE 5 Find the area under the curve $y = x$ from $x = 1$ to $x = 2$.

Solution The region whose area we must calculate is the shaded region in Fig. 2(a). Here $f(x) = x$, $a = 1$, and $b = 2$. The value of the definite integral is then

$$\int_1^2 x\, dx = \tfrac{1}{2}x^2 \Big|_1^2 = \tfrac{1}{2}\cdot 2^2 - \tfrac{1}{2}\cdot 1^2 = \tfrac{3}{2}.$$

Therefore the area is $\tfrac{3}{2}$.

In Example 5 it is also possible to compute the area by elementary geometry. Let us break up the region into a triangle and a square, as pictured in Fig. 2(b). The area of the triangle is $\tfrac{1}{2}\cdot$ (base) $\cdot$ (height) $= \tfrac{1}{2}\cdot 1 \cdot 1 = \tfrac{1}{2}$. The area of the square is $1 \cdot 1 = 1$. Thus the area of the region is $\tfrac{1}{2} + 1 = \tfrac{3}{2}$, the same result that we obtained by using the definite integral.

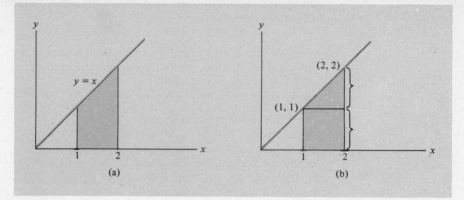

(a)

(b)

FIGURE 2

EXAMPLE 6 Compute the area under the curve $y = x^2 - 3x + 5$ from $x = 1$ to $x = 4$.

Solution Using our curve-sketching techniques ($y' = 2x - 3$, $y' = 0$ when $x = \frac{3}{2}$), we obtain the graph shown in Fig. 3. Since the function $f(x) = x^2 - 3x + 5$ is nonnegative for all x between 1 and 4, the area under the curve is given by the integral

$$\int_1^4 (x^2 - 3x + 5)\, dx = \left(\frac{x^3}{3} - \frac{3}{2}x^2 + 5x\right)\Bigg|_1^4$$

$$= \left(\frac{64}{3} - \frac{48}{2} + 20\right) - \left(\frac{1}{3} - \frac{3}{2} + 5\right) = 13\tfrac{1}{2}.$$

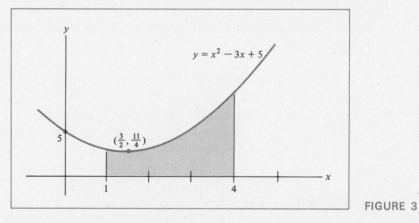

FIGURE 3

Historically, calculus centered about two problems: finding the slope of a curve at a point (differentiation) and finding the area of a region under the graph of a function. The fundamental theorem of calculus is so named because it connects these two problems. Computing an area can be accomplished by finding an antiderivative, the reverse process of differentiation.

To explain why the Fundamental Theorem of Calculus is true, we need the following theorem.

Theorem III Let $f(x)$ be a continuous nonnegative function for $a \leq x \leq b$. Let $A(x)$ be the area under the graph of the function from a to the number x. (See Fig. 4.) Then $A(x)$ is an antiderivative of $f(x)$.

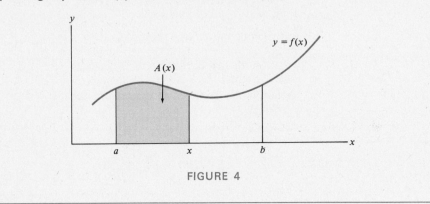

FIGURE 4

Verification of the Fundamental Theorem of Calculus Note that the "area function" $A(x)$ in Theorem III has the property that $A(a)$ is zero and $A(b)$ equals the area under the curve $y = f(x)$ from a to b. Therefore, we may conclude from Theorem III that

$$\int_a^b f(x)\, dx = A(b) - A(a) = A(b) - 0$$

$$= [\text{Area under } y = f(x) \text{ from } a \text{ to } b].$$

This is the Fundamental Theorem of Calculus.

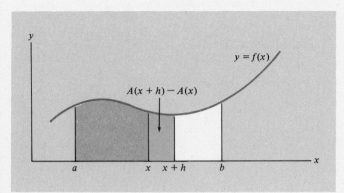

FIGURE 5

It is not difficult to explain why Theorem III is true. Let h be a small positive number. Then $A(x + h) - A(x)$ is the area of the shaded region in Fig. 5. This shaded region is approximately a rectangle of width h, height $f(x)$, and area $h \cdot f(x)$. Thus

$$A(x + h) - A(x) \approx h \cdot f(x),$$

where the approximation becomes better as h approaches zero. Dividing by h, we have

$$\frac{A(x + h) - A(x)}{h} \approx f(x).$$

Since the approximation becomes more exact as h approaches zero, the quotient must approach $f(x)$. However, the limit definition of the derivative tells us that the quotient approaches $A'(x)$ as h approaches zero. Therefore, we have $A'(x) = f(x)$. Since x represented any number between a and b, this shows that $A(x)$ is an antiderivative of $f(x)$.

PRACTICE PROBLEMS 2

1. Find the area under the curve $y = e^{x/2}$ from $x = -3$ to $x = 2$.

2. Let $C(x)$ be the cost of producing x units of a certain commodity. A common notation for the marginal cost function is $MC(x)$ (or just MC). What economic interpretation can be given to the area of the shaded region in Fig. 6?

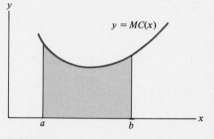

FIGURE 6

EXERCISES 2

Calculate the following definite integrals.

1. $\displaystyle\int_{-1}^{1} x\,dx$

2. $\displaystyle\int_{4}^{5} e^2\,dx$

3. $\displaystyle\int_{1}^{2} 5\,dx$

4. $\displaystyle\int_{-1}^{-1/2} \frac{1}{x^2}\,dx$

5. $\displaystyle\int_{1}^{2} 8x^3\,dx$

6. $\displaystyle\int_{0}^{1} e^{x/3}\,dx$

7. $\displaystyle\int_{0}^{1} 4e^{-3x}\,dx$

8. $\displaystyle\int_{1}^{3} \frac{5}{x}\,dx$

9. $\displaystyle\int_{1}^{4} 3\sqrt{x}\,dx$

10. $\displaystyle\int_{1}^{8} 2x^{1/3}\,dx$

11. $\displaystyle\int_{0}^{5} e^{\cdot 2t}\,dt$

12. $\displaystyle\int_{0}^{1} \frac{4}{e^{3t}}\,dt$

13. $\displaystyle\int_{3}^{6} x^{-1}\,dx$

14. $\displaystyle\int_{1}^{3} (5t-1)^3\,dt$

15. $\displaystyle\int_{-1}^{1} \frac{4}{(t+2)^3}\,dt$

16. $\displaystyle\int_{-3}^{0} \sqrt{25+3t}\,dt$

17. $\displaystyle\int_{2}^{3} (5-2t)^4\,dt$

18. $\displaystyle\int_{4}^{9} \frac{3}{t-2}\,dt$

19. $\displaystyle\int_{0}^{3} (x^3+x-7)\,dx$

20. $\displaystyle\int_{-5}^{5} (e^{x/10}-x^2-1)\,dx$

21. $\displaystyle\int_{2}^{4} \left(x^2 + \frac{2}{x^2} - \frac{1}{x+5}\right)dx$

22. $\displaystyle\int_{1}^{2} (4x^3 + 3x^{-4} - 5)\,dx$

23. Suppose that the velocity of a car at time t is $40 + 8/(t+1)^2$ kilometers per hour. How far does the car travel from $t = 1$ to $t = 9$ hours?

24. A helicopter is rising straight up in the air. Its velocity at time t is $2t + 1$ feet per second for the first five seconds. How high does it rise in that time period?

25. After t hours of operation an assembly line is assembling power lawn mowers at the rate of $21 - \frac{1}{3}t$ mowers per hour. How many mowers are assembled during the time from $t = 2$ to $t = 6$ hours?

26. Some food is placed in a freezer. After t hours the temperature of the food is dropping at a rate of $12 + 4/(t + 3)^2$ degrees Fahrenheit per hour. How many degrees will the temperature of the food change during the first two hours?

27. Suppose the marginal cost function of a handbag manufacturer is $\frac{3}{32}x^2 - x + 200$ dollars per unit at production level x (where x is measured in units of 100 handbags). Find the total cost of producing 6 additional units if 2 units are currently being produced.

28. If $C(x)$ is the cost of producing x units of some commodity, then $C(0)$ represents the fixed costs. Suppose that the marginal cost function is $C'(x) = .06x^2 - 8x + 250$ dollars per unit. Find the total cost of producing the first 100 units, not counting the fixed costs.

29. Let $P(x)$ denote the profit earned from the sale of x units of some commodity. Then $P'(x)$ is the marginal profit. Let a and b be two values of x, with $a < b$, and give an economic interpretation to the area under the marginal profit curve $y = P'(x)$ from $x = a$ to $x = b$.

30. Suppose the marginal profit function for a company is $100 + 50x - 3x^2$. Find the total profit earned from the sale of 3 additional units if 5 units are currently being produced.

Find the area under each of the given curves.

31. $y = 4x$; $x = 2$ to $x = 3$ 32. $y = 3x^2$; $x = -1$ to $x = 1$

33. $y = e^{x/2}$; $x = 0$ to $x = 1$ 34. $y = \sqrt{x}$; $x = 0$ to $x = 4$

35. $y = (x - 3)^4$; $x = 1$ to $x = 4$ 36. $y = e^{3x}$; $x = -\frac{1}{3}$ to $x = 0$

Find the area under each of the given curves by antidifferentiation and use elementary geometry to check the answer.

37. $y = 5$; $x = -1$ to $x = 2$ 38. $y = x + 1$; $x = 2$ to $x = 4$

39. $y = 2x$; $x = 0$ to $x = 3$ 40. $y = 2x + 1$; $x = 0$ to $x = 3$

41. Verify the formula

$$\ln b = \int_1^b \frac{1}{t}\, dt, \qquad b > 0.$$

42. For $x > 1$, show how $\ln x$ may be interpreted as the area under an appropriate curve $y = f(x)$. (This interpretation of $\ln x$ is used in some texts as the *definition* of the natural logarithm function.)

43. For each positive number x, let $A(x)$ be the area under the curve $y = x^2 + 1$ from 0 to x. Find $A'(3)$.

44. For each number $x > 2$, let $A(x)$ be the area under the curve $y = x^3$ from 2 to x. Find $A'(6)$.

1. The desired area is

$$\int_{-3}^{2} e^{x/2}\, dx = 2e^{x/2}\Big|_{-3}^{2} = 2e - 2e^{-3/2} \approx 4.99.$$

2. The area shaded is the area under the curve MC from $x = a$ to $x = b$. This area can be computed as the definite integral $\int_a^b MC(x)\, dx$. Since $C(x)$ is an antiderivative of $MC(x)$,

$$\int_a^b MC(x)\, dx = C(x)\Big|_a^b = C(b) - C(a).$$

So the shaded area equals the additional cost of increasing production from a units to b units.

7.3. Areas in the x-y Plane

In Section 2 we introduced the concept of the definite integral and showed how the definite integral is used to compute the area of the region lying above the x-axis and below the graph of the nonnegative function $y = f(x)$. In this section we will show how to use the definite integral to compute areas of more general regions.

Three simple but important properties of the integral will be used repeatedly.

> Let $f(x)$ and $g(x)$ be functions and a, b, k any constants. Then
>
> $$\int_a^b f(x)\, dx + \int_a^b g(x)\, dx = \int_a^b [f(x) + g(x)]\, dx \qquad (1)$$
>
> $$\int_a^b f(x)\, dx - \int_a^b g(x)\, dx = \int_a^b [f(x) - g(x)]\, dx \qquad (2)$$
>
> $$\int_a^b kf(x)\, dx = k\int_a^b f(x)\, dx. \qquad (3)$$

Verification of (1) Let $F(x)$ and $G(x)$ be antiderivatives of $f(x)$ and $g(x)$, respectively. Then $F(x) + G(x)$ is an antiderivative of $f(x) + g(x)$, and so

$$\int_a^b [f(x) + g(x)]\, dx = [F(x) + G(x)]\Big|_a^b$$
$$= [F(b) + G(b)] - [F(a) + G(a)]$$
$$= [F(b) - F(a)] + [G(b) - G(a)]$$
$$= \int_a^b f(x)\, dx + \int_a^b g(x)\, dx.$$

The verifications of (2) and (3) are similar and use the facts that $F(x) - G(x)$ is an antiderivative of $f(x) - g(x)$ and $kF(x)$ is an antiderivative of $kf(x)$.

Let us now consider regions that are bounded both above and below by graphs of functions. Referring to Fig. 1, we would like to find a simple expression for the area of the shaded region under the graph of $y = f(x)$ and above the graph of $y = g(x)$ from $x = a$ to $x = b$. It is the region under the graph of $f(x)$ with the region under the graph of $g(x)$ taken away. Therefore

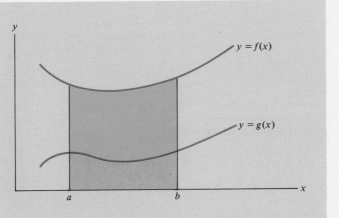

FIGURE 1

[area of shaded region] = [area under $f(x)$] − [area under $g(x)$]

$$= \int_a^b f(x)\, dx - \int_a^b g(x)\, dx$$

$$= \int_a^b [f(x) - g(x)]\, dx \quad \text{[by property (2)].}$$

Area Between Two Curves If $y = f(x)$ lies above $y = g(x)$ from $x = a$ to $x = b$, the area of the region between $f(x)$ and $g(x)$ from $x = a$ to $x = b$ is

$$\int_a^b [f(x) - g(x)]\, dx.$$

EXAMPLE 1 Find the area of the region between $y = 2x^2 - 4x + 6$ and $y = -x^2 + 2x + 1$ from $x = 1$ to $x = 2$.

Solution Upon sketching the two graphs (Fig. 2), we see that $f(x) = 2x^2 - 4x + 6$ lies above $g(x) = -x^2 + 2x + 1$ for $1 \le x \le 2$. Therefore our formula gives the

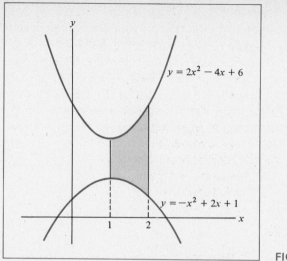

FIGURE 2

area of the shaded region as

$$\int_1^2 [(2x^2 - 4x + 6) - (-x^2 + 2x + 1)]\, dx = \int_1^2 (3x^2 - 6x + 5)\, dx$$

$$= (x^3 - 3x^2 + 5x)\Big|_1^2$$

$$= 6 - 3 = 3.$$

EXAMPLE 2 Find the area of the region between $y = x^2$ and $y = (x - 2)^2 = x^2 - 4x + 4$ from $x = 0$ to $x = 3$.

Solution Upon sketching the graphs (Fig. 3), we see that the two graphs cross; and by setting

FIGURE 3

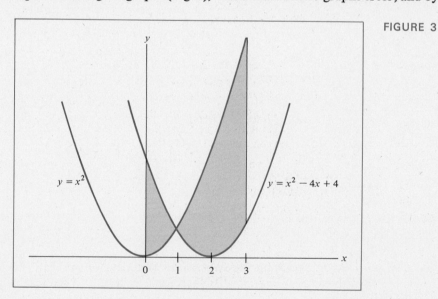

$x^2 = x^2 - 4x + 4$, we find that they cross when $x = 1$. Thus one graph does not always lie above the other from $x = 0$ to $x = 3$, so that we cannot directly apply our rule for finding the area between two curves. However, the difficulty is easily surmounted if we break the region into two parts, namely, the area from $x = 0$ to $x = 1$ and the area from $x = 1$ to $x = 3$. For from $x = 0$ to $x = 1$, $y = x^2 - 4x + 4$ is on top; from $x = 1$ to $x = 3$, $y = x^2$ is on top. Consequently,

$$[\text{Area from } x = 0 \text{ to } x = 1] = \int_0^1 [(x^2 - 4x + 4) - (x^2)]\, dx$$

$$= \int_0^1 (-4x + 4)\, dx$$

$$= (-2x^2 + 4x)\Big|_0^1$$

$$= 2 - 0 = 2.$$

$$[\text{Area from } x = 1 \text{ to } x = 3] = \int_1^3 [(x^2) - (x^2 - 4x + 4)]\, dx$$

$$= \int_1^3 (4x - 4)\, dx$$

$$= (2x^2 - 4x)\Big|_1^3$$

$$= 6 - (-2) = 8.$$

Thus the total area is $2 + 8 = 10$.

In our derivation of the formula for the area between two curves, we examined functions that are nonnegative. However, the statement of the rule does not contain this stipulation, and rightfully so. Consider the case where $f(x)$ and $g(x)$ are not always positive. Let us determine the area of the shaded region in Fig. 4(a). Select some constant c such that the graphs of the functions $f(x) + c$ and

FIGURE 4

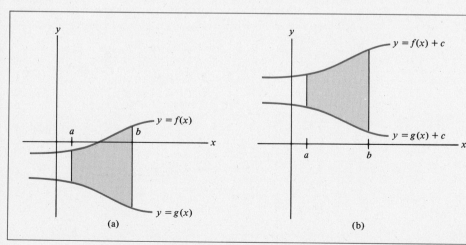

(a)

(b)

$g(x) + c$ lie completely above the x-axis [Fig. 4(b)]. The region between them will have the same area as the original region. Using the rule as applied to non-negative functions, we have

$$[\text{area of the region}] = \int_a^b [(f(x) + c) - (g(x) + c)]\, dx$$

$$= \int_a^b [f(x) - g(x)]\, dx.$$

Therefore we see that our rule is valid for any functions $f(x)$ and $g(x)$ as long as the graph of $f(x)$ lies above the graph of $g(x)$ for all x from $x = a$ to $x = b$.

EXAMPLE 3 Set up the integral that gives the area between the curves $y = x^2 - 2x$ and $y = -e^x$ from $x = -1$ to $x = 2$.

Solution Since $y = x^2 - 2x$ lies above $y = -e^x$ (Fig. 5), the rule for finding the area between two curves can be applied directly. The area between the curves is

$$\int_{-1}^2 (x^2 - 2x + e^x)\, dx.$$

Sometimes we are asked to find the area between two curves without being given the values of a and b. In these cases there is a region that is completely enclosed by the two curves. As the next examples illustrate, we must first find the

FIGURE 5

FIGURE 6

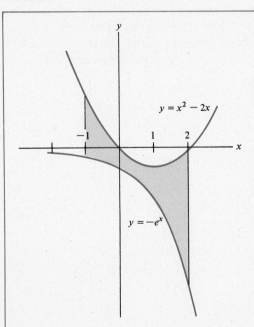

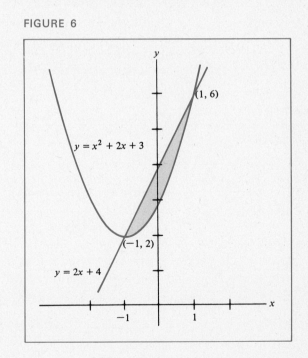

points of intersection of the two curves in order to obtain the values of a and b. In such problems careful curve sketching is especially important.

EXAMPLE 4   Set up the integral that gives the area bounded by the curves $y = x^2 + 2x + 3$ and $y = 2x + 4$.

Solution   The two curves are sketched in Fig. 6, and the region bounded by them is shaded. In order to find the points of intersection, we set $x^2 + 2x + 3 = 2x + 4$ and solve for x. We obtain $x^2 = 1$ or $x = -1$ and $x = +1$. When $x = -1$, $2x + 4 = 2(-1) + 4 = 2$. When $x = 1$, $2x + 4 = 2(1) + 4 = 6$. Thus the curves intersect at the points $(1, 6)$ and $(-1, 2)$.

Since $y = 2x + 4$ lies above $y = x^2 + 2x + 3$ from $x = -1$ to $x = 1$, the area between the curves is given by

$$\int_{-1}^{1} [(2x + 4) - (x^2 + 2x + 3)]\, dx = \int_{-1}^{1} (1 - x^2)\, dx.$$

EXAMPLE 5   Set up the integral that gives the area bounded by the two curves $y = 2x^2$ and $y = x^3 - 3x$.

Solution   First we make a rough sketch of the two curves as in Fig. 7. The curves intersect where $x^3 - 3x = 2x^2$ or $x^3 - 2x^2 - 3x = 0$. Note that $x^3 - 2x^2 - 3x = x(x^2 - 2x - 3) = x(x - 3)(x + 1)$. So the solutions to $x^3 - 2x^2 - 3x = 0$ are $x = 0, 3, -1$, and the curves intersect at $(-1, 2)$, $(0, 0)$, and $(3, 18)$. From $x = -1$ to $x = 0$, the curve $y = x^3 - 3x$ lies above $y = 2x^2$. But from $x = 0$ to $x = 3$, the reverse is true. Thus the area between the curves is given by

$$\int_{-1}^{0} (x^3 - 3x - 2x^2)\, dx + \int_{0}^{3} (2x^2 - x^3 + 3x)\, dx.$$

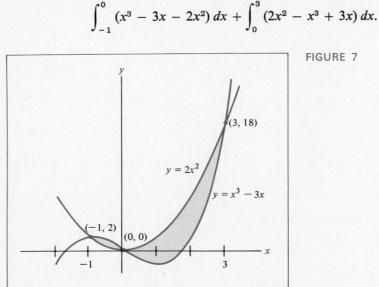

FIGURE 7

EXAMPLE 6 Beginning in 1974, with the advent of dramatically higher oil prices, the exponential rate of growth of world oil consumption slowed down from a growth constant of 7% to a growth constant of 4% per year. A fairly good model for the annual rate of oil consumption since 1974 is given by

$$C_1(t) = 21.3e^{.04(t-4)}, \quad t \geq 4,$$

where $t = 0$ corresponds to 1970. Determine the total amount of oil saved between 1976 and 1980 by not consuming oil at the rate predicted by the model of Example 1, Section 1, namely

$$C(t) = 16.1e^{.07t}, \quad t \geq 0.$$

Solution If oil consumption had continued to grow as it did prior to 1974, then the total oil consumed between 1976 and 1980 would have been

$$\int_6^{10} C(t)\, dt. \tag{3}$$

However, taking into account the slower increase in the rate of oil consumption since 1974, we find that the total oil consumed between 1976 and 1980 was approximately

$$\int_6^{10} C_1(t)\, dt. \tag{4}$$

The integrals in (3) and (4) may be interpreted as the areas under the curves $y = C(t)$ and $y = C_1(t)$, respectively, from $t = 6$ to $t = 10$. (See Fig. 8.) By

FIGURE 8

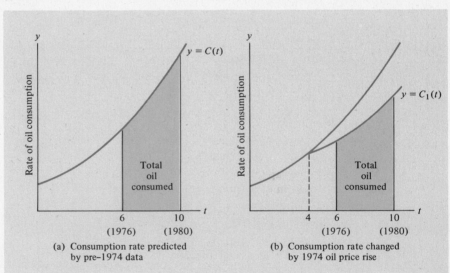

(a) Consumption rate predicted by pre-1974 data

(b) Consumption rate changed by 1974 oil price rise

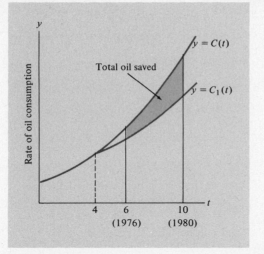

FIGURE 9

superimposing the two curves we see that the area between them from $t = 6$ to $t = 10$ represents the total oil that was saved by consuming oil at the rate given by $C_1(t)$ instead of $C(t)$. (See Fig. 9.) The area between the two curves equals

$$\int_6^{10} [C(t) - C_1(t)]\, dt = \int_6^{10} [16.1e^{.07t} - 21.3e^{.04(t-4)}]\, dt$$

$$= \left(\frac{16.1}{.07} e^{.07t} - \frac{21.3}{.04} e^{.04(t-4)} \right) \Big|_6^{10}$$

$$\approx 13.02.$$

Thus about 13 billion barrels of oil were saved between 1976 and 1980.

PRACTICE PROBLEMS 3

1. Find the area between the curves $y = x + 3$ and $y = x^2 + x - 13$ from $x = 1$ to $x = 3$.

2. A company plans to increase its production from 10 to 15 units per day. The present marginal cost function is $MC_1(x) = x^2 - 20x + 108$. By redesigning the production process and purchasing new equipment the company can change the marginal cost function to $MC_2(x) = \frac{1}{2}x^2 - 12x + 75$. Determine the area between the graphs of the two marginal cost curves from $x = 10$ to $x = 15$. Interpret this area in economic terms.

EXERCISES 3

Find the area of the region between the curves:

1. $y = 2x^2$ and $y = 8$ (a horizontal line) from $x = -2$ to $x = 2$.

2. $y = 13 - 3x^2$ and $y = 1$ from $x = -2$ to $x = 2$.

3. $y = x^2 - 6x + 12$ and $y = 1$ from $x = 0$ to $x = 4$.

4. $y = x(2 - x)$ and $y = 4$ from $x = 0$ to $x = 2$.

5. $y = x^2$ and $y = (x - 4)^2$ from $x = 0$ to $x = 3$.

6. $y = x^2$ and $y = (x + 3)^2$ from $x = -3$ to $x = 0$.

7. $y = 3x^2$ and $y = -3x^2$ from $x = -1$ to $x = 2$.

8. $y = e^{2x}$ and $y = -e^{2x}$ from $x = -1$ to $x = 1$.

Find the area of the region bounded by the curves:

9. $y = x^2 + x$ and $y = 3 - x$.

10. $y = 3x - x^2$ and $y = 4 - 2x$.

11. $y = -x^2 + 6x - 5$ and $y = 2x - 5$.

12. $y = 2x^2 + x - 7$ and $y = x + 1$.

13. Find the area bounded by $y = x^2 - 3x$ and the x-axis

    (a) between $x = 0$ and $x = 3$.

    (b) between $x = 0$ and $x = 4$.

    (c) between $x = -2$ and $x = 3$.

14. Find the area bounded by $y = x^2$ and $y = 1/x^2$

    (a) between $x = 1$ and $x = 4$.

    (b) between $x = \frac{1}{2}$ and $x = 4$.

15. Find the area bounded by $y = 1/x^2$, $y = x$, and $y = 8x$, for $x \geq 0$. (Make a careful sketch.)

16. Find the area bounded by $y = 1/x$, $y = 4x$, and $y = \frac{1}{4}x$, for $x \geq 0$.

17. Find the area bounded by $y = 12/x$, $y = \frac{3}{2}\sqrt{x}$, and $y = \frac{1}{3}x$, for $0 \leq x \leq 6$.

18. Find the area of the region in the first quadrant of the x-y plane that is bounded by $y = 12 - x^2$, $y = 4x$ and $y = x$.

19. Refer to the oil-consumption data in Example 1, Section 1. Suppose that in 1970 the growth constant for the annual rate of oil consumption had been held to .04. What effect would this action have had on oil consumption from 1970 to 1974?

20. The marginal profit function for a certain company is $MP_1(x) = -x^2 + 14x - 24$. The company expects the daily production level to rise from $x = 6$ to $x = 8$ units. The management is considering a plan that would have the effect of changing the marginal profit to $MP_2(x) = -x^2 + 12x - 20$. Should the company adopt the plan? Determine the area between the graphs of the two marginal cost functions from $x = 6$ to $x = 8$. Interpret this area in economic terms.

1. First graph the two curves, as shown in the accompanying sketch. The curve $y = x + 3$ lies on top. So the area between the curves is

$$\int_1^3 [(x + 3) - (x^2 + x - 13)]\, dx = \int_1^3 [-x^2 + 16]\, dx$$

$$= (-\tfrac{1}{3}x^3 + 16x)\Big|_1^3$$

$$= 23\tfrac{1}{3}.$$

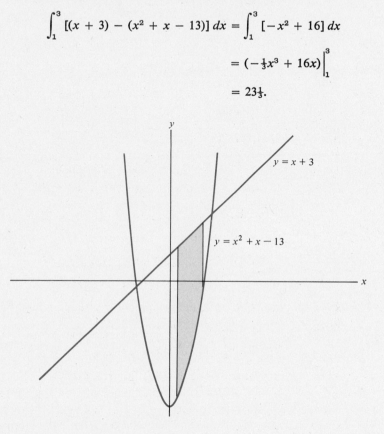

2. Graphing the two marginal cost functions yields the results shown on the next page. So the area between the curves equals

$$\int_{10}^{15} [MC_1(x) - MC_2(x)]\, dx = \int_{10}^{15} (x^2 - 20x + 108) - (\tfrac{1}{2}x^2 - 12x + 75)]\, dx$$

$$= \int_{10}^{15} [\tfrac{1}{2}x^2 - 8x + 33]\, dx$$

$$= (\tfrac{1}{6}x^3 - 4x^2 + 33x)\Big|_{10}^{15}$$

$$= 60\tfrac{5}{6}.$$

The area under a marginal cost curve is total cost. So the area between the curves represents the difference in the cost of increasing production from 10 to 15 units using the old production process versus the new production process. Using the new production process allows the company to perform the increased production for $60\frac{2}{3}$ dollars less than using the old process.

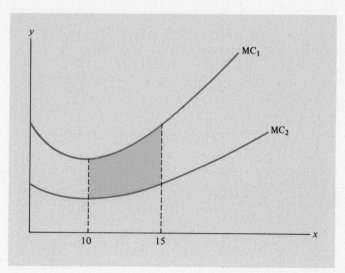

7.4. Riemann Sums

We have seen many applied problems which require us to calculate a definite integral $\int_a^b f(x)\,dx$. So far, our only technique for doing so requires us to determine an antiderivative of $f(x)$. However, as we shall now demonstrate, it is possible to numerically approximate a definite integral without first determining an anti-derivative. The numerical approximation is in terms of so-called *Riemann sums*.

Let us begin by defining Riemann sums for a continuous function $f(x)$ over the interval from $x = a$ to $x = b$. Divide the interval on the x-axis from $x = a$ to $x = b$ into n equal subintervals (Fig. 1). (Here n may be any positive integer.) Since the entire interval is of length $b - a$, each subinterval has length $(1/n)$th of the total, or $(b - a)/n$. For brevity, let us set $\Delta x = (b - a)/n$. In each subinterval choose a point (any point will do). Label these points $x_1, x_2, x_3, \ldots, x_n$. (See Fig. 1.) Now form the sum

$$f(x_1)\,\Delta x + f(x_2)\,\Delta x + \ldots + f(x_n)\,\Delta x.$$

FIGURE 1

Such a sum is called a *Riemann sum* for the function $f(x)$ over the interval from $x = a$ to $x = b$.

The significance of Riemann sums is that they approximate the definite integral $\int_a^b f(x)\,dx$. More precisely, we can state the following fact.

> Suppose that the interval from $x = a$ to $x = b$ is divided into n equal subintervals of length $\Delta x = (b - a)/n$. Let x_i be any point of the ith subinterval. Then
> $$\int_a^b f(x)\,dx \approx f(x_1)\,\Delta x + f(x_2)\,\Delta x + \ldots + f(x_n)\,\Delta x, \tag{1}$$
> where the approximation becomes exact as n gets large.

Justification of Approximation (1)

We shall justify (1) only for the case when $f(x)$ is nonnegative for x between a and b. [Our argument can easily be modified to cover general $f(x)$.] Note that by the Fundamental Theorem of Calculus the left side of (1) equals the area under the curve $y = f(x)$ from $x = a$ to $x = b$.

Let us approximate this area by the area enclosed by n rectangles, as in Fig. 2. The first rectangle has height $f(x_1)$ and width Δx, so its area is $f(x_1)\,\Delta x$. The second rectangle has height $f(x_2)$ and width Δx, so its area is $f(x_2)\,\Delta x$. And so on. Each term on the right side of (1) corresponds to the area of one of the rectangles in Fig. 2. The sum of the areas of these rectangles closely approximates the area under the curve $y = f(x)$ from $x = a$ to $x = b$. This explains why (1) is true. Moreover, as n increases, the width of the rectangles decreases and the rectangular area approximates the area under the curve more closely. This completes the justification.

The approximation (1) may be used in a number of ways. First of all, we may apply it to estimate definite integrals. For this purpose, it is often convenient to choose $x_1, x_2, \ldots, x_n$ to be the midpoints of their respective subintervals, in which case approximation (1) is called the *rectangle rule*.

FIGURE 2

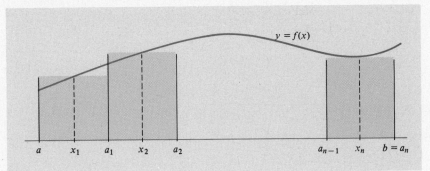

EXAMPLE 1   Approximate $\int_0^1 x^2\, dx$, using the rectangle rule with $n = 5$.

Note: We know, of course, that

$$\int_0^1 x^2\, dx = \frac{x^3}{3}\bigg|_0^1 = \frac{1}{3},$$

so the rectangle rule is not really necessary here. But this example will give us some assurance that the rectangle rule works. Also, it will give us some idea of the accuracy we obtain for a given amount of calculation.

Solution   (a) If $n = 5$, $\Delta x = (1 - 0)/5 = .2$. Therefore, $x_1 = .1$, $x_2 = .3$, $x_3 = .5$, $x_4 = .7$, and $x_5 = .9$. Thus we have

$$\int_0^1 x^2\, dx \approx (.1)^2(.2) + (.3)^2(.2) + (.5)^2(.2) + (.7)^2(.2) + (.9)^2(.2)$$

$$= [.01 + .09 + .25 + .49 + .81](.2)$$

$$= .33.$$

Consequently, we get two-place accuracy with $n = 5$. The error is $.00333\ldots$.

Approximate techniques of integration become manageable for most examples by using one of the currently available calculators. (The examples in this book were done that way.) However, for very large n (such as $n = 100$), even the minicalculator must make way for a high-speed digital computer or a programmable calculator.

The fundamental approximation (1) can be applied in a manner completely different from that described above. Namely, it may be used to express certain quantities as definite integrals. The principle behind such applications of (1) is as follows: Suppose that we are interested in finding the value of a certain quantity, call it Q. We observe that we can obtain an approximation of Q by subdividing an interval from $x = a$ to $x = b$ into n equal subintervals and forming the sum $f(x_1)\,\Delta x + \ldots + f(x_n)\,\Delta x$, where $f(x)$ is some function (not given originally) and where x_i, Δx are as before. Also, we observe that this approximation to Q becomes exact as n gets large. We then invoke our fundamental formula to conclude that $Q = \int_a^b f(x)\, dx$.

Examples 2–5 illustrate such applications of (1).

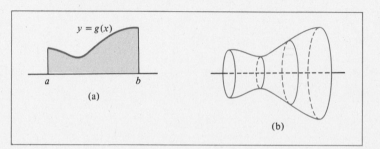

$y = g(x)$

a b

(a)

(b)

FIGURE 3

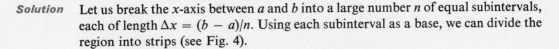

EXAMPLE 2   Revolve the region of Fig. 3(a) about the x-axis so that it sweeps out a solid [Fig. 3(b)]. Derive a formula for the volume of this *solid of revolution*.

Solution   Let us break the x-axis between a and b into a large number n of equal subintervals, each of length $\Delta x = (b - a)/n$. Using each subinterval as a base, we can divide the region into strips (see Fig. 4).

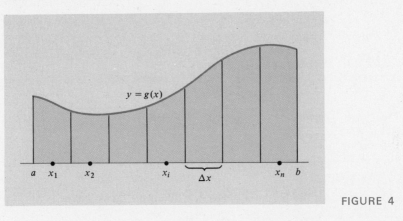

FIGURE 4

Let x_i be a point in the ith subinterval. Then the volume swept out by revolving the ith strip is approximately the same as the volume of the cylinder swept out by revolving the rectangle of height $g(x_i)$ and base Δx around the x-axis (Fig. 5). The volume of the cylinder is

$$[\text{area of circular side}] \cdot [\text{width}] = \pi[g(x_i)]^2 \cdot \Delta x.$$

The total volume swept out by all the strips is approximated by the total volume swept out by the rectangles, which is

$$\text{Volume} \approx \pi[g(x_1)]^2 \, \Delta x + \pi[g(x_2)]^2 \, \Delta x + \ldots + \pi[g(x_n)]^2 \, \Delta x.$$

As n gets larger and larger, this approximation becomes arbitrarily close to the true volume. The expression on the right is a Riemann sum with $f(x) = \pi[g(x)]^2$.

FIGURE 5

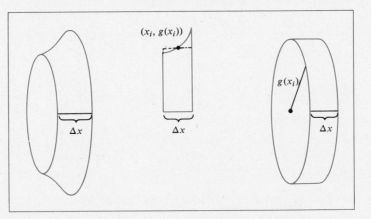

Therefore we conclude that

$$\text{Volume} = \int_a^b \pi[g(x)]^2 \, dx.$$

EXAMPLE 3   Find the volume of the solid of revolution obtained by revolving the region of Fig. 6 about the x-axis.

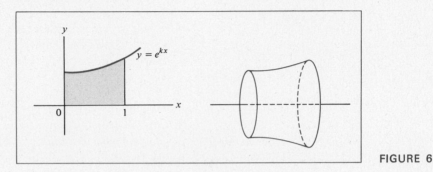

FIGURE 6

Solution   Using the formula derived in Example 2, we have

$$\text{Volume} = \int_0^1 \pi(e^{kx})^2 \, dx$$

$$= \int_0^1 \pi e^{2kx} \, dx = \frac{\pi}{2k} e^{2kx} \Big|_0^1 = \frac{\pi}{2k}(e^{2k} - 1).$$

EXAMPLE 4   Find the volume of a right circular cone of radius r and height h.

Solution   The cone [Fig. 7(a)] is the solid of revolution swept out when the shaded region in Fig. 7(b) is revolved about the x-axis. Using the formula developed in Example 2, the volume of the cone is

$$\int_0^h \pi\left(\frac{r}{h}x\right)^2 dx = \frac{\pi r^2}{h^2} \int_0^h x^2 \, dx = \frac{\pi r^2}{h^2} \frac{x^3}{3} \Big|_0^h = \frac{\pi r^2 h}{3}.$$

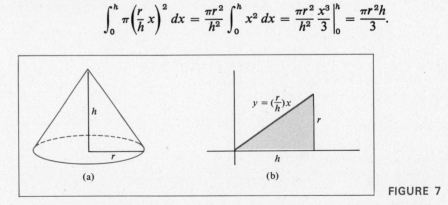

(a)   (b)

FIGURE 7

The next example provides an application of Riemann sums to economics.

EXAMPLE 5 (Consumers' Surplus) Using a demand curve of economics, derive a formula showing the amount that consumers benefit from an open system that has no price discrimination.

Solution Figure 8(a) is a *demand curve* for a commodity. It is determined by complex economic factors and gives a relationship between the quantity sold and the unit price of a commodity. Specifically, it says that, in order to sell x units, the price must be set at $f(x)$ dollars per unit. Since, for most commodities, selling larger quantities requires a lowering of the price, demand functions are usually decreasing. Interactions between supply and demand determine the amount of a quantity available. Let A designate the amount of the commodity currently available and $B = f(A)$ the current selling price.

Divide the interval from 0 to A into n subintervals, each of length $\Delta x = (A - 0)/n$, and take x_i to be the right-hand endpoint of the ith interval. Consider the first subinterval, from 0 to x_1. [See Fig. 8(b).] Suppose that only x_1 units had been available. Then the price per unit could have been set at $f(x_1)$ dollars and these x_1 units sold. Of course, at this price we could not have sold any more units. However, those people who paid $f(x_1)$ dollars had a great demand for the commodity. It is extremely valuable to them, and there is no advantage to substituting another commodity at that price. They are actually paying what the commodity is worth to them. Under price discrimination the commodity could be sold to these people at $f(x_1)$ dollars per unit, yielding (price per unit) · (number of units) = $f(x_1) \cdot (x_1) = f(x_1) \cdot \Delta x$ dollars.

After selling the first x_1 units, suppose that more units become available, so that now a total of x_2 units have been produced. Setting the price at $f(x_2)$, the remaining $x_2 - x_1 = \Delta x$ units can be sold, yielding $f(x_2) \cdot \Delta x$ dollars. Here, again, the second group of buyers would have paid as much for the commodity as it is worth to them. Continuing this process of price discrimination, the amount of money paid by the consumers would be

$$f(x_1)\,\Delta x + f(x_2)\,\Delta x + \ldots + f(x_n)\,\Delta x.$$

Taking n large, we see that this amount approaches $\int_0^A f(x)\,dx$.

FIGURE 8 Consumers' surplus.

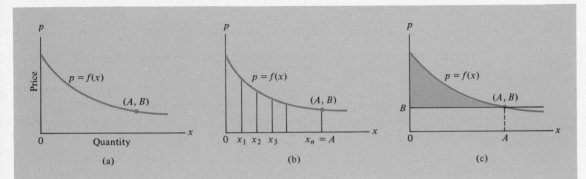

(a)   (b)   (c)

Of course, in our open system, without price discrimination, everyone pays the same price, B, and so the total amount paid is [price per unit] · [number of units] $= B \cdot A$. The amount of money saved by the consumers is

$$\int_0^A f(x)\,dx - B \cdot A = \int_0^A [f(x) - B]\,dx$$

and is called the *consumers' surplus*. Referring to Fig. 8(c), we see that the consumers' surplus is the area of the shaded region. It gives a numerical value to one benefit of a modern efficient economy.

The next example shows how the definite integral can be used to approximate the sum of a large number of terms.

EXAMPLE 6 Suppose that money is deposited daily into a savings account at an annual rate of $1000. The account pays 6% interest compounded continuously. Approximate the amount of money in the account at the end of five years.

Solution Divide the time interval from 0 to 5 years into daily subintervals. Each subinterval is then of duration $\Delta t = \frac{1}{365}$ years. Let $t_1, t_2, \ldots, t_n$ be points chosen from these subintervals. Since we deposit money at an annual rate of $1000, the amount deposited during one of the subintervals is $1000\,\Delta t$ dollars. If this amount is deposited at time t_i, the $1000\,\Delta t$ dollars will earn interest for the remaining $5 - t_i$ years. The total amount resulting from this one deposit at time t_i is then

$$1000\,\Delta t e^{.06(5-t_i)}.$$

Add up the effects of the deposits at times $t_1, t_2, \ldots, t_n$ to arrive at the total balance in the account:

$$A = 1000e^{.06(5-t_1)}\,\Delta t + 1000e^{.06(5-t_2)}\,\Delta t + \ldots + 1000e^{.06(5-t_n)}\,\Delta t.$$

This sum is a Riemann sum for the integral

$$\int_0^5 1000e^{.06(5-t)}\,dt = \frac{1000}{-.06}\,e^{.06(5-t)}\Big|_0^5$$

$$= \frac{1000}{-.06}\,(1 - e^{.3})$$

$$\approx 5831.$$

That is, the approximate balance in the account at the end of five years is $5831.

Average Value of a Function   Let $f(x)$ be a function. The definite integral may be used to define the *average value of* $f(x)$ *from* $x = a$ *to* $x = b$. To calculate the average of a collection of numbers $y_1, y_2, \ldots, y_n$, we add up the numbers and divide by n to obtain

$$\frac{y_1 + y_2 + \ldots + y_n}{n}.$$

FIGURE 9

To determine the average value of $f(x)$ from $x = a$ to $x = b$ we proceed similarly. Choose n values of x, say $x_1, x_2, \ldots, x_n$, and calculate the corresponding function values $f(x_1), f(x_2), \ldots, f(x_n)$. The average of these values is

$$\frac{f(x_1) + f(x_2) + \ldots + f(x_n)}{n}. \tag{*}$$

If the points $x_1, x_2, \ldots, x_n$ are "evenly" spread out from a to b, we would expect this sum to closely approximate the average value of $f(x)$ from $x = a$ to $x = b$. In fact, as n becomes large the average (*) should approximate the average value of $f(x)$ to any arbitrary degree of accuracy. To guarantee that the points $x_1, x_2, \ldots, x_n$ are "evenly" spread out from a to b, let us divide the interval from $x = a$ to $x = b$ into n subintervals of equal length $\Delta x = (b - a)/n$ (Fig. 9). Then choose x_1 from the first subinterval, x_2 from the second, and so forth. Thus, we see that for n large,

[average value of $f(x)$ from $x = a$ to $x = b$]

$$\approx f(x_1) \cdot \frac{1}{n} + f(x_2) \cdot \frac{1}{n} + \ldots + f(x_n) \cdot \frac{1}{n}$$

$$= \frac{1}{b - a}\left[f(x_1) \cdot \frac{b - a}{n} + f(x_2) \cdot \frac{b - a}{n} + \ldots + f(x_n) \cdot \frac{b - a}{n}\right]$$

$$= \frac{1}{b - a}[f(x_1) \Delta x + f(x_2) \Delta x + \ldots + f(x_n) \Delta x]$$

$$\approx \frac{1}{b - a}\int_a^b f(x)\,dx,$$

where the approximation becomes exact as n becomes large. Thus, we have motivated the following important definition:

The *average value* of $f(x)$ from $x = a$ to $x = b$ is defined as the quantity

$$\frac{1}{b - a}\int_a^b f(x)\,dx. \tag{2}$$

EXAMPLE 7   Find the average value of $f(x) = \sqrt{x}$ over the interval from 0 to 9.

Solution   Using (2) with $a = 0$ and $b = 9$, the average value of $f(x) = \sqrt{x}$ from $x = 0$ to $x = 9$ is

$$\frac{1}{9 - 0}\int_0^9 \sqrt{x}\,dx.$$

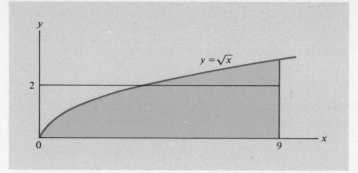

FIGURE 10

Since $\sqrt{x} = x^{1/2}$, an antiderivative of $\sqrt{x}$ is $\frac{2}{3}x^{3/2}$. Therefore

$$\frac{1}{9}\int_0^9 \sqrt{x}\,dx = \frac{1}{9}\left(\frac{2}{3}x^{3/2}\Big|_0^9\right) = \frac{1}{9}\left(\frac{2}{3}\cdot 9^{3/2} - 0\right) = \frac{1}{9}\left(\frac{2}{3}\cdot 27\right) = 2,$$

so that the average value of $\sqrt{x}$ over the interval from 0 to 9 is 2. The area of the shaded region is the same as the area of the pictured rectangle (Fig. 10).

EXAMPLE 8 Suppose that the current world population is 4.2×10^9 and the population t years from now is given by the exponential growth law

$$P(t) = (4.2 \times 10^9)e^{.023t}.$$

Find the average population of the earth during the next 30 years. (This number is important in long-range planning for agricultural production and for the allocation of goods and services.)

Solution The average value of the population $P(t)$ from $t = 0$ to $t = 30$ is

$$\frac{1}{30 - 0}\int_0^{30} P(t)\,dt = \frac{1}{30}\int_0^{30} (4.2 \times 10^9)e^{.023t}\,dt$$

$$= \left(\frac{1}{30}\right)\frac{4.2 \times 10^9}{.023}\,e^{.023t}\Big|_0^{30}$$

$$= \frac{4.2 \times 10^9}{.69}\,(e^{.69} - 1)$$

$$\approx 6 \times 10^9.$$

PRACTICE PROBLEMS 4

1. Approximate $\displaystyle\int_1^3 \frac{1}{x}\,dx$ using the rectangle rule with $n = 4$.

2. Suppose that the interval $0 \le x \le 2$ is divided into 100 subintervals, each of width Δx. Show that the sum,

    $$[3(\tfrac{1}{100})^2 - \tfrac{1}{100}]\Delta x + [3(\tfrac{3}{100})^2 - \tfrac{3}{100}]\Delta x$$
    $$+ [3(\tfrac{5}{100})^2 - \tfrac{5}{100}]\Delta x + \ldots + [3(\tfrac{199}{100})^2 - \tfrac{199}{100}]\Delta x,$$

    is close to 6.

Approximate the following integrals by the rectangle rule using the given value of n. Also, find the exact value of the integral.

1. $\displaystyle\int_0^1 x^3\,dx;\ n = 2, 4$

2. $\displaystyle\int_1^2 \frac{1}{(4x-3)^2}\,dx;\ n = 4$

3. $\displaystyle\int_0^1 e^{-x}\,dx;\ n = 5$

4. $\displaystyle\int_0^4 (x^2-5)\,dx;\ n = 2, 4$

Find the consumers' surplus for each of the following demand curves at the given sales level x.

5. $p = 3 - (x/10);\ x = 20$

6. $p = \dfrac{x^2}{200} - x + 50;\ x = 20$

7. $p = -.01x + 5;\ x = 100$

8. $p = \sqrt{16 - .02x};\ x = 350$

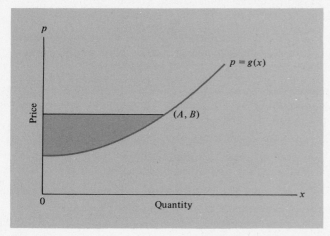

FIGURE 11   Producers' surplus.

Figure 11 shows a supply curve for a commodity. It gives the relationship between the selling price of the commodity and the quantity that producers will manufacture. At a higher selling price, a greater quantity will be produced. Therefore the curve is increasing. If (A, B) is a point on the curve, then, in order to stimulate the production of A units of the commodity, the price per unit must be B dollars. Of course, some producers will be willing to produce the commodity even with a lower selling price. Since everyone receives the same price in an open efficient economy, most producers are receiving more than their minimal required price. The excess is called the *producers' surplus*. Using an argument analogous to that of the *consumers' surplus*, one can show that the total producers' surplus when the price is B is the area of the shaded region in Fig. 11. Find the producers' surplus for each of the following supply curves at the given sales level x.

9. $p = .01x + 3;\ x = 200$

10. $p = (x^2/9) + 1;\ x = 3$

11. $p = (x/2) + 7;\ x = 10$

12. $p = 1 + \tfrac{1}{2}\sqrt{x};\ x = 36$

For a particular commodity, the quantity produced and the unit price are given by the coordinates of the point where the supply and demand curves intersect. For each pair of supply and demand curves, determine the point of intersection (A, B) and the consumers' and producers' surplus. (See Fig. 12.)

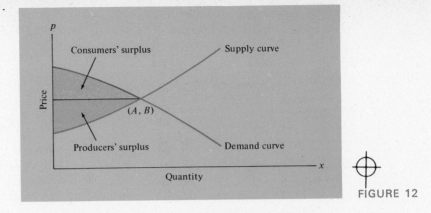

FIGURE 12

13. Demand curve: $p = 12 - (x/50)$; supply curve; $p = (x/20) + 5$.

14. Demand curve: $p = \sqrt{25 - .1x}$; supply curve: $p = \sqrt{.1x + 9} - 2$.

15. Suppose money is deposited daily into a savings account at an annual rate of $1000. If the account pays 5% interest compounded continuously, estimate the balance in the account at the end of three years.

16. Suppose money is deposited daily into a savings account at an annual rate of $2000. If the account pays 6% interest compounded continuously, approximately how much will be in the account at the end of two years?

17. Suppose money is deposited daily into a savings account at the rate of P dollars per year. The savings account pays 6% interest compounded continuously. Estimate the balance at the end of four years.

18. Suppose money is deposited daily into a savings account at the rate of P dollars per year for T years. The account pays 5% interest compounded continuously. Find a definite integral that gives the approximate balance at the end of the T years.

Find the average value of $f(x)$ over the interval from $x = a$ to $x = b$, where:

19. $f(x) = x^2$; $a = 0$, $b = 3$

20. $f(x) = e^{x/3}$; $a = 0$, $b = 3$

21. $f(x) = x^3$; $a = -1$, $b = 1$

22. $f(x) = 5$; $a = 1$, $b = 10$

23. $f(x) = 1/x^2$; $a = \frac{1}{4}$, $b = \frac{1}{2}$

24. $f(x) = 2x - 6$; $a = 2$, $b = 4$

25. During a certain 12-hour period the temperature at time t (measured in hours from the start of the period) was $47 + 4t - \frac{1}{3}t^2$ degrees. What was the average temperature during that period?

26. Assuming that a country's population is now 3 million and growing exponentially with growth constant .02, what will be the average population during the next 50 years?

27. One hundred grams of radioactive radium having a half-life of 1690 years is placed in a concrete vault. What will be the average amount of radium in the vault during the next 1000 years?

28. One hundred dollars is deposited in the bank at 5% interest compounded continuously. What will be the average value of the money in the account during the next 20 years?

29. Suppose the interval $0 \le x \le 3$ is divided into 100 subintervals of width Δx. Let $x_1, \ldots, x_{100}$ be points in these subintervals. Suppose that in some application one needs to estimate the sum

$$(3 - x_1)^2 \, \Delta x + (3 - x_2)^2 \, \Delta x + \ldots + (3 - x_{100})^2 \, \Delta x.$$

Show that this sum is close to 9.

30. Suppose the interval $0 \le t \le 3$ is divided into 1000 subintervals of width Δt. Let $t_1, t_2, \ldots, t_{1000}$ denote the right endpoints of these subintervals. Suppose that in some problem one needs to estimate the sum

$$5000e^{-.1t_1} \, \Delta t + 5000e^{-.1t_2} \, \Delta t + \ldots + 5000e^{-.1t_{1000}} \, \Delta t.$$

Show that this sum is close to 13,000. (Note: A sum such as this would arise if one wanted to compute the present value of a continuous stream of income of $5000 per year for three years, with interest compounded continuously at 10%.)

31. Suppose that the interval $0 \le x \le 1$ is divided into 100 subintervals of width $\Delta x = \frac{1}{100}$. Show that the sum

$$[2 \cdot \tfrac{1}{100} + (\tfrac{1}{100})^3] \, \Delta x + [2 \cdot \tfrac{2}{100} + (\tfrac{2}{100})^3] \, \Delta x$$
$$+ [2 \cdot \tfrac{3}{100} + (\tfrac{3}{100})^3] \, \Delta x + \ldots + [2 \cdot \tfrac{100}{100} + (\tfrac{100}{100})^3] \, \Delta x$$

is close to $\frac{5}{4}$.

32. Suppose that water is flowing into a tank at a rate of $r(t)$ gallons per hour, where the rate depends on the time t according to the formula

$$r(t) = 20 - 4t, \quad 0 \le t \le 5.$$

(a) Consider a brief period of time, say, from t_1 to t_2. The length of this time period is $\Delta t = t_2 - t_1$. During this period the rate of flow does not change much and is approximately $20 - 4t_1$ (the rate at the beginning of the brief time interval). Approximately how much water flows into the tank during the time from t_1 to t_2?

(b) Explain why the total amount of water added to the tank during the time interval from $t = 0$ to $t = 5$ is given by $\int_0^5 r(t) \, dt$.

33. Suppose the world consumes oil at the rate of $C_1(t)$ billion barrels per year at time t, where $t = 0$ corresponds to 1970 and $C_1(t) = 21.3e^{.04(t-4)}$ for $t \ge 4$. Use the rectangle rule with $n = 4$ to approximate $\int_6^{10} C_1(t) \, dt$. What interpretation can you give to each of the four terms in the approximation?

34. Find the average rate of oil consumption between 1976 and 1980, where the rate $C_1(t)$ is given in Exercise 33.

Find the volume of the solid of revolution generated by revolving about the x-axis the region under each of the following curves.

35. $y = \sqrt{r^2 - x^2}$ from $x = -r$ to $x = r$ (generates a sphere of radius r)

36. $y = r$ from $x = 0$ to $x = h$ (generates a cylinder)

37. $y = x^2$ from $x = 1$ to $x = 2$

38. $y = \dfrac{1}{x}$ from $x = 1$ to $x = 100$

39. $y = \sqrt{x}$ from $x = 0$ to $x = 4$ (The solid generated is called a *paraboloid*.)

40. $y = 2x - x^2$ from $x = 0$ to $x = 2$

41. $y = e^{-x}$ from $x = 0$ to $x = r$

42. $y = 2x + 1$ from $x = 0$ to $x = 1$ (The solid generated is called a *truncated cone*.)

SOLUTIONS TO PRACTICE PROBLEMS 4

1. Here $a = 1$, $b = 3$, $n = 4$, so $\Delta x = (3 - 1)/4 = \frac{1}{2}$. Moreover, by drawing the interval from $x = 1$ to $x = 3$, we see that $x_1 = 1.25$, $x_2 = 1.75$, $x_3 = 2.25$, $x_4 = 2.75$. Thus,

$$\int_1^3 \frac{1}{x}\,dx \approx \left[\frac{1}{1.25} + \frac{1}{1.75} + \frac{1}{2.25} + \frac{1}{2.75}\right] \cdot \frac{1}{2}$$

$$\approx 1.0898.$$

$$x_1 = 1.25 \quad x_2 = 1.75 \quad x_3 = 2.25 \quad x_4 = 2.75$$

We note that

$$\int_1^3 \frac{1}{x}\,dx = \ln x \Big|_1^3 = \ln 3.$$

To four decimal places, $\ln 3 = 1.0986$, so the error from the rectangle rule is .0088. When $n = 10$, the error is .0015.

2. Here $\Delta x = (2 - 0)/100 = .02$. The subdivision is shown in Fig. 13 where the respective midpoints of the subintervals are seen to be $x_1 = .01$, $x_2 = .03$, $x_3 = .05, \ldots, x_{100} = 1.99$. Therefore, the sum in question is

$$[3x_1^2 - x_1]\,\Delta x + [3x_2^2 - x_2]\,\Delta x + [3x_3^2 - x_3]\,\Delta x + \ldots + [3x_{100}^2 - x_{100}]\,\Delta x,$$

a Riemann sum for the function $f(x) = 3x^2 - x$ over the interval from $x = 0$ to $x = 2$. By (1), this sum is approximately equal to $\int_0^2 (3x^2 - x)\,dx$. Now,

$$\int_0^2 (3x^2 - x)\,dx = \left(x^3 - \frac{x^2}{2}\right)\Big|_0^2 = 6.$$

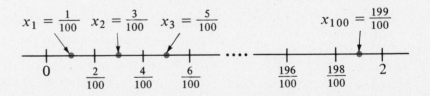

$$x_1 = \frac{1}{100} \quad x_2 = \frac{3}{100} \quad x_3 = \frac{5}{100} \qquad\qquad x_{100} = \frac{199}{100}$$

Chapter 7: CHECKLIST

- □ Antiderivative
- □ Indefinite integral
- □ $\displaystyle\int x^r \, dx = \frac{1}{r+1} x^{r+1} + C, \qquad r \neq -1$
- □ $\displaystyle\int e^{kx} \, dx = \frac{1}{k} e^{kx} + C, \qquad k \neq 0$
- □ $\displaystyle\int \frac{1}{x} \, dx = \ln|x| + C, \qquad x \neq 0$
- □ $\displaystyle\int [f(x) + g(x)] \, dx = \int f(x) \, dx + \int g(x) \, dx$
- □ $\displaystyle\int kf(x) \, dx = k \int f(x) \, dx$
- □ Net change: $F(x)\big|_a^b$
- □ $\displaystyle\int_a^b f(x) \, dx$
- □ Fundamental theorem of calculus
- □ Area between two curves: $\displaystyle\int_a^b [f(x) - g(x)] \, dx$
- □ Riemann sum approximation to $\displaystyle\int_a^b f(x) \, dx$
- □ Rectangle rule
- □ Average value of $f(x)$: $\displaystyle\frac{1}{b-a} \int_a^b f(x) \, dx$

Chapter 7: SUPPLEMENTARY EXERCISES

Calculate the following integrals.

1. $\displaystyle\int e^{-x/2} \, dx$

2. $\displaystyle\int \frac{5}{\sqrt{x-7}} \, dx$

3. $\displaystyle\int (3x^4 - 4x^3) \, dx$

4. $\displaystyle\int (2x + 3)^7 \, dx$

5. $\displaystyle\int \sqrt{4 - x} \, dx$

6. $\displaystyle\int \left(\frac{5}{x} - \frac{x}{5}\right) dx$

7. $\displaystyle\int_1^4 \frac{1}{x^2} \, dx$

8. $\displaystyle\int_3^6 e^{2-(x/3)} \, dx$

9. $\displaystyle\int_0^5 (5 + 3x)^{-1} \, dx$

10. Find the area under the curve $y = 1 + \sqrt{x}$ from $x = 1$ to $x = 9$.

11. Find the area under the curve $y = (3x - 2)^{-3}$ from $x = 1$ to $x = 2$.

12. Find the area of the region bounded by the curves $y = 16 - x^2$ and $y = 10 - x$.

13. Find the area of the region bounded by the curves $y = x^3 - 3x + 1$ and $y = x + 1$.

14. Find the area of the region between the curves $y = 2x^2 + x$ and $y = x^2 + 2$ from $x = 0$ to $x = 2$.

15. Find the function $f(x)$ for which $f'(x) = (x - 5)^2, f(8) = 2$.

16. Find the function $f(x)$ for which $f'(x) = e^{-5x}, f(0) = 1$.

17. Let k be a constant, and let $y = f(t)$ be a function such that $y' = kty$. Show that $y = Ce^{kt^2/2}$, for some constant C. [*Hint*: Use the product rule to evaluate $d/dt\,[f(t)e^{-kt^2/2}]$, and then apply Theorem II of Section 1.]

18. Describe all solutions of the following differential equations, where y represents a function of t.

    (a) $y' = 4t$ (b) $y' = 4y$ (c) $y' = e^{4t}$

19. An airplane tire plant finds that its marginal cost of producing tires is $.04x + 150$ dollars at a production level of x tires per day. If fixed costs are \$500 per day, find the cost of producing x tires per day.

20. Suppose the marginal revenue function for a company is $400 - 3x^2$. Find the additional revenue received from doubling production if currently 10 units are being produced.

21. A drug is injected into a patient at the rate of $f(t)$ cubic centimeters per minute at time t. What does the area under the graph of $y = f(t)$ from $t = 0$ to $t = 4$ represent?

22. The rate of production of copper ore in North America has been $2.6e^{.034t}$ million metric tons per year at time t, where $t = 0$ corresponds to 1975. Find a formula for the total production from 1980 to time t.

23. The known U.S. natural gas reserves are decreasing at the rate of $R(t)$ quadrillion Btu per year at time t, where $t = 0$ corresponds to 1980 and $R(t) = 22.3e^{-.14t}$.

    If the known reserves in 1980 are 159 quadrillion Btu, what will be the reserves in 1986, assuming no new gas sources are discovered?

24. The United States has been consuming sulfur at the rate of $C(t)$ million long tons per year at time t, where $t = 0$ corresponds to 1980 and $C(t) = 12.4e^{.035t}$.

    (a) How much sulfur will the U.S. consume between 1981 and 1985?

    (b) How much sulfur did the U.S. consume between 1975 and 1980?

25. Use the rectangle rule with $n = 2$ to approximate $\displaystyle\int_2^4 \frac{1}{x + 2}\, dx$. Then find the exact value to five decimal places.

26. Use the rectangle rule with $n = 5$ to approximate $\displaystyle\int_0^1 e^{x^2}\, dx$.

27. Find the consumer's surplus for the demand curve $p = \sqrt{25 - .04x}$ at the sales level $x = 400$.

28. Three thousand dollars is deposited in the bank at 6% interest compounded continuously. What will be the average value of the money in the account during the next 10 years?

29. Find the average value of $f(x) = 1/x^3$ from $x = \frac{1}{3}$ to $x = \frac{1}{2}$.

30. Suppose that the interval $0 \leq x \leq 1$ is divided into 100 subintervals of width $\Delta x = .01$. Show that the sum $[3e^{-.01}]\Delta x + [3e^{-.02}]\Delta x + [3e^{-.03}]\Delta x + \dots + [3e^{-1}]\Delta x$ is close to $3(1 - e^{-1})$.

31. Use an appropriate definite integral to estimate to two decimal places the value of the following sum.

$$\left(e^2 - \frac{3}{2}\right)(.01) + \left(e^{2.01} - \frac{3}{2.01}\right)(.01) + \left(e^{2.02} - \frac{3}{2.02}\right)(.01) + \dots$$

$$+ \left(e^{2.99} - \frac{3}{2.99}\right)(.01)$$

32. Find the volume of the solid of revolution generated by revolving about the x-axis the region under the curve $y = 1 - x^2$ from $x = 0$ to $x = 1$.

33. A store has an inventory of Q units of a certain product at time $t = 0$. The store sells the product at the steady rate of Q/A units per week, and exhausts the inventory in A weeks.
    (a) Find a formula $f(t)$ for the amount of product in inventory at time t.
    (b) Find the average inventory level during the period $0 \leq t \leq A$.

34. A retail store sells a certain product at the rate of $g(t)$ units per week at time t, where $g(t) = rt$. At time $t = 0$, the store has Q units of the product in inventory.
    (a) Find a formula $f(t)$ for the amount of product in inventory at time t.
    (b) Determine the value of r in (a) such that the inventory is exhausted in A weeks.
    (c) Using $f(t)$, with r as in (b), find the average inventory level during the period $0 \leq t \leq A$.

35. Let x be any positive number, and define $g(x)$ to be the number determined by the definite integral

$$g(x) = \int_0^x \frac{1}{1 + t^2}\, dt$$

    (a) Give a geometric interpretation of the number $g(3)$.
    (b) Find the derivative $g'(x)$.

36. For each number x satisfying $-1 \leq x \leq 1$, define $h(x)$ by

$$h(x) = \int_{-1}^x \sqrt{1 - t^2}\, dt$$

    (a) Give a geometric interpretation of the values $h(0)$ and $h(1)$.
    (b) Find the derivative $h'(x)$.

chapter 8

FUNCTIONS OF SEVERAL VARIABLES

8.1. Examples of Functions of Several Variables

In Chapters 1 through 7 we were concerned with calculus for functions of one variable. In the present chapter we briefly show how calculus can be extended to functions of several variables.

A function $f(x, y)$ of the two variables x and y is a rule that associates to each set of values for the variables a number. Examples of functions of two variables are

$$f(x, y) = x^2 + 2y$$
$$g(x, y) = e^x(x + y).$$

An example of a function of three variables is

$$f(x, y, z) = 5xyz.$$

It is possible to consider functions of any number of variables.

EXAMPLE 1 Let $f(x, y) = xy + 5y - x^2$.

(a) Calculate $f(2, 3)$. That is, calculate the value of $f(x, y)$ when $x = 2$ and $y = 3$.

(b) Calculate $f(1, 0)$ and $f(0, 1)$.

(c) Calculate $f(x, y)$ when $(x, y) = (-4, 6)$—that is, when $x = -4$ and $y = 6$.

Solution

(a) $f(2, 3) = 2 \cdot 3 + 5 \cdot 3 - (2)^2 = 17$.

(b) $f(1, 0) = 1 \cdot 0 + 5 \cdot 0 - 1^2 = -1$.

$f(0, 1) = 0 \cdot 1 + 5 \cdot 1 - 0^2 = 5$.

(c) $f(-4, 6) = (-4) \cdot 6 + 5 \cdot 6 - (-4)^2 = -10$.

A function $f(x, y)$ of two variables may be graphed in a manner analogous to that for functions of one variable. It is necessary to use a three-dimensional coordinate system, where each point is identified by three coordinates (x, y, z). For each choice of x, y, the graph of $f(x, y)$ includes the point $(x, y, f(x, y))$. The graph of $f(x, y)$ is thus a surface in three-dimensional space. (See Fig. 1.) We will not employ graphical analysis to study $f(x, y)$ for two reasons. First, it is fairly difficult to draw good graphs in three-dimensional space. This fact makes graphical analysis inconvenient for those not artistically inclined. (See Fig. 2.*) Second, and

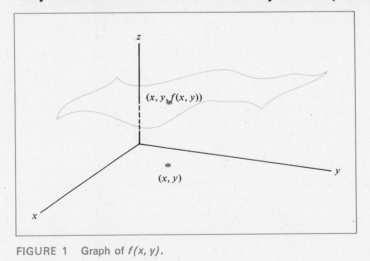

FIGURE 1   Graph of $f(x, y)$.

---

*These graphs were drawn by Norton Starr at the University of Waterloo Computing Centre.

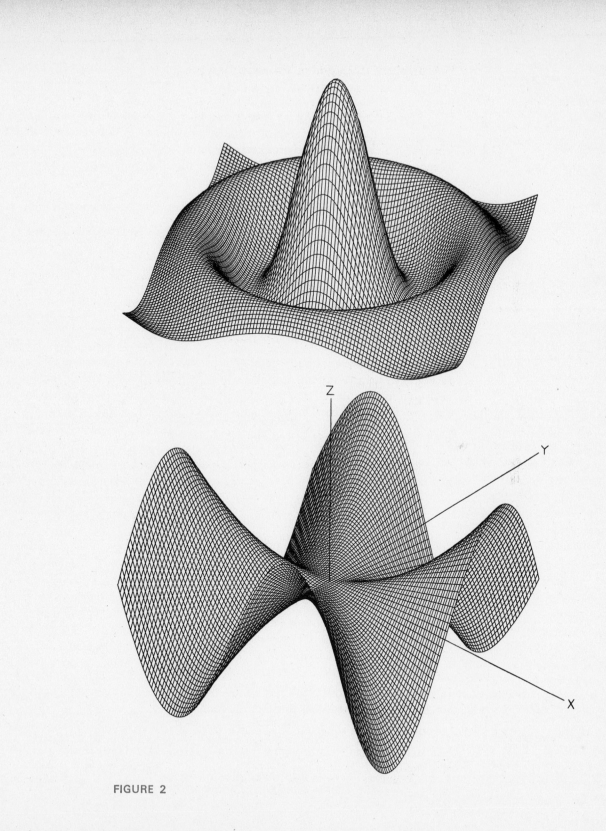

FIGURE 2

more significantly, however, we will not use graphical analysis because it is a tool that exists only for functions of one or two variables. It is not possible to graph a function of three or more variables.

Throughout this book we have shown how functions of one variable arise in applications. Functions of several variables are equally important. In Example 2 we present an application to architectural design and in Example 3 an application to economics. These two examples will recur throughout the chapter to illustrate the various techniques that will be developed.

In designing a building, it is necessary to know, at least approximately, how much heat the building loses per day. The heat loss affects many aspects of the design, such as the size of the heating plant, the size and location of duct work, and so on. A building loses heat through its sides, roof, and floor. How much heat is lost will generally differ for each face of the building and will depend on such factors as insulation, materials used in construction, exposure (north, south, east, or west), and climate. It is possible to estimate how much heat is lost per square foot of each face. Using this data, one can construct a heat-loss function as in the following example.

EXAMPLE 2   A rectangular industrial building of dimensions x, y, and z is shown in Fig. 3(a). In Fig. 3(b) we give the amount of heat lost per day by each side of the building, measured in suitable units of heat per square foot. Let $f(x, y, z)$ be the total daily heat loss for such a building.

(a) Find a formula for $f(x, y, z)$.

(b) Find the total daily heat loss if the building has length 100 feet, width 70 feet, and height 50 feet.

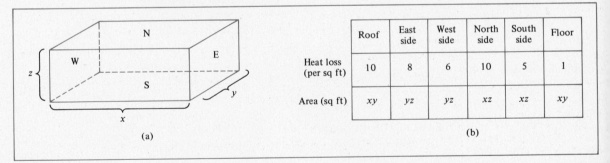

	Roof	East side	West side	North side	South side	Floor
Heat loss (per sq ft)	10	8	6	10	5	1
Area (sq ft)	xy	yz	yz	xz	xz	xy

(a)   (b)

FIGURE 3

Solution   (a) The total heat loss is the sum of the amount of heat loss through each side of the building. The heat loss through the roof is

[heat loss per square foot of roof] · [area of roof in square feet] $= 10xy$.

Similarly, the heat loss through the east side is $8yz$. Continuing in this way, we see that the total daily heat loss is

$$f(x, y, z) = 10xy + 8yz + 6yz + 10xz + 5xz + 1 \cdot xy.$$

We collect terms to obtain

$$f(x, y, z) = 11xy + 14yz + 15xz.$$

(b) The amount of heat loss when $x = 100$, $y = 70$, and $z = 50$ is given by $f(100, 70, 50)$, which equals

$$f(100, 70, 50) = 11(100)(70) + 14(70)(50) + 15(100)(50)$$

$$= 77,000 + 49,000 + 75,000 = 201,000.$$

In Section 3 we will determine the dimensions x, y, z that minimize the heat loss for a building of specified volume.

The costs of a manufacturing process can generally be classified as one of two types: cost of labor and cost of capital. The meaning of the cost of labor is clear. By the cost of capital, we mean the cost of buildings, tools, machines, and similar items used in the production process. A manufacturer usually has some control over the relative portions of labor and capital utilized in his production process. He can completely automate production so that labor is at a minimum, or he can utilize mostly labor and little capital. Suppose that x units of labor and y units of capital are used.* Let $f(x, y)$ denote the number of units of finished product that are manufactured. Economists have found that $f(x, y)$ is often a function of the form

$$f(x, y) = Cx^A y^{1-A},$$

where A and C are constants, $0 < A < 1$. Such a function is called a *Cobb-Douglas production function*.

EXAMPLE 3 (Production in a Firm) Suppose that during a certain time period the number of units of goods produced when utilizing x units of labor and y units of capital is $f(x, y) = 60x^{3/4}y^{1/4}$.

(a) How many units of goods will be produced by using 81 units of labor and 16 units of capital?

(b) Show that whenever the amounts of labor and capital being used are doubled, so is the production. (Economists say that the production function has "constant returns to scale.")

Solution (a) $f(81, 16) = 60(81)^{3/4} \cdot (16)^{1/4} = 60 \cdot 27 \cdot 2 = 3240$. There will be 3240 units of goods produced.

(b) Utilization of a units of labor and b units of capital results in the production of $f(a, b) = 60a^{3/4}b^{1/4}$ units of goods. Utilizing $2a$ and $2b$ units of labor and capital,

---

* Economists normally use L and K, respectively, for labor and capital. However, for simplicity, we use x and y.

respectively, results in $f(2a, 2b)$ units produced. Set $x = 2a$ and $y = 2b$. Then we see that

$$f(2a, 2b) = 60(2a)^{3/4}(2b)^{1/4}$$

$$= 60 \cdot 2^{3/4} \cdot a^{3/4} \cdot 2^{1/4} \cdot b^{1/4}$$

$$= 60 \cdot 2^{(3/4 + 1/4)} \cdot a^{3/4}b^{1/4}$$

$$= 2^1 \cdot 60a^{3/4}b^{1/4}$$

$$= 2f(a, b).$$

PRACTICE PROBLEMS 1

1. Let $f(x, y, z) = x^2 + y/(x - z) - 4$. Compute $f(3, 5, 2)$.

2. Suppose that in a certain country the daily demand for coffee is given by $f(p_1, p_2) = 16p_1/p_2$ thousand pounds, where p_1 and p_2 are the respective prices of tea and coffee per pound. Compute and interpret $f(3, 4)$.

EXERCISES 1

1. Let $f(x, y) = x^2 + 8y$. Compute $f(1, 0)$, $f(0, 1)$, $f(3, 2)$.

2. Let $g(x, y) = 3xe^y$. Compute $g(2, 1)$, $g(1, 0)$, $g(0, 0)$.

3. Let $f(L, K) = 3\sqrt{LK}$. Compute $f(0, 1)$, $f(3, 12)$, $f(a, b)$.

4. Let $f(p, q) = pe^{q/p}$. Compute $f(1, 0)$, $f(3, 12)$, $f(a, b)$.

5. Let $f(x, y, z) = x/(y - z)$. Compute $f(2, 3, 4)$, $f(7, 46, 44)$.

6. Let $f(x, y, z) = x^2 e^{\sqrt{y^2 + z^2}}$. Compute $f(1, 0, 1)$, $f(5, 2, 3)$.

7. Let $f(x, y) = xy$. Show that $f(2 + h, 3) - f(2, 3) = 3h$.

8. Let $f(x, y) = xy$. Show that $f(2, 3 + K) - f(2, 3) = 2K$.

9. Let $f(x, y) = \dfrac{x^2 + 3xy + 3y^2}{x + y}$. Show that $f(2a, 2b) = 2f(a, b)$.

10. Let $f(x, y) = 75x^A y^{1-A}$, where $0 < A < 1$. Show that $f(2a, 2b) = 2f(a, b)$.

11. The present value of A dollars to be paid t years in the future (assuming a 5% continuous interest rate) is $P(A, t) = Ae^{-.05t}$. Find and interpret $P(100, 13.8)$.

12. Refer to Example 3. Suppose that labor costs $100 per unit and capital costs $200 per unit. Express as a function of two variables, $C(x, y)$, the cost of utilizing x units of labor and y units of capital.

1. Substitute 3 for x, 5 for y, and 2 for z.

$$f(3, 5, 2) = 3^2 + \frac{5}{3 - 2} - 4 = 10.$$

2. To compute $f(3, 4)$, substitute 3 for p_1 and 4 for p_2 into $f(p_1, p_2) = 16p_1/p_2$. Thus,

$$f(3, 4) = 16 \cdot \tfrac{3}{4} = 12.$$

Therefore, if the price of tea is \$3 per pound and the price of coffee is \$4 per pound, then 12,000 pounds of coffee will be sold each day. (Notice that as the price of coffee increases, the demand decreases.)

8.2. Partial Derivatives

In Chapter 1 we introduced the notion of a derivative to measure the rate at which a function $f(x)$ is changing with respect to changes in the variable x. Let us now study the analog of the derivative for functions of two (or more) variables.

Let $f(x, y)$ be a function of the two variables x and y. Since we want to know how $f(x, y)$ changes both with respect to changes in the variable x and changes in the variable y, we shall define two derivatives of $f(x, y)$ (to be called "partial derivatives"), one with respect to each of the variables. The *partial derivative of* $f(x, y)$ *with respect to* x, written $\dfrac{\partial f}{\partial x}$, is the derivative of $f(x, y)$, where y is treated as a constant and $f(x, y)$ is considered as a function of x alone. The *partial derivative of* $f(x, y)$ *with respect to* y, written $\dfrac{\partial f}{\partial y}$, is the derivative of $f(x, y)$, where x is treated as a constant.

EXAMPLE 1   Let $f(x, y) = 5x^3y^2$. Compute $\dfrac{\partial f}{\partial x}$ and $\dfrac{\partial f}{\partial y}$.

Solution   To compute $\dfrac{\partial f}{\partial x}$, we think of $f(x, y)$ written as

$$f(x, y) = [5y^2]x^3,$$

where the brackets emphasize that $5y^2$ is to be treated as a constant. Therefore when differentiating with respect to x, $f(x, y)$ is just a constant times x^3. Recall that if k is any constant, then

$$\frac{d}{dx}(kx^3) = 3 \cdot k \cdot x^2.$$

Thus

$$\frac{\partial f}{\partial x} = 3 \cdot [5y^2] \cdot x^2 = 15x^2y^2.$$

After some practice, it is unnecessary to place the y^2 in front of the x^3 before differentiating.

Now, in order to compute $\frac{\partial f}{\partial y}$, we think of

$$f(x, y) = [5x^3]y^2.$$

When differentiating with respect to y, $f(x, y)$ is simply a constant, $5x^3$, times y^2.. Hence

$$\frac{\partial f}{\partial y} = 2 \cdot [5x^3] \cdot y = 10x^3y.$$

EXAMPLE 2   Let $f(x, y) = 3x^2 + 2xy + 5y$. Compute $\frac{\partial f}{\partial x}$ and $\frac{\partial f}{\partial y}$.

Solution   To compute $\frac{\partial f}{\partial x}$, we think of

$$f(x, y) = 3x^2 + [2y]x + [5y].$$

Now we differentiate $f(x, y)$ as if it were a quadratic polynomial in x:

$$\frac{\partial f}{\partial x} = 6x + [2y] + 0 = 6x + 2y.$$

Note that $5y$ is treated as a constant when differentiating with respect to x, so the partial derivative of $5y$ with respect to x is zero.

To compute $\frac{\partial f}{\partial y}$, we think of

$$f(x, y) = [3x^2] + [2x]y + 5y.$$

Then

$$\frac{\partial f}{\partial y} = 0 + [2x] + 5 = 2x + 5.$$

Note that $3x^2$ is treated as a constant when differentiating with respect to y, so the partial derivative of $3x^2$ with respect to y is zero.

EXAMPLE 3   Compute $\frac{\partial f}{\partial x}$ and $\frac{\partial f}{\partial y}$ for each of the following:

(a) $f(x, y) = (4x + 3y - 5)^8$

(b) $f(x, y) = e^{xy^2}$

(c) $f(x, y) = y/(x + 3y)$.

Solution   (a) To compute $\frac{\partial f}{\partial x}$, we think of

$$f(x, y) = (4x + [3y - 5])^8.$$

By the general power rule,

$$\frac{\partial f}{\partial x} = 8 \cdot (4x + [3y - 5])^7 \cdot 4 = 32(4x + 3y - 5)^7.$$

Here we used the fact that the derivative of $4x + 3y - 5$ with respect to x is just 4.
To compute $\frac{\partial f}{\partial y}$, we think of

$$f(x, y) = ([4x] + 3y - 5)^8.$$

Then

$$\frac{\partial f}{\partial y} = 8 \cdot ([4x] + 3y - 5)^7 \cdot 3 = 24(4x + 3y - 5)^7.$$

(b) To compute $\frac{\partial f}{\partial x}$, we observe that

$$f(x, y) = e^{x[y^2]},$$

so that

$$\frac{\partial f}{\partial x} = [y^2]e^{x[y^2]} = y^2 e^{xy^2}.$$

To compute $\frac{\partial f}{\partial y}$, we think of

$$f(x, y) = e^{[x]y^2}.$$

Thus

$$\frac{\partial f}{\partial y} = e^{[x]y^2} \cdot 2[x]y = 2xye^{xy^2}.$$

(c) To compute $\frac{\partial f}{\partial x}$, we use the general power rule to differentiate $[y](x + [3y])^{-1}$ with respect to x:

$$\frac{\partial f}{\partial x} = (-1) \cdot [y](x + [3y])^{-2} \cdot 1 = -\frac{y}{(x + 3y)^2}.$$

To compute $\frac{\partial f}{\partial y}$, we use the quotient rule to differentiate

$$f(x, y) = \frac{y}{[x] + 3y}$$

with respect to y. We find that

$$\frac{\partial f}{\partial y} = \frac{([x] + 3y) \cdot 1 - y \cdot 3}{([x] + 3y)^2} = \frac{x}{(x + 3y)^2}.$$

The use of brackets to highlight constants is helpful initially in order to compute partial derivatives. From now on we shall merely form a mental picture of those terms to be treated as constants and dispense with brackets.

A partial derivative of a function of several variables is also a function of several variables and hence can be evaluated at specific values of the variables. We write

$$\frac{\partial f}{\partial x}(a, b)$$

for $\frac{\partial f}{\partial x}$ evaluated at $x = a$, $y = b$. Similarly,

$$\frac{\partial f}{\partial y}(a, b)$$

denotes the function $\frac{\partial f}{\partial y}$ evaluated at $x = a$, $y = b$.

EXAMPLE 4 Let $f(x, y) = 3x^2 + 2xy + 5y$.

(a) Calculate $\frac{\partial f}{\partial x}(1, 4)$.

(b) Evaluate $\frac{\partial f}{\partial y}$ at $(x, y) = (1, 4)$.

Solution (a) $\frac{\partial f}{\partial x} = 6x + 2y$, $\frac{\partial f}{\partial x}(1, 4) = 6 \cdot 1 + 2 \cdot 4 = 14$.

(b) $\frac{\partial f}{\partial y} = 2x + 5$, $\frac{\partial f}{\partial y}(1, 4) = 2 \cdot 1 + 5 = 7$.

Since $\frac{\partial f}{\partial x}$ is simply the ordinary derivative with y held constant, $\frac{\partial f}{\partial x}$ gives the rate of change of $f(x, y)$ with respect to x for y held constant. In other words, keeping y constant and increasing x by one (small) unit produces a change in $f(x, y)$ that is approximately given by $\frac{\partial f}{\partial x}$. An analogous interpretation holds for $\frac{\partial f}{\partial y}$.

EXAMPLE 5 Interpret the partial derivatives of $f(x, y) = 3x^2 + 2xy + 5y$ calculated in Example 4.

Solution We showed in Example 4 that

$$\frac{\partial f}{\partial x}(1, 4) = 14, \qquad \frac{\partial f}{\partial y}(1, 4) = 7.$$

The fact that $\frac{\partial f}{\partial x}(1, 4) = 14$ means that if y is kept constant at 4 and x is allowed to vary near 1, then $f(x, y)$ changes at a rate equal to 14 times the change in x.

That is, if x increases by one small unit, then $f(x, y)$ increases by approximately 14 units. If x increases by h units (h small), then $f(x, y)$ increases approximately $14 \cdot h$ units. That is, we have

$$f(1 + h, 4) - f(1, 4) \approx 14 \cdot h.$$

Similarly, the fact that $\dfrac{\partial f}{\partial y}(1, 4) = 7$ means that if we keep x constant at 1 and let y vary near 4, then $f(x, y)$ changes at a rate equal to seven times the change in y. So for a small value of k, we have

$$f(1, 4 + k) - f(1, 4) \approx 7 \cdot k.$$

We can generalize the interpretations of $\dfrac{\partial f}{\partial x}$ and $\dfrac{\partial f}{\partial y}$ given in Example 5 to yield the following general fact.

> Let $f(x, y)$ be a function of two variables. Then if h and k are small, we have
>
> $$f(a + h, b) - f(a, b) \approx \frac{\partial f}{\partial x}(a, b) \cdot h$$
>
> $$f(a, b + k) - f(a, b) \approx \frac{\partial f}{\partial y}(a, b) \cdot k.$$

Partial derivatives can be computed for functions of any number of variables. When taking the partial with respect to one variable, we treat the other variables as constants.

EXAMPLE 6 Let $f(x, y, z) = x^2yz - 3z$.

(a) Compute $\dfrac{\partial f}{\partial x}, \dfrac{\partial f}{\partial y}$, and $\dfrac{\partial f}{\partial z}$.

(b) Calculate $\dfrac{\partial f}{\partial z}(2, 3, 1)$.

Solution (a) $\dfrac{\partial f}{\partial x} = 2xyz, \qquad \dfrac{\partial f}{\partial y} = x^2z, \qquad \dfrac{\partial f}{\partial z} = x^2y - 3$.

(b) $\dfrac{\partial f}{\partial z}(2, 3, 1) = 2^2 \cdot 3 - 3 = 12 - 3 = 9$.

EXAMPLE 7 Let $f(x, y, z)$ be the heat loss function computed in Example 2 of Section 1. That is, $f(x, y, z) = 11xy + 14yz + 15xz$. Calculate and interpret $\dfrac{\partial f}{\partial x}(10, 7, 5)$.

Solution   We have

$$\frac{\partial f}{\partial x} = 11y + 15z$$

$$\frac{\partial f}{\partial x}(10, 7, 5) = 11 \cdot 7 + 15 \cdot 5 = 77 + 75 = 152.$$

The quantity $\frac{\partial f}{\partial x}$ is commonly referred to as the *marginal heat loss with respect to change in x*. Specifically, if x is changed from 10 by h units (where h is small) and the values of y and z remain fixed at 7 and 5, then the amount of heat loss will change by approximately $152 \cdot h$ units.

EXAMPLE 8   (Production)   Consider the production function $f(x, y) = 60x^{3/4}y^{1/4}$, which gives the number of units of goods produced when utilizing x units of labor and y units of capital.

(a) Find $\frac{\partial f}{\partial x}$ and $\frac{\partial f}{\partial y}$.

(b) Evaluate $\frac{\partial f}{\partial x}$ and $\frac{\partial f}{\partial y}$ at $x = 81$, $y = 16$.

(c) Interpret the numbers computed in part (b).

Solution   (a) $\frac{\partial f}{\partial x} = 60 \cdot \frac{3}{4} x^{-1/4} \cdot y^{1/4} = 45x^{-1/4}y^{1/4} = 45\frac{y^{1/4}}{x^{1/4}}$

$$\frac{\partial f}{\partial y} = 60 \cdot \frac{1}{4} x^{3/4}y^{-3/4} = 15x^{3/4}y^{-3/4} = 15\frac{x^{3/4}}{y^{3/4}}.$$

(b) $\frac{\partial f}{\partial x}(81, 16) = 45 \cdot \frac{16^{1/4}}{81^{1/4}} = 45 \cdot \frac{2}{3} = 30$

$$\frac{\partial f}{\partial y}(81, 16) = 15 \cdot \frac{(81)^{3/4}}{(16)^{3/4}} = 15 \cdot \frac{27}{8} = \frac{405}{8} = 50\tfrac{5}{8}.$$

(c) The quantities $\frac{\partial f}{\partial x}$ and $\frac{\partial f}{\partial y}$ are referred to as the *marginal productivity of labor* and the *marginal productivity of capital*. If the amount of capital is held fixed at $y = 16$ and the amount of labor increases by one unit, then the quantity of goods produced will increase by approximately 30 units. Similarly, an increase in capital of one unit (with labor fixed at 81) results in an increase in production of approximately $50\tfrac{5}{8}$ units of goods.

Just as we formed second derivatives in the case of one variable, we can form second partial derivatives of a function $f(x, y)$ of two variables. Since $\frac{\partial f}{\partial x}$ is a function of x and y, we can differentiate it with respect to x or y. The partial derivative of $\frac{\partial f}{\partial x}$ with respect to x is denoted by $\frac{\partial^2 f}{\partial x^2}$. The partial derivative of $\frac{\partial f}{\partial x}$ with respect to y is denoted by $\frac{\partial^2 f}{\partial y\, \partial x}$. Similarly, the partial derivative of the function $\frac{\partial f}{\partial y}$ with respect to x is denoted by $\frac{\partial^2 f}{\partial x\, \partial y}$, and the partial derivative of $\frac{\partial f}{\partial y}$ with respect to y is denoted by $\frac{\partial^2 f}{\partial y^2}$. Almost all functions $f(x, y)$ encountered in applications [and all functions $f(x, y)$ in this text] have the property that

$$\frac{\partial^2 f}{\partial y\, \partial x} = \frac{\partial^2 f}{\partial x\, \partial y}.$$

EXAMPLE 9   Let $f(x, y) = x^2 + 3xy + 2y^2$. Calculate $\dfrac{\partial^2 f}{\partial x^2}, \dfrac{\partial^2 f}{\partial y^2}, \dfrac{\partial^2 f}{\partial x\, \partial y}, \dfrac{\partial^2 f}{\partial y\, \partial x}.$

Solution   First we compute $\frac{\partial f}{\partial x}$ and $\frac{\partial f}{\partial y}$.

$$\frac{\partial f}{\partial x} = 2x + 3y, \qquad \frac{\partial f}{\partial y} = 3x + 4y.$$

To compute $\frac{\partial^2 f}{\partial x^2}$, we differentiate $\frac{\partial f}{\partial x}$ with respect to x:

$$\frac{\partial^2 f}{\partial x^2} = 2.$$

Similarly, to compute $\frac{\partial^2 f}{\partial y^2}$, we differentiate $\frac{\partial f}{\partial y}$ with respect to y:

$$\frac{\partial^2 f}{\partial y^2} = 4.$$

To compute $\frac{\partial^2 f}{\partial x\, \partial y}$, we differentiate $\frac{\partial f}{\partial y}$ with respect to x:

$$\frac{\partial^2 f}{\partial x\, \partial y} = 3.$$

Finally, to compute $\frac{\partial^2 f}{\partial y\, \partial x}$, we differentiate $\frac{\partial f}{\partial x}$ with respect to y:

$$\frac{\partial^2 f}{\partial y\, \partial x} = 3.$$

PRACTICE PROBLEMS 2

1.   The number of TV sets sold per week by an appliance store is given by a function of two variables, $f(x, y)$, where x is the price per TV set and y is the amount of

money spent weekly on advertising. Suppose that the current price is \$400 per set and that currently \$2,000 per week is being spent for advertising.

(a) Would you expect $\dfrac{\partial f}{\partial x}$ (400, 2000) to be positive or negative?

(b) Would you expect $\dfrac{\partial f}{\partial y}$ (400, 2000) to be positive or negative?

2. The monthly mortgage payment for a house is a function of two variables, $f(A, r)$, where A is the amount of the mortgage and the interest rate is $r\%$. For a 30-year mortgage, $f(72{,}000, 9) = 579.33$ and $\dfrac{\partial f}{\partial r}(72{,}000, 9) = 51.80$. What is the significance of the number 51.80?

EXERCISES 2

Find $\dfrac{\partial f}{\partial x}$ and $\dfrac{\partial f}{\partial y}$ for each of the following functions.

1. $f(x, y) = 5xy$

2. $f(x, y) = 3x^2 + 2y + 1$

3. $f(x, y) = 2x^2 e^y$

4. $f(x, y) = x + e^{xy}$

5. $f(x, y) = y^2/x$

6. $f(x, y) = x/(1 + e^y)$

7. $f(x, y) = (2x - y + 5)^2$

8. $f(x, y) = (9x^2 y + 3x)^{12}$

9. $f(x, y) = x^2 e^{3x} \ln y$

10. $f(x, y) = (x - \ln y)e^{xy}$

11. $f(x, y) = (x - y)/(x + y)$

12. $f(x, y) = 2xy/e^x$

13. Let $f(L, K) = 3\sqrt{LK}$. Compute $\dfrac{\partial f}{\partial L}$.

14. Let $f(p, q) = e^{q/p}$. Compute $\dfrac{\partial f}{\partial q}, \dfrac{\partial f}{\partial p}$.

15. Let $f(x, y, z) = (1 + x^2 y)/z$. Compute $\dfrac{\partial f}{\partial x}, \dfrac{\partial f}{\partial y}, \dfrac{\partial f}{\partial z}$.

16. Let $f(x, y, z) = x^2 y + 3yz - z^2$. Compute $\dfrac{\partial f}{\partial x}, \dfrac{\partial f}{\partial y}, \dfrac{\partial f}{\partial z}$.

17. Let $f(x, y, z) = xze^{yz}$. Find $\dfrac{\partial f}{\partial x}, \dfrac{\partial f}{\partial y}, \dfrac{\partial f}{\partial z}$.

18. Let $f(x, y, z) = ze^{z/xy}$. Find $\dfrac{\partial f}{\partial x}, \dfrac{\partial f}{\partial y}, \dfrac{\partial f}{\partial z}$.

19. Let $f(x, y) = x^2 + 2xy + y^2 + 3x + 5y$. Compute $\dfrac{\partial f}{\partial x}$ (2, −3) and $\dfrac{\partial f}{\partial y}$ (2, −3).

20. Let $f(x, y) = xye^{2x-y}$. Evaluate $\dfrac{\partial f}{\partial x}$ and $\dfrac{\partial f}{\partial y}$ at $(x, y) = (1, 2)$.

21. Let $f(x, y, z) = xy^2z + 5$. Evaluate $\dfrac{\partial f}{\partial y}$ at $(x, y, z) = (2, -1, 3)$.

22. Let $f(x, y, z) = \dfrac{x}{y - z}$. Compute $\dfrac{\partial f}{\partial y}(2, -1, 3)$.

23. Let $f(x, y) = x^3y + 2xy^2$. Find $\dfrac{\partial^2 f}{\partial x^2}, \dfrac{\partial^2 f}{\partial y^2}, \dfrac{\partial^2 f}{\partial x\, \partial y}, \dfrac{\partial^2 f}{\partial y\, \partial x}$.

24. Let $f(x, y) = xe^y + x^4y + y^3$. Find $\dfrac{\partial^2 f}{\partial x^2}, \dfrac{\partial^2 f}{\partial y^2}, \dfrac{\partial^2 f}{\partial x\, \partial y}, \dfrac{\partial^2 f}{\partial y\, \partial x}$.

25. A farmer can produce $f(x, y) = 100(6x^2 + y^2)$ units of produce by utilizing x units of labor and y units of capital. (The capital is used to rent or purchase land, materials, and equipment.)

    (a) Calculate the marginal productivities of labor and capital when $x = 10$ and $y = 5$.

    (b) Use the result of (a) to determine the approximate effect on production of utilizing 5 units of capital but cutting back to $9\frac{1}{2}$ units of labor.

26. The productivity of a country is given by $f(x, y) = 300x^{2/3}y^{1/3}$, where x and y are the amounts of labor and capital.

    (a) Compute the marginal productivities of labor and capital when $x = 125$ and $y = 64$.

    (b) What would be the approximate effect of utilizing 125 units of labor but cutting back to 62 units of capital?

27. In a certain suburban community commuters have the choice of getting into the city by bus or train. The demand for these modes of transportation varies with their cost. Let $f(p_1, p_2)$ be the number of people who will take the bus when p_1 is the price of the bus ride and p_2 is the price of the train ride. Explain why $\dfrac{\partial f}{\partial p_1} < 0$ and $\dfrac{\partial f}{\partial p_2} > 0$.

28. The demand for a certain gas-guzzling car is given by $f(p_1, p_2)$, where p_1 is the price of the car and p_2 is the price of gasoline. Explain why $\dfrac{\partial f}{\partial p_1} < 0$ and $\dfrac{\partial f}{\partial p_2} < 0$.

29. Using data collected from 1929–1941, Richard Stone* determined that the yearly quantity Q of beer consumed in the United Kingdom was approximately given by the formula $Q = f(m, p, r, s)$, where

$$f(m, p, r, s) = (1.058)m^{.136}p^{-.727}r^{.914}s^{.816}$$

and m is the aggregate real income (personal income after direct taxes, adjusted for retail price changes), p is the average retail price of the commodity (in this case, beer), r is the average retail price level of all other consumer goods and services, and s is a measure of the strength of the beer. Determine which partial

---

* Richard Stone, "The analysis of market demand," *Journal of the Royal Statistical Society*, CVIII (1945), 286–391.

derivatives are positive and which are negative and give interpretations. (For example, since $\frac{\partial f}{\partial r} > 0$, people buy more beer when the prices of other goods increase and the other factors remain constant.)

30. Richard Stone (see Exercise 29) determined that the yearly consumption of food in the United States was given by

$$f(m, p, r) = (2.186)m^{.595}p^{-.543}r^{.922}.$$

Determine which partial derivatives are positive and which are negative and give interpretations of these facts.

31. The volume (V) of a certain amount of a gas is determined by the temperature (T) and the pressure (P) by the formula, $V = .08(T/P)$. Calculate and interpret $\frac{\partial V}{\partial P}$ and $\frac{\partial V}{\partial T}$ when $P = 20$, $T = 300$.

32. For the production function, $f(x, y) = 60x^{3/4}y^{1/4}$, considered in Example 8, think of $f(x, y)$ as the revenue when utilizing x units of labor and y units of capital. Under actual operating conditions, say $x = a$ and $y = b$, $\frac{\partial f}{\partial x}(a, b)$ is referred to as the *wage per unit of labor* and $\frac{\partial f}{\partial y}(a, b)$ is referred to as the *wage per unit of capital*. Show that

$$f(a, b) = a \cdot \left[\frac{\partial f}{\partial x}(a, b)\right] + b \cdot \left[\frac{\partial f}{\partial y}(a, b)\right].$$

(This equation shows how the revenue is distributed between labor and capital.)

33. Compute $\frac{\partial^2 f}{\partial x^2}$ where $f(x, y) = 60x^{3/4}y^{1/4}$, a production function (where x is units of labor). Explain why $\frac{\partial^2 f}{\partial x^2}$ is always negative.

34. Compute $\frac{\partial^2 f}{\partial y^2}$ where $f(x, y) = 60x^{3/4}y^{1/4}$, a production function (where y is units of capital). Explain why $\frac{\partial^2 f}{\partial y^2}$ is always negative.

35. Let $f(x, y) = 3x^2 + 2xy + 5y$, as in Example 5. Show that
$$f(1 + h, 4) - f(1, 4) = 14h + 3h^2.$$

Thus the error in approximating $f(1 + h, 4) - f(1, 4)$ by $14h$ is $3h^2$. (If $h = .01$, for instance, the error is only .0003.)

36. Physicians, particularly pediatricians, sometimes need to know the body surface area of a patient. For instance, the surface area is used to adjust the results of certain tests of kidney performance. Tables are available that give the approximate body surface A in square meters of a person who weighs W kilograms and is H centimeters tall. The following empirical formula* is also used:

$$A = .007W^{.425}H^{.725}.$$

---

*See J. Routh, *Mathematical Preparation for Laboratory Technicians* (Philadelphia: W. B. Saunders Co., 1971), p. 92.

Evaluate $\dfrac{\partial A}{\partial W}$ and $\dfrac{\partial A}{\partial H}$ when $W = 54$, $H = 165$, and give a physical interpretation of your answers. You may use the approximations: $(54)^{.425} \approx 5.4$, $(54)^{-.575} \approx .10$, $(165)^{.725} \approx 40.5$, $(165)^{-.275} \approx .25$.

SOLUTIONS TO PRACTICE PROBLEMS 2

1. (a) Negative. $\dfrac{\partial f}{\partial x}$ (400, 2000) is approximately the change in sales due to a $1 increase in x (price). Since raising prices lowers sales, we would expect $\dfrac{\partial f}{\partial x}$ (400, 2000) to be negative.

   (b) Positive. $\dfrac{\partial f}{\partial y}$ (400, 2000) is approximately the change in sales due to a $1 increase in advertising. Since spending more money on advertising brings in more customers, we would expect sales to increase; that is, $\dfrac{\partial f}{\partial y}$ (400, 2000) is most likely positive.

2. If the interest rate is raised from 9% to 10%, then the monthly payment will increase by about $51.80. (An increase to $9\frac{1}{2}\%$ causes an increase in the monthly payment of $\frac{1}{2} \cdot (51.80)$ or $25.90, and so on.)

8.3. Maxima and Minima of Functions of Several Variables

Previously, we studied how to determine the maxima and minima of functions of a single variable. Let us extend that discussion to functions of several variables.

If $f(x, y)$ is a function of two variables, then we say that $f(x, y)$ has a *maximum* when $x = a$, $y = b$ if $f(x, y)$ is at most equal to $f(a, b)$ whenever x is near a and y is near b. Geometrically, the graph of $f(x, y)$ has a peak at the point (a, b). [See Fig. 1(a).] Similarly, we say that $f(x, y)$ has a *minimum* when $x = a$, $y = b$

FIGURE 1   Maximum and minimum points.

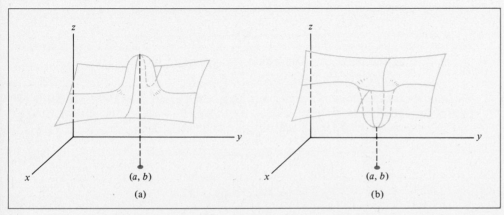

(a)

(b)

if $f(x, y)$ is at least equal to $f(a, b)$ whenever x is near a and y is near b. Geometrically, the graph of $f(x, y)$ has a pit with bottom at the point (a, b). [See Fig. 1(b).]

Suppose that $f(x, y)$ has a maximum at $(x, y) = (a, b)$. Then, keeping y constant at b, $f(x, y)$ becomes a function of the variable x with a maximum at $x = a$. Therefore its derivative with respect to x is zero at $x = a$. That is,

$$\frac{\partial f}{\partial x}(a, b) = 0.$$

Similarly, keeping x constant at a, $f(x, y)$ becomes a function of the variable y with a maximum at $y = b$. Therefore its derivative with respect to y is zero at $y = b$. That is,

$$\frac{\partial f}{\partial y}(a, b) = 0.$$

Similar considerations apply if $f(x, y)$ has a minimum at $(x, y) = (a, b)$. Thus we have the following test for extrema in two variables.

First Derivative Test for Extrema If $f(x, y)$ has either a maximum or a minimum at $(x, y) = (a, b)$, then

$$\frac{\partial f}{\partial x}(a, b) = 0$$

and

$$\frac{\partial f}{\partial y}(a, b) = 0.$$

This result is quite helpful in finding maximum and minimum points, especially when additional information about the function is known.

EXAMPLE 1 The function $f(x, y) = 3x^2 - 4xy + 3y^2 + 8x - 17y + 5$ has the graph pictured in Fig. 2. Find the point (x, y) at which $f(x, y)$ attains its minimum.

Solution We look for those values of x and y at which both partial derivatives are zero. The partial derivatives are

$$\frac{\partial f}{\partial x} = 6x - 4y + 8$$

$$\frac{\partial f}{\partial y} = -4x + 6y - 17.$$

Setting $\frac{\partial f}{\partial x} = 0$ and $\frac{\partial f}{\partial y} = 0$, we obtain

$$6x - 4y + 8 = 0 \quad \text{or} \quad y = \frac{6x + 8}{4}$$

$$-4x + 6y - 17 = 0 \quad \text{or} \quad y = \frac{4x + 17}{6}.$$

By equating these two expressions for y, we have

$$\frac{6x + 8}{4} = \frac{4x + 17}{6}.$$

Cross-multiplying, we see that

$$36x + 48 = 16x + 68$$

$$20x = 20$$

$$x = 1.$$

When we substitute this value for x into our first equation for y in terms of x, we obtain

$$y = \frac{6x + 8}{4} = \frac{6 \cdot 1 + 8}{4} = \frac{7}{2}.$$

If $f(x, y)$ has a minimum, it must occur where $\frac{\partial f}{\partial x} = 0$ and $\frac{\partial f}{\partial y} = 0$. We have determined that the partial derivatives are zero only when $x = 1$, $y = \frac{7}{2}$. From Fig. 2 we know that $f(x, y)$ has a minimum, so it must be at $(x, y) = (1, \frac{7}{2})$.

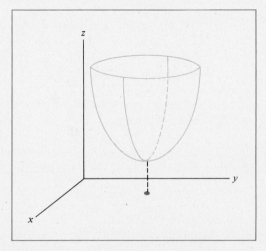

FIGURE 2 Graph of $f(x, y) = 3x^2 - 4xy + 3y^2 + 8x - 17y + 5$.

EXAMPLE 2 (Price Discrimination)   A monopolist markets his product in two countries and can charge different amounts in each country. Let x be the number of units to be sold in the first country and y the number of units to be sold in the second country. Due to the laws of demand, the monopolist must set the price at $97 - (x/10)$ dollars in the first country and $83 - (y/20)$ dollars in the second country in order to sell all the units. The cost of producing these units is $20,000 + 3(x + y)$. Find the values of x and y that maximize the profit.

Solution   Let $f(x, y)$ be the profit derived from selling x units in the first country and y in the second. Then

$f(x, y)$ = [revenue from first country] + [revenue from second country] − [cost]

$$= \left(97 - \frac{x}{10}\right)x + \left(83 - \frac{y}{20}\right)y - [20,000 + 3(x + y)]$$

$$= 97x - \frac{x^2}{10} + 83y - \frac{y^2}{20} - 20,000 - 3x - 3y$$

$$= 94x - \frac{x^2}{10} + 80y - \frac{y^2}{20} - 20,000.$$

To find where $f(x, y)$ has its maximum value, we look for those values of x and y at which both partial derivatives are zero.

$$\frac{\partial f}{\partial x} = 94 - \frac{x}{5}$$

$$\frac{\partial f}{\partial y} = 80 - \frac{y}{10}.$$

We set $\dfrac{\partial f}{\partial x} = 0$ and $\dfrac{\partial f}{\partial y} = 0$ to obtain

$$94 - \frac{x}{5} = 0 \quad \text{or} \quad x = 470$$

$$80 - \frac{y}{10} = 0 \quad \text{or} \quad y = 800.$$

Therefore the firm should adjust its prices to levels where it will sell 470 units in the first country and 800 units in the second country.

EXAMPLE 3   Suppose that we want to design a rectangular building having volume 147,840 cubic feet. Assuming that the daily loss of heat is given by

$$11xy + 14yz + 15xz,$$

where x, y, and z are, respectively, the length, width, and height of the building, find the dimensions of the building for which the daily heat loss is minimal.

Solution   We must minimize the function

$$11xy + 14yz + 15xz,\qquad(1)$$

where x, y, z satisfy the constraint equation

$$xyz = 147{,}840.$$

For simplicity, let us denote 147,840 by V. Then $xyz = V$, so that $z = V/xy$. We substitute this expression for z into the objective function (1) to obtain a heat-loss function $g(x, y)$ of two variables—namely,

$$g(x, y) = 11xy + 14y\,\frac{V}{xy} + 15x\,\frac{V}{xy}$$

$$= 11xy + \frac{14V}{x} + \frac{15V}{y}.$$

To minimize this function, we first compute the partial derivatives with respect to x and y; then we equate them to zero.

$$\frac{\partial g}{\partial x} = 11y - \frac{14V}{x^2} = 0$$

$$\frac{\partial g}{\partial y} = 11x - \frac{15V}{y^2} = 0.$$

These two equations yield

$$y = \frac{14V}{11x^2}\qquad(2)$$

$$11xy^2 = 15V.\qquad(3)$$

If we substitute the value of y from (2) into (3), we see that

$$11x\left(\frac{14V}{11x^2}\right)^2 = 15V$$

$$\frac{14^2 V^2}{11x^3} = 15V$$

$$x^3 = \frac{14^2 \cdot V^2}{11 \cdot 15 \cdot V} = \frac{14^2 \cdot V}{11 \cdot 15}$$

$$= \frac{14^2 \cdot 147{,}840}{11 \cdot 15}$$

$$= 175{,}616.$$

Therefore we see that (using a calculator or a table of cube roots)

$$x = 56.$$

From equation (2) we find that

$$y = \frac{14 \cdot V}{11x^2} = \frac{14 \cdot 147{,}840}{11 \cdot 56^2} = 60.$$

Finally,

$$z = \frac{V}{xy} = \frac{147,840}{56 \cdot 60} = 44.$$

Thus the building should be 56 feet long, 60 feet wide, and 44 feet high in order to minimize the heat loss.*

When considering a function of two variables, we find points (x, y) at which $f(x, y)$ has a potential maximum or minimum by setting $\frac{\partial f}{\partial x}$ and $\frac{\partial f}{\partial y}$ equal to zero and solving for x and y. However, if we are given no additional information about $f(x, y)$, it may be difficult to determine whether we have found a maximum or a minimum (or neither). In the case of functions of one variable, we studied concavity and deduced the second derivative test. There is an analog of the second derivative test for functions of two variables, but it is much more complicated than the one-variable test. We state it without proof.

Second Derivative Test for Maxima and Minima Suppose that $f(x, y)$ is a function and (a, b) is a point at which

$$\frac{\partial f}{\partial x}(a, b) = 0 \quad \text{and} \quad \frac{\partial f}{\partial y}(a, b) = 0,$$

and let

$$D(x, y) = \frac{\partial^2 f}{\partial x^2} \cdot \frac{\partial^2 f}{\partial y^2} - \left(\frac{\partial^2 f}{\partial x \, \partial y}\right)^2.$$

1. If

$$D(a, b) > 0 \quad \text{and} \quad \frac{\partial^2 f}{\partial x^2}(a, b) > 0,$$

then $f(x, y)$ has a minimum at (a, b).

2. If

$$D(a, b) > 0 \quad \text{and} \quad \frac{\partial^2 f}{\partial x^2}(a, b) < 0,$$

then $f(x, y)$ has a maximum at (a, b).

3. If

$$D(a, b) < 0,$$

then $f(x, y)$ has neither a maximum nor a minimum at (a, b).

4. If $D(a, b) = 0$, then no conclusion is possible.

---

* For further discussion of this heat-loss problem, as well as other examples of optimization in architectural design, see L. March, "Elementary Models of Built Forms," Chapter 3 in *Urban Space and Structures* by L. Martin and L. March, eds. (Cambridge: Cambridge University Press, 1972).

The saddle-shaped graph in Fig. 3 illustrates a function $f(x, y)$ for which $D(a, b) < 0$. Both partial derivatives are zero at $(x, y) = (a, b)$ and yet the function has neither a maximum nor a minimum there. (Observe that the function has a maximum with respect to x when y is held constant and a minimum with respect to y when x is held constant.)

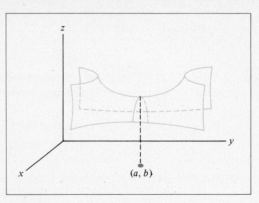

FIGURE 3

EXAMPLE 4 Let $f(x, y) = y^3 - x^2 + 6x - 12y + 5$. Find all possible maximum and minimum points of $f(x, y)$. Use the second derivative test to determine the nature of each such point.

Solution Since

$$\frac{\partial f}{\partial x} = -2x + 6, \qquad \frac{\partial f}{\partial y} = 3y^2 - 12,$$

we find that $f(x, y)$ has a potential extreme point when

$$-2x + 6 = 0$$
$$3y^2 - 12 = 0.$$

The first equation implies that $x = 3$. From the second equation we have

$$3y^2 = 12$$
$$y^2 = 4$$
$$y = \pm 2.$$

Thus $\frac{\partial f}{\partial x}$ and $\frac{\partial f}{\partial y}$ are both zero when $(x, y) = (3, 2)$ and when $(x, y) = (3, -2)$. To apply the second derivative test, we compute

$$\frac{\partial^2 f}{\partial x^2} = -2, \qquad \frac{\partial^2 f}{\partial y^2} = 6y, \qquad \frac{\partial^2 f}{\partial x \, \partial y} = 0,$$

and

$$D(x, y) = \frac{\partial^2 f}{\partial x^2} \cdot \frac{\partial^2 f}{\partial y^2} - \left(\frac{\partial^2 f}{\partial x \, \partial y} \right)^2 = (-2)(6y) - 0 = -12y. \qquad (4)$$

Since $D(3, 2) = -24$ is negative, case 3 of the second derivative test says that $f(x, y)$ has neither a maximum nor a minimum at $(3, 2)$. Next, note that

$$D(3, -2) = 24 > 0, \qquad \frac{\partial^2 f}{\partial x^2} (3, -2) = -2 < 0.$$

Thus, by case 2 of the second derivative test, the function $f(x, y)$ has a maximum at $(3, -2)$.

In this section we have restricted ourselves to functions of two variables, but the case of three or more variables is handled in a similar fashion. For instance, here is the first derivative test for a function of three variables.

If $f(x, y, z)$ has a maximum or a minimum at $(x, y, z) = (a, b, c)$, then

$$\frac{\partial f}{\partial x}(a, b, c) = 0$$

$$\frac{\partial f}{\partial y}(a, b, c) = 0$$

$$\frac{\partial f}{\partial z}(a, b, c) = 0.$$

PRACTICE PROBLEMS 3

1. Find all points (x, y) where $f(x, y) = x^3 - 3xy + \frac{1}{2}y^2 + 8$ has a possible maximum or minimum.

2. Apply the second derivative test to the function $g(x, y)$ of Example 3 to confirm that a minimum point actually occurs when $x = 56$ and $y = 60$.

EXERCISES 3

Find all points (x, y) where $f(x, y)$ has a possible maximum or minimum.

1. $f(x, y) = x^2 - 3y^2 + 4x + 6y + 8$

2. $f(x, y) = \frac{1}{2}x^2 + y^2 - 3x + 2y - 5$

3. $f(x, y) = x^2 - 5xy + 6y^2 + 3x - 2y + 4$

4. $f(x, y) = -3x^2 + 7xy - 4y^2 + x + y$

5. $f(x, y) = x^3 + y^2 - 3x + 6y$

6. $f(x, y) = x^2 - y^3 + 5x + 12y + 1$

7. $f(x, y) = \frac{1}{3}x^3 - 2y^3 - 5x + 6y - 5$

8. $f(x, y) = x^4 - 8xy + 2y^2 - 3$

9. The function $f(x, y) = 2x + 3y + 9 - x^2 - xy - y^2$ has a maximum at some point (x, y). Find the values of x and y where this maximum occurs.

10. The function $f(x, y) = \frac{1}{2}x^2 + 2xy + 3y^2 - x + 2y$ has a minimum at some point (x, y). Find the values of x and y where this minimum occurs.

   Find all points (x, y) where $f(x, y)$ has a possible maximum or minimum. Then use the second derivative test to determine, if possible, the nature of $f(x, y)$ at each of these points. If the second derivative test is inconclusive, so state.

11. $f(x, y) = x^2 - 2xy + 4y^2$

12. $f(x, y) = 2x^2 + 3xy + 5y^2$

13. $f(x, y) = -2x^2 + 2xy - y^2 + 4x - 6y + 5$

14. $f(x, y) = -x^2 - 8xy - y^2$

15. $f(x, y) = x^2 + 2xy + 5y^2 + 2x + 10y - 3$

16. $f(x, y) = x^2 - 2xy + 3y^2 + 4x - 16y + 22$

17. $f(x, y) = x^3 - y^2 - 3x + 4y$

18. $f(x, y) = x^3 - 2xy + 4y$

19. $f(x, y) = 2x^2 + y^3 - x - 12y + 7$

20. $f(x, y) = x^2 + 4xy + 2y^4$

21. Find the values of x, y, z at which
$$f(x, y, z) = 2x^2 + 3y^2 + z^2 - 2x - y - z$$
assumes its minimum value.

22. Find the values of x, y, z at which
$$f(x, y, z) = 5 + 8x - 4y + x^2 + y^2 + z^2$$
assumes its maximum value.

23. U.S. postal rules require that the length plus the girth of a package cannot exceed 84 inches in order to be mailed. Find the dimensions of the rectangular package of greatest volume that can be mailed. [Note: From Fig. 4 we see that $84 = $ (length) $+$ (girth) $= l + (2x + 2y)$ and volume $= xyl$.]

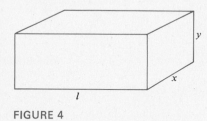

FIGURE 4

24. Find the dimensions of the rectangular box of least surface area that has a volume of 100 cubic inches. [Note: From Fig. 4 we see that the surface area $= 2xy + 2xl + 2yl$ and volume $= 100 = xyl$.]

25. A company manufactures and sells two products, call them I and II, that sell for $10 and $9 per unit, respectively. The cost of producing x units of product I and y units of product II is
$$400 + 2x + 3y + .01(3x^2 + xy + 3y^2).$$
Find the values of x and y that maximize the company's profit. [Note: Profit $=$ (revenue) $-$ (cost).]

26. A monopolist manufactures and sells two competing products, call them I and II, that cost \$30 and \$20 per unit, respectively, to produce. The revenue from marketing x units of product I and y units of product II is $98x + 112y - .04xy - .1x^2 - .2y^2$. Find the values of x and y that maximize the monopolist's profits.

27. A company manufactures and sells two products, call them I and II, that sell for $\$p_{\mathrm{I}}$ and $\$p_{\mathrm{II}}$ per unit, respectively. Let $C(x, y)$ be the cost of producing x units of product I and y units of product II. Show that if the company's profit is maximized when $x = a, y = b$, then

$$\frac{\partial C}{\partial x}(a, b) = p_{\mathrm{I}} \quad \text{and} \quad \frac{\partial C}{\partial y}(a, b) = p_{\mathrm{II}}.$$

28. A monopolist manufactures and sells two competing products, call them I and II, that cost $\$p_{\mathrm{I}}$ and $\$p_{\mathrm{II}}$ per unit, respectively, to produce. Let $R(x, y)$ be the revenue from marketing x units of product I and y units of product II. Show that if the monopolist's profit is maximized when $x = a, y = b$, then

$$\frac{\partial R}{\partial x}(a, b) = p_{\mathrm{I}} \quad \text{and} \quad \frac{\partial R}{\partial y}(a, b) = p_{\mathrm{II}}.$$

SOLUTIONS TO PRACTICE PROBLEMS 3

1. Compute the first partial derivatives of $f(x, y)$ and solve the system of equations that results from setting the partials equal to zero.

$$\frac{\partial f}{\partial x} = 3x^2 - 3y = 0$$

$$\frac{\partial f}{\partial y} = -3x + y = 0.$$

Solve each equation for y in terms of x.

$$\begin{cases} y = x^2 \\ y = 3x. \end{cases}$$

Equate expressions for y and solve for x.

$$x^2 = 3x$$
$$x^2 - 3x = 0$$
$$x(x - 3) = 0$$
$$x = 0 \quad \text{or} \quad x = 3.$$

When $x = 0, y = 0^2 = 0$. When $x = 3, y = 3^2 = 9$. Therefore, the possible maximum or minimum points are $(0, 0)$ and $(3, 9)$.

2. We have

$$g(x, y) = 11xy + \frac{14V}{x} + \frac{15V}{y},$$

$$\frac{\partial g}{\partial x} = 11y - \frac{14V}{x^2}, \quad \text{and} \quad \frac{\partial g}{\partial y} = 11x - \frac{15V}{y^2}.$$

Now,

$$\frac{\partial^2 g}{\partial x^2} = \frac{28V}{x^3}, \qquad \frac{\partial^2 g}{\partial y^2} = \frac{30V}{y^3}, \quad \text{and} \quad \frac{\partial^2 g}{\partial x\, \partial y} = 11.$$

Therefore,

$$D(x, y) = \frac{28V}{x^3} \cdot \frac{30V}{y^3} - (11)^2$$

$$D(56, 60) = \frac{28(147{,}840)}{(56)^3} \cdot \frac{30(147{,}840)}{(60)^3} - 121$$

$$= 484 - 121 = 363 > 0,$$

and

$$\frac{\partial^2 g}{\partial x^2}(56, 60) = \frac{28(147{,}840)}{(56)^3} > 0.$$

It follows that $g(x, y)$ has a minimum at $x = 56$, $y = 60$.

8.4. Lagrange Multipliers and Constrained Optimization

We have seen a number of optimization problems in which we were required to minimize (or maximize) an objective function where the variables were subject to a constraint equation. For instance, in Example 4 of Section 2.5, we minimized the cost of a rectangular enclosure by minimizing the objective function $21x + 14y$, where x and y were subject to the constraint equation $600 - xy = 0$. In the preceding section (Example 3) we minimized the daily heat loss from a building by minimizing the objective function $11xy + 14yz + 15xz$, subject to the constraint equation $147{,}840 - xyz = 0$.

In this section we introduce a powerful technique for solving problems of this type. Let us begin with the following general problem, which involves two variables.

PROBLEM Let $f(x, y)$ and $g(x, y)$ be functions of two variables. Find values of x and y that maximize (or minimize) the objective function $f(x, y)$ and that also satisfy the constraint equation $g(x, y) = 0$.

Of course, if we can solve the equation $g(x, y)$ for one variable in terms of the other and substitute the resulting expression into $f(x, y)$, we arrive at a function of a single variable that can be maximized (or minimized) by using the methods of Chapter 2. However, this technique can be unsatisfactory for two reasons. First, it may be difficult to solve the equation $g(x, y) = 0$ for x or for y. For example, if $g(x, y) = x^4 + 5x^3y + 7x^2y^3 + y^5 - 17 = 0$, then it is difficult to write y as a function of x or x as a function of y. Second, even if $g(x, y) = 0$ can be solved for one variable in terms of the other, substitution of the result into $f(x, y)$ may yield a complicated function.

One clever idea for handling the preceding problem was discovered by the eighteenth-century mathematician Lagrange, and the technique that he pioneered today bears his name—the method of *Lagrange multipliers*. The basic idea of this method is to replace $f(x, y)$ by an auxiliary function of three variables $F(x, y, \lambda)$, defined as

$$F(x, y, \lambda) = f(x, y) + \lambda g(x, y).$$

The new variable λ (lambda) is called a *Lagrange multiplier* and always multiplies the constraint function $g(x, y)$. The following theorem is stated without proof.

> Theorem   Suppose that, subject to the constraint $g(x, y) = 0$, the function $f(x, y)$ has a maximum or a minimum at $(x, y) = (a, b)$. Then there is a value of λ, say $\lambda = c$, such that the partial derivatives of $F(x, y, \lambda)$ all equal zero at $(x, y, \lambda) = (a, b, c)$.

The theorem implies that if we locate all points (x, y, λ) where the partial derivatives of $F(x, y, \lambda)$ are all zero, then among the corresponding points (x, y) we will find all possible places where $f(x, y)$ may have a constrained maximum or minimum. Thus, the first step in the method of Lagrange Multipliers is to set the partial derivatives of $F(x, y, \lambda)$ equal to zero and solve for x, y, and λ:

$$\frac{\partial F}{\partial x} = 0 \tag{L-1}$$

$$\frac{\partial F}{\partial y} = 0 \tag{L-2}$$

$$\frac{\partial F}{\partial \lambda} = 0. \tag{L-3}$$

From the definition of $F(x, y, \lambda)$, we see that $\dfrac{\partial F}{\partial \lambda} = g(x, y)$. Thus, the third equation (L-3) is just the original constraint equation $g(x, y) = 0$. So when we find a point (x, y, λ) that satisfies (L-1), (L-2), and (L-3), the coordinates x and y will automatically satisfy the constraint equation.

Let us see how this method works in practice.

EXAMPLE 1   Minimize $x^2 + y^2$, subject to the constraint $2x + 3y - 4 = 0$.

Solution   Here we have $f(x, y) = x^2 + y^2$, $g(x, y) = 2x + 3y - 4$, and

$$F(x, y, \lambda) = x^2 + y^2 + \lambda(2x + 3y - 4).$$

Equations (L-1) to (L-3) read

$$\frac{\partial F}{\partial x} = 2x + 2\lambda = 0 \tag{1}$$

$$\frac{\partial F}{\partial y} = 2y + 3\lambda = 0 \tag{2}$$

$$\frac{\partial F}{\partial \lambda} = 2x + 3y - 4 = 0. \tag{3}$$

Let us solve the first two equations for λ:

$$\lambda = -x \tag{4}$$

$$\lambda = -\tfrac{2}{3}y.$$

If we equate these two expressions for λ, we obtain

$$-x = -\tfrac{2}{3}y$$

$$x = \tfrac{2}{3}y. \tag{5}$$

Then, substituting this expression for x into the equation (3), we have

$$2(\tfrac{2}{3}y) + 3y - 4 = 0$$

$$\tfrac{13}{3}y = 4$$

$$y = \tfrac{12}{13}.$$

Using this value for y, equations (4) and (5) give us the value of x and λ:

$$x = \tfrac{2}{3}y = \tfrac{2}{3}(\tfrac{12}{13}) = \tfrac{8}{13}$$

$$\lambda = -x = -\tfrac{8}{13}.$$

Therefore the minimum value of $F(x, y, \lambda)$ occurs when $x = \tfrac{8}{13}$, $y = \tfrac{12}{13}$, and $\lambda = -\tfrac{8}{13}$. And so the minimum value of $x^2 + y^2$, subject to the constraint $2x + 3y - 4 = 0$, is

$$(\tfrac{8}{13})^2 + (\tfrac{12}{13})^2 = \tfrac{16}{13}.$$

The preceding technique for solving three equations in the three variables x, y, λ can usually be applied to solve Lagrange multiplier problems. Here is the basic procedure.

1. Solve (L-1) and (L-2) for λ in terms of x and y; then equate the resulting expressions for λ.
2. Solve the resulting equation for one of the variables.
3. Substitute the expression so derived into the equation (L-3) and solve the resulting equation of one variable.
4. Use the one known variable and the equations of steps (1) and (2) to determine the other two variables.

EXAMPLE 2 Using Lagrange multipliers, minimize $21x + 14y$, subject to the constraint $600 - xy = 0$, where x and y are restricted to positive values. (This problem arose in Example 4 of Section 2.5, where $21x + 14y$ was the cost of building a 600 square-foot enclosure having dimensions x and y.)

Solution We have $f(x, y) = 21x + 14y$, $g(x, y) = 600 - xy$, and

$$F(x, y, \lambda) = 21x + 14y + \lambda(600 - xy).$$

The equations (L-1) to (L-3), in this case, are

$$\frac{\partial F}{\partial x} = 21 - \lambda y = 0$$

$$\frac{\partial F}{\partial y} = 14 - \lambda x = 0$$

$$\frac{\partial F}{\partial \lambda} = 600 - xy = 0.$$

From the first two equations we see that

$$\lambda = \frac{21}{y} = \frac{14}{x} \qquad \text{(step 1)}.$$

Therefore

$$21x = 14y$$

and

$$x = \tfrac{2}{3}y \qquad \text{(step 2)}.$$

Substituting this expression for x into the third equation, we derive

$$600 - (\tfrac{2}{3}y)y = 0$$

$$y^2 = \tfrac{3}{2} \cdot 600 = 900$$

$$y = \pm 30 \qquad \text{(step 3)}.$$

We discard the case $y = -30$ because we are interested only in positive values of x and y. Using $y = 30$, we find that

$$\left. \begin{array}{l} x = \tfrac{2}{3}(30) = 20 \\ \lambda = \tfrac{14}{20} = \tfrac{7}{10}. \end{array} \right\} \qquad \text{(step 4)}$$

So the minimum value of $21x + 14y$ with x and y subject to the constraint occurs when $x = 20$, $y = 30$, and $\lambda = \tfrac{7}{10}$. And the minimum value of $21x + 14y$, subject to the constraint $600 - xy = 0$, is

$$21 \cdot (20) + 14 \cdot (30) = 840.$$

EXAMPLE 3 (Production) Suppose that x units of labor and y units of capital can produce $f(x, y) = 60x^{3/4}y^{1/4}$ units of a certain product. Also suppose that each unit of labor costs \$100, whereas each unit of capital costs \$200. Assume that \$30,000 is available to spend on production. How many units of labor and how many of capital should be utilized in order to maximize production?

Solution   The cost of x units of labor and y units of capital equals $100x + 200y$. Therefore since we want to use all the available money ($30,000), we must satisfy the constraint equation

$$100x + 200y = 30,000$$

or

$$g(x, y) = 30,000 - 100x - 200y = 0.$$

Our objective function is $f(x, y) = 60x^{3/4}y^{1/4}$. In this case, we have

$$F(x, y, \lambda) = 60x^{3/4}y^{1/4} + \lambda(30,000 - 100x - 200y).$$

The equations (L-1) to (L-3) read

$$\frac{\partial F}{\partial x} = 45x^{-1/4}y^{1/4} - 100\lambda = 0 \qquad \text{(L-1)}$$

$$\frac{\partial F}{\partial y} = 15x^{3/4}y^{-3/4} - 200\lambda = 0 \qquad \text{(L-2)}$$

$$\frac{\partial F}{\partial \lambda} = 30,000 - 100x - 200y = 0. \qquad \text{(L-3)}$$

By solving the first two equations for λ, we see that

$$\lambda = \tfrac{45}{100}x^{-1/4}y^{1/4} = \tfrac{9}{20}x^{-1/4}y^{1/4}$$

$$\lambda = \tfrac{15}{200}x^{3/4}y^{-3/4} = \tfrac{3}{40}x^{3/4}y^{-3/4}.$$

Therefore we must have

$$\tfrac{9}{20}x^{-1/4}y^{1/4} = \tfrac{3}{40}x^{3/4}y^{-3/4}.$$

To solve for y in terms of x, let us multiply both sides of this equation by $x^{1/4}y^{3/4}$:

$$\tfrac{9}{20}y = \tfrac{3}{40}x$$

or

$$y = \tfrac{1}{6}x.$$

Inserting this result in (L-3), we find that

$$100x + 200(\tfrac{1}{6}x) = 30,000$$

$$\frac{400x}{3} = 30,000$$

$$x = 225.$$

Hence

$$y = \tfrac{225}{6} = 37.5.$$

And so maximum production is achieved by using 225 units of labor and 37.5 units of capital.

In Example 3 it turns out that, at the optimum value of x and y,

$$\lambda = \tfrac{9}{20}x^{-1/4}y^{1/4} = \tfrac{9}{20}(225)^{-1/4}(37.5)^{1/4} \approx .2875$$

$$\frac{\partial f}{\partial x} = 45x^{-1/4}y^{1/4} = 45(225)^{-1/4}(37.5)^{1/4} \tag{6}$$

$$\frac{\partial f}{\partial y} = 15x^{3/4}y^{-3/4} = 15(225)^{3/4}(37.5)^{-3/4}. \tag{7}$$

It can be shown that the Lagrange multiplier λ can be interpreted as the marginal productivity of money. That is, if one additional dollar is available, then approximately .2875 additional units of the product can be produced.

Recall that the partial derivatives $\dfrac{\partial f}{\partial x}$ and $\dfrac{\partial f}{\partial y}$ are called the marginal productivity of labor and capital, respectively. From (6) and (7) we have

$$\frac{[\text{marginal productivity of labor}]}{[\text{marginal productivity of capital}]} = \frac{45(225)^{-1/4}(37.5)^{1/4}}{15(225)^{3/4}(37.5)^{-3/4}}$$

$$= \frac{45}{15}(225)^{-1}(37.5)^{1}$$

$$= \frac{3(37.5)}{225} = \frac{37.5}{75} = \frac{1}{2}.$$

On the other hand,

$$\frac{[\text{cost per unit of labor}]}{[\text{cost per unit of capital}]} = \frac{100}{200} = \frac{1}{2}.$$

This result illustrates the following law of economics. *If labor and capital are at their optimal levels, then the ratio of their marginal productivities equals the ratio of their unit costs.*

The method of Lagrange multipliers generalizes to functions of any number of variables. For instance, we can maximize $f(x, y, z)$, subject to the constraint equation $g(x, y, z) = 0$, by forming the Lagrange function

$$F(x, y, z, \lambda) = f(x, y, z) + \lambda g(x, y, z).$$

The analogues of equations (L-1) to (L-3) are

$$\frac{\partial F}{\partial x} = 0$$

$$\frac{\partial F}{\partial y} = 0$$

$$\frac{\partial F}{\partial z} = 0$$

$$\frac{\partial F}{\partial \lambda} = 0.$$

Let us now show how we can solve the heat-loss problem of Section 3 by using this method.

EXAMPLE 4   Use Lagrange multipliers to find the values of x, y, z that minimize the objective function

$$f(x, y, z) = 11xy + 14yz + 15xz,$$

subject to the constraint

$$xyz = 147{,}840.$$

Solution   The Lagrange function is

$$F(x, y, z, \lambda) = 11xy + 14yz + 15xz + \lambda(147{,}840 - xyz).$$

The conditions for a minimum are

$$\frac{\partial F}{\partial x} = 11y + 15z - \lambda yz = 0$$

$$\frac{\partial F}{\partial y} = 11x + 14z - \lambda xz = 0$$

$$\frac{\partial F}{\partial z} = 14y + 15x - \lambda xy = 0$$

$$\frac{\partial F}{\partial \lambda} = 147{,}840 - xyz = 0. \tag{8}$$

From the first three equations we have

$$\left. \begin{aligned}
\lambda &= \frac{11y + 15z}{yz} = \frac{11}{z} + \frac{15}{y} \\[2mm]
\lambda &= \frac{11x + 14z}{xz} = \frac{11}{z} + \frac{14}{x} \\[2mm]
\lambda &= \frac{14y + 15x}{xy} = \frac{14}{x} + \frac{15}{y}
\end{aligned} \right\}. \tag{9}$$

Let us equate the first two expressions for λ:

$$\frac{11}{z} + \frac{15}{y} = \frac{11}{z} + \frac{14}{x}$$

$$\frac{15}{y} = \frac{14}{x}$$

$$x = \frac{14}{15}y.$$

Next, we equate the second and third expressions for λ in (9):

$$\frac{11}{z} + \frac{14}{x} = \frac{14}{x} + \frac{15}{y}$$

$$\frac{11}{z} = \frac{15}{y}$$

$$z = \frac{11}{15} y.$$

We now substitute the expressions for x and z into the constraint equation (8) and obtain

$$\tfrac{14}{15}y \cdot y \cdot \tfrac{11}{15}y = 147{,}840$$

$$y^3 = \frac{(147{,}840)(15)^2}{(14)(11)} = 216{,}000$$

$$y = 60.$$

From this, we find that

$$x = \tfrac{14}{15}(60) = 56 \quad \text{and} \quad z = \tfrac{11}{15}(60) = 44.$$

We conclude that the heat loss is minimized when $x = 56$, $y = 60$, and $z = 44$.

In the solution of Example 4, we found that at the optimal values of x, y, and z,

$$\frac{14}{x} = \frac{15}{y} = \frac{11}{z}.$$

Referring to Example 2 of Section 1, we see that 14 is the combined heat loss through the east and west sides of the building, 15 is the heat loss through the north and south sides of the building, and 11 is the heat loss through the floor and roof. Thus we have that under optimal conditions

$$\frac{[\text{heat loss through east and west sides}]}{[\text{distance between east and west sides}]} = \frac{[\text{heat loss through north and south sides}]}{[\text{distance between north and south sides}]}$$

$$= \frac{[\text{heat loss through floor and roof}]}{[\text{distance between floor and roof}]}.$$

This is a principle of optimal design: minimal heat loss occurs when the distance between each pair of opposite sides is some fixed constant times the heat loss from the pair of sides.

The value of λ in Example 4 corresponding to the optimal values of x, y, and z is

$$\lambda = \frac{11}{z} + \frac{15}{y} = \frac{11}{44} + \frac{15}{60} = \frac{1}{2}.$$

One can show that the Lagrange multiplier λ is the marginal heat loss with respect to volume. That is, if a building of volume slightly more than 147,840 cubic feet is optimally designed, then $\frac{1}{2}$ unit of additional heat will be lost for each additional cubic foot of volume.

PRACTICE PROBLEMS 4

1. Let $F(x, y, \lambda) = 2x + 3y + \lambda(90 - 6x^{1/3}y^{2/3})$. Find $\dfrac{\partial F}{\partial x}$.

2. Refer to Exercise 23 of Section 3. What is the function $F(x, y, \lambda)$ when the exercise is solved using the method of Lagrange multipliers?

EXERCISES 4

Solve the following exercises by the method of Lagrange multipliers.

1. Minimize the function $x^2 + 3y^2 + 10$, subject to the constraint $8 - x - y = 0$.

2. Maximize the function $x^2 - y^2$, subject to the constraint $2x + y - 3 = 0$.

3. Maximize $x^2 + xy - 3y^2$, subject to the constraint $2 - x - 2y = 0$.

4. Minimize $\frac{1}{2}x^2 - 3xy + y^2 + \frac{1}{2}$, subject to the constraint $3x - y - 1 = 0$.

5. Find the values of x, y that maximize the function

$$-2x^2 - 2xy - \tfrac{3}{2}y^2 + x + 2y,$$

subject to the constraint $x + y - \frac{5}{2} = 0$.

6. Find the values of x, y that minimize the function

$$x^2 + xy + y^2 - 2x - 5y,$$

subject to the constraint $1 - x + y = 0$.

7. Find the values of x, y, z that maximize the function

$$3x + 5y + z - x^2 - y^2 - z^2,$$

subject to the constraint $6 - x - y - z = 0$.

8. Find the values of x, y, z that minimize the function

$$x^2 + y^2 + z^2 - 3x - 5y - z,$$

subject to the constraint $20 - 2x - y - z = 0$.

9. The material for a rectangular box costs $2 per square foot for the top and $1 per square foot for the sides and bottom. Using Lagrange multipliers, find the dimensions for which the volume of the box is 12 cubic feet and the cost of the materials is minimized. [Referring to Fig. 2(a), the cost will be $3xy + 2xz + 2yz$.]

10. Use Lagrange multipliers to find the three positive numbers whose sum is 15 and whose product is as large as possible.

11. Find the dimensions of the rectangle of maximum area that can be inscribed in the unit circle. [See Fig. 1(a).] That is, find the values of x, y that maximize $4xy$, subject to the constraint $x^2 + y^2 = 1$.

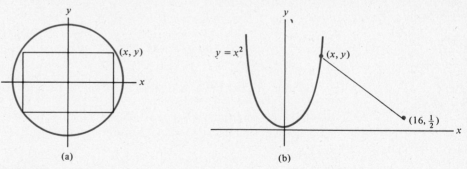

FIGURE 1

12. Find the point on the parabola $y = x^2$ that has minimal distance from point $(16, \frac{1}{2})$. [See Fig. 1(b).] Suggestion: If d denotes the distance from (x, y) to $(16, \frac{1}{2})$, then $d^2 = (x - 16)^2 + (y - \frac{1}{2})^2$. If d^2 is minimized, then d will be minimized. Thus it suffices to minimize $(x - 16)^2 + (y - \frac{1}{2})^2$, subject to the constraint $y - x^2 = 0$.

13. Find the dimensions of an open rectangular glass tank of volume 32 cubic feet for which the amount of material needed to construct the tank is minimized. [See Fig. 2(a).] That is, find the values of x, y, z that minimize $xy + 2xz + 2yz$, subject to the constraint $32 - xyz = 0$.

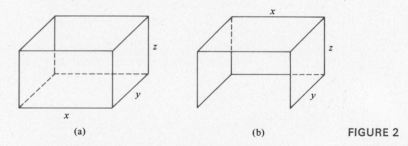

FIGURE 2

14. A shelter for use at the beach has a back, two sides, and a top made of canvas. [See Fig. 2(b).] Find the dimensions that maximize the volume and require 96 square feet of canvas. That is, find the values of x, y, z that maximize xyz, subject to the constraint $xz + xy + 2yz - 96 = 0$.

15. The production function for a firm is $f(x, y) = 64x^{3/4}y^{1/4}$, where x and y are the number of units of labor and capital utilized. Suppose that labor costs $96 per unit and capital costs $162 per unit and that the firm decides to produce 3456 units of goods.

(a) Determine the amounts of labor and capital that should be utilized in order to minimize the cost. That is, find the values of x, y that minimize $96x + 162y$, subject to the constraint $3456 - 64x^{3/4}y^{1/4} = 0$.

(b) Find the value of λ at the optimal level of production.

(c) Show that, at the optimal level of production, we have

$$\frac{[\text{marginal productivity of labor}]}{[\text{marginal productivity of capital}]} = \frac{[\text{unit price of labor}]}{[\text{unit price of capital}]}.$$

16. The production function for a firm is $f(x, y) = 1000\sqrt{6x^2 + y^2}$, where x and y are, respectively, the number of units of labor and capital utilized. Suppose that labor costs \$480 per unit and capital costs \$40 per unit and that the firm has \$5,000 to spend.

    (a) Determine the amounts of labor and capital that should be utilized in order to maximize production.

    (b) Find the value of λ.

    (c) Repeat part (c) of Exercise 15 for $f(x, y) = 1000\sqrt{6x^2 + y^2}$.

17. Suppose that a firm makes two products A and B that use the same raw materials. Given a fixed amount of raw materials and a fixed amount of manpower, the firm must decide how much of its resources should be allocated to the production of A and how much to B. If x units of A and y units of B are produced, suppose that x and y must satisfy

    $$9x^2 + 4y^2 = 18,000.$$

    The graph of this equation (for $x \geq 0$, $y \geq 0$) is called a *production possibilities curve* (Fig. 3). A point (x, y) on

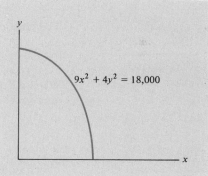

FIGURE 3

    this curve represents a *production schedule* for the firm, committing it to produce x units of A and y units of B. The reason for the relationship between x and y involves the limitations on manpower and raw materials available to the firm. Suppose that each unit of A yields a \$3 profit, whereas each unit of B yields a \$4 profit. Then the profit of the firm is

    $$P(x, y) = 3x + 4y.$$

    Find the production schedule that maximizes the profit function $P(x, y)$.

18. Consider the monopolist of Example 2, Section 3, who sells his goods in two countries. Suppose that he must set the same price in each country. That is, $97 - (x/10) = 83 - (y/20)$. Find the values of x and y that maximize profits under this new restriction.

19. Let $f(x, y)$ be any production function where x represents labor (costing \$$a$ per unit) and y represents capital (costing \$$b$ per unit). Assuming that \$$c$ is available, show that, at the values of x, y that maximize production,

    $$\frac{\frac{\partial f}{\partial x}}{\frac{\partial f}{\partial y}} = \frac{a}{b}.$$

[Note: Let $F(x, y, \lambda) = f(x, y) + \lambda(c - ax - by)$. The result follows from (L-1) and (L-2).]

20. By applying the result in Exercise 19 to the production function $f(x, y) = kx^\alpha y^\beta$, show that, for the values of x, y that maximize production, we have

$$\frac{y}{x} = \frac{a\beta}{b\alpha}.$$

(This tells us that the ratio of capital to labor does not depend on the amount of money available nor on the level of production but only on the numbers a, b, α, and β.)

SOLUTIONS TO PRACTICE PROBLEMS 4

1. The function can be written as

$$F(x, y, \lambda) = 2x + 3y + \lambda \cdot 90 - \lambda \cdot 6x^{1/3}y^{2/3}.$$

When differentiating with respect to x, both y and λ should be treated as constants (and so $\lambda \cdot 90$ and $\lambda \cdot 6$ are also regarded as constants).

$$\frac{\partial F}{\partial x} = 2 - \lambda \cdot 6 \cdot \tfrac{1}{3}x^{-2/3} \cdot y^{2/3}$$

$$= 2 - \lambda \cdot 2 \cdot x^{-2/3}y^{2/3}.$$

[Note: It is not necessary to write out the multiplication by λ as we did. Most people just do this mentally and then differentiate.]

2. The quantity to be maximized is the volume xyl. The constraint is that length plus girth is 84. This translates to $84 = l + 2x + 2y$ or $84 - l - 2x - 2y = 0$. Therefore,

$$F(x, y, l, \lambda) = xyl + \lambda(84 - l - 2x - 2y).$$

8.5. Total Differentials and Their Applications

In Section 2 we saw that the partial derivatives of $f(x, y)$ indicate how much the function changes with respect to small changes in its variables. In particular, if h and k are small, then, at the point (a, b), a change in x of h units produces a change in $f(x, y)$ of approximately $\left[\dfrac{\partial f}{\partial x}(a, b)\right] \cdot h$ units, and a change in y of k units produces a change in $f(x, y)$ of approximately $\left[\dfrac{\partial f}{\partial y}(a, b)\right] \cdot k$ units. Therefore it is not surprising that when both coordinates are changed, the change in $f(x, y)$ is approximated by the sum of these terms. Precisely, we have

$$f(a + h, b + k) - f(a, b) \approx \left[\frac{\partial f}{\partial x}(a, b)\right] \cdot h + \left[\frac{\partial f}{\partial y}(a, b)\right] \cdot k, \qquad (1)$$

where the approximation improves as h, k approach zero.

The expression on the right side of (1) is usually called a *total differential*. Its value depends on h and k, as well as the partial derivatives at $x = a$, $y = b$.

Formula (1) generalizes to functions of any number of variables. For example, for a function of three variables, $f(x, y, z)$, we have

$$f(a + h, b + k, c + l) - f(a, b, c) \approx \frac{\partial f}{\partial x} \cdot h + \frac{\partial f}{\partial y} \cdot k + \frac{\partial f}{\partial z} \cdot l, \qquad (2)$$

where each of the partial derivatives is evaluated at $x = a$, $y = b$, $z = c$.

We shall use total differentials to approximate function values, estimate errors, and interpret Lagrange multipliers.

EXAMPLE 1   Approximate $60 \cdot (80)^{3/4} \cdot (17)^{1/4}$.

Solution   Let $f(x, y) = 60x^{3/4}y^{1/4}$, $a = 81$, $b = 16$, $h = -1$, and $k = 1$. Then, by formula (1), we have

$$f(81 - 1, 16 + 1) - f(81, 16) \approx \left[\frac{\partial f}{\partial x}(81, 16)\right] \cdot (-1) + \left[\frac{\partial f}{\partial y}(81, 16)\right] \cdot 1.$$

Using the values computed in Example 3 of Section 1 and Example 8 of Section 2, we have

$$f(80, 17) - 3240 \approx 30(-1) + 50\tfrac{5}{8}(1) = 20\tfrac{5}{8}$$

$$f(80, 17) \approx 20\tfrac{5}{8} + 3240 = 3260\tfrac{5}{8}.$$

Therefore $60 \cdot (80)^{3/4} \cdot (17)^{1/4} \approx 3260\tfrac{5}{8}$. [The actual value of $60 \cdot (80)^{3/4} \cdot (17)^{1/4}$ is $3258.97\ldots$, so that our approximation yields an error of about 1.65.]

EXAMPLE 2   A rectangle is measured and found to have dimensions of 3.3 and 4.6 millimeters, with an error of at most .1 millimeter in each measurement. Approximate the maximum error that might occur if these numbers are used to compute the area of the rectangle.

Solution   The true dimensions are $3.3 + h$ and $4.6 + k$, where $|h|$ and $|k|$ are at most .1 millimeter. Let $f(x, y) = x \cdot y$. Then

$$[\text{error}] = [\text{true area}] - [\text{computed area}]$$

$$= f(3.3 + h, 4.6 + k) - f(3.3, 4.6)$$

$$\approx \left[\frac{\partial f}{\partial x}(3.3, 4.6)\right] \cdot h + \left[\frac{\partial f}{\partial y}(3.3, 4.6)\right] \cdot k.$$

Since $\frac{\partial f}{\partial x} = y$ and $\frac{\partial f}{\partial y} = x$,

$$[\text{error}] \approx (4.6) \cdot h + (3.3) \cdot k.$$

In addition, since h and k might be positive or negative, the safest estimate is

$$|\text{error}| \leq |(4.6)h| + |(3.3)k|$$
$$\leq (4.6)(.1) + (3.3)(.1)$$
$$\leq .46 + .33 = .79$$

Consequently, the maximum error in the area of the rectangle is approximately .79 square millimeter.

EXAMPLE 3   (Cardiac Output)   The quantity of blood pumped through the lungs per minute can be computed by measuring the amount of oxygen (O_2) used by the body per minute and the concentration of oxygen in well-chosen blood samples. Let

C = number of liters of blood pumped into lungs per minute,

x = number of milliliters of O_2 utilized by the body per minute (computed by analyzing expelled air),

y = number of milliliters of O_2 per liter of arterial blood (blood sampled from an artery after absorbing O_2 from the lungs),

z = number of milliliters of O_2 per liter of mixed venous blood (blood sampled from veins before entering the lungs).

Then, for each minute,

[amount of O_2 utilized] = [amount of O_2 gained per liter of blood]
· [number of liters of blood pumped];

that is,

$$x = (y - z) \cdot C.$$

Equivalently,

$$C = \frac{x}{y - z}.$$

Suppose that, for a particular person at rest, the measurements $x = 250$, $y = 180$, $z = 140$ are taken and used to compute $C = 6.25$. Approximate the maximum error in C if each measurement might be in error by at most 1%.

Solution   Let $f(x, y, z) = x/(y - z)$. The partial derivatives of $f(x, y, z)$ are

$$\frac{\partial f}{\partial x} = \frac{1}{y - z},$$

$$\frac{\partial f}{\partial y} = \frac{-x}{(y - z)^2},$$

and

$$\frac{\partial f}{\partial z} = \frac{x}{(y - z)^2}.$$

The values of the partial derivatives when $x = 250$, $y = 180$, and $z = 140$ are

$$\frac{\partial f}{\partial x} = \frac{1}{180 - 140} = \frac{1}{40},$$

$$\frac{\partial f}{\partial y} = \frac{-250}{(40)^2} = -\frac{5}{32},$$

and

$$\frac{\partial f}{\partial z} = \frac{250}{(40)^2} = \frac{5}{32}.$$

By formula (2), the error in C is approximately

$$\frac{1}{40} h - \frac{5}{32} k + \frac{5}{32} l,$$

where h, k, and l are the errors in the measurements of x, y, and z, respectively. Since each error is at most 1%,

$$|h| \leq 2.5$$
$$|k| \leq 1.8$$

and

$$|l| \leq 1.4.$$

Moreover, since h, k, and l could each be positive or negative, the safest estimate is

$$|\text{error}| \leq \left| \frac{1}{40} h \right| + \left| \frac{-5}{32} k \right| + \left| \frac{5}{32} l \right|$$

$$\leq \frac{1}{40} (2.5) + \frac{5}{32} (1.8) + \frac{5}{32} (1.4) = .5625.$$

Since .5625 is 9% of 6.25, we see that small errors in the measurements of the oxygen can result in large errors in the computation of cardiac output.

EXAMPLE 4   (Production)   In Example 3 of Section 4 we found that, with \$30,000 to spend, the production function $f(x, y) = 60x^{3/4}y^{1/4}$ was maximized when $x = 225$ units of labor (costing \$100 per unit) and $y = 37.5$ units of capital (costing \$200 per unit) were employed. We also found that $\lambda \approx .2875$. From (6) and (7) of Section 4 it follows that

$$\frac{\partial f}{\partial x}(225, 37.5) = 100\lambda, \qquad \frac{\partial f}{\partial y}(225, 37.5) = 200\lambda.$$

Show that λ is the marginal productivity of money. That is, if one extra dollar is spent, then approximately λ additional units can be produced.

Solution   With one additional dollar, we can purchase h units of labor and k units of capital, where

$$100 \cdot h + 200 \cdot k = 1.$$

By (1), the additional number of units of goods produced is

$$f(225 + h, 37.5 + k) - f(225, 37.5) \approx \left[\frac{\partial f}{\partial x}(225, 37.5)\right] \cdot h + \left[\frac{\partial f}{\partial y}(225, 37.5)\right] \cdot k$$

$$\approx (100\lambda)h + (200\lambda)k$$

$$= \lambda(100h + 200k) = \lambda \cdot 1 = \lambda.$$

Therefore approximately an additional .2875 unit of goods can be produced.

PRACTICE PROBLEMS 5

1. Let $f(x, y) = x^4 - (y/x^2)$. Use the total differential to approximate $f(1.02, 1.99)$.

2. Consider the production function $f(x, y) = 60x^{3/4}y^{1/4}$ of Example 8, Section 2. Suppose that currently 81 units of labor and 16 units of capital are being employed. Use the total differential to estimate the increase in production from employing two more units of labor and one more unit of capital.

EXERCISES 5

1. Let $f(x, y) = 2x^3 + 3xy$. Approximate $f(1.001, 1)$.

2. Let $f(x, y) = x^{100}y^3$. Approximate $f(.99, 2.02)$.

3. Let $f(x, y) = ye^x$. Approximate $f(.01, .98)$.

4. Let $f(x, y) = xe^{-y}$. Approximate $f(7.06, .03)$.

5. Let $f(x, y) = y/(1 - x)$. Approximate $f(-.01, 5)$.

6. Let $f(x, y) = \sqrt{x}/y$. Approximate $f(3.98, .99)$.

7. Let $f(x, y) = \sqrt{xy}$. Approximate $f(11.9, 3.2)$.

8. Let $f(x, y) = x^{2/3}y^3$. Approximate $f(7.9, -1.2)$.

9. Let $f(x, y, z) = x^{10}y^2z^3$. Approximate $f(1.01, 2.97, 1.98)$.

10. Let $f(x, y, z) = \sqrt{z}e^{x-y^2}$. Approximate $f(4.1, 1.9, 9.1)$.

11. Suppose that the profit for a firm selling two products at prices p and q, respectively, is given by

$$P(p, q) = -100,000 + 5000p + 10,000q - 50p^2 - 100q^2 + 10pq.$$

Thus when $p = 100$ and $q = 50$, the total profit of the firm is $200,000. Use the total differential to approximate the effect of a $1 increase in the price p and a $.20 increase in the price q.

12. Let S denote the sales of a certain product, p the unit price, and a the number of dollars spent per unit in advertising. Studies have shown that

$$S = 100,000 + 20,000\left(1 - \frac{1}{a+1}\right)e^{-p}.$$

(a) Calculate the total sales when $a = 1, $p = 3.

(b) Approximate the effect of reducing a to $.95 and increasing p to $3.10.

13. The volume of a cylinder is $\pi x^2 y$, where x is the radius of the base and y is the height. A cylindrical container is measured and found to have radius 20 millimeters and height 100 millimeters. Find the approximate error in calculating the volume of the container if each measurement is in error by at most 1 millimeter.

14. The volume of a certain gas is related to the temperature and pressure of the gas by the formula $V = .08\dfrac{T}{P}$, where V is the volume in liters, P is the pressure in atmospheres, and T is the temperature in degrees Kelvin ($^\circ K = {}^\circ C + 273$). A container of the gas is found to have temperature $293^\circ K$ and pressure 20 atmospheres. Find the approximate maximum error in computing the volume if the temperature and pressure readings might be in error by at most 1 unit.

15. The solution to Exercise 15 of Section 4 is $x = 81$, $y = 16$, $\lambda = 3$. Show that λ is the marginal cost of production. That is, the cost of producing one additional unit of goods is approximately $3.

16. The solution to Exercise 16 of Section 4 is $x = 10$, $y = 5$, $\lambda = 5$. Show that λ is the marginal productivity of money. That is, if $1 additional is available, then approximately five more units of goods can be produced.

17. (Renal clearance)   An important function of a person's kidneys is to remove urea (a nitrogenous end product of protein decomposition) from the blood. A standard clinical measure of the health of the kidneys is the so-called maximum clearance rate at which the blood is cleared of urea:

$$C = \frac{UV}{S},$$

where C is expressed in milliliters of blood cleared per minute, U is the concentration of urea in the urine (in mg/100 ml), S is the concentration of urea in the blood (in mg/100 ml), and V is the volume of urine excreted per minute (ml/min).
    Suppose that laboratory tests for a patient indicate that $U = 200$, $V = 3.00$, and $S = 12.0$, so that $C = 50$ ml/min. Estimate the maximum error in C if the error in each measurement does not exceed 1%.

18. (Renal clearance, continued) For certain physiological reasons, physicians use the formula

$$f(U, V, S) = \frac{U\sqrt{V}}{S}$$

to calculate the "standard urea clearance" whenever the value of V (see Exercise 17) drops below 2 ml/min. Suppose that laboratory tests for a patient indicate that $U = 500$, $V = 1.44$, $S = 20$. Estimate the maximum error in $f(U, V, S)$ if the error in each measurement does not exceed 1%.

SOLUTIONS TO PRACTICE PROBLEMS 5

1. Evaluating the function at $(1.02, 1.99)$ is not as easy as evaluating it at $(1, 2)$, a nearby point.

$$f(1, 2) = 1^4 - \frac{2}{1^2} = 1 - 2 = -1$$

$$\frac{\partial f}{\partial x} = 4x^3 + 2\frac{y}{x^3}; \qquad \frac{\partial f}{\partial y} = -\frac{1}{x^2}$$

$$\frac{\partial f}{\partial x}(1, 2) = 4 + 4 = 8; \qquad \frac{\partial f}{\partial y}(1, 2) = -1.$$

$(1.02, 1.99) = (1 + h, 2 + k)$ where $h = .02$, $k = -.01$. By (1),

$$f(1.02, 1.99) - f(1, 2) \approx \left[\frac{\partial f}{\partial x}(1, 2)\right] \cdot h + \left[\frac{\partial f}{\partial y}(1, 2)\right] \cdot k$$

$$f(1.02, 1.99) - (-1) \approx (8) \cdot (.02) + (-1)(-.01)$$

$$f(1.02, 1.99) + 1 \approx .16 + .01 = .17$$

$$f(1.02, 1.99) \approx .17 - 1 = -.83.$$

2. We have shown in Example 8 of Section 2 that

$$\frac{\partial f}{\partial x}(81, 16) = 30 \quad \text{and} \quad \frac{\partial f}{\partial y}(81, 16) = 50\tfrac{5}{8}.$$

By (1),

$$f(83, 17) - f(81, 16) \approx \left[\frac{\partial f}{\partial x}(81, 16)\right] \cdot 2 + \left[\frac{\partial f}{\partial y}(81, 16)\right] \cdot 1$$

$$\approx 30 \cdot 2 + 50\tfrac{5}{8} \cdot 1$$

$$= 110\tfrac{5}{8}.$$

That is, approximately $110\tfrac{5}{8}$ additional units can be produced.

8.6. The Method of Least Squares

Modern man compiles graphs of literally thousands of different quantities: the purchasing value of the dollar as a function of time, the pressure of a fixed volume of air as a function of temperature, the average income of people as a function of their years of formal education, or the incidence of strokes as a function of blood pressure. The observed points on such graphs tend to be irregularly distributed due both to the complicated nature of the phenomena underlying them as well as to errors made in observation (for example, a given procedure for measuring average income may not count certain groups). In spite of the imperfect nature of the data, we are often faced with the problem of making assessments and predictions based on them. Roughly speaking, this problem amounts to filtering the sources of errors in the data and isolating the basic underlying trend. Frequently, on the basis of a suspicion or a working hypothesis, we may suspect that the underlying trend is linear—that is, the data should lie on a straight line. But which straight line? This is the problem that the *method of least squares* attempts to answer. To be more specific, let us consider the following problem:

PROBLEM   Given observed data points $(x_1, y_1), (x_2, y_2), \ldots, (x_N, y_N)$ on a graph, find the straight line that "best" fits these points.

In order to completely understand the statement of the problem being considered, we must define what it means for a line to "best" fit a set of points. If (x_i, y_i) is one of our observed points, then we will measure how far it is from a given line $y = Ax + B$ by the vertical distance from the point to the line. Since the point on the line with x-coordinate x_i is $(x_i, Ax_i + B)$, this vertical distance is the distance between the y-coordinates $Ax_i + B$ and y_i (see Fig. 1). If $E_i =$

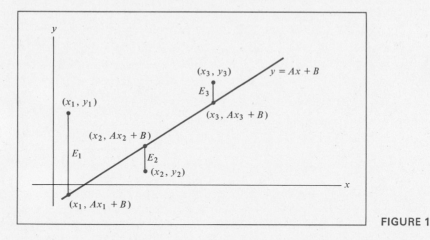

FIGURE 1

$(Ax_i + B) - y_i$, then either E_i or $-E_i$ is the vertical distance from (x_i, y_i) to the line. It turns out to be convenient to work with the square of this vertical distance, namely,

$$E_i^2 = (Ax_i + B - y_i)^2.$$

The total error in approximating the data points $(x_1, y_1), \ldots, (x_N, y_N)$ by the line $y = Ax + B$ is usually measured by the sum E of the squares of the vertical distances from the points to the line,

$$E = E_1^2 + E_2^2 + \ldots + E_N^2. \tag{1}$$

E is called the *least squares error* of the observed points with respect to the line. If all the observed points lie on the line $y = Ax + B$, then all E_i are zero and the error E is zero. If a given observed point is far away from the line, the corresponding E_i^2 is large and hence makes a large contribution to the error E.

In general, we cannot expect to find a line $y = Ax + B$ that fits the observed points so well that the error E is zero. Actually, this situation will occur only if the observed points lie on a straight line. However, we can rephrase our original problem as follows:

PROBLEM   Given observed data points $(x_1, y_1), (x_2, y_2), \ldots, (x_N, y_N)$, find a straight line $y = Ax + B$ for which the error E is as small as possible.

It turns out that this problem is a minimization problem in the two variables A and B and so can be solved by using the methods of Section 3. Let us consider an example.

EXAMPLE 1   Find the straight line that minimizes the least-squares error for the points $(1, 4)$ $(2, 5), (3, 8)$.

Solution   Let the straight line be $y = Ax + B$. When $x = 1, 2, 3$, the y-coordinate of the corresponding point of the line is $A + B, 2A + B, 3A + B$, respectively. Therefore the squares of the vertical distances from the points $(1, 4), (2, 5), (3, 8)$ are, respectively,

$$E_1^2 = (A + B - 4)^2$$

$$E_2^2 = (2A + B - 5)^2$$

$$E_3^2 = (3A + B - 8)^2.$$

(See Fig. 2.)
Thus the least-squares error is

$$E_1^2 + E_2^2 + E_3^2 = (A + B - 4)^2 + (2A + B - 5)^2 + (3A + B - 8)^2.$$

This error obviously depends on the choice of A and B. Let $f(A, B)$ denote this least-squares error. We want to find values of A and B that minimize $f(A, B)$. To do so, we take partial derivatives with respect to A and B and set the partial derivatives equal to zero:

$$\frac{\partial f}{\partial A} = 2(A + B - 4) + 2(2A + B - 5) \cdot 2 + 2(3A + B - 8) \cdot 3$$

$$= 28A + 12B - 76 = 0,$$

$$\frac{\partial f}{\partial B} = 2(A + B - 4) + 2(2A + B - 5) + 2(3A + B - 8)$$

$$= 12A + 6B - 34 = 0.$$

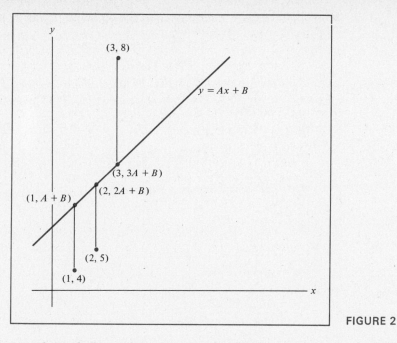

FIGURE 2

In order to find A and B, we must solve the system of simultaneous linear equations

$$28A + 12B = 76$$

$$12A + 6B = 34.$$

Multiplying the second equation by 2 and subtracting from the first equation, we have $4A = 8$ or $A = 2$. Therefore $B = \frac{5}{3}$, and the straight line that minimizes the least squares error is $y = 2x + \frac{5}{3}$.

We may follow a similar procedure for finding the line $y = Ax + B$ when more observed points are given.

EXAMPLE 2   The following table* gives the crude male death rate for lung cancer in 1950 and the per capita consumption of cigarettes in 1930 in various countries.

	Cigarette consumption (per capita)	Lung cancer deaths (per million males)
Norway	250	95
Sweden	300	120
Denmark	350	165
Australia	470	170

(a) Use the method of least squares to obtain the straight line that best fits these data.

---

* These data were obtained from *Smoking and Health*, Report of the Advisory Committee to the Surgeon General of the Public Health Service, U.S. Dept of Health, Education, and Welfare, Washington, D.C., Public Health Service Publication No. 1103, p. 176.

(b) In 1930 the per capita cigarette consumption in Finland was 1100. Use the straight line found in part (a) to estimate the male lung cancer death rate in Finland in 1950.

Solution

(a) The points are plotted in Fig. 3. Let the line that best fits these points have equation $y = Ax + B$. The least-squares error of the given data points is

$$E = (250A + B - 95)^2 + (300A + B - 120)^2$$

$$+ (350A + B - 165)^2 + (470A + B - 170)^2.$$

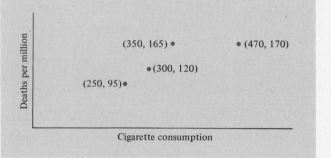

FIGURE 3

We determine the values of A and B for which E is minimized by taking the partial derivatives with respect to A and B and setting them equal to zero. Computing the partial derivatives, we get

$$\frac{\partial E}{\partial A} = 2(250A + B - 95) \cdot 250 + 2(300A + B - 120) \cdot 300$$

$$+ 2(350A + B - 165) \cdot 350 + 2(470A + B - 170) \cdot 470$$

$$\frac{\partial E}{\partial B} = 2(250A + B - 95) + 2(300A + B - 120)$$

$$+ 2(350A + B - 165) + 2(470A + B - 170).$$

If we set $\frac{\partial E}{\partial A} = 0$ and $\frac{\partial E}{\partial B} = 0$, the following system of equations results:

$$495{,}900A + 1370B = 197{,}400 \tag{1}$$

$$1370A + 4B = 550. \tag{2}$$

There are several ways to solve this system of equations. We will solve equation (2) for B and then insert this expression for B into equation (1) so as to obtain an equation in A alone. Solving (2) for B gives

$$B = \frac{550 - 1370A}{4} = 137.5 - 342.5A. \tag{3}$$

Now substitute (3) into (1) to obtain

$$495900A + 1370(137.5 - 342.5A) = 197,400$$
$$[495,900 - (1370)(342.5)]A = 197,400 - (1370)(137.5)$$
$$26,675A = 9025$$
$$A = \frac{9025}{26,675} \approx .338.$$

Then substitute this value for A into (3) to find

$$B = 137.5 - 342.5(.338) = 21.735.$$

Therefore the straight line that best fits these points is

$$y = .338x + 21.735.$$

(b) We use the straight line to estimate the lung cancer death rate in Finland by setting $x = 1100$. Then we get

$$y = .338(1100) + 21.735 = 393.535 \approx 394.$$

Therefore we estimate the lung cancer death rate in Finland to be 394.
(Note: The actual rate was 350.)

PRACTICE PROBLEMS 6

1. Let $E = (A + B + 2)^2 + (3A + B)^2 + (6A + B - 8)^2$. What is $\frac{\partial E}{\partial A}$?

2. Find the formula (of the type in Problem 1) that gives the least-squares error for the points (1, 10), (5, 8), (7, 0).

EXERCISES 6

Find the straight line that best fits the following data points, where "best" is meant in the sense of least squares.

1. $(1, 0), (2, -1), (3, -4)$
2. $(2, 2), (3, 0), (7, -1)$
3. $(1, 6), (3, 0), (6, 10)$
4. $(1, 1), (3, 2), (5, 4)$
5. $(0, 1), (1, 1), (2, 2), (3, 2)$
6. $(0, 2), (1, 2), (2, 2), (3, 3)$
7. $(1, 0), (2, 1), (4, 2), (5, 3)$
8. $(2, 3), (3, 2), (5, 1), (6, 0)$
9. $(1, 0), (2, 1), (3, 5), (15, 11), (-1, 4)$
10. $(0, 7), (-1, 5), (1, 9), (3, 14), (-2, 0)$

11. An ecologist wished to know whether certain species of aquatic insects have their ecological range limited by temperature. He collected the following data, relating the average daily temperature at different portions of a creek with the elevation of that portion of the creek (above sea level).*

---

* The authors express their thanks to Dr. J. David Allen, Department of Zoology, University of Maryland, for suggesting this exercise.

Elevation (kilometers)	Average temperature (Celsius)
2.7	11.2
2.8	10
3.0	8.5
3.5	7.5

(a) Find the straight line that provides the best least squares fit to these data.

(b) Use the linear function to estimate the average daily temperature for this creek at altitude 3.2 kilometers.

12. A soap manufacturer hopes to determine the demand curve for his soap by setting different prices in each of three cities of the same size and observing the volume of sales in each city. The results are given below.

Price (in cents)	Volume of sales (in thousands of cases)
30	120
35	100
40	90

(a) Let x be the price and y the volume. Find the straight line that best fits these data.

(b) Use the demand curve of part (a) to estimate the volume of sales if the price is 50 cents.

13. The trend of sales of new cars by a car dealer is as shown in the table.

Week number	Number of cars sold
1	38
2	40
3	41
4	39
5	45

Make a prediction of the number of cars sold during the seventh week.

14. A company analyzes production and profit figures and determines that in recent years productivity and profits have been related as shown in the table below.

Productivity (in thousands of units)	Profits
78	52,000
83	55,000
100	68,000
110	77,000
129	97,000

For the current year, their plant has been increased to allow production of 150,000 units. Estimate the profits.

1. $\dfrac{\partial E}{\partial A} = 2(A + B + 2)\cdot 1 + 2(3A + B)\cdot 3 + 2(6A + B - 8)\cdot 6$

   $= (2A + 2B + 4) + (18A + 6B) + (72A + 12B - 96)$

   $= 92A + 20B - 92.$

   [Notice that we used the general power rule when differentiating and so had to always multiply by the derivative of the quantity inside the parenthesis. Also, you might be tempted to first square the terms in the expression for E and then differentiate. We recommend that you resist that temptation.]

2. $E = (A + B - 10)^2 + (5A + B - 8)^2 + (7A + B)^2$. In general, E is a sum of squares, one for each point being fitted. The point (a, b) gives rise to the term $(aA + B - b)^2$.

8.7. Double Integrals

Up to this point our discussion of the calculus of functions of several variables has been confined to differentiation. Let us now take up the topic of integration of functions of several variables. As has been the case throughout most of this chapter, we will limit our discussion to functions $f(x, y)$ of two variables.

Our goal is to define for $f(x, y)$ an analogue of the definite integral. As motivation, let us briefly review the approximation of the definite integral $\int_a^b g(x)\, dx$ by Riemann sums. Let n be a positive integer. We subdivide the interval from $x = a$ to $x = b$ into n equal subintervals, each of length $\Delta x = (b - a)/n$. (See Fig. 1.) In each subinterval we choose a point, say x_1 in the first, x_2 in the

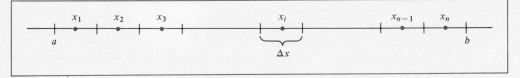

FIGURE 1

second, and so forth. Then the definite integral $\int_a^b g(x)\, dx$ may be approximated by a Riemann sum.

$$\int_a^b g(x)\, dx \approx g(x_1)\,\Delta x + g(x_2)\,\Delta x + \ldots + g(x_n)\,\Delta x.$$

As n increases, and consequently Δx becomes arbitrarily small, this approximation becomes increasingly accurate.

Based on the discussion above, let us first define Riemann sums for a function $f(x, y)$ of two variables and use them to define an analogue of the definite integral.

The two-variable analogue of the interval from $x = a$ to $x = b$ is a region R in the x-y plane [Fig. 2(a)]. Let us cover the region R with small rectangles of size Δx by Δy, where Δx and Δy are small [Fig. 2(b)]. Note that we include in our sub-

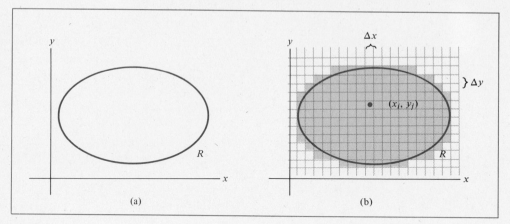

(a)   (b)

FIGURE 2

division any rectangle which touches the region. In each rectangle we choose a point (x_i, y_j) which is also in R [Fig. 2(b)]. Then a *Riemann sum for $f(x, y)$ over the region R* is a sum of all expressions of the form

$$f(x_i, y_j) \, \Delta x \, \Delta y,$$

where the sum is over all rectangles touching the region R.

Analogous to the one variable case, we allow Δx and Δy to become arbitrarily small. The shaded rectangles then approximate R more closely, and the Riemann sums approach a specific number. This number is called *the double integral of $f(x, y)$ over R* and is denoted

$$\iint_R f(x, y) \, dx \, dy.$$

It is fortunate that, in order to calculate double integrals, it is not usually necessary to resort to Riemann sums. Rather, we may utilize what are called *iterated integrals*—that is, integrals of the form

$$\int_c^d \left(\int_a^b f(x, y) \, dx \right) dy. \tag{1}$$

To explain the meaning of such a collection of symbols, we proceed from the inside out. To evaluate the integral

$$\int_a^b f(x, y) \, dx,$$

we consider $f(x, y)$ as a function of x alone, with y viewed as a constant. The value of this integral is then a function of y, say $h(y)$, which we may integrate:

$$\int_c^d h(y) \, dy.$$

The result is a number, the value of the iterated integral (1).

EXAMPLE 1 Evaluate the iterated integral $\int_1^2 \left(\int_2^3 x^2 y \, dx \right) dy$.

Solution We treat the inner integral first. The variable is x, and so y is treated as a constant.

$$\int_2^3 x^2 y \, dx = \tfrac{1}{3} x^3 y \bigg|_{x=2}^{x=3} = \tfrac{1}{3}(3)^3 y - \tfrac{1}{3}(2)^3 y$$

$$= \tfrac{19}{3} y.$$

Now carry out the integration with respect to y:

$$\int_1^2 \left(\int_2^3 x^2 y \, dx \right) dy = \int_1^2 \tfrac{19}{3} y \, dy = \tfrac{19}{3} \cdot \tfrac{1}{2} y^2 \bigg|_1^2$$

$$= \tfrac{19}{6} \cdot 4 - \tfrac{19}{6} \cdot 1$$

$$= \tfrac{19}{2}.$$

Here is the relationship between iterated integrals and double integrals.

Let R be the rectangle in the x-y plane defined by the conditions $a \le x \le b$, $c \le y \le d$. Then

$$\iint_R f(x, y) \, dx \, dy = \int_c^d \left(\int_a^b f(x, y) \, dx \right) dy.$$

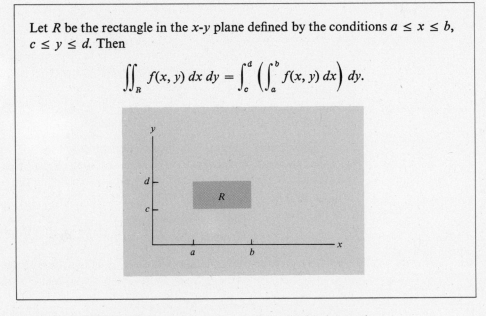

We may use this fundamental fact to calculate double integrals.

EXAMPLE 2   Evaluate $\iint_R (x^2 - 3y^2)\, dx\, dy$, where R is the rectangle $0 \le x \le 1$, $0 \le y \le 2$.

Solution   The double integral equals an iterated integral.

$$\iint_R (x^2 - 3y^2)\, dx\, dy = \int_0^2 \left(\int_0^1 (x^2 - 3y^2)\, dx \right) dy$$

$$= \int_0^2 \left(\frac{x^3}{3} - 3y^2 x \right) \Big|_{x=0}^{x=1} dy$$

$$= \int_0^2 \left[\left(\frac{1}{3} - 3y^2 \right) - 0 \right] dy$$

$$= \left(\frac{y}{3} - y^3 \right) \Big|_0^2$$

$$= \frac{2}{3} - 8$$

$$= -\frac{22}{3}.$$

EXAMPLE 3   Evaluate the double integral $\iint_R e^{x+y}\, dx\, dy$, where R is the rectangle $0 \le x \le 3$, $1 \le y \le 2$.

Solution   The double integral equals

$$\int_1^2 \left(\int_0^3 e^{x+y}\, dx \right) dy = \int_1^2 \left(e^{x+y} \Big|_{x=0}^{x=3} \right) dy$$

$$= \int_1^2 (e^{y+3} - e^y)\, dy$$

$$= (e^{y+3} - e^y) \Big|_1^2$$

$$= (e^5 - e^2) - (e^4 - e^1)$$

$$\approx 89.14.$$

The double integral has many applications, most of which are beyond the scope of this text. However, let us quote one such. Recall that the definite integral of a function $f(x)$ of one variable can be used to define the average value of $f(x)$ from $x = a$ to $x = b$ via the formula

$$\begin{bmatrix} \text{average value of } f(x) \\ \text{from } x = a \text{ to } x = b \end{bmatrix} = \frac{1}{b-a} \int_a^b f(x)\, dx.$$

Similarly, double integrals may be used to define the average value of a function $f(x, y)$ as x and y range independently from a to b and c to d, respectively:

$$\begin{bmatrix} \text{average value of } f(x, y), \\ a \le x \le b,\ c \le y \le d \end{bmatrix} = \frac{1}{(b-a)(d-c)} \iint_R f(x, y)\, dx\, dy.$$

1. Calculate the iterated integral,

$$\int_2^4 \left(\int_0^3 (x^2 y - 2y) \, dx \right) dy.$$

2. Let R be the rectangle consisting of all points (x, y) such that $0 \le x \le 3, 2 \le y \le 4$. Calculate

$$\iint_R (x^2 y - 2y) \, dx \, dy.$$

EXERCISES 7

Calculate the following iterated integrals.

1. $\displaystyle\int_0^1 \left(\int_0^1 e^{x+y} \, dx \right) dy$ 2. $\displaystyle\int_{-1}^1 \left(\int_0^2 (x^2 + 3y^3) \, dx \right) dy$

3. $\displaystyle\int_{-2}^0 \left(\int_{-1}^1 y e^{xy} \, dx \right) dy$ 4. $\displaystyle\int_1^3 \left(\int_2^5 (2x - 3y) \, dx \right) dy$

Let R be the rectangle consisting of all points (x, y) such that $0 \le x \le 2$, $2 \le y \le 3$. Calculate the following double integrals.

5. $\displaystyle\iint_R xy^2 \, dx \, dy$ 6. $\displaystyle\iint_R (xy - y^2) \, dx \, dy$

7. $\displaystyle\iint_R e^{-x-y} \, dx \, dy$ 8. $\displaystyle\iint_R e^{y-x} \, dx \, dy$

SOLUTIONS TO PRACTICE PROBLEMS 7

1. $\displaystyle\int_2^4 \left(\int_0^3 (x^2 y - 2y) \, dx \right) dy = \int_2^4 \left(\frac{x^3}{3} y - 2yx \right) \Big|_{x=0}^{x=3} dy$

$$= \int_2^4 \left[\left(\frac{27}{3} y - 2y \cdot 3 \right) - (0 - 0) \right] dy$$

$$= \int_2^4 3y \, dy$$

$$= \frac{3}{2} y^2 \Big|_2^4$$

$$= \frac{3}{2} \cdot 4^2 - \frac{3}{2} \cdot 2^2 = 18.$$

26. Using the method of Lagrange multipliers, find the values of x, y that minimize the function $-x^2 - 3xy - \frac{1}{2}y^2 + y + 10$, subject to the constraint $10 - x - y = 0$.

27. Using the method of Lagrange multipliers, find the values of x, y, z that minimize the function $3x^2 + 2y^2 + z^2 + 4x + y + 3z$, subject to the constraint $4 - x - y - z = 0$.

28. Using the method of Lagrange multipliers, find the dimensions of the rectangular box of volume 1000 cubic inches for which the sum of the dimensions is minimized.

29. A person wants to plant a rectangular garden along one side of a house and put a fence on the other three sides of the garden. (See Fig. 1). Using the method of Lagrange multipliers, find the dimensions of the garden of greatest area that can be enclosed by using 40 feet of fencing.

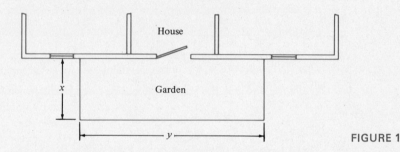

FIGURE 1

30. The solution to Exercise 29 is $x = 10$, $y = 20$, $\lambda = 10$. Suppose that one additional foot of fencing becomes available and that now the optimal dimensions for the garden are $10 + h$ and $20 + k$, where $2h + k = 1$. Use the total differential to show that the increased area is approximately $\lambda = 10$ square feet. In other words, λ is the marginal change in area with respect to change in the length of fencing.

31. Let $f(x, y) = x\sqrt{y}$. Use the total differential to approximate $f(12.1, 3.8)$.

32. Let $f(x, y) = x^5 + (y/x)$. Use the total differential to approximate $f(.98, 2.99)$.

33. Let $f(x, y, z) = (x^2 + y)/(z - 1)$. Use the total differential to approximate $f(1.98, 1.99, 2.01)$.

34. The surface area of a square-ended rectangular box is $2x^2 + 4xy$, where x is the dimension of the square side. Such a box is measured and found to have $x = 30$ millimeters and $y = 50$ millimeters. Estimate the maximum error in calculating the surface area of the box if each measurement is in error by at most 1 millimeter.

In Exercises 35–37 find the straight line that best fits the following data points, where "best" is meant in the sense of least squares.

35. $(1, 1), (2, 3), (3, 6)$ 36. $(1, 1), (3, 4), (5, 7)$

37. $(0, 1), (1, -1), (2, -3), (3, -5)$

In Exercises 38 and 39 calculate the iterated integral.

38. $\int_0^4 \left(\int_0^1 x\sqrt{y} + y \, dx \right) dy$

39. $\int_1^2 \left(\int_0^5 2xy^4 + 3 \, dx \right) dy$

In Exercises 40 and 41 let R be the rectangle consisting of all points (x, y) such that $0 \le x \le 4$, $1 \le y \le 3$, and calculate the double integral.

40. $\iint_R 2x + 3y \, dx \, dy$

41. $\iint_R 5 \, dx \, dy$

THE TRIGONOMETRIC FUNCTIONS

In this chapter we expand the collection of functions to which we can apply calculus by introducing the trigonometric functions. As we shall see, these functions are *periodic*. That is, after a certain point their graphs repeat themselves. This repetitive phenomenon is not displayed by any of the functions that we have considered until now. Yet many natural phenomena are repetitive or cyclical—for example, the motion of the planets in our solar system, earthquake vibrations, and the natural rhythm of the heart. Thus the functions introduced in this chapter add considerably to our capacity to describe physical processes.

9.1. Radian Measure of Angles

The ancient Babylonians introduced angle measurement in terms of degrees, minutes, and seconds, and these units are still generally used today for navigation and practical measurements. In calculus, however, it is more convenient to measure angles in terms of *radians*, for in this case the differentiation formulas for the trigonometric functions are easier to remember and use. Also, the radian is becoming more widely used today in scientific work because it is the unit of angle measurement in the international metric system (Système International d'Unites).

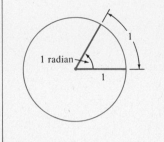

In order to define a radian, we consider a circle of radius 1 and measure angles in terms of distances around the circumference. The central angle determined by an arc of length 1 along the circumference is said to have a measure of *one radian*. (See Fig. 1.) Since the circumference of the circle of radius 1 has length 2π, there are 2π radians in one full revolution of the circle. Equivalently,

$$360° = 2\pi \text{ radians.} \tag{1}$$

FIGURE 1

The following important relations should be memorized (see Fig. 2):

$$90° = \frac{\pi}{2} \text{ radians} \qquad \text{(one quarter-revolution)}$$

$$180° = \pi \text{ radians} \qquad \text{(one half-revolution)}$$

$$270° = \frac{3\pi}{2} \text{ radians} \qquad \text{(three quarter-revolutions)}$$

$$360° = 2\pi \text{ radians} \qquad \text{(one full revolution).}$$

FIGURE 2

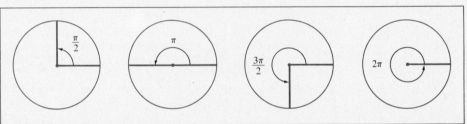

From (1) we see that

$$1° = \frac{2\pi}{360} \text{ radians} = \frac{\pi}{180} \text{ radians}.$$

If d is any number, then

$$d° = d \times \frac{\pi}{180} \text{ radians}. \qquad (2)$$

That is, in order to convert degrees to radians, multiply the number of degrees by $\pi/180$.

EXAMPLE 1   Convert 45°, 60°, and 135° into radians.

Solution
$$45° = \overset{1}{\cancel{45}} \times \frac{\pi}{\underset{4}{\cancel{180}}} \text{ radians} = \frac{\pi}{4} \text{ radians}$$

$$60° = \overset{1}{\cancel{60}} \times \frac{\pi}{\underset{3}{\cancel{180}}} \text{ radians} = \frac{\pi}{3} \text{ radians}$$

$$135° = \overset{3}{\cancel{135}} \times \frac{\pi}{\underset{4}{\cancel{180}}} \text{ radians} = \frac{3\pi}{4} \text{ radians}.$$

These three angles are shown in Fig. 3.

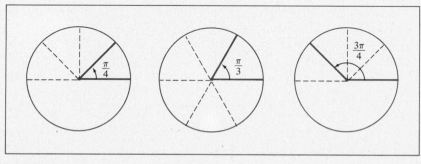

FIGURE 3

We usually omit the word "radian" when measuring angles, because all our angle measurements will be in radians unless degrees are specifically indicated.

For our purposes, it is important to be able to speak of negative as well as positive angles, so let us define what we mean by a negative angle. We shall usually consider angles that are in *standard position* on a coordinate system, with the vertex of the angle at $(0, 0)$ and one side, called the "initial side," along the positive x-axis. We measure such an angle from the initial side to the "terminal side," where a *counterclockwise angle is positive* and a *clockwise angle is negative*. Some examples are given in Fig. 4.

Notice in Fig. 4(a) and (b) how essentially the same picture can describe more than one angle.

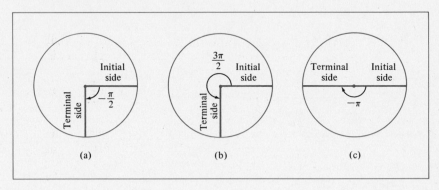

FIGURE 4

By considering angles formed from more than one revolution (in the positive or negative direction), we can construct angles whose measure is of arbitrary size (i.e., not necessarily between -2π and 2π). Some examples are illustrated in Fig. 5.

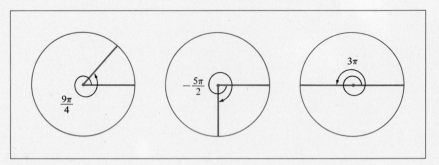

FIGURE 5

EXAMPLE 2 (a) What is the radian measure of the angle in Fig. 6?

(b) Construct an angle of $5\pi/2$ radians.

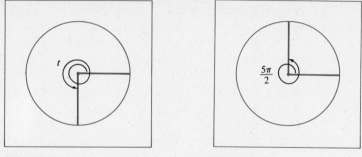

FIGURE 6   FIGURE 7

Solution (a) The angle described in Fig. 6 consists of one full revolution (2π radians) plus three quarter-revolutions [$3 \times (\pi/2)$ radians]. That is,

$$t = 2\pi + 3 \times \frac{\pi}{2} = 4 \times \frac{\pi}{2} + 3 \times \frac{\pi}{2} = \frac{7\pi}{2}.$$

(b) Think of $5\pi/2$ radians as $5 \times (\pi/2)$ radians—that is, five quarter-revolutions of the circle. This is one full revolution plus one quarter-revolution. An angle of $5\pi/2$ radians is shown in Fig. 7.

PRACTICE PROBLEMS 1

1. A right triangle has one angle of $\pi/3$ radians. What are the other angles?

2. How many radians in an angle of $-780°$? Draw the angle.

EXERCISES 1

Convert the following to radian measure.

1. 30°, 120°, 315°   2. 18°, 72°, 150°

3. 450°, $-210°$, $-90°$ 4. 990°, $-270°$, $-540°$

Give the radian measure of each angle described below.

5.   6.   7.

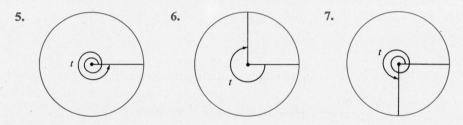

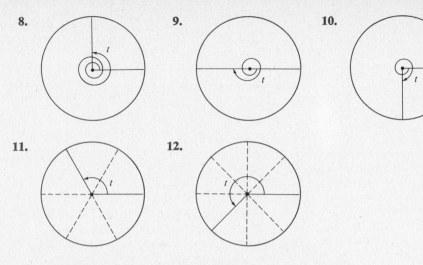

8. **9.** **10.**

11. **12.**

Construct angles with the following radian measure.

13. $3\pi/2$, $3\pi/4$, 5π **14.** $\pi/3$, $5\pi/2$, 6π

15. $-\pi/3$, $-3\pi/4$, $-7\pi/2$ **16.** $-\pi/4$, $-3\pi/2$, -3π

17. $\pi/6$, $-2\pi/3$, $-\pi$ **18.** $2\pi/3$, $-\pi/6$, $7\pi/2$

SOLUTIONS TO PRACTICE PROBLEMS 1

1. The sum of the angles of a triangle is $180°$ or π radians. Since a right angle is $\pi/2$ radians and one angle is $\pi/3$ radians, the remaining angle is $\pi - (\pi/2 + \pi/3) = \pi/6$ radians.

2. $-780° = -780 \times (\pi/180)$ radians $= -13\pi/3$ radians. Since $-13\pi/3 = -4\pi - \pi/3$, we draw the angle by making first two revolutions in the negative direction and then a rotation of $\pi/3$ in the negative direction.

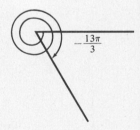

$-\dfrac{13\pi}{3}$

9.2. The Sine and the Cosine

Given a number t, we consider an angle of t radians placed in standard position, as in Fig. 1, and we let P be a point on the terminal side of this angle. Denote the coordinates of P by (x, y) and let r be the length of the segment OP; that is,

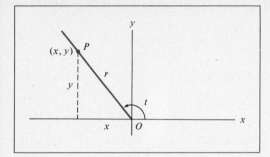

FIGURE 1

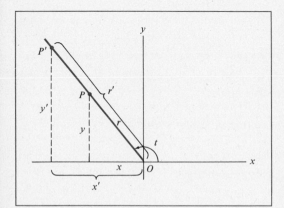

FIGURE 2

$r = \sqrt{x^2 + y^2}$. The *sine* and *cosine* of t, denoted by $\sin t$ and $\cos t$, respectively, are defined by the ratios

$$\sin t = \frac{y}{r}$$

$$\cos t = \frac{x}{r}.$$

(1)

It does not matter which point on the ray through P we use to define $\sin t$ and $\cos t$. If $P' = (x', y')$ is another point on the same ray and if r' is the length of OP' (Fig. 2), then, by properties of similar triangles, we have

$$\frac{y'}{r'} = \frac{y}{r} = \sin t$$

$$\frac{x'}{r'} = \frac{x}{r} = \cos t.$$

Three examples that illustrate the definition of $\sin t$ and $\cos t$ are shown in Fig. 3. We have included a table of approximate values of $\sin t$ and $\cos t$ for various values of t between 0 and 2π. (See Table 3, page A 7)

When $0 < t < \pi/2$, the values of $\sin t$ and $\cos t$ may be expressed as ratios of the lengths of the sides

FIGURE 3

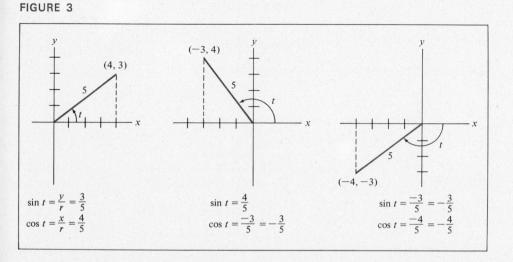

of a right triangle. Indeed, if we are given a right triangle as in Fig. 4, then we have

$$\sin t = \frac{\text{opposite}}{\text{hypotenuse}}, \qquad \cos t = \frac{\text{adjacent}}{\text{hypotenuse}}. \tag{2}$$

A typical application of (2) appears in Example 1.

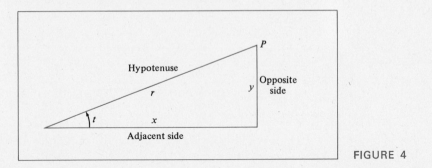

FIGURE 4

EXAMPLE 1   The hypotenuse of a right triangle is four units and one angle is .7 radian. Determine the length of the side opposite this angle.

Solution   See Fig. 5. Since $y/4 = \sin .7$, we have (using Table 3, page A7),

$$y = 4 \sin .7$$
$$= 4(.64422) = 2.57688.$$

Another way to describe the sine and cosine functions is to choose the point P in Fig. 1 so that $r = 1$. That is, choose P on the unit circle. (See Fig. 6.) In this case

$$\sin t = \frac{y}{1} = y$$

$$\cos t = \frac{x}{1} = x. \tag{3}$$

FIGURE 5

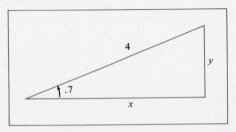

FIGURE 6

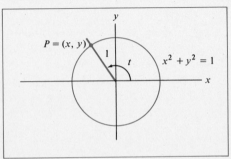

So the y-coordinate of P is $\sin t$ and the x-coordinate of P is $\cos t$. Thus we have the following result.

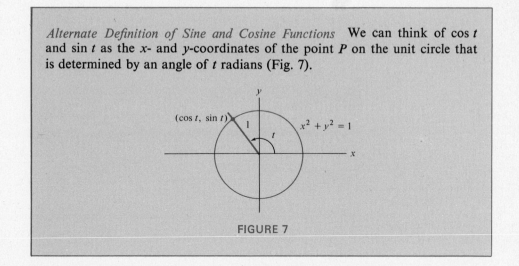

Alternate Definition of Sine and Cosine Functions We can think of $\cos t$ and $\sin t$ as the x- and y-coordinates of the point P on the unit circle that is determined by an angle of t radians (Fig. 7).

FIGURE 7

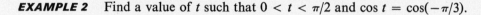

EXAMPLE 2 Find a value of t such that $0 < t < \pi/2$ and $\cos t = \cos(-\pi/3)$.

Solution On the unit circle we locate the point P that is determined by an angle of $-\pi/3$ radians. The x-coordinate of P is $\cos(-\pi/3)$. There is another point Q on the unit circle with the same x-coordinate. (See Fig. 8.) Let t be the radian measure of the angle determined by Q. Then

$$\cos t = \cos\left(-\frac{\pi}{3}\right)$$

because Q and P have the same x-coordinate. Also $0 < t < \pi/2$. From the symmetry of the diagram it is clear that $t = \pi/3$.

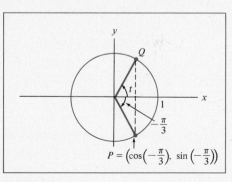

FIGURE 8

Properties of the Sine and Cosine Functions Each number t determines a point $(\cos t, \sin t)$ on the unit circle $x^2 + y^2 = 1$ as in Fig. 7. Therefore $(\cos t)^2 + (\sin t)^2 = 1$. It is convenient (and traditional) to write $\sin^2 t$ instead of $(\sin t)^2$ and $\cos^2 t$ instead of $(\cos t)^2$. Thus we can write the last formula as follows:

$$\cos^2 t + \sin^2 t = 1. \tag{4}$$

The numbers t and $t \pm 2\pi$ determine the same point on the unit circle (because 2π represents a full revolution of the circle). But $t + 2\pi$ and $t - 2\pi$ correspond to the points $(\cos(t + 2\pi), \sin(t + 2\pi))$ and $(\cos(t - 2\pi), \sin(t - 2\pi))$, respectively. Hence

$$\cos(t \pm 2\pi) = \cos t, \qquad \sin(t \pm 2\pi) = \sin t. \tag{5}$$

Figure 9(a) illustrates another property of the sine and cosine—namely,

$$\cos(-t) = \cos t, \qquad \sin(-t) = -\sin t. \tag{6}$$

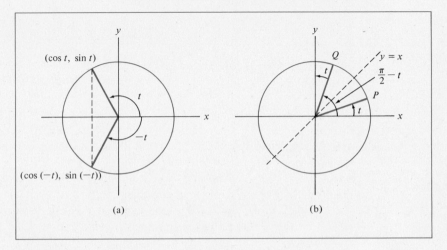

FIGURE 9

Figure 9(b) shows that the points P and Q corresponding to t and to $\pi/2 - t$ are reflections of each other through the line $y = x$. Consequently, the coordinates of Q are obtained by interchanging the coordinates of P. This means that

$$\cos\left(\frac{\pi}{2} - t\right) = \sin t, \qquad \sin\left(\frac{\pi}{2} - t\right) = \cos t. \tag{7}$$

The equations in (4) to (7) are called *identities* because they hold for all values of t. Another identity that holds for all numbers s and t is

$$\sin(s + t) = \sin s \cos t + \cos s \sin t. \tag{8}$$

A proof of (8) may be found in any introductory book on trigonometry. There are a number of other identities concerning trigonometric functions, but we shall not need them here.

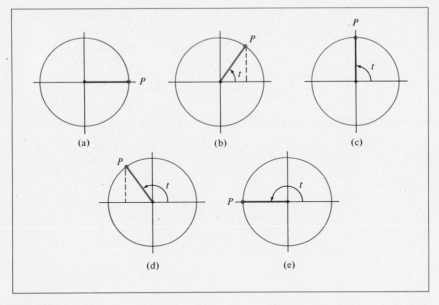

FIGURE 10

The Graph of sin t Let us analyze what happens to sin t as t increases from 0 to π. When $t = 0$, the point $P = (\cos t, \sin t)$ is at $(1, 0)$, as in Fig. 10(a). As t increases, P moves counterclockwise around the unit circle. The y-coordinate of P—that is, sin t—increases until $t = \pi/2$, where $P = (0, 1)$. [See Fig. 10(c).] As t increases from $\pi/2$ to π, the y-coordinate of P—that is, sin t—decreases from 1 to 0. See Fig. 10(d) and (e).

Part of the graph of sin t is sketched in Fig. 11, using Table 3. Notice that,

FIGURE 11

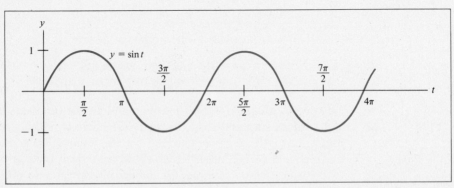

for t between 0 and π, the values of $\sin t$ increase from 0 to 1 and then decrease back to 0, just as we predicted from Fig. 10. For t between π and 2π, the values of $\sin t$ are negative. Can you explain why? The graph of $y = \sin t$ for t between 2π and 4π is exactly the same as the graph for t between 0 and 2π. This result follows from formula (5). We say that the sine function is *periodic with period 2π* because the graph repeats itself every 2π units. We can use this fact to make a quick sketch of part of the graph for negative values of t (Fig. 12).

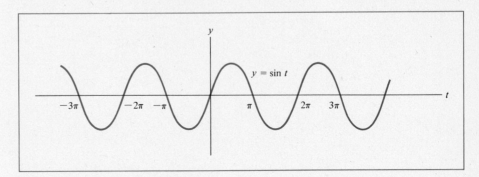

FIGURE 12   Graph of the sine function.

The Graph of cos t   By analyzing what happens to the first coordinate of the point $(\cos t, \sin t)$ as t varies, we obtain the graph of $\cos t$. Note from Fig. 13 that the graph of the cosine function is also periodic with period 2π.

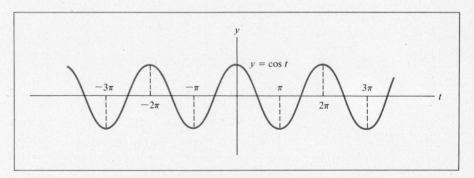

FIGURE 13   Graph of the cosine function.

Note that since the sine and cosine are periodic functions, the tables need not give $\sin t$ and $\cos t$ for values of t larger than 2π.

A Remark about Notation: The sine and cosine functions assign to each number t the values $\sin t$ and $\cos t$, respectively. There is nothing special, however, about the letter t. Although we chose to use the letters t, x, y, and r in the *definition* of the sine and cosine, other letters could have been used as well. Now that the sine and cosine of every number are defined, we are free to use *any* letter to represent the independent variable.

PRACTICE PROBLEMS 2

Use Table 3 on page A7 to solve the following problems.

1. Find $\sin(-2.5)$ and $\cos(-2.5)$.

2. Estimate the value of t in Fig. 14.

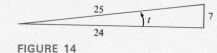

FIGURE 14

EXERCISES 2

In Exercises 1 to 12 give the values of $\sin t$ and $\cos t$, where t is the radian measure of the angle shown.

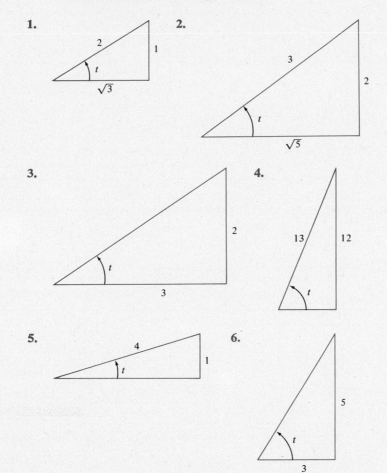

1.

2.

3.

4.

5.

6.

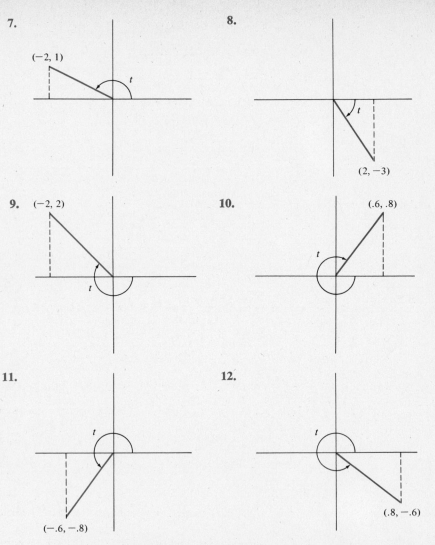

7. (−2, 1)   *t*

8. (2, −3)   *t*

9. (−2, 2)   *t*

10. (.6, .8)   *t*

11. (−.6, −.8)   *t*

12. (.8, −.6)   *t*

Exercises 13 through 20 refer to various right triangles whose sides and angles are labeled as in the accompanying sketch. Round off all lengths of sides to one decimal place.

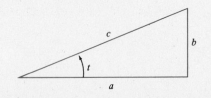

13. Estimate *t* if $a = 12$, $b = 5$, and $c = 13$. (Use Table 3, page A7.)

14. If $t = 1.1$ and $c = 10.0$, find *b*.   **15.** If $t = 1.1$ and $b = 3.2$, find *c*.

16. If $t = .4$ and $c = 5.0$, find a.

17. If $t = .4$ and $a = 10.0$, find c.

18. If $t = .9$ and $c = 20.0$, find a and b.

19. If $t = .5$ and $a = 2.4$, find b and c.

20. If $t = 1.1$ and $b = 3.5$, find a and c.

Find t such that $0 \le t \le \pi$ and t satisfies the stated condition.

21. $\cos t = \cos(-\pi/6)$ 22. $\cos t = \cos(3\pi/2)$

23. $\cos t = \cos(5\pi/4)$ 24. $\cos t = \cos(-4\pi/6)$

25. $\cos t = \cos(-5\pi/8)$ 26. $\cos t = \cos(-3\pi/4)$

Find t such that $-\pi/2 \le t \le \pi/2$ and t satisfies the stated condition.

27. $\sin t = \sin(3\pi/4)$ 28. $\sin t = \sin(7\pi/6)$

29. $\sin t = \sin(-4\pi/3)$ 30. $\sin t = -\sin(3\pi/8)$

31. $\sin t = -\sin(\pi/6)$ 32. $\sin t = -\sin(-\pi/3)$

33. $\sin t = \cos t$ 34. $\sin t = -\cos t$

35. By referring to Fig. 10, describe what happens to $\cos t$ as t increases from 0 to π.

36. Use the unit circle to describe what happens to $\sin t$ as t increases from π to 2π.

37. Determine the value of $\sin t$ when $t = 5\pi, -2\pi, 17\pi/2, -13\pi/2$.

38. Determine the value of $\cos t$ when $t = 5\pi, -2\pi, 17\pi/2, -13\pi/2$.

SOLUTIONS TO PRACTICE PROBLEMS 2

1. Table 3 does not give the values of the trigonometric functions for negative values of t. However, by (6),

$$\cos(-2.5) = \cos 2.5, \qquad \sin(-2.5) = -\sin 2.5.$$

Therefore, $\cos(-2.5) = -.80114$ and $\sin(-2.5) = -.59847$.

2. By (2),

$$\sin t = \frac{\text{opposite}}{\text{hypotenuse}} = \frac{7}{25} = .28.$$

Look down the $\sin t$ column in Table 3 to find the number that is closest to .28. The best we can do is .29552, which is $\sin(.3)$. Therefore, $t \approx .3$. [We could get a more accurate estimate of t by using a more extensive table or an electronic calculator. To three decimal places t is .284.]

9.3. Differentiation of sin *t* and cos *t*

In this section we study the two differentiation rules

$$\frac{d}{dt} \sin t = \cos t \qquad (1)$$

$$\frac{d}{dt} \cos t = -\sin t. \qquad (2)$$

It is not difficult to see why these rules might be true. Formula (1) says that the slope of the curve $y = \sin t$ at a particular value of t is given by the corresponding value of $\cos t$. To check it, we draw a careful graph of $y = \sin t$ and estimate the slope at various points, as indicated in Fig. 1. Let us plot the slope as a function

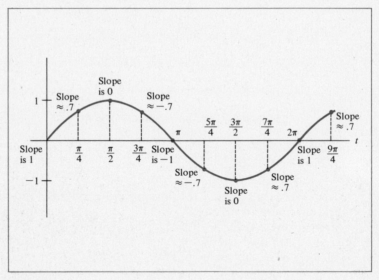

FIGURE 1

of t (Fig. 2). As can be seen, the "slope function" (i.e., the derivative) of $\sin t$ has a graph similar to the curve $y = \cos t$. Thus formula (1) seems to be reasonable. A similar analysis of the graph of $y = \cos t$ would show why (2) might be true. Proofs of these differentiation rules are outlined in an appendix at the end of this section.

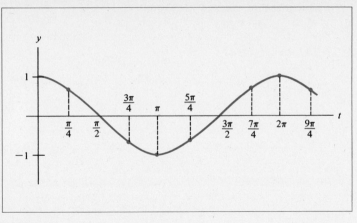

FIGURE 2

Combining (1), (2), and the chain rule, we obtain the following general rules.

$$\frac{d}{dt} \sin g(t) = [\cos g(t)]g'(t) \qquad (3)$$

$$\frac{d}{dt} \cos g(t) = [-\sin g(t)]g'(t). \qquad (4)$$

EXAMPLE 1 Differentiate

(a) $\sin 3t$.

(b) $(t^2 + 3 \sin t)^5$.

Solution (a) $\frac{d}{dt} (\sin 3t) = (\cos 3t) \frac{d}{dt} (3t) = (\cos 3t) \cdot 3 = 3 \cos 3t$.

(b) $\frac{d}{dt} (t^2 + 3 \sin t)^5 = 5(t^2 + 3 \sin t)^4 \cdot \frac{d}{dt} (t^2 + 3 \sin t)$

$$= 5(t^2 + 3 \sin t)^4(2t + 3 \cos t).$$

EXAMPLE 2 Differentiate

(a) $\cos(t^2 + 1)$.

(b) $\cos^2 t$.

Solution (a) $\frac{d}{dt} \cos(t^2 + 1) = -\sin(t^2 + 1) \frac{d}{dt} (t^2 + 1) = -\sin(t^2 + 1) \cdot (2t)$

$$= -2t \sin(t^2 + 1).$$

(b) The notation $\cos^2 t$ means $(\cos t)^2$.

$$\frac{d}{dt}\cos^2 t = \frac{d}{dt}(\cos t)^2 = 2(\cos t)\frac{d}{dt}\cos t$$

$$= -2\cos t \sin t.$$

EXAMPLE 3 Differentiate

(a) $t^2 \cos 3t$.

(b) $(\sin 2t)/t$.

Solution (a) From the product rule we have

$$\frac{d}{dt}(t^2 \cos 3t) = t^2 \frac{d}{dt}\cos 3t + (\cos 3t)\frac{d}{dt}t^2$$

$$= t^2(-3\sin 3t) + (\cos 3t)(2t)$$

$$= -3t^2 \sin 3t + 2t \cos 3t.$$

(b) From the quotient rule we have

$$\frac{d}{dt}\left(\frac{\sin 2t}{t}\right) = \frac{t\dfrac{d}{dt}\sin 2t - (\sin 2t)\cdot 1}{t^2}$$

$$= \frac{2t \cos 2t - \sin 2t}{t^2}$$

EXAMPLE 4 A V-shaped trough is to be constructed with sides that are 200 centimeters long and 30 centimeters wide (Fig. 3). Find the angle t between the sides that maximizes the capacity of the trough.

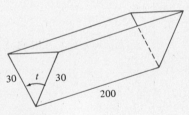

FIGURE 3

Solution The volume of the trough is its length times its cross-sectional area. Since the length is constant, it suffices to maximize the cross-sectional area. Let us rotate the diagram of a cross section so that one side is horizontal (Fig. 4). Note that $h/30 = \sin t$, and so $h = 30 \sin t$. Thus the area A of the cross section is

$$A = \tfrac{1}{2}\cdot\text{base}\cdot\text{height}$$

$$= \tfrac{1}{2}(30)(h) = 15(30 \sin t) = 450 \sin t.$$

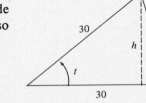

FIGURE 4

To find where A is a maximum, we set the derivative equal to zero and solve for t,

$$\frac{dA}{dt} = 0$$

$$450 \cos t = 0.$$

Physical considerations force us to consider only values of t between 0 and π. From the graph of $y = \cos t$ we see that $t = \pi/2$ is the only value of t between 0 and π that makes $\cos t = 0$. So in order to maximize the volume of the trough, the two sides should be perpendicular to one another.

EXAMPLE 5 Calculate the following indefinite integrals.

(a) $\displaystyle\int \sin t \, dt$ (b) $\displaystyle\int \sin 3t \, dt$.

Solution (a) Since $\dfrac{d}{dt}(-\cos t) = \sin t$, we have

$$\int \sin t \, dt = -\cos t + C.$$

where C is an arbitrary constant.

(b) From part (a) we guess that an antiderivative of $\sin 3t$ should resemble the function $-\cos 3t$. However, if we differentiate, we find that

$$\frac{d}{dt}(-\cos 3t) = (\sin 3t) \cdot \frac{d}{dt}(3t)$$

$$= 3 \sin 3t,$$

which is three times too much. So we multiply this last equation by $\frac{1}{3}$ to derive that

$$\frac{d}{dt}\left(-\frac{1}{3}\cos 3t\right) = \sin 3t,$$

so

$$\int \sin 3t \, dt = -\frac{1}{3}\cos 3t + C.$$

EXAMPLE 6 Find the area under the curve $y = \sin 3t$ from $t = 0$ to $t = \pi/3$.

Solution The area is shaded in the accompanying figure.

[Shaded area]

$$= \int_0^{\pi/3} \sin 3t \, dt$$

$$= -\tfrac{1}{3}\cos 3t \,\Big|_0^{\pi/3}$$

$$= -\tfrac{1}{3}\cos 3 \cdot \frac{\pi}{3} - (-\tfrac{1}{3}\cos 0)$$

$$= -\tfrac{1}{3}\cos \pi + \tfrac{1}{3}\cos 0$$

$$= \tfrac{1}{3} + \tfrac{1}{3} = \tfrac{2}{3}.$$

As we mentioned earlier, the trigonometric functions are required to model situations which are repetitive (or periodic). The next example illustrates such a situation.

EXAMPLE 7   In many mathematical models used to study the interaction between predators and prey, both the number of predators and the number of prey are described by periodic functions. Suppose that in one such model, the number of predators (in a particular geographical region) at time t is given by $N(t) = 5000 + 2000 \cos(2\pi t/36)$, where t is measured in months from June 1, 1980.

(a) At what rate is the number of predators changing on August 1, 1980?

(b) What is the average number of predators during the time interval from June 1, 1980, to June 1, 1983?

Solution   (a) The date August 1, 1980, corresponds to $t = 2$. The rate of change of $N(t)$ is given by the derivative $N'(t)$:

$$N'(t) = \frac{d}{dt}\left[5000 + 2000 \cos\left(\frac{2\pi t}{36}\right)\right]$$

$$= 2000\left[-\sin\left(\frac{2\pi t}{36}\right) \cdot \left(\frac{2\pi}{36}\right)\right]$$

$$= -\frac{1000\pi}{9}\sin\left(\frac{2\pi t}{36}\right),$$

$$N'(2) = -\frac{1000\pi}{9}\sin\left(\frac{\pi}{9}\right)$$

$$\approx -119.$$

Thus, on August 1, 1980, the number of predators is decreasing at the rate of 119 per month.

(b) The time interval from June 1, 1980, to June 1, 1983, corresponds to $t = 0$ to $t = 36$. The average value of $N(t)$ over this interval is

$$\frac{1}{36 - 0}\int_0^{36} N(t)\,dt = \frac{1}{36}\int_0^{36}\left[5000 + 2000\cos\left(\frac{2\pi t}{36}\right)\right]dt$$

$$= \frac{1}{36}\left[5000t + \frac{2000}{2\pi/36}\sin\left(\frac{2\pi t}{36}\right)\right]\Bigg|_0^{36}$$

$$= \frac{1}{36}\left[5000 \cdot 36 + \frac{2000}{2\pi/36}\sin(2\pi)\right]$$

$$- \frac{1}{36}\left[5000 \cdot 0 + \frac{2000}{2\pi/36}\sin(0)\right]$$

$$= 5000.$$

We have sketched the graph of $N(t)$ in Fig. 5. Note how $N(t)$ oscillates around 5000, the average value.

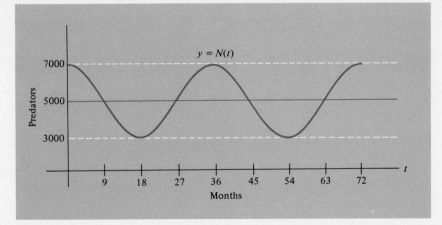

FIGURE 5   Periodic fluctuation of a predator population.

APPENDIX   Proofs of the Differentiation Rules for sin t and cos t

First, let us examine the derivatives of cos t and sin t at $t = 0$. The function cos t has a maximum at $t = 0$; consequently, its derivative there must be zero. See Fig. 6(a). If we approximate the tangent line at $t = 0$ by a secant line, as in Fig. 6(b), then the slope of the secant line must approach 0 as $h \to 0$. Since the slope of the secant line is $(\cos h - 1)/h$, we conclude that

$$\lim_{h \to 0} \frac{\cos h - 1}{h} = 0. \tag{5}$$

FIGURE 6

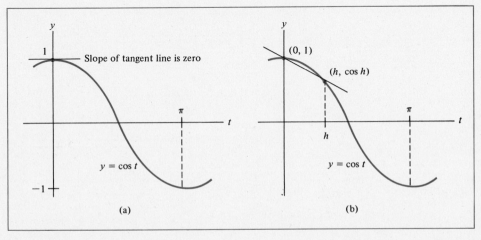

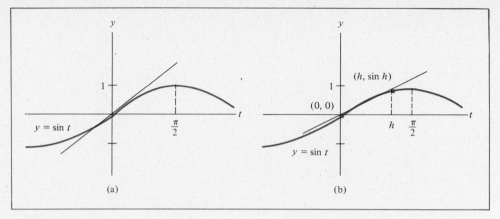

FIGURE 7

It appears from the graph of $y = \sin t$ that the tangent line at $t = 0$ has slope 1 [Fig. 7(a)]. If it does, then the slope of the approximating secant line [Fig. 7(b)] must approach 1. Since the slope of the secant line in Fig. 7(b) is $(\sin h)/h$, this would imply

$$\lim_{h \to 0} \frac{\sin h}{h} = 1. \tag{6}$$

We can check (6) for small values of h with a calculator.

h	.1	.01	.001
$\sin h$	.099833417	.009999833	.0009999998
$\dfrac{\sin h}{h}$	.99833417	.9999833	.9999998

The numerical evidence does not prove (6), but it should be sufficiently convincing for our purposes.

To obtain the differentiation formula for $\sin t$, we approximate the slope of a tangent line by the slope of a secant line. (See Fig. 8). The slope of a secant line is

$$\frac{\sin(t + h) - \sin t}{h}.$$

From formula (8) of Section 2 we note that $\sin(t + h) = \sin t \cos h + \cos t \sin h$. Thus

$$[\text{slope of secant line}] = \frac{(\sin t \cos h + \cos t \sin h) - \sin t}{h}$$

$$= \frac{\sin t(\cos h - 1) + \cos t \sin h}{h}$$

$$= (\sin t)\,\frac{\cos h - 1}{h} + (\cos t)\,\frac{\sin h}{h}.$$

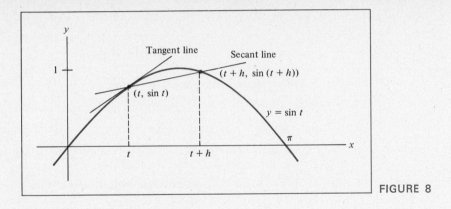

FIGURE 8

From (5) and (6) it follows that

$$\frac{d}{dt}\sin t = \lim_{h \to 0}\left[(\sin t)\frac{\cos h - 1}{h} + (\cos t)\frac{\sin h}{h}\right]$$

$$= (\sin t)\lim_{h \to 0}\frac{\cos h - 1}{h} + (\cos t)\lim_{h \to 0}\frac{\sin h}{h}$$

$$= (\sin t)\cdot 0 + (\cos t)\cdot 1$$

$$= \cos t.$$

A similar argument may be given to verify the formula for the derivative of cos t. Here is a shorter proof that uses the chain rule and the two identities

$$\cos t = \sin\left(\frac{\pi}{2} - t\right), \qquad \sin t = \cos\left(\frac{\pi}{2} - t\right).$$

[See formula (7) of Section 2.] We have

$$\frac{d}{dt}\cos t = \frac{d}{dt}\sin\left(\frac{\pi}{2} - t\right)$$

$$= \cos\left(\frac{\pi}{2} - t\right)\cdot\frac{d}{dt}\left(\frac{\pi}{2} - t\right)$$

$$= \cos\left(\frac{\pi}{2} - t\right)\cdot(-1)$$

$$= -\sin t.$$

PRACTICE PROBLEMS 3

1. Differentiate $y = 2\sin[t^2 + (\pi/6)]$.

2. Differentiate $y = e^t\sin 2t$.

Differentiate (with respect to t or x).

1. $\sin 4t$
2. $-3 \cos t$
3. $4 \sin t$
4. $\cos(-3t)$
5. $2 \cos 3t$
6. $2 \sin \pi t$
7. $t + \cos \pi t$
8. $t^2 - 2 \sin 4t$
9. $\sin(\pi - t)$
10. $2 \cos(t + \pi)$
11. $\cos^3 t$
12. $\sin t^3$
13. $\sin \sqrt{x - 1}$
14. $\cos(e^x)$
15. $\sqrt{\sin(x - 1)}$
16. $e^{\cos x}$
17. $(1 + \cos t)^8$
18. $\sqrt[3]{\sin \pi t}$
19. $\cos^2 x^3$
20. $\sin^3 x + 4 \sin^2 x$
21. $e^x \sin x$
22. $x\sqrt{\cos x}$
23. $\sin 2x \cos 3x$
24. $\sin^3 x \cos x$
25. $\sin t / \cos t$
26. $e^t / \cos 2t$
27. $\ln(\cos t)$
28. $\ln(\sin 2t)$
29. $\sin(\ln t)$
30. $(\cos t)\ln t$

31. Find the slope of the line tangent to the graph of $y = \cos 3x$ at $x = 13\pi/6$.

32. Find the slope of the line tangent to the graph of $y = \sin 2x$ at $x = 5\pi/4$.

33. Find the equation of the line tangent to the graph of $y = 3 \sin x + \cos 2x$ at $x = \pi/2$.

34. Find the equation of the line tangent to the graph of $y = 3 \sin 2x - \cos 2x$ at $x = 3\pi/4$.

Find the following indefinite integrals.

35. $\displaystyle\int \cos 2x \, dx$

36. $\displaystyle\int \sin \frac{x}{3} \, dx$

37. $\displaystyle\int \sin(4x + 1) \, dx$

38. $\displaystyle\int \cos(5 - x) \, dx$

39. Suppose that a person's blood pressure P at time t is given by $P = 100 + 20 \cos 6t$.

   (a) Find the maximum value of P (called the systolic pressure) and the minimum value of P (called the diastolic pressure) and give one or two values of t where these maximum and minimum values of P occur.

   (b) If time is measured in seconds, approximately how many heartbeats per minute are predicted by the equation for P?

40. The *basal metabolism* (BM) of an organism over a certain time period may be described as the total amount of heat in kilocalories (kcal) the organism produces during that period, assuming that the organism is at rest and not subject to stress. The *basal metabolic rate* (BMR) is the rate in kcal per hour at which the organism produces heat. The BMR of an animal such as a desert rat fluctuates in response to changes in temperature and other environmental factors. The BMR generally follows a *diurnal* cycle—rising at night during low temperatures and decreasing

during the warmer daytime temperatures. Find the BM for one day if $BMR(t) = .4 + .2 \sin(\pi t/12)$ kcal per hour.

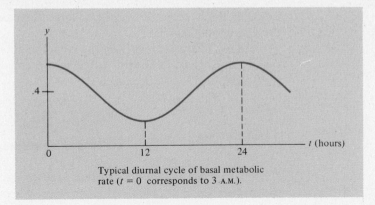

Typical diurnal cycle of basal metabolic rate ($t = 0$ corresponds to 3 A.M.).

SOLUTIONS TO PRACTICE PROBLEMS 3

1. By the chain rule,

$$y' = 2 \cos\left(t^2 + \frac{\pi}{6}\right) \frac{d}{dt}\left(t^2 + \frac{\pi}{6}\right)$$

$$= 2 \cos\left(t^2 + \frac{\pi}{6}\right) \cdot 2t$$

$$= 4t \cos\left(t^2 + \frac{\pi}{6}\right).$$

2. By the product rule,

$$y' = e^t \frac{d}{dt}[\sin 2t] + (\sin 2t) \cdot \frac{d}{dt} e^t$$

$$= 2e^t \cos 2t + e^t \sin 2t.$$

9.4. The Tangent and Other Trigonometric Functions

Certain functions involving the sine and cosine functions occur so frequently in applications that they have been given special names. The *tangent* (tan), *cotangent* (cot), *secant* (sec), and *cosecant* (csc) are such functions and are defined as follows:

$$\tan t = \frac{\sin t}{\cos t}, \qquad \cot t = \frac{\cos t}{\sin t},$$

$$\sec t = \frac{1}{\cos t}, \qquad \csc t = \frac{1}{\sin t}.$$

They are defined only for t such that the denominators in the preceding quotients are not zero. These four functions, together with the sine and cosine, are called *the trigonometric functions*. Our main interest in this section is with the tangent function. Some properties of the cotangent, secant, and cosecant are developed in the Exercises.

Many identities involving the trigonometric functions can be deduced from the identities given in Section 2. We shall mention just one:

$$\tan^2 t + 1 = \sec^2 t. \tag{1}$$

[Here $\tan^2 t$ means $(\tan t)^2$ and $\sec^2 t$ means $(\sec t)^2$.] This identity follows from the identity $\sin^2 t + \cos^2 t = 1$ when we divide everything by $\cos^2 t$.

An important interpretation of the tangent function can be given in terms of the diagram used to define the sine and cosine. For a given t, let us construct an angle of t radians, as in Fig. 1. Since $\sin t = y/r$ and $\cos t = x/r$, we have

$$\frac{\sin t}{\cos t} = \frac{y/r}{x/r} = \frac{y}{x},$$

where this formula holds provided that $x \neq 0$. Thus

$$\tan t = \frac{y}{x}. \tag{2}$$

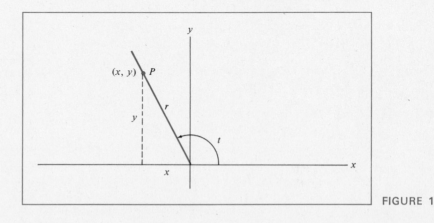

FIGURE 1

Three examples that illustrate this property of the tangent appear in Fig. 2.

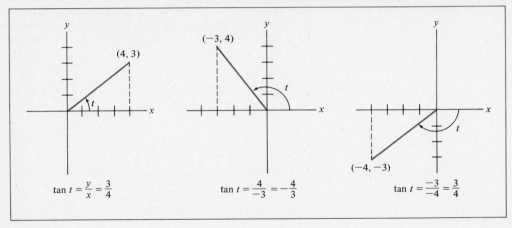

$$\tan t = \frac{y}{x} = \frac{3}{4} \qquad\qquad \tan t = \frac{4}{-3} = -\frac{4}{3} \qquad\qquad \tan t = \frac{-3}{-4} = \frac{3}{4}$$

FIGURE 2

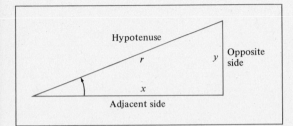

FIGURE 3

When $0 < t < \pi/2$, the value of $\tan t$ is a ratio of the lengths of the sides of a right triangle. In other words, suppose that we are given a triangle as in Fig. 3. Then we have

$$\tan t = \frac{\text{opposite}}{\text{adjacent}}. \qquad (3)$$

EXAMPLE 1 The angle of elevation from an observer to the top of a building is 29° (Fig. 4). If the observer is 100 meters from the base of the building, how high is the building?

Solution Let h denote the height of the building. Then (3) implies that

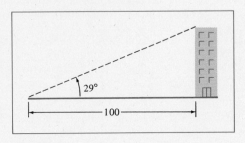

FIGURE 4

$$\frac{h}{100} = \tan 29°$$

$$h = 100 \tan 29°.$$

In order to find $\tan 29°$ from Table 3 on page A7, we convert 29° into radians. We find that $29° = (\pi/180) \cdot 29$ radians $\approx .5$ radian, and $\tan .5 = .54630$. Hence

$$h \approx 100(.54630) = 54.63 \text{ meters.}$$

The Derivative of tan t Since $\tan t$ is defined in terms of $\sin t$ and $\cos t$, we can compute the derivative of $\tan t$ from our rules of differentiation. That is, by applying the quotient rule for differentiation, we have

$$\frac{d}{dt}(\tan t) = \frac{d}{dt}\left(\frac{\sin t}{\cos t}\right) = \frac{(\cos t)(\cos t) - (\sin t)(-\sin t)}{(\cos t)^2}$$

$$= \frac{\cos^2 t + \sin^2 t}{\cos^2 t} = \frac{1}{\cos^2 t}.$$

Now

$$\frac{1}{\cos^2 t} = \frac{1}{(\cos t)^2} = \left(\frac{1}{\cos t}\right)^2 = (\sec t)^2 = \sec^2 t.$$

And so the derivative of $\tan t$ can be expressed in two equivalent ways:

$$\frac{d}{dt}(\tan t) = \frac{1}{\cos^2 t} = \sec^2 t. \tag{4}$$

Combining (4) with the chain rule, we have

$$\frac{d}{dt}(\tan g(t)) = [\sec^2 g(t)]g'(t). \tag{5}$$

EXAMPLE 2 Differentiate

(a) $\tan(t^3 + 1)$.

(b) $\tan^3 t$.

Solution (a) From (5) we find that

$$\frac{d}{dt}[\tan(t^3 + 1)] = \sec^2(t^3 + 1) \cdot \frac{d}{dt}(t^3 + 1)$$

$$= 3t^2 \sec^2(t^3 + 1).$$

(b) We write $\tan^3 t$ as $(\tan t)^3$ and use the chain rule (in this case, the general power rule):

$$\frac{d}{dt}(\tan t)^3 = (3\tan^2 t) \cdot \frac{d}{dt}\tan t$$

$$= 3\tan^2 t \sec^2 t.$$

The Graph of tan t Recall that $\tan t$ is defined for all t except where $\cos t = 0$. (We cannot have zero in the denominator of $\sin t/\cos t$.) The graph of $\tan t$ is sketched in Fig. 5, using Table 3. Note that $\tan t$ is periodic with period π.

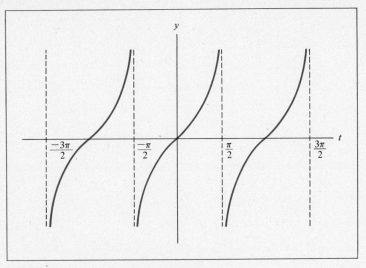

FIGURE 5   Graph of the tangent function.

PRACTICE PROBLEMS 4

1. Show that the slope of a straight line is equal to the tangent of the angle that the line makes with the x-axis.

2. Calculate $\int_0^{\pi/4} \sec^2 t \, dt$.

EXERCISES 4

1. Suppose that $0 < t < \pi/2$ and use Fig. 3 to describe $\sec t$ as a ratio of the lengths of the sides of a right triangle.

2. Describe $\cot t$ for $0 < t < \pi/2$ as a ratio of the lengths of the sides of a right triangle.

    In Exercises 3 to 10 give the value of $\tan t$ and $\sec t$, where t is the radian measure of the angle shown.

3.   4.

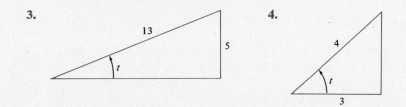

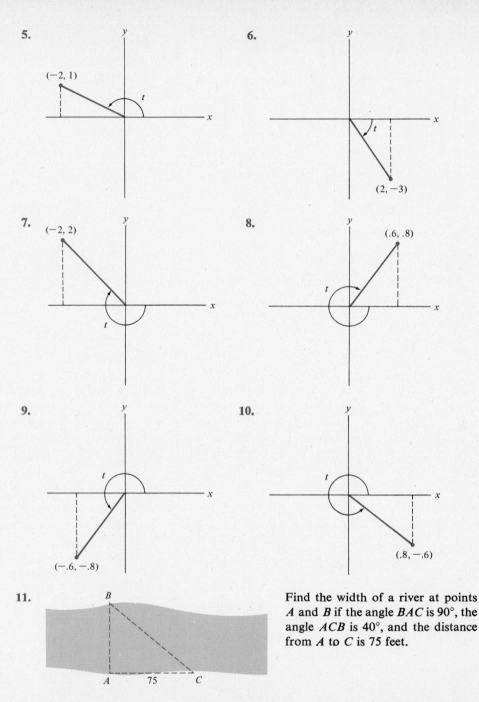

5. (−2, 1) t

6. (2, −3) t

7. (−2, 2) t

8. (.6, .8) t

9. (−.6, −.8) t

10. (.8, −.6) t

11.

Find the width of a river at points *A* and *B* if the angle *BAC* is 90°, the angle *ACB* is 40°, and the distance from *A* to *C* is 75 feet.

12. The angle of elevation from an observer to the top of a church is .3 radian, while the angle of elevation from the observer to the top of the church spire is .4 radian. If the observer is 70 meters from the church, how tall is the spire on top of the church?

Differentiate (with respect to t or x).

13. $f(t) = \sec t$

14. $f(t) = \csc t$

15. $f(t) = \cot t$

16. $f(t) = \cot 3t$

17. $f(t) = \tan 4t$

18. $f(t) = \tan \pi t$

19. $f(x) = 3 \tan(\pi - x)$

20. $f(x) = 5 \tan(2x + 1)$

21. $f(x) = 4 \tan(x^2 + x + 3)$

22. $f(x) = 3 \tan(1 - x^2)$

23. $y = \tan \sqrt{x}$

24. $y = 2 \tan \sqrt{x^2 - 4}$

25. $y = x \tan x$

26. $y = e^{3x} \tan 2x$

27. $y = \tan^2 x$

28. $y = \sqrt{\tan x}$

29. $y = (1 + \tan 2t)^3$

30. $y = \tan^4 3t$

31. $y = \ln(\tan t + \sec t)$

32. $y = \ln(\tan t)$

SOLUTIONS TO PRACTICE PROBLEMS 4

1. A line of positive slope m is shown in Fig. 6(a). Here, $\tan \theta = m/1 = m$. Suppose $y = mx + b$ where the slope m is negative. The line $y = mx$ has the same slope and makes the same angle with the x-axis. (See Fig. 6(b).) We see that $\tan \theta = -m/-1 = m$.

2. $\int_0^{\pi/4} \sec^2 t \, dt = \tan t \Big|_0^{\pi/4} = \tan \frac{\pi}{4} - \tan 0 = 1 - 0 = 1.$

FIGURE 6

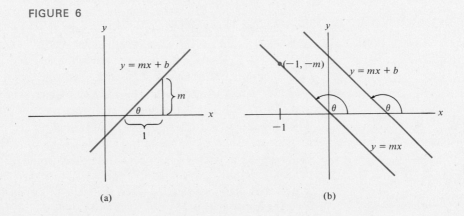

(a)

(b)

Chapter 9: CHECKLIST

- ☐ Radian
- ☐ 2π radians $= 360°$
- ☐ $\sin t$, $\cos t$, $\tan t$
- ☐ Triangle interpretation of $\sin t$, $\cos t$, $\tan t$ for $0 < t < \pi/2$
- ☐ $\sin^2 t + \cos^2 t = 1$
- ☐ $\sin t$ and $\cos t$ are periodic with period 2π
- ☐ $\dfrac{d}{dt}\sin t = \cos t$
- ☐ $\dfrac{d}{dt}\cos t = -\sin t$
- ☐ $\dfrac{d}{dt}\tan t = \sec^2 t$

Chapter 9: SUPPLEMENTARY EXERCISES

Determine the radian measure of the angles shown in Exercises 1 to 3.

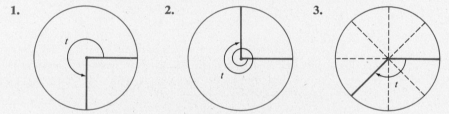

1.   2.   3.

Construct angles with the following radian measure.

4. $-\pi$ 5. $\dfrac{5\pi}{4}$ 6. $-\dfrac{9\pi}{2}$

In Exercises 7 to 10 the point with the given coordinates determines an angle of t radians, where $0 \le t \le 2\pi$. Find $\sin t$, $\cos t$, $\tan t$.

7. $(3, 4)$ 8. $(-.6, .8)$ 9. $(-.6, -.8)$ 10. $(3, -4)$

11. If $\sin t = \frac{1}{5}$, what are the possible values for $\cos t$?

12. If $\cos t = -\frac{2}{3}$, what are the possible values for $\sin t$?

13. Find four values of t between -2π and 2π at which $\sin t = \cos t$.

14. Find four values of t between -2π and 2π at which $\sin t = -\cos t$.

15. When $-\pi/2 < t < 0$, is $\tan t$ positive or negative?

16. When $\pi/2 < t < \pi$, is $\sin t$ positive or negative?

17. A gabled roof is to be built on a house that is 30 feet wide, so that the roof rises at a pitch of $23°$. Determine the length of the rafters needed to support the roof.

18. A tree casts a 60-foot shadow when the angle of elevation of the sun (measured from the horizontal) is $53°$. How tall is the tree?

Differentiate (with respect to t or x).

19. $f(t) = 3 \sin t$

20. $f(t) = \sin 3t$

21. $f(t) = \sin \sqrt{t}$

22. $f(t) = \cos t^3$

23. $g(x) = x^3 \sin x$

24. $g(x) = \sin(-2x) \cos 5x$

25. $f(x) = \dfrac{\cos 2x}{\sin 3x}$

26. $f(x) = \dfrac{\cos x - 1}{x^3}$

27. $f(x) = \cos^3 4x$

28. $f(x) = \tan^3 2x$

29. $y = \tan(x^4 + x^2)$

30. $y = \tan e^{-2x}$

31. $y = \sin(\tan x)$

32. $y = \tan(\sin x)$

33. $y = \sin x \tan x$

34. $y = \ln x \cos x$

35. $y = \ln(\sin x)$

36. $y = \ln(\cos x)$

37. $y = e^{3x} \sin^4 x$

38. $y = \sin^4 e^{3x}$

39. $f(t) = \dfrac{\sin t}{\tan 3t}$

40. $f(t) = \dfrac{\tan 2t}{\cos t}$

41. $f(t) = e^{\tan t}$

42. $f(t) = e^t \tan t$

43. If $f(t) = \sin^2 t$, find $f''(t)$.

44. Show that $y = 3 \sin 2t + \cos 2t$ satisfies the differential equation $y'' = -4y$.

45. If $f(s, t) = \sin s \cos 2t$, find $\dfrac{\partial f}{\partial s}$ and $\dfrac{\partial f}{\partial t}$.

46. If $z = \sin wt$, find $\dfrac{\partial z}{\partial w}$ and $\dfrac{\partial z}{\partial t}$.

47. If $f(s, t) = t \sin st$, find $\dfrac{\partial f}{\partial s}$ and $\dfrac{\partial f}{\partial t}$.

48. The identity
$$\sin(s + t) = \sin s \cos t + \cos s \sin t$$
was given in Section 2. Compute the partial derivative of each side with respect to t and obtain an identity involving $\cos(s + t)$.

49. Find the equation of the line tangent to the graph of $y = \tan t$ at $t = \pi/4$.

50. Sketch the graph of $f(t) = \sin t + \cos t$ for $-2\pi \le t \le 2\pi$, using the following steps:

    (a) Find all t (between -2π and 2π) such that $f'(t) = 0$. Plot the corresponding points on the graph of $y = f(t)$.

    (b) Check the concavity of $f(t)$ at the points in part (a). Make sketches of the graph near these points.

    (c) Determine any inflection points and plot them. Then complete the sketch of the graph.

51. Sketch the graph of $y = t + \sin t$ for $0 \le t \le 2\pi$.

52. Find the area under the curve $y = 2 + \sin 3t$ from $t = 0$ to $t = \pi/2$.

53. Find the area of the region between the curve $y = \sin t$ and the t-axis from $t = 0$ to $t = 2\pi$.

54. Find the area of the region between the curve $y = \cos t$ and the t-axis from $t = 0$ to $t = 3\pi/2$.

55. Find the area of the region bounded by the curves $y = x$ and $y = \sin x$ from $x = 0$ to $x = \pi$.

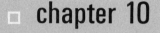

chapter 10

TECHNIQUES
OF
INTEGRATION

In this chapter we develop techniques for calculating integrals, both indefinite and definite. The need for these techniques has been justified in the preceding chapters. In addition to adding to our fund of applications, we will see even more clearly how the need to calculate integrals arises in physical problems.

As noted, integration is the reverse process of differentiation. However, integration is much harder to carry out. If a function is an expression involving elementary functions (such as x^r, $\sin x$, e^x, ...) then so is its derivative. Moreover, we were able to develop methods of calculation that enable us to differentiate, with comparative ease, almost any function that we can write down. Although many integration problems have these characteristics, certain ones do not. There are some elementary functions (e.g., e^{x^2}) for which an antiderivative cannot be expressed in terms of elementary functions. Even where an elementary antiderivative exists, the techniques for finding the antiderivative are often complicated.

For this reason, we must be prepared with a broad range of tools in order to cope with the problem of calculating integrals. Among the ideas to be discussed are

1. Techniques for evaluating indefinite integrals. We will concentrate on two methods—integration by substitution and integration by parts.
2. Evaluation of definite integrals.
3. Approximation of definite integrals. We will develop a new technique for obtaining numerical approximations to $\int_a^b f(x)\,dx$. This technique is especially useful in those cases in which we cannot find an antiderivative for $f(x)$.

Let us review the most elementary facts about integration. The indefinite integral

$$\int f(x)\,dx$$

is, by definition, a function whose derivative is $f(x)$. If $F(x)$ is one such function, then the most general function whose derivative is $f(x)$ is simply $F(x) + C$, where C is any constant. We write

$$\int f(x)\,dx = F(x) + C$$

to mean that all antiderivatives of $f(x)$ are precisely the functions $F(x) + C$, where C is any constant.

Each time we differentiate a function, we also derive an integration formula. For example, the fact that

$$\frac{d}{dx}(3x^2) = 6x$$

can be turned into the integration formula

$$\int 6x\,dx = 3x^2 + C.$$

Some of the formulas that follow immediately from differentiation formulas are reviewed in the table on the next page.

This table illustrates the need for techniques of integration. For although $\sin x$, $\cos x$, and $\sec^2 x$ occur as derivatives of simple trigonometric functions, the functions $\tan x$ and $\cot x$ are not on our list. In fact, if we experiment with various elementary combinations of the trigonometric functions, it is easy to convince ourselves that antiderivatives of $\tan x$ and $\cot x$ are not easy to compute. In this chapter we develop techniques for calculating such antiderivatives (among others).

Differentiation formula	Corresponding integration formula

$$\frac{d}{dx}(x^r) = rx^{r-1} \qquad \int rx^{r-1}\, dx = x^r + C \quad \text{or}$$

$$\int x^r\, dx = \frac{x^{r+1}}{r+1} + C, \quad r \neq -1$$

$$\frac{d}{dx}(e^x) = e^x \qquad \int e^x\, dx = e^x + C$$

$$\frac{d}{dx}(\ln|x|) = \frac{1}{x} \qquad \int \frac{1}{x}\, dx = \ln|x| + C$$

$$\frac{d}{dx}(\sin x) = \cos x \qquad \int \cos x\, dx = \sin x + C$$

$$\frac{d}{dx}(\cos x) = -\sin x \qquad \int \sin x\, dx = -\cos x + C$$

$$\frac{d}{dx}(\tan x) = \sec^2 x \qquad \int \sec^2 x\, dx = \tan x + C$$

10.1. Integration by Substitution

As noted earlier, every differentiation formula can be turned into a corresponding integration formula. This point is true even for the chain rule. The resulting formula is called *integration by substitution* and is often used to transform a complicated integral into a simpler one.

Let $f(x)$ and $g(x)$ be two given functions and let $F(x)$ be an antiderivative for $f(x)$. The chain rule asserts that

$$\frac{d}{dx}[F(g(x))] = F'(g(x))g'(x)$$
$$= f(g(x))g'(x) \quad (\text{since } F'(x) = f(x)).$$

Turning this formula into an integration formula, we have

$$\int f(g(x))g'(x)\, dx = F(g(x)) + C, \tag{1}$$

where C is any constant.

EXAMPLE 1   Determine $\int (x^2 + 1)^3 \cdot 2x\, dx$.

Solution   If we set $f(x) = x^3$, $g(x) = x^2 + 1$, then $f(g(x)) = (x^2 + 1)^3$ and $g'(x) = 2x$. Therefore we can apply formula (1). An antiderivative $F(x)$ of $f(x)$ is given by

$$F(x) = \tfrac{1}{4}x^4,$$

so that, by formula (1), we have

$$\int (x^2 + 1)^3 \cdot 2x\, dx = F(g(x)) + C$$
$$= \tfrac{1}{4}(x^2 + 1)^4 + C.$$

Formula (1) can be elevated from the status of a sometimes useful formula to a technique of integration by the introduction of a simple mnemonic device. Suppose that we are faced with integrating a function of the form $f(g(x))g'(x)$. Of course, we know the answer from formula (1). However, let us proceed somewhat differently. Replace the expression $g(x)$ by a new variable u, and replace $g'(x)\,dx$ by du. Such a substitution has the advantage that it reduces the generally complex expression $f(g(x))$ to the simpler form $f(u)$. In terms of u, the integration problem may be written

$$\int f(g(x))g'(x)\,dx = \int f(u)\,du.$$

However, the integral on the right is easy to evaluate, since

$$\int f(u)\,du = F(u) + C.$$

Since $u = g(x)$, we then obtain

$$\int f(g(x))g'(x)\,dx = F(u) + C = F(g(x)) + C,$$

which is the correct answer by (1). Remember, however, that replacing $g'(x)\,dx$ by du only has status as a correct mathematical statement because doing so leads to the correct answers. We do not, in this book, seek to explain in any deeper way what this replacement means.

Let us rework Example 1 using this method.

Second solution of Example 1   Set $u = x^2 + 1$. Then $du = \dfrac{d}{dx}(x^2 + 1)\,dx = 2x\,dx$, and

$$\int (x^2 + 1)^3 \cdot 2x\,dx = \int u^3\,du$$

$$= \tfrac{1}{4}u^4 + C$$

$$= \tfrac{1}{4}(x^2 + 1)^4 + C \quad (\text{since } u = x^2 + 1).$$

EXAMPLE 2   Evaluate $\int 2xe^{x^2}\,dx$.

Solution   Let $u = x^2$, so that $du = \dfrac{d}{dx}(x^2)\,dx = 2x\,dx$. Therefore

$$\int 2xe^{x^2}\,dx = \int e^{x^2} \cdot 2x\,dx$$

$$= \int e^u\,du$$

$$= e^u + C$$

$$= e^{x^2} + C.$$

From Examples 1 and 2 we can deduce the following method for integration of functions of the form $f'(g(x))g'(x)$.

Integration by Substitution

1. Define a new variable $u = g(x)$, where $g(x)$ is chosen in such a way that, when written in terms of u, the integrand is simpler than when written in terms of x.

2. Transform the integral with respect to x into an integral with respect to u by replacing $g(x)$ everywhere by u and $g'(x)\, dx$ by du.

3. Integrate the resulting function of u.

4. Rewrite the answer in terms of x by replacing u by $g(x)$.

Let us try a few more examples.

EXAMPLE 3   Evaluate $\int 3x^2 \sqrt{x^3 + 1}\, dx$.

Solution   The first problem facing us is to find an appropriate substitution that will simplify the integral. An immediate possibility is offered by setting $u = x^3 + 1$. Then $\sqrt{x^3 + 1}$ will become $\sqrt{u}$, a significant simplification. If $u = x^3 + 1$, then $du = \frac{d}{dx}(x^3 + 1)\, dx = 3x^2\, dx$, so that

$$\int 3x^2 \sqrt{x^3 + 1}\, dx = \int \sqrt{u}\, du$$

$$= \tfrac{2}{3} u^{3/2} + C$$

$$= \tfrac{2}{3}(x^3 + 1)^{3/2} + C.$$

EXAMPLE 4   Find $\int \dfrac{(\ln x)^2}{x}\, dx$.

Solution   Let $u = \ln x$. Then $du = (1/x)\, dx$ and

$$\int \frac{(\ln x)^2}{x}\, dx = \int (\ln x)^2 \cdot \frac{1}{x}\, dx$$

$$= \int u^2\, du$$

$$= \frac{u^3}{3} + C$$

$$= \frac{(\ln x)^3}{3} + C \quad \text{(since } u = \ln x\text{)}.$$

Knowing the correct substitution to make is a skill that develops through practice. Basically, we look for an occurrence of function composition, $f(g(x))$, where $f(x)$ is a function that we know how to integrate and where $g'(x)$ also appears in the integrand. Sometimes $g'(x)$ does not appear exactly but can be obtained by multiplying by a constant. Such a shortcoming is easily remedied, as is illustrated in Examples 5 and 6.

EXAMPLE 5   Find $\int x^2 e^{x^3}\, dx$.

Solution   Let $u = x^3$; then $du = 3x^2\, du$. A simple trick allows us to introduce the needed factor of 3 into the integrand. Note that

$$\int x^2 e^{x^3}\, dx = \int \tfrac{1}{3} \cdot 3x^2 e^{x^3}\, dx = \tfrac{1}{3}\int 3x^2 e^{x^3}\, dx.$$

Substituting, we obtain

$$\int x^2 e^{x^3}\, dx = \tfrac{1}{3}\int e^u\, du = \tfrac{1}{3}e^u + C$$
$$= \tfrac{1}{3}e^{x^3} + C \quad (\text{since } u = x^3).$$

EXAMPLE 6   Find $\int (\cos 5x)\sqrt{1 - \sin 5x}\, dx$.

Solution   Let $u = 1 - \sin 5x$; then $du = -5 \cos 5x\, dx$.

$$\int (\cos 5x)\sqrt{1 - \sin 5x}\, dx = \frac{1}{-5}\int (-5 \cos 5x)\sqrt{1 - \sin 5x}\, dx$$

$$= -\tfrac{1}{5}\int \sqrt{u}\, du = -\tfrac{1}{5}\int u^{1/2}\, du$$

$$= -\tfrac{1}{5} \cdot \tfrac{2}{3}u^{3/2} + C = -\tfrac{2}{15}u^{3/2} + C$$

$$= -\tfrac{2}{15}(1 - \sin 5x)^{3/2} + C.$$

You should be on the alert for integrals of the form

$$\int \frac{f'(x)}{f(x)}\, dx.$$

If we make the substitution $u = f(x)$, $du = f'(x)\, dx$, the integral becomes

$$\int \frac{1}{u}\, du = \ln |u| + C = \ln |f(x)| + C.$$

EXAMPLE 7   Find $\int \frac{2x}{x^2 + 1}\, dx$.

Solution   We note that the derivative of $x^2 + 1$ is $2x$. Thus we make the substitution $u = x^2 + 1$, $du = 2x\, dx$, in order to derive

$$\int \frac{2x}{x^2 + 1}\, dx = \int \frac{1}{u}\, du = \ln |u| + C = \ln |x^2 + 1| + C.$$

EXAMPLE 8   Evaluate $\int \tan x \, dx$.

Solution   Since $\tan x = \dfrac{\sin x}{\cos x}$, we have

$$\int \tan x \, dx = \int \frac{\sin x}{\cos x} \, dx.$$

Let $u = \cos x$, so that $du = -\sin x \, dx$. Then

$$\int \frac{\sin x}{\cos x} \, dx = -\int \frac{-\sin x}{\cos x} \, dx$$

$$= -\int \frac{1}{u} \, du$$

$$= -\ln |u| + C$$

$$= -\ln |\cos x| + C.$$

PRACTICE PROBLEMS 1

Evaluate the following integrals.

1. $\int \dfrac{\sin(\ln x)}{x} \, dx.$

2. $\int \dfrac{e^{3/x}}{x^2} \, dx.$

EXERCISES 1

Calculate each of the following indefinite integrals.

1. $\int 2x(x^2 + 4)^5 \, dx$ 2. $\int 3x^2(x^3 + 1)^2 \, dx$

3. $\int (x^2 - 5x)^3(2x - 5) \, dx$ 4. $\int 2x\sqrt{x^2 + 3} \, dx$

5. $\int \dfrac{3x^2}{x^3 - 1} \, dx$ 6. $\int 2xe^{-x^2} \, dx$

7. $\int \dfrac{\ln(2x)}{x} \, dx$ 8. $\int \dfrac{1}{x} (\ln x)^3 \, dx$

9. $\int \cos^3 x \sin x \, dx$ 10. $\int 2x \cos x^2 \, dx$

11. $\int \sin x \cos x \, dx$ 12. $\int (\cos x)e^{\sin x} \, dx$

13. $\displaystyle\int \frac{1}{\sqrt{2x+1}}\, dx$

14. $\displaystyle\int (x^3 - 4x)^7 (3x^2 - 4)\, dx$

15. $\displaystyle\int x^2 \sqrt{x^3 - 1}\, dx$

16. $\displaystyle\int 2\cos(2x+1)\, dx$

Calculate each of the following integrals by making the indicated substitutions.

17. $\displaystyle\int (x+5)^{-1/2} e^{\sqrt{x+5}}\, dx,\; u = \sqrt{x+5}.$

18. $\displaystyle\int [1 + \ln x] \sin[x \ln x]\, dx;\; u = x \ln x.$

19. $\displaystyle\int \sqrt{\sin 2x}\, \cos 2x\, dx;\; u = \sin 2x.$

20. $\displaystyle\int (\cos^2 x + 3\cos x + 1)\sin x\, dx;\; u = \cos x.$

21. $\displaystyle\int \frac{1}{x(\ln x)^3}\, dx;\; u = \ln x.$

22. $\displaystyle\int \frac{e^{1/x}}{x^2}\, dx;\; u = \frac{1}{x}.$

Determine each of the following integrals by making an appropriate substitution.

23. $\displaystyle\int \frac{x}{\sqrt{x^2+1}}\, dx$

24. $\displaystyle\int e^{\sin(x^2)} \cdot x \cos(x^2)\, dx$

25. $\displaystyle\int \frac{x^4}{x^5 - 7}\ln(x^5 - 7)\, dx$

26. $\displaystyle\int \frac{1 + \ln x}{[x \ln x]^2}\, dx$

27. $\displaystyle\int \cot x\, dx$

28. $\displaystyle\int (x^{10} - 2x^5 + 10)(x^9 - x^4)\, dx$

29. $\displaystyle\int \frac{x}{e^{x^2}}\, dx$

30. $\displaystyle\int \frac{e^{\sqrt{x}}}{\sqrt{x}}\, dx$

31. $\displaystyle\int \frac{x^2 - 2x}{x^3 - 3x^2 + 1}\, dx$

32. $\displaystyle\int \frac{1}{x \ln x}\, dx$

33. $\displaystyle\int \frac{x^3}{\sqrt{x^4 + 4}}\, dx$

34. $\displaystyle\int \frac{x + 4}{1 - 8x - x^2}\, dx$

35. $\displaystyle\int \frac{e^{2x} - e^{-2x}}{e^{2x} + e^{-2x}}\, dx$

36. $\displaystyle\int \frac{\sin x}{(2 + \cos x)^3}\, dx$

SOLUTIONS TO PRACTICE PROBLEMS 1

1. Let $u = \ln x$, $du = (1/x)\, dx$. Then

$$\int \frac{\sin(\ln x)}{x}\, dx = \int \sin u\, du$$

$$= -\cos u + C$$

$$= -\cos(\ln x) + C.$$

2. Let $u = 3/x$, $du = -3/x^2 \, dx$. Then

$$\int \frac{e^{3/x}}{x^2} \, dx = -\tfrac{1}{3} \int e^{3/x} \cdot \left(-\frac{3}{x^2} \right) dx$$

$$= -\tfrac{1}{3} \int e^u \, du$$

$$= -\tfrac{1}{3} e^u + C$$

$$= -\tfrac{1}{3} e^{3/x} + C.$$

10.2. Integration by Parts

In the preceding section we developed the method of integration by substitution by turning the chain rule into an integration formula. Let us do the same for the product rule. Let $f(x)$ and $g(x)$ be any two functions and let $G(x)$ be an anti-derivative of $g(x)$. The product rule asserts that

$$\frac{d}{dx} [f(x)G(x)] = f(x)G'(x) + f'(x)G(x)$$

$$= f(x)g(x) + f'(x)G(x) \quad \text{[since } G'(x) = g(x)\text{]}.$$

Therefore

$$f(x)G(x) = \int f(x)g(x) \, dx + \int f'(x)G(x) \, dx.$$

This last formula can be rewritten in the following more useful form.

$$\int f(x)g(x) \, dx = f(x)G(x) - \int f'(x)G(x) \, dx. \tag{1}$$

Equation (1) is the principle of *integration by parts* and is one of the most important techniques of integration.

EXAMPLE 1   Evaluate $\int xe^x \, dx$.

Solution   Set $f(x) = x$, $g(x) = e^x$. Then $f'(x) = 1$, $G(x) = e^x$, and equation (1) yields

$$\int xe^x \, dx = xe^x - \int 1 \cdot e^x \, dx$$

$$= xe^x - e^x + C.$$

The following principles underlie Example 1 and also illustrate general features of situations to which integration by parts may be applied:

1. The integrand is the product of two functions $f(x) = x$ and $g(x) = e^x$.
2. It is easy to compute $f'(x)$ and $G(x)$. That is, we can differentiate $f(x)$ and integrate $g(x)$.
3. The integral $\int f'(x)G(x)\, dx$ can be calculated.

Let us consider another example to see how these three principles work.

EXAMPLE 2 Evaluate $\int x(x + 5)^8\, dx$.

Solution Our calculations can be set up as follows:

$$f(x) = x, \qquad g(x) = (x + 5)^8,$$
$$f'(x) = 1, \qquad G(x) = \tfrac{1}{9}(x + 5)^9.$$

Then

$$\int x(x + 5)^8\, dx = x \cdot \tfrac{1}{9}(x + 5)^9 - \int 1 \cdot \tfrac{1}{9}(x + 5)^9\, dx$$

$$= \tfrac{1}{9}x(x + 5)^9 - \tfrac{1}{9}\int (x + 5)^9\, dx$$

$$= \tfrac{1}{9}x(x + 5)^9 - \tfrac{1}{9} \cdot \tfrac{1}{10}(x + 5)^{10} + C$$

$$= \tfrac{1}{9}x(x + 5)^9 - \tfrac{1}{90}(x + 5)^{10} + C.$$

We were led to try integration by parts because our integrand is the product of two functions. We were led to choose $f(x) = x$ [and not $(x + 5)^9$] because $f'(x) = 1$, so that the factor x in the integrand is made to disappear, thereby simplifying the integral.

EXAMPLE 3 Evaluate $\int x \sin x\, dx$.

Solution Let us set

$$f(x) = x, \qquad g(x) = \sin x,$$
$$f'(x) = 1, \qquad G(x) = -\cos x.$$

Then

$$\int x \sin x\, dx = -x \cos x - \int 1 \cdot (-\cos x)\, dx$$

$$= -x \cos x + \int \cos x\, dx$$

$$= -x \cos x + \sin x + C.$$

EXAMPLE 4   Evaluate $\int x^2 \ln x \, dx$.

Solution   Set

$$f(x) = \ln x, \qquad g(x) = x^2,$$
$$f'(x) = \frac{1}{x}, \qquad G(x) = \frac{x^3}{3}.$$

Then

$$\int x^2 \ln x \, dx = \frac{x^3}{3} \ln x - \int \frac{1}{x} \cdot \frac{x^3}{3} \, dx$$

$$= \frac{x^3}{3} \ln x - \frac{1}{3} \int x^2 \, dx$$

$$= \frac{x^3}{3} \ln x - \frac{1}{9} x^3 + C.$$

The next example shows how integration by parts can be used to compute a reasonably complicated integral.

EXAMPLE 5   Find $\int \dfrac{xe^x}{(x + 1)^2} \, dx$.

Solution   Let $f(x) = xe^x$, $g(x) = \dfrac{1}{(x + 1)^2}$. Then

$$f'(x) = xe^x + e^x \cdot 1 = (x + 1)e^x, \qquad G(x) = \frac{-1}{x + 1}.$$

As a result, we have

$$\int \frac{xe^x}{(x + 1)^2} \, dx = xe^x \cdot \frac{-1}{x + 1} - \int (x + 1)e^x \cdot \frac{-1}{x + 1} \, dx$$

$$= -\frac{xe^x}{x + 1} + \int e^x \, dx$$

$$= -\frac{xe^x}{x + 1} + e^x + C = \frac{e^x}{x + 1} + C.$$

Sometimes we must use integration by parts more than once.

EXAMPLE 6   Find $\int x^2 \sin x \, dx$.

Solution   Let $f(x) = x^2$, $g(x) = \sin x$. Then $f'(x) = 2x$ and $G(x) = -\cos x$. Applying our formula for integration by parts, we have

$$\int x^2 \sin x \, dx = -x^2 \cos x - \int 2x \cdot (-1) \cos x \, dx$$

$$= -x^2 \cos x + 2 \int x \cos x \, dx. \qquad (2)$$

The integral $\int x \cos x$ can itself be handled by integration by parts. Let $f(x) = x$, $g(x) = \cos x$. Then $f'(x) = 1$ and $G(x) = \sin x$, so that

$$\int x \cos x \, dx = x \sin x - \int 1 \cdot \sin x \, dx$$

$$= x \sin x + \cos x. \tag{3}$$

Combining (2) and (3), we see that

$$\int x^2 \sin x \, dx = -x^2 \cos x + 2(x \sin x + \cos x) + C.$$

$$= -x^2 \cos x + 2x \sin x + 2 \cos x + C.$$

EXAMPLE 7   Evaluate $\int \ln x \, dx$.

Solution   Since $\ln x = 1 \cdot \ln x$, we may view $\ln x$ as a product $f(x)g(x)$, where $f(x) = \ln x$, $g(x) = 1$. Then

$$f'(x) = \frac{1}{x}, \qquad G(x) = x.$$

Finally,

$$\int \ln x \, dx = x \ln x - \int \frac{1}{x} \cdot x \, dx$$

$$= x \ln x - \int 1 \, dx$$

$$= x \ln x - x + C.$$

PRACTICE PROBLEMS 2

Evaluate the following integrals.

1. $\int x^3 \ln x \, dx$

2. $\int x^5 \sin x^3 \, dx$

EXERCISES 2

Evaluate the following integrals.

1. $\int xe^{5x} \, dx$ 2. $\int xe^{-x/2} \, dx$ 3. $\int x(2x + 1)^4 \, dx$

4. $\int (x + 1)e^x \, dx$ 5. $\int xe^{-x} \, dx$ 6. $\int x(x + 5)^{-3} \, dx$

7. $\displaystyle\int \frac{x}{\sqrt{x+1}}\, dx$

8. $\displaystyle\int \frac{x}{\sqrt{3+2x}}\, dx$

9. $\displaystyle\int x(1-3x)^4\, dx$

10. $\displaystyle\int x(1+x)^{10}\, dx$

11. $\displaystyle\int \frac{3x}{e^x}\, dx$

12. $\displaystyle\int \frac{x+1}{e^x}\, dx$

13. $\displaystyle\int x\sqrt{x+1}\, dx$

14. $\displaystyle\int x\sqrt{2-x}\, dx$

15. $\displaystyle\int \sqrt{x}\, \ln\sqrt{x}\, dx$

16. $\displaystyle\int x^5 \ln x\, dx$

17. $\displaystyle\int x^2 \cos x\, dx$

18. $\displaystyle\int x \sin 8x\, dx$

19. $\displaystyle\int x \ln 5x\, dx$

20. $\displaystyle\int x^{-3} \ln x\, dx$

21. $\displaystyle\int x^3 e^{x^2}\, dx$

22. $\displaystyle\int x^3 \sin x^2\, dx$

23. $\displaystyle\int \frac{(\ln x)^5}{x}\, dx$

24. $\displaystyle\int x^3\sqrt{4-x^2}\, dx$

25. $\displaystyle\int \frac{\ln x}{\sqrt{x}}\, dx$

26. $\displaystyle\int x^2(e^x-1)\, dx$

27. $\displaystyle\int x^2 e^{-x}\, dx$

28. $\displaystyle\int x^2 \sin 3x\, dx$

SOLUTIONS TO PRACTICE PROBLEMS 2

1. Set $f(x) = \ln x$, $g(x) = x^3$. Then

$$f'(x) = \frac{1}{x}, \qquad G(x) = \frac{x^4}{4},$$

so that

$$\int x^3 \ln x\, dx = \frac{x^4}{4} \ln x - \int \frac{1}{x}\cdot\frac{x^4}{4}\, dx$$

$$= \frac{x^4}{4} \ln x - \int \frac{1}{4} x^3\, dx$$

$$= \frac{x^4}{4} \ln x - \frac{x^4}{16} + C.$$

2. $\displaystyle\int x^5 \sin x^3\, dx = \int \underbrace{\frac{x^3}{3}}_{f(x)} \cdot \underbrace{3x^2 \sin x^3}_{g(x)}\, dx,$

$$f'(x) = x^2, \qquad G(x) = -\cos x^3,$$

$$\int x^5 \sin x^3\, dx = -\tfrac{1}{3}x^3 \cos x^3 - \int x^2[-\cos x^3]\, dx$$

$$= -\tfrac{1}{3}x^3 \cos x^3 + \int x^2 \cos x^3\, dx$$

$$= -\tfrac{1}{3}x^3 \cos x^3 + \tfrac{1}{3} \sin x^3 + C.$$

10.3. Evaluation of Definite Integrals

Earlier we discussed techniques for determining antiderivatives (indefinite integrals). One of the most important applications of such techniques concerns the computation of definite integrals. For if $F(x)$ is an antiderivative of $f(x)$, then

$$\int_a^b f(x) \, dx = F(b) - F(a).$$

Thus the techniques of the previous sections can be used to evaluate definite integrals. Here we will simplify the method of evaluating definite integrals in those cases where the antiderivative is found by integration by substitution or parts.

EXAMPLE 1   Evaluate $\int_0^1 2x(x^2 + 1)^5 \, dx$.

Solution—
First Method   Let $u = x^2 + 1$, $du = 2x \, dx$. Then

$$\int 2x(x^2 + 1)^5 \, dx = \int u^5 \, du$$

$$= \frac{u^6}{6} + C$$

$$= \frac{(x^2 + 1)^6}{6} + C.$$

Consequently,

$$\int_0^1 2x(x^2 + 1)^5 \, dx = \frac{(x^2 + 1)^6}{6} \Big|_0^1 = \frac{2^6}{6} - \frac{1^6}{6} = \frac{21}{2}.$$

Solution—
Second Method   Again we make the substitution $u = x^2 + 1$, $du = 2x \, dx$; however, we also apply the substitution to the limits of integration. When $x = 0$ (the lower limit of integration), we have $u = 0^2 + 1 = 1$; and when $x = 1$ (the upper limit of integration), we have $u = 1^2 + 1 = 2$. Therefore

$$\int_0^1 2x(x^2 + 1)^5 \, dx = \int_1^2 u^5 \, du$$

$$= \frac{u^6}{6} \Big|_1^2$$

$$= \frac{2^6}{6} - \frac{1^6}{6} = \frac{21}{2}.$$

Notice that, in utilizing the second method, we did not need to reexpress the function $u^6/6$ in terms of x.

The foregoing computation is an example of a general computational method, which can be expressed as follows:

Change of Limits Rule Suppose that the integral $\int f(g(x))g'(x)\,dx$ is subjected to the substitution $u = g(x)$, so that $\int f(g(x))g'(x)\,dx$ becomes $\int f(u)\,du$. Then

$$\int_a^b f(g(x))g'(x)\,dx = \int_{g(a)}^{g(b)} f(u)\,du.$$

Justification of Change of Limits Rule If $F(x)$ is an antiderivative of $f(x)$, then

$$\frac{d}{dx}[F(g(x))] = F'(g(x))g'(x) = f(g(x))g'(x).$$

Therefore

$$\int_a^b f(g(x))g'(x)\,dx = F(g(x))\Big|_a^b = F(g(b)) - F(g(a)) = \int_{g(a)}^{g(b)} f(u)\,du.$$

EXAMPLE 2 Evaluate $\int_3^5 x\sqrt{x^2 - 9}\,dx$.

Solution Let $u = x^2 - 9$, then $du = 2x\,dx$. When $x = 3$, we have $u = 3^2 - 9 = 0$. When $x = 5$, we have $u = 5^2 - 9 = 16$. Thus

$$\int_3^5 x\sqrt{x^2 - 9}\,dx = \tfrac{1}{2}\int_3^5 2x\sqrt{x^2 - 9}\,dx$$

$$= \tfrac{1}{2}\int_0^{16} \sqrt{u}\,du$$

$$= \tfrac{1}{2}\cdot\tfrac{2}{3}u^{3/2}\Big|_0^{16}$$

$$= \tfrac{1}{3}\cdot[16^{3/2} - 0] = \tfrac{1}{3}\cdot 16^{3/2}$$

$$= \tfrac{1}{3}\cdot 64 = \tfrac{64}{3}.$$

EXAMPLE 3 Determine the area of the ellipse $x^2/a^2 + y^2/b^2 = 1$. (See Fig. 1.)

FIGURE 1

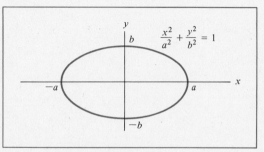

Solution   Owing to the symmetry of the ellipse, the area is equal to twice the area of the upper half of the ellipse. Solving for y,

$$\frac{y^2}{b^2} = 1 - \frac{x^2}{a^2}$$

$$\frac{y}{b} = \sqrt{1 - \left(\frac{x}{a}\right)^2}$$

$$y = b\sqrt{1 - \left(\frac{x}{a}\right)^2}.$$

Since the area of the upper half-ellipse is the area under the curve

$$y = b\sqrt{1 - \left(\frac{x}{a}\right)^2},$$

the area of the ellipse is given by the integral

$$2\int_{-a}^{a} b\sqrt{1 - \left(\frac{x}{a}\right)^2}\, dx.$$

Let $u = x/a$; then $du = 1/a\, dx$. When $x = -a$, we have $u = -a/a = -1$. When $x = a$, we have $u = a/a = 1$.

$$2\int_{-a}^{a} b\sqrt{1 - \left(\frac{x}{a}\right)^2}\, dx = 2b \cdot a \int_{-a}^{a} \frac{1}{a}\sqrt{1 - \left(\frac{x}{a}\right)^2}\, dx$$

$$= 2ba \int_{-1}^{1} \sqrt{1 - u^2}\, du.$$

Although we cannot evaluate this integral by our existing techniques, we obtain its value immediately by recognizing that since the area under the curve $y = \sqrt{1 - x^2}$ is the area of the top half of the unit circle, this integral has value $\pi/2$. And so the area of the ellipse is $2ba \cdot (\pi/2) = \pi ab$.

Integration by Parts in Definite Integrals

EXAMPLE 4   Evaluate $\int_0^{\pi/2} x \cos x \, dx$.

Solution—   We use integration by parts to find an antiderivative of $x \cos x$. Let $f(x) = x$,
First Method $g(x) = \cos x, f'(x) = 1, G(x) = \sin x$. Then

$$\int x \cos x \, dx = x \sin x - \int 1 \cdot \sin x \, dx = x \sin x + \cos x + C.$$

Hence

$$\int_0^{\pi/2} x \cos x \, dx = (x \sin x + \cos x)\Big|_0^{\pi/2}$$

$$= \left(\frac{\pi}{2} \sin \frac{\pi}{2} + \cos \frac{\pi}{2}\right) - (0 + \cos 0)$$

$$= \frac{\pi}{2} - 1.$$

Note that the antiderivative of $x \cos x$ consists of two terms. Instead of evaluating the entire antiderivative of $x \cos x$ at $\pi/2$ and then subtracting its value at 0, we may evaluate each term of the antiderivative separately. The computations may be written in the form

$$\int_0^{\pi/2} x \cos x \, dx = x \sin x \Big|_0^{\pi/2} - \int_0^{\pi/2} 1 \cdot \sin x \, dx$$

$$= x \sin x \Big|_0^{\pi/2} - (-\cos x) \Big|_0^{\pi/2}$$

$$= \left(\frac{\pi}{2} \sin \frac{\pi}{2} - 0\right) - \left(-\cos \frac{\pi}{2} + \cos 0\right) = \frac{\pi}{2} - 1.$$

The second method involved less writing because we never had to consider the *in*definite integral $\int x \cos x \, dx$.

The second method in Example 4 illustrates the following rule for integrating a definite integral by parts.

Integration by Parts If $G(x)$ is an antiderivative of $g(x)$, then

$$\int_a^b f(x)g(x) \, dx = f(x)G(x) \Big|_a^b - \int_a^b f'(x)G(x) \, dx.$$

Using the integration-by-parts formula, we have

$$\int_a^b f(x)g(x) \, dx = \left[f(x)G(x) - \int f'(x)G(x) \, dx\right]\Big|_a^b$$

$$= f(x)G(x) \Big|_a^b - \left[\int f'(x)G(x) \, dx\right]\Big|_a^b$$

$$= f(x)G(x) \Big|_a^b - \int_a^b f'(x)G(x) \, dx.$$

EXAMPLE 5 Evaluate $\int_0^5 \frac{x}{\sqrt{x+4}} \, dx$.

Solution Let $f(x) = x$, $g(x) = (x+4)^{-1/2}$, $f'(x) = 1$, $G(x) = 2(x+4)^{1/2}$. Then

$$\int_0^5 \frac{x}{\sqrt{x+4}} \, dx = 2x(x+4)^{1/2} \Big|_0^5 - \int_0^5 1 \cdot 2(x+4)^{1/2} \, dx$$

$$= 2x(x+4)^{1/2} \Big|_0^5 - \tfrac{4}{3}(x+4)^{3/2} \Big|_0^5$$

$$= [10(9)^{1/2} - 0] - [\tfrac{4}{3}(9)^{3/2} - \tfrac{4}{3}(4)^{3/2}]$$

$$= [30] - [36 - \tfrac{32}{3}] = 4\tfrac{2}{3}.$$

PRACTICE PROBLEMS 3

Evaluate the following definite integrals.

1. $\displaystyle\int_0^1 (2x + 3)e^{x^2+3x+6}\, dx$

2. $\displaystyle\int_e^{e^{\pi/2}} \frac{\sin(\ln x)}{x}\, dx$

EXERCISES 3

Evaluate the following definite integrals.

1. $\displaystyle\int_{3/2}^2 2(2x - 3)^{17}\, dx$
2. $\displaystyle\int_{\sqrt{3}}^2 [(x^2 - 3)^{17} - (x^2 - 3)]x\, dx$

3. $\displaystyle\int_1^3 (4 - 2x) \sin(4x - x^2)\, dx$
4. $\displaystyle\int_5^{13} x\sqrt{x^2 - 25}\, dx$

5. $\displaystyle\int_1^e \frac{\ln x}{x}\, dx$
6. $\displaystyle\int_0^\pi e^{(\sin x)^2} \cdot \cos x \sin x\, dx$

7. $\displaystyle\int_3^5 x\sqrt{x^2 - 9}\, dx$
8. $\displaystyle\int_1^{e^{\pi/2}} \frac{\sin(\ln x)}{x}\, dx$

9. $\displaystyle\int_{\ln 1}^{\ln 2} [(e^x)^2 + e^x]e^x\, dx$
10. $\displaystyle\int_{\pi/4}^{\pi/2} \ln(\sin x) \cos x\, dx$

11. $\displaystyle\int_{-1}^2 (x^3 - 2x - 1)^2 \cdot (3x^2 - 2)\, dx$

12. $\displaystyle\int_{-1}^1 x \sin x^4\, dx$ (Let $u = x^2$.)

13. $\displaystyle\int_1^3 x^2 e^{x^3}\, dx$
14. $\displaystyle\int_0^1 (2x - 1)(x^2 - x)^{10}\, dx$

15. $\displaystyle\int_{-\pi}^{2\pi} \sin(8x - \pi)\, dx$
16. $\displaystyle\int_{-2}^2 2x \sin(x^2)\, dx$

17. $\displaystyle\int_0^2 xe^{x/2}\, dx$
18. $\displaystyle\int_0^4 8x(x + 4)^{-3}\, dx$

19. $\displaystyle\int_0^1 x \sin \pi x\, dx$
20. $\displaystyle\int_1^4 \ln x\, dx$

Use substitutions and the fact that a circle of radius r has area πr^2 to evaluate the following integrals.

21. $\displaystyle\int_{-\pi/2}^{\pi/2} \sqrt{1 - \sin^2 x} \cos x\, dx$
22. $\displaystyle\int_0^{\sqrt{2}} \sqrt{4 - x^4} \cdot 2x\, dx$

23. $\displaystyle\int_{-6}^0 \sqrt{-x^2 - 6x}\, dx$ [Complete the square: $-x^2 - 6x = 9 - (x + 3)^2$.]

1. Let $u = x^2 + 3x + 6$ so that $du = 2x + 3$. When $x = 0$, $u = 6$; when $x = 1$, $u = 10$. Thus,

$$\int_0^1 (2x + 3)e^{x^2+3x+6}\,dx = \int_6^{10} e^u\,du$$

$$= e^u \Big|_6^{10}$$

$$= e^{10} - e^6.$$

2. Let $u = \ln x$, $du = (1/x)\,dx$. When $x = e$, $u = \ln e = 1$; when $x = e^{\pi/2}$, $u = \ln e^{\pi/2} = \pi/2$. Thus,

$$\int_e^{e^{\pi/2}} \frac{\sin(\ln x)}{x}\,dx = \int_1^{\pi/2} \sin u\,du$$

$$= -\cos u \Big|_1^{\pi/2}$$

$$= -\cos \frac{\pi}{2} + \cos 1$$

$$\approx .54030.$$

10.4. The Trapezoidal Rule

Many applications involve a definite integral

$$\int_a^b f(x)\,dx. \tag{1}$$

Earlier sections provided us with the tools necessary to calculate the value of such an integral if we can determine an antiderivative for $f(x)$. However, as already noted, certain simple functions (e.g., e^{x^2}) have no antiderivative that can be expressed in terms of elementary functions. In this section we discuss a method for numerically approximating the value of (1)—the *trapezoidal rule*. This rule is applicable to arbitrary functions $f(x)$ and can easily be implemented on a computer. In fact, as we shall see below, the trapezoidal rule can be used to approximate (1) even when $f(x)$ is not explicitly known but is only known for certain values of x, as is the case, for example, when $f(x)$ is obtained from a set of experimental data.

To begin, we subdivide the x-axis from $x = a$ to $x = b$ into n equal subintervals, each of length $(b - a)/n$. For brevity, let us write Δx in place of $(b - a)/n$. Then denote the endpoints of the subintervals by $a_0, a_1, a_2, \ldots, a_n$. Since each subinterval is of length Δx, we see that

$$a_0 = a$$
$$a_1 = a + \Delta x$$
$$a_2 = a + 2\,\Delta x$$
$$\vdots$$
$$a_{n-1} = a + (n-1)\,\Delta x$$
$$a_n = b.$$

Our subdivision of the x-axis is illustrated in Fig. 1.

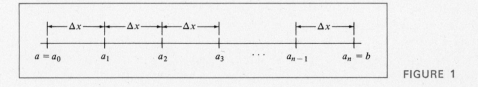

FIGURE 1

Suppose that x_1 is any point selected from the first subinterval, x_2 any point selected from the second, and so forth. Then the Riemann sum approximation of the integral yields the approximation

$$\int_a^b f(x)\,dx \approx f(x_1)\,\Delta x + f(x_2)\,\Delta x + \ldots + f(x_n)\,\Delta x \qquad (2)$$

This approximation arises from replacing the area under the curve $y = f(x)$ with a series of n rectangular areas, one lying over each subinterval, the first of height $f(x_1)$, the second of height $f(x_2)$, and so forth. (See Fig. 2.)

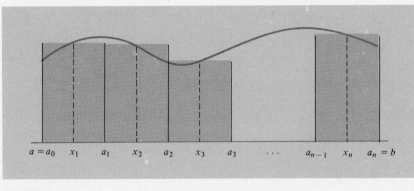

FIGURE 2

The basic principle behind the approximation (2) may be described as follows: On each small subinterval into which we divide our interval from a to b we approximate the function $f(x)$ by a constant function—namely, the value of $f(x)$ at the selected point of the subinterval. (See Fig. 3.) There is another way of

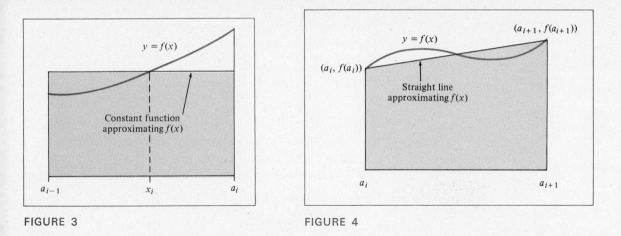

FIGURE 3

FIGURE 4

approximating the integral (1) for which computation is easier: On each sub-interval, let us approximate $f(x)$ by a straight line. This simple idea leads to the *trapezoidal rule*.

We approximate $f(x)$ on the subinterval from a_i to a_{i+1} by the straight line connecting the points $(a_i, f(a_i))$ and $(a_{i+1}, f(a_{i+1}))$. (See Fig. 4.) The function $f(x)$ is then approximated by a broken line. (See Fig. 5.) Let us approximate the area under the curve by the area under the broken line.

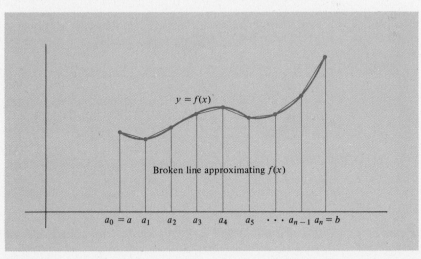

FIGURE 5

To find the area under the broken line, we find the areas of each of the figures like the one shaded in Fig. 4 and then add them up. The geometrical figure shaded in Fig. 4 is called a *trapezoid*. We will use the fact (from high school geometry)

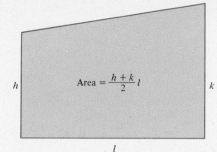

$$\text{Area} = \frac{h+k}{2} l$$

h k

l

FIGURE 6

that the area of the trapezoid of Fig. 6 is just $[(h + k)/2] \cdot l$. Hence we see that the area of the trapezoid of Fig. 4 is

$$\frac{f(a_i) + f(a_{i+1})}{2} (a_{i+1} - a_i). \tag{3}$$

However, $a_{i+1} - a_i$ is the length of the subinterval, so

$$a_{i+1} - a_i = (b - a)/n = \Delta x.$$

Thus, by using formula (3), we find that the area of the trapezoid of Fig. 5 resting on the subinterval from a_0 to a_1 is simply

$$\frac{f(a_0) + f(a_1)}{2} \Delta x.$$

Similarly, the area of the next trapezoid is

$$\frac{f(a_1) + f(a_2)}{2} \Delta x.$$

The area of the trapezoid on the right is

$$\frac{f(a_{n-1}) + f(a_n)}{2} \Delta x.$$

Therefore our approximation for the area under the curve is

$$\left[\frac{f(a_0) + f(a_1)}{2} + \frac{f(a_1) + f(a_2)}{2} + \ldots + \frac{f(a_{n-1}) + f(a_n)}{2} \right] \Delta x$$

$$= [f(a_0) + f(a_1) + f(a_1) + f(a_2) + \ldots + f(a_{n-1}) + f(a_n)] \frac{\Delta x}{2}$$

$$= [f(a_0) + 2f(a_1) + 2f(a_2) + \ldots + 2f(a_{n-1}) + f(a_n)] \frac{\Delta x}{2}.$$

> *Trapezoidal Rule*   Let $f(x)$ be a given function, a, b numbers, n a positive integer. Furthermore, let $\Delta x = (b - a)/n$, $a_0 = a$, $a_1 = a + \Delta x, \ldots, a_n = b$. Then
>
> $$\int_a^b f(x)\, dx \approx [f(a_0) + 2f(a_1) + 2f(a_2) + \ldots + 2f(a_{n-1}) + f(a_n)] \frac{\Delta x}{2}.$$

EXAMPLE 1   Approximate $\displaystyle\int_0^1 \frac{1}{1 + x}\, dx$ by using the trapezoidal rule with $n = 10$.

Solution   Since $n = 10$, $\Delta x = (1 - 0)/10 = \frac{1}{10}$. Then

$$a_0 = 0, \quad a_1 = \tfrac{1}{10}, \quad a_2 = \tfrac{2}{10}, \quad a_3 = \tfrac{3}{10}, \quad \ldots, \quad a_9 = \tfrac{9}{10}, \quad a_{10} = 1,$$

$$f(a_0) = 1, \quad f(a_1) = \tfrac{10}{11}, \quad f(a_2) = \tfrac{10}{12},$$

$$f(a_3) = \tfrac{10}{13}, \quad \ldots, \quad f(a_9) = \tfrac{10}{19}, \quad f(a_{10}) = \tfrac{1}{2}.$$

Therefore the trapezoidal rule yields

$$\int_0^1 \frac{1}{1+x}\,dx \approx [1 + 2\cdot\tfrac{10}{11} + 2\cdot\tfrac{10}{12} + 2\cdot\tfrac{10}{13} + 2\cdot\tfrac{10}{14} + 2\cdot\tfrac{10}{15}$$

$$+ 2\cdot\tfrac{10}{16} + 2\cdot\tfrac{10}{17} + 2\cdot\tfrac{10}{18} + 2\cdot\tfrac{10}{19} + \tfrac{1}{2}]\cdot\tfrac{1}{20}$$

$$= .69377.$$

The actual value of the integral is ln 2 = .69315, and so the error of approximation is .00062.

Using a larger value of n improves the estimate. Basically, doubling the value of n yields an error of one-fourth the previous error.

Distribution of IQs   Psychologists use various standardized tests to measure intelligence. The method most commonly used to describe the results of such tests is an intelligence quotient (or IQ). An IQ is a positive number that, in theory, indicates how a person's mental age compares with the person's chronological age. The median IQ is arbitrarily set at 100, so that half the population has an IQ less than 100 and half greater. IQs are distributed according to a bell-shaped curve called the *normal curve*, pictured in Fig. 7. The proportion of all people having IQs between A and B is given by the area under the curve from A to B, that is by the integral

$$\frac{1}{16\sqrt{2\pi}}\int_A^B e^{-(1/2)[(x-100)/16]^2}\,dx.$$

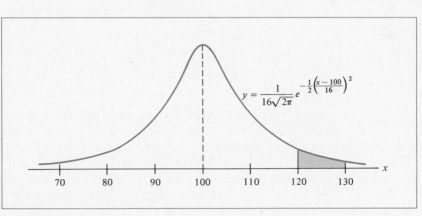

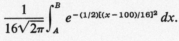

FIGURE 7

EXAMPLE 2   Find the proportion of all people having IQs between 120 and 130.

Solution   We have seen that this proportion is given by

$$\frac{1}{16\sqrt{2\pi}}\int_{120}^{130} e^{-(1/2)[(x-100)/16]^2}\,dx.$$

Let us evaluate this integral approximately, using the trapezoidal rule with $n = 10$. Let $f(x) = e^{-(1/2)[(x-100)/16]^2}$. The integral approximately equals

$$\frac{1}{16\sqrt{2\pi}} \, [f(120) + 2f(121) + 2f(122) + 2f(123) + 2f(124) + 2f(125)$$

$$+ \, 2f(126) + 2f(127) + 2f(128) + 2f(129) + f(130)] \cdot \frac{\Delta x}{2}$$

$$= \frac{1}{16\sqrt{2\pi}} \, [.4578 + .8452 + .7771 + .7117 + .6493 + .5900$$

$$+ \, .5341 + .4816 + .4325 + .3870 + .1724] \cdot \tfrac{1}{2}$$

$$= (.0249)(6.0387) \cdot \tfrac{1}{2}$$

$$= .0752.$$

Thus 7.52% of the population have IQs between 120 and 130.

As mentioned at the beginning of the section, the trapezoidal rule can be used to estimate the integral of a function whose value is known only at certain points.

EXAMPLE 3 In a drive along a country road, the speedometer readings are recorded each minute during a five-minute interval.

Time (minutes)	Velocity (mph)
0	33
1	32
2	28
3	30
4	32
5	35

That is, when first observed the speedometer read 33. After one minute, the speedometer read 32 and so on. Using these data, estimate the distance traveled during the 5 minutes.

Solution Let $s(t)$ be the distance traveled by time t. Then the distance traveled during the five minutes (or $\frac{1}{12}$ hour) is $s(\frac{1}{12})$. Since $s'(t) = v(t)$, the velocity at time t,

$$\int_0^{1/12} v(t) \, dt = s(t) \Big|_0^{1/12} = s(\tfrac{1}{12}) - s(0) = s(\tfrac{1}{12}).$$

We cannot evaluate $\int_0^{1/12} v(t) \, dt$ directly, since we only know the value of $v(t)$ when $t = 0, \frac{1}{60}, \frac{2}{60}, \frac{3}{60}, \frac{4}{60}, \frac{5}{60}$ hours. However, we can approximate the integral by using the trapezoidal rule. Letting $n = 5$, we have

$$\Delta x = \frac{\frac{1}{12} - 0}{5} = \frac{1}{60}.$$

Then

$$a_0 = 0, \quad a_1 = \tfrac{2}{60}, \quad a_2 = \tfrac{1}{60}, \quad a_3 = \tfrac{3}{60}, \quad a_4 = \tfrac{4}{60}, \quad a_5 = \tfrac{5}{60},$$

$$v(a_0) = 33, \quad v(a_1) = 32, \quad v(a_2) = 28, \quad v(a_3) = 30,$$

$$v(a_4) = 32, \quad v(a_5) = 35.$$

From the trapezoidal rule we have

$$\int_0^{1/12} v(t)\, dt \approx [33 + 2 \cdot (32) + 2 \cdot (28) + 2 \cdot (30) + 2 \cdot (32) + 35]\tfrac{1}{120}$$

$$= 2.6.$$

Therefore the car traveled approximately 2.6 miles during the five minutes.

PRACTICE PROBLEMS 4

1. Calculate the definite integral $\int_0^1 e^x\, dx$ using the trapezoidal rule with $n = 5$.

2. Calculate the exact value of the definite integral in Problem 1 and compare it with the answer derived in Problem 1.

EXERCISES 4

Approximate the following integrals by the trapezoidal rule and then find the exact value by integrating.

1. $\displaystyle\int_0^1 x^2\, dx; \ n = 5$

2. $\displaystyle\int_1^5 (x - 1)\, dx; \ n = 3$

3. $\displaystyle\int_0^{\pi/2} \sin x\, dx; \ n = 3$

4. The area of the semicircle of radius 1 is given by $\int_{-1}^1 \sqrt{1 - x^2}\, dx$. Use the trapezoidal rule with $n = 4$ to approximate this area. (Note: $\sqrt{.75} \approx .866$.)

Approximate the following integrals by the trapezoidal rule for each value n and then find the exact value by integrating.

5. $\displaystyle\int_0^4 (x^2 - 5)\, dx; \ n = 2, 4$

6. $\displaystyle\int_1^3 \frac{1}{x}\, dx; \ n = 3, 5$

7. $\displaystyle\int_0^1 x^3\, dx; \ n = 2, 4$

8. $\displaystyle\int_1^5 \frac{1}{x^2}\, dx; \ n = 3, 6$

Use the trapezoidal and rectangle rules to estimate the following integrals. Compare the estimates obtained with the exact values of the integrals.

9. $\int_4^6 \left(\frac{1}{2}x - 2 \right) dx; \; n = 2$ 10. $\int_1^2 \frac{1}{4x - 3} dx; \; n = 4$

11. $\int_0^1 e^{-x} dx; \; n = 5$ 12. $\int_2^3 e^{x+2} dx; \; n = 3$

13. Upon surveying a piece of oceanfront property, measurements of the distance to the water were made every 50 feet along a 200-foot side (Fig. 8). Use the trapezoidal rule to estimate the area of the property.

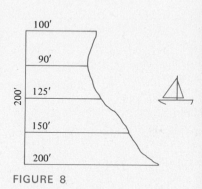

FIGURE 8

14. In order to determine the amount of water flowing down a certain 100-yard-wide river, engineers need to know the area of a vertical cross section of the river. Measurements of the depth of the river were made every 20 yards from one bank to the other. The readings in fathoms were 0, 1, 2, 3, 1, 0. (One fathom equals two yards.) Use the trapezoidal rule to estimate the area of the cross section.

15. Upon takeoff, the velocity readings of a rocket noted every second for 10 seconds were 0, 20, 50, 80, 110, 140, 170, 200, 230, 250, 280 mph. Use the trapezoidal rule to estimate the distance traveled by the rocket during the first 10 seconds.

SOLUTIONS TO PRACTICE PROBLEMS 4

1. If $n = 5$, then $\Delta x = (b - a)/n = \frac{1}{5} = .2$. By the trapezoidal rule,

$$\int_0^1 e^x \, dx = [e^0 + 2 \cdot e^{.2} + 2 \cdot e^{.4} + 2 \cdot e^{.6} + 2 \cdot e^{.8} + e^1] \frac{\Delta x}{2}$$

$$\approx [1 + 2.44280 + 2.98364 + 3.64424 + 4.45108 + 2.71828](.1)$$

$$\approx 1.72400.$$

2. The exact value of the integral (to 5 decimal places) is

$$\int_0^1 e^x \, dx = e^x \Big|_0^1 = e - 1 = 1.71828.$$

The error in the trapezoidal rule with $n = 5$ is thus

$$1.72400 - 1.71828 = .00572.$$

10.5. Some Applications of the Integral

Recall that the integral

$$\int_a^b f(t)\, dt$$

can be approximated by a sum as follows: We divide the t-axis from a to b into n subintervals by adding intermediate points $t_0 = a, t_1, \ldots, t_{n-1}, t_n = b$.

We assume that the points are equally spaced, so that each subinterval has length $\Delta t = (b - a)/n$. For large n, the integral is very closely approximated by the sum

$$\int_a^b f(t)\, dt \approx f(t_0)\, \Delta t + f(t_1)\, \Delta t + \ldots + f(t_{n-1})\, \Delta t. \tag{1}$$

The approximation in (1) works both ways. If we encounter a sum like the one in (1), then we can approximate it by the corresponding integral. The approximation becomes better as the number of subintervals increases—that is, as n gets large. Thus as n gets large, the sum approaches the value of the integral. This will be our approach in the examples below.

Our first two examples involve the concept of the present value of money. Suppose that we make an investment that promises to repay K dollars at time t (measuring the present as time 0). How much should we be willing to pay for such an investment? Clearly, we would not want to pay as much as K dollars. For if we had K dollars now, we could invest it at the current rate of interest and, at time t, we would get back our original K dollars plus the accrued interest. Instead, we should only be willing to pay an amount C that, if invested for t years, would yield K dollars. We call C the *present value of K dollars in t years*. We shall assume continuous compounding of interest. If the current (annual) rate of interest is r, then C dollars invested for t years will yield Ce^{rt} dollars (see Section 6.2). That is,

$$Ce^{rt} = K.$$

Thus the formula for the present value C of K dollars in t years at interest rate r is

$$C = Ke^{-rt}.$$

EXAMPLE 1   (*Present Value of an Income Stream*)   Consider a small printing company that does most of its work on one printing press. The firm's profits are directly influenced by the amount of material that the press can produce (assuming that other factors, such as wages, are held constant). We may say that the press is producing a continuous stream of income for the company. Of course, the efficiency of the press may decline as it gets older. At time t, let $K(t)$ be the annual rate of income from the press. [This means that the press is producing $K(t) \cdot \frac{1}{365}$ dollars per day at time t.] Find a model for the present value of the income generated by the printing press over the next T years, assuming an interest rate r (with interest compounded continuously).

Solution   Let us divide the T-year period into n small subintervals of time, each of duration Δt years. (If each subinterval were one day, for example, then Δt would equal $\frac{1}{365}$.)

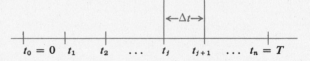

We now consider the income produced by the printing press during a small time interval from t_j to t_{j+1}. Since Δt is small, the rate $K(t)$ of income production changes by only a negligible amount in that interval and can be considered approximately equal to $K(t_j)$. Since $K(t_j)$ gives an annual rate of income, the actual income produced during the period of Δt years is $K(t_j) \, \Delta t$. This income will be produced at approximately time t_j (i.e., t_j years from $t = 0$), and so its present value is

$$[K(t_j) \, \Delta t]e^{-rt_j}.$$

The present value of the total income produced over the T-year period is the sum of the present values of the amounts produced during each time subinterval; that is,

$$K(t_0)e^{-rt_0} \, \Delta t + K(t_1)e^{-rt_1} \, \Delta t + \ldots + K(t_{n-1})e^{-rt_{n-1}} \, \Delta t. \tag{2}$$

As the number of subintervals gets large, the length Δt of each subinterval becomes small, and the sum in (2) approaches the integral

$$\int_0^T K(t)e^{-rt} \, dt. \tag{3}$$

We call this quantity the *present value of the stream of income* produced by the printing press over the period from $t = 0$ to $t = T$ years.

The concept of the present value of a continuous stream of income is an important tool in management decision processes involving the selection or replacement of equipment. It is also useful when analyzing various investment opportunities. Even when $K(t)$ is a simple function, the evaluation of the integral in (3) usually requires special techniques, such as integration by parts, as we see in the next example.

EXAMPLE 2   A company estimates that the rate of revenue produced by a machine at time t will be $5000 - 100t$ dollars per year. Find the present value of this continuous stream of income over the next four years at a 6% interest rate.

Solution   We use (3) with $K(t) = 5000 - 100t$, $T = 4$, and $r = .06$. The present value of this income stream is

$$\int_0^4 (5000 - 100t)e^{-.06t}\, dt.$$

Using integration by parts, with $f(t) = 5000 - 100t$ and $g(t) = e^{-.06t}$, we find that the preceding integral equals

$$(5000 - 100t)\frac{1}{-.06} e^{-.06t}\Big|_0^4 - \int_0^4 (-100)\frac{1}{-.06} e^{-.06t}\, dt$$

$$\approx 23{,}025 - \frac{100}{.06}\cdot\frac{1}{-.06} e^{-.06t}\Big|_0^4$$

$$\approx 23{,}025 - 5{,}927 = 17{,}098.$$

EXAMPLE 3   (*A Demographic Model*)   Let $P(t)$ denote the number of people living at time t in a certain metropolitan area. Suppose that the dynamics of population change are governed by two rules:

1.   At time t, new people are arriving in the area (by birth and by moving van) at a rate $r(t)$ per year.

2.   During any time interval, say from T_1 to T_2, a certain fraction of the original $P(T_1)$ people present at time T_1 will still be present at time T_2 (the others having died or moved away). Suppose that we write the number still present as $h(T_2 - T_1)P(T_1)$, where $h(T_2 - T_1)$ is a fraction between 0 and 1 that depends only on the length $T_2 - T_1$ of the time interval.

Construct a model for the size of the population at an arbitrary time T.

Solution   We divide the time interval from 0 to T into n equal subintervals, each of length Δt.

Of the original population of $P(0)$ people, only $h(T)P(0)$ will remain at time T, by rule 2. Let us analyze what happens during the time interval from t_j to t_{j+1}. If Δt is small, then the rate of new arrivals in the time interval from t_j to t_{j+1} will be nearly constant and equal to $r(t_j)$. The number of arrivals in the time period of length Δt will be $r(t_j)\,\Delta t$. How many of these people will remain from time t_j until the time T? By rule 2., this number is

$$h(T - t_j)r(t_j)\,\Delta t.$$

Therefore the total population at time T is approximately

$$h(T)P(0) + h(T - t_0)r(t_0)\,\Delta t + h(T - t_1)r(t_1)\,\Delta t + \ldots$$

$$+ h(T - t_{n-1})r(t_{n-1})\,\Delta t \approx h(T)P(0) + \int_0^T h(T - t)r(t)\, dt.$$

And so, letting n get large, we see that

$$P(T) = h(T)P(0) + \int_0^T h(T - t)r(t)\, dt. \tag{4}$$

EXAMPLE 4   Suppose that in Example 3 we have $r(t) = 50{,}000 + 2000t$ and $h(t) = e^{-t/40}$. Determine the population at the end of 10 years if the initial population is 10^6.

Solution   Using (4), we have

$$P(10) = h(10)P(0) + \int_0^{10} h(10 - t)r(t)\, dt$$

$$= e^{-1/4} \cdot 10^6 + \int_0^{10} e^{-(10-t)/40}(50{,}000 + 2000t)\, dt$$

$$\approx 778{,}800 + \int_0^{10} e^{(t-10)/40}(50{,}000 + 2000t)\, dt.$$

We evaluate the integral by using integration by parts, with $f(t) = 50{,}000 + 2000t$ and $g(t) = e^{(t-10)/40}$. After some calculations we obtain

$$P(10) \approx 778{,}800 + 534{,}560 = 1{,}313{,}360.$$

EXAMPLE 5   It has been determined that* in 1940 the population density r miles from the center of New York City was approximately $120e^{-.2r}$ thousands of people per square mile. Estimate the number of people who lived within two miles of the center of New York in 1940.

Solution   Let us choose a fixed line emanating from the center of the city along which to measure distance. Subdivide this line from $r = 0$ to $r = 2$ into a large number of subintervals, each of length Δr. Each subinterval determines a ring. (See Fig. 1.) Let us determine the population of each ring and add up all the populations so derived. If the inner circle of the $(i + 1)$st ring is at distance r_i from the center, the outer circle of that ring is at distance $r_{i+1} = r_i + \Delta r$ from the center. The area of the $(i + 1)$st ring is

$$\pi r_{i+1}^2 - \pi r_i^2 = \pi(r_i + \Delta r)^2 - \pi r_i^2$$

$$= \pi(r_i^2 + 2r_i\, \Delta r + \Delta r^2) - \pi r_i^2$$

$$= 2\pi r_i\, \Delta r + \pi(\Delta r)^2.$$

Assume that Δr is very small. Then $(\Delta r)^2$ is much smaller than $2\pi r_i\, \Delta r$. Hence, the area of this ring is very close to $2\pi r_i\, \Delta r$.

---

* Clark, C., "Urban Population Densities," *J. of the Royal Statistical Society*, Series A, **114** (1951), 490–496. Also see White, M. J., "On Cumulative Urban Growth and Urban Density Functions," *J. of Urban Economics*, **4** (1977), 104–112.

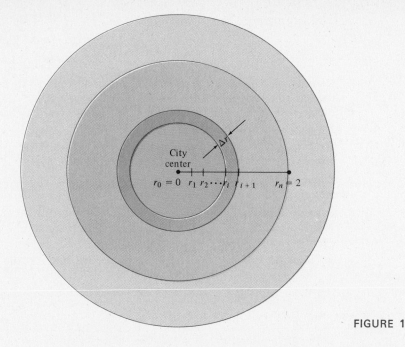

City center

$r_0 = 0 \quad r_1 \, r_2 \cdots r_i \; r_{i+1} \qquad r_n = 2$

Δr

FIGURE 1

Within the $(i + 1)$st ring, the density of people is about $120e^{-.2r_i}$ thousand people per square mile. So the number of people in this ring is approximately

$$\text{(population density)} \cdot \text{(area of ring)} = 120e^{-.2r_i} \cdot 2\pi r_i \, \Delta r$$
$$= 240\pi r_i e^{-.2r_i} \, \Delta r.$$

Adding up the populations of all the rings gives a total of

$$240\pi r_0 e^{-.2r_0} \, \Delta r + 240\pi r_1 e^{-.2r_1} \, \Delta r + \ldots + 240\pi r_{n-1} e^{-.2r_{n-1}} \, \Delta r,$$

which is a Riemann sum for the function $f(r) = 240\pi r e^{-.2r}$ over the interval from $r = 0$ to $r = 2$. Thus, the number of people (in thousands) who lived within two miles of the center of the city was

$$\int_0^2 240\pi r e^{-.2r} \, dr = 240\pi \int_0^2 r e^{-.2r} \, dr.$$

The last integral can be computed using integration by parts to obtain

$$240\pi \int_0^2 r e^{-.2r} \, dr = 240\pi \left. \frac{r e^{-.2r}}{-.2} \right|_0^2 - 240\pi \int_0^2 \frac{e^{-.2r}}{-.2} \, dr$$

$$= -2400\pi e^{-.4} + 1200\pi \left(\frac{e^{-.2r}}{-.2} \right) \Bigg|_0^2$$

$$= -2400\pi e^{-.4} + (-6000\pi e^{-.4} + 6000\pi)$$

$$\approx 1160.$$

Thus, in 1940 approximately 1,160,000 people lived within two miles of the center of the city.

1. Consider a continuous stream of income that is produced by some investment such that, at any given time t, the annual rate of income is $K(t) = -500 + 300t$. Find the present value of the income over the next ten years, using a 5% interest rate.

2. Suppose that $K(t) = 5000$ in Example 1: that is, suppose that the rate of income is a constant \$5000 per year. Find the present value of the stream of income produced over the next four years if the interest rate is 6%.

3. Suppose that in Example 3 we have $r(t) = 10,000 - 100t$ and $h(t) = e^{-t/50}$. Determine the population at the end of 10 years if the initial population is 250,000.

4. Suppose that in Example 3 the rate $r(t)$ of new arrivals is a constant 10,000 per year (no matter how large the population becomes) and also suppose that $h(t) = e^{-t/40}$.

    (a) Suppose that the initial population is 300,000 and use (4) to determine a formula for $P(T)$, where T is arbitrary.

    (b) What happens to the size of the population in part (a) as T gets larger and larger?

5. In 1900 the population density of Philadelphia r miles from the city center was $120e^{-.65r}$ thousand people per square mile. Calculate the number of people who lived within five miles of the city center.

6. In 1940, the population density of Philadelphia was given by the function $60e^{-.4r}$. Sketch the graphs of the population densities for 1900 and 1940 on a common graph. What trend do the graphs exhibit?

10.6. Improper Integrals

In applications of calculus, especially to statistics, it is often necessary to consider the area of a region that extends infinitely far to the right or left along the x-axis. We have drawn several such regions in Fig. 1. The areas of such "infinite" regions may be computed using *improper integrals*.

FIGURE 1

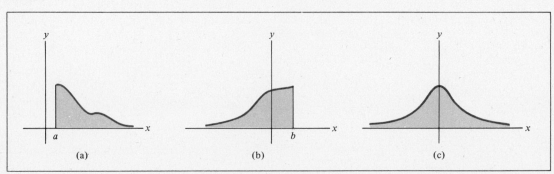

(a)   (b)   (c)

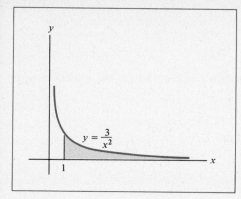

FIGURE 2

In order to motivate the idea of an improper integral, let us attempt to calculate the area under the curve $y = 3/x^2$ to the right of $x = 1$ (Fig. 2).

First, we shall compute the area under the graph of this function from $x = 1$ to $x = b$, where b is some number greater than 1. [See Fig. 3(a).] Then we shall examine how the area increases as we let b get larger, as in Figs. 3(b) and 3(c). The area from 1 to b is given by

$$\int_1^b \frac{3}{x^2}\, dx = -\frac{3}{x}\Big|_1^b = \left(-\frac{3}{b}\right) - \left(-\frac{3}{1}\right) = 3 - \frac{3}{b}.$$

When b is large, $3/b$ is small and the integral nearly equals 3. That is, the area under the curve from 1 to b nearly equals 3. (See Table 1.) In fact, the area gets arbitrarily close to 3 as b gets larger. Thus it is reasonable to say that the region under the curve $y = 3/x^2$ for $x \geq 1$ has area 3.

TABLE 1

b	Area $= \int_1^b \dfrac{3}{x^2}\, dx = 3 - \dfrac{3}{b}$
10	2.7000
100	2.9700
1000	2.9970
10,000	2.9997

Recall from Chapter 1 that we write $b \to \infty$ as a shorthand for "b gets arbitrarily large, without bound." Then, to express the fact that the value of $\int_1^b \frac{3}{x^2}\, dx$ approaches 3 as $b \to \infty$, we write

$$\int_1^\infty \frac{3}{x^2}\, dx = 3.$$

FIGURE 3

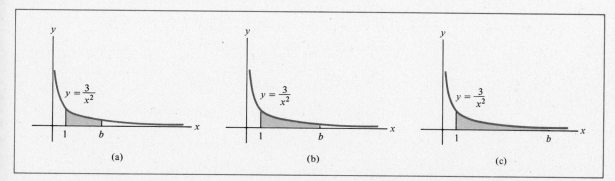

(a)   (b)   (c)

We call $\int_1^\infty \frac{3}{x^2}\,dx$ an *improper* integral because the upper limit of the integral is ∞ (infinity) rather than a finite number.

Definition   Let a be fixed and suppose $f(x)$ is a nonnegative function for $x \geq a$. If $\lim_{b \to \infty} \int_a^b f(x)\,dx = L$, then we say that

$$\int_a^\infty f(x)\,dx = L.$$

We say that the improper integral $\int_a^\infty f(x)\,dx$ is *convergent* and that the region under the curve $y = f(x)$ for $x \geq a$ has area L. (See Fig. 4.)

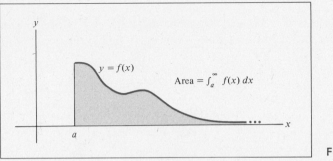

FIGURE 4

It is possible to consider improper integrals in which $f(x)$ is both positive and negative. However, we shall consider only nonnegative functions, since this is the case occurring in most applications.

EXAMPLE 1   Find the area under the curve $y = e^{-x}$ for $x \geq 0$ (Fig. 5).

Solution   We must calculate the improper integral

$$\int_0^\infty e^{-x}\,dx.$$

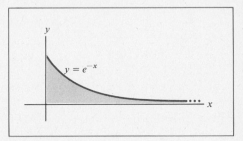

FIGURE 5

We take $b > 0$ and compute

$$\int_0^b e^{-x}\,dx = -e^{-x}\Big|_0^b = (-e^{-b}) - (-e^0) = 1 - e^{-b} = 1 - \frac{1}{e^b}.$$

We now consider the limit as $b \to \infty$ and note that $1/e^b$ approaches zero. Thus,

$$\int_0^\infty e^{-x}\,dx = \lim_{b \to \infty} \int_0^b e^{-x}\,dx = \lim_{b \to \infty} \left(1 - \frac{1}{e^b}\right) = 1.$$

Therefore, the region in Fig. 5 has area 1.

EXAMPLE 2 Evaluate the improper integral $\int_7^\infty \dfrac{1}{(x-5)^2}\,dx$.

Solution
$$\int_7^b \frac{1}{(x-5)^2}\,dx = -\frac{1}{x-5}\bigg|_7^b = -\frac{1}{b-5} - \left(-\frac{1}{7-5}\right) = \frac{1}{2} - \frac{1}{b-5}.$$

As $b \to \infty$, the fraction $1/(b-5)$ approaches zero, and so

$$\int_7^\infty \frac{1}{(x-5)^2}\,dx = \lim_{b\to\infty} \int_7^b \frac{1}{(x-5)^2}\,dx = \lim_{b\to\infty}\left(\frac{1}{2} - \frac{1}{b-5}\right) = \frac{1}{2}.$$

Not every improper integral is convergent. If the value of $\int_a^b f(x)\,dx$ does not have a limit as $b \to \infty$, then we cannot assign any numerical value to $\int_a^\infty f(x)\,dx$, and we say that the improper integral $\int_a^\infty f(x)\,dx$ is *divergent*.

EXAMPLE 3 Show that $\int_1^\infty \dfrac{1}{\sqrt{x}}\,dx$ is divergent.

Solution For $b > 1$ we have

$$\int_1^b \frac{1}{\sqrt{x}}\,dx = 2\sqrt{x}\,\bigg|_1^b = 2\sqrt{b} - 2. \qquad (1)$$

As $b \to \infty$, the quantity $2\sqrt{b} - 2$ increases without bound. That is, $2\sqrt{b} - 2$ can be made larger than any specific number. Therefore, $\int_1^b \dfrac{1}{\sqrt{x}}\,dx$ has no limit as $b \to \infty$, and so $\int_0^\infty \dfrac{1}{\sqrt{x}}\,dx$ is divergent.

In some cases it is necessary to consider improper integrals of the form

$$\int_{-\infty}^b f(x)\,dx.$$

Let b be fixed and examine the value of $\int_a^b f(x)\,dx$ as $a \to -\infty$—that is, as a moves arbitrarily far to the left on the number line. If $\lim\limits_{a\to-\infty} \int_a^b f(x)\,dx = L$, we say that the improper integral $\int_{-\infty}^b f(x)\,dx$ is *convergent* and we write

$$\int_{-\infty}^b f(x)\,dx = L.$$

Otherwise, the improper integral is divergent. An integral of the form $\int_{-\infty}^b f(x)\,dx$ may be used to compute the area of a region like that shown in Fig. 1(b).

EXAMPLE 4 Determine if $\int_{-\infty}^0 e^{5x}\,dx$ is convergent. If convergent, find its value.

Solution

$$\int_{-\infty}^{0} e^{5x}\,dx = \lim_{a \to -\infty} \int_{a}^{0} e^{5x}\,dx$$

$$= \lim_{a \to -\infty} \tfrac{1}{5}e^{5x}\Big|_{a}^{0}$$

$$= \lim_{a \to -\infty} (\tfrac{1}{5} - \tfrac{1}{5}e^{5a}).$$

As $a \to -\infty$, e^{5a} approaches 0 so that $\tfrac{1}{5} - \tfrac{1}{5}e^{5a}$ approaches $\tfrac{1}{5}$. Thus, the improper integral converges and has value $\tfrac{1}{5}$.

Areas of regions that extend infinitely far to the left *and* right, such as the region in Fig. 1(c), are calculated using improper integrals of the form

$$\int_{-\infty}^{\infty} f(x)\,dx.$$

We define such an integral to have the value

$$\int_{-\infty}^{0} f(x)\,dx + \int_{0}^{\infty} f(x)\,dx,$$

provided both of the latter improper integrals are convergent.

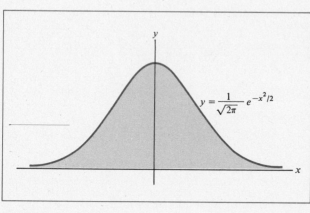

FIGURE 6

An important area that arises in probability theory is the area under the so-called normal curve, whose equation is

$$y = \frac{1}{\sqrt{2\pi}}\,e^{-x^2/2}.$$

(See Fig. 6.) It is of fundamental importance for probability theory that this area is 1. In terms of an improper integral, this fact may be written

$$\int_{-\infty}^{\infty} \frac{1}{\sqrt{2\pi}}\,e^{-x^2/2}\,dx = 1.$$

The proof of this result is beyond the scope of this book.

PRACTICE PROBLEMS 6

1. Does $1 + 2a^{-3}$ approach a limit as $a \to -\infty$?

2. Evaluate the improper integral, $\displaystyle\int_{-\infty}^{-2} \frac{1}{x^4}\,dx.$

EXERCISES 6

In Exercises 1–14 determine if the given expression approaches a limit as $b \to \infty$, and find that number when it exists.

1. $5/b$ 2. b^2 3. $-3e^{2b}$ 4. $(1/b) + \frac{1}{3}$ 5. $\frac{1}{4} - (1/b^2)$

6. $\frac{1}{2}\sqrt{b}$ 7. $2 - (b + 1)^{-1/2}$ 8. $(2/b) - (3/b^{3/2})$

9. $5 - 1/(b - 1)$ 10. $5(b^2 + 3)^{-1}$ 11. $6 - 3b^{-2}$

12. $e^{-b/2} + 5$ 13. $2(1 - e^{-3b})$ 14. $4(1 - b^{-3/4})$

15. Find the area under the graph of $y = 1/x^2$ for $x \geq 2$.

16. Find the area under the graph of $y = (x + 1)^{-2}$ for $x \geq 0$.

17. Find the area under the graph of $y = e^{-x/2}$ for $x \geq 0$.

18. Find the area under the graph of $y = 4e^{-4x}$ for $x \geq 0$.

19. Find the area under the graph of $y = (x + 1)^{-3/2}$ for $x \geq 3$.

20. Find the area under the graph of $y = (3x + 1)^{-2}$ for $x \geq 1$.

21. Show that the region under the graph of $y = x^{-3/4}$ for $x \geq 1$ cannot be assigned any finite number as its area.

22. Show that the region under the graph of $y = (x - 1)^{-1/3}$ for $x \geq 2$ cannot be assigned any finite number as its area.

Evaluate the following improper integrals whenever they are convergent.

23. $\displaystyle\int_1^\infty \frac{1}{x^3}\, dx$ 24. $\displaystyle\int_1^\infty \frac{2}{x^{3/2}}\, dx$ 25. $\displaystyle\int_0^\infty e^{2x}\, dx$

26. $\displaystyle\int_0^\infty x^2 + 1\, dx$ 27. $\displaystyle\int_0^\infty \frac{1}{(2x + 3)^2}\, dx$ 28. $\displaystyle\int_0^\infty e^{-3x}\, dx$

29. $\displaystyle\int_2^\infty \frac{1}{(x - 1)^{5/2}}\, dx$ 30. $\displaystyle\int_2^\infty e^{2-x}\, dx$ 31. $\displaystyle\int_0^\infty .01e^{-.01x}\, dx$

32. $\displaystyle\int_0^\infty \frac{4}{(2x + 1)^3}\, dx$ 33. $\displaystyle\int_0^\infty 6e^{1-3x}\, dx$ 34. $\displaystyle\int_1^\infty e^{-.2x}\, dx$

35. $\displaystyle\int_{-\infty}^0 e^{4x}\, dx$ 36. $\displaystyle\int_{-\infty}^0 \frac{8}{(x - 5)^2}\, dx$ 37. $\displaystyle\int_{-\infty}^0 \frac{6}{(1 - 3x)^2}\, dx$

38. $\displaystyle\int_{-\infty}^0 \frac{1}{\sqrt{4 - x}}\, dx$ 39. If $k > 0$, show that $\displaystyle\int_0^\infty ke^{-kx}\, dx = 1$.

40. If $k > 0$, show that $\displaystyle\int_1^\infty \frac{k}{x^{k+1}}\, dx = 1$.

1. The expression $1 + 2a^{-3}$ can also be written as $1 + (2/a^3)$. If a is large and negative, then $(2/a^3)$ is a small negative number. For instance, when $a = -100$, $(2/a^3) = (2/-1,000,000) = -.000002$. As $a \to -\infty$, $(2/a^3)$ approaches 0 and $1 + (2/a^3)$ approaches the limit 1.

2. $\displaystyle \int_a^{-2} \frac{1}{x^4}\, dx = \int_a^{-2} x^{-4}\, dx = \frac{x^{-3}}{-3}\Big|_a^{-2} = \frac{1}{-3x^3}\Big|_a^{-2}$

$$= \frac{1}{-3(-2)^3} - \left(\frac{1}{-3 \cdot a^3}\right)$$

$$= \frac{1}{24} + \frac{1}{3a^3}.$$

$\displaystyle \int_{-\infty}^{-2} \frac{1}{x^4}\, dx = \lim_{a \to -\infty} \int_a^{-2} \frac{1}{x^4}\, dx = \lim_{a \to -\infty} \left(\frac{1}{24} + \frac{1}{3a^3}\right) = \frac{1}{24}.$

Chapter 10: CHECKLIST

☐ Integration by substitution:

$$\int f(g(x))g'(x)\, dx = \int f(u)\, du, \quad u = g(x)$$

☐ Integration by parts:

$$\int f(x)g(x)\, dx = f(x)G(x) - \int f'(x)G(x)\, dx, \, G'(x) = g(x)$$

☐ Change of limits of integration:

$$\int_a^b f(g(x))g'(x)\, dx = \int_{g(a)}^{g(b)} f(u)\, du$$

☐ Trapezoidal rule:

$$\int_a^b f(x)\, dx \approx [f(a_0) + 2f(a_1) + \ldots + 2f(a_{n-1}) + f(a_n)]\frac{\Delta x}{2}$$

☐ Improper integrals, convergence and divergence

Chapter 10: SUPPLEMENTARY EXERCISES

Determine the following indefinite integrals.

1. $\displaystyle \int x \sin(3x^2)\, dx$ 2. $\displaystyle \int \sqrt{2x + 1}\, dx$ 3. $\displaystyle \int x(1 - 3x^2)^5\, dx$

4. $\displaystyle \int \frac{(\ln x)^5}{x}\, dx$ 5. $\displaystyle \int \frac{(\ln x)^2}{x}\, dx$ 6. $\displaystyle \int \frac{1}{\sqrt{4x + 3}}\, dx$

7. $\displaystyle \int x^3\sqrt{4 - x^2}\, dx$ 8. $\displaystyle \int x \sin 3x\, dx$ 9. $\displaystyle \int x^2 e^{-x^3}\, dx$

10. $\displaystyle \int \frac{x \ln(x^2 + 1)}{x^2 + 1}\, dx$ 11. $\displaystyle \int x^2 \cos 3x\, dx$ 12. $\displaystyle \int \frac{\ln(\ln x)}{x \ln x}\, dx$

13. $\displaystyle\int \ln x^2 \, dx$ 14. $\displaystyle\int x\sqrt{x+1} \, dx$ 15. $\displaystyle\int \frac{x}{\sqrt{3x-1}} \, dx$

16. $\displaystyle\int x^2 \ln x^3 \, dx$ 17. $\displaystyle\int \frac{x}{(1-x)^5} \, dx$ 18. $\displaystyle\int \frac{\ln x}{(1+\ln x)^2} \, dx$

Evaluate the following definite integrals.

19. $\displaystyle\int_0^1 \frac{2x}{(x^2+1)^3} \, dx$ 20. $\displaystyle\int_0^{\pi/2} x \sin 8x \, dx$

21. $\displaystyle\int_0^2 xe^{-(1/2)x^2} \, dx$ 22. $\displaystyle\int_{1/2}^1 \frac{\ln(2x+3)}{2x+3} \, dx$

23. $\displaystyle\int_1^2 xe^{-2x} \, dx$ 24. $\displaystyle\int_1^2 x^{-3/2} \ln x \, dx$

25. $\displaystyle\int_1^7 \frac{(\ln x)^3}{x} \, dx$ 26. $\displaystyle\int_0^1 xe^{-2x} \, dx$

Approximate the following integrals, using the trapezoidal rule with the indicated value of n.

27. $\displaystyle\int_0^{\pi/2} \sin x^2 \, dx, \, n = 4$ 28. $\displaystyle\int_0^{10} e^{\sqrt{x}} \, dx, \, n = 5$

29. $\displaystyle\int_1^3 \frac{1}{x} \, dx, \, n = 4$ 30. $\displaystyle\int_0^1 \frac{1}{1+x^2} \, dx, \, n = 5$

In Exercises 31 to 46 decide whether integration by parts or a substitution should be used to compute the indefinite integral. If substitution, indicate the substitution to be made. If by parts, indicate the functions $f(x)$ and $g(x)$ to be used in formula (1) of Section 2.

31. $\displaystyle\int xe^{2x} \, dx$ 32. $\displaystyle\int (x-3)e^{-x} \, dx$

33. $\displaystyle\int (x+1)^{-1/2} e^{\sqrt{x+1}} \, dx$ 34. $\displaystyle\int x^2 \sin(x^3-1) \, dx$

35. $\displaystyle\int \frac{x-2x^3}{x^4-x^2+4} \, dx$ 36. $\displaystyle\int \ln\sqrt{5-x} \, dx$

37. $\displaystyle\int e^{-x}(3x-1)^2 \, dx$ 38. $\displaystyle\int xe^{3-x^2} \, dx$

39. $\displaystyle\int (500-4x)e^{-x/2} \, dx$ 40. $\displaystyle\int (2+8x-2x^2)(x^3-6x^2-3x) \, dx$

41. $\displaystyle\int (x+3)e^{x^2+6x} \, dx$ 42. $\displaystyle\int (x+1)^2 e^{3x} \, dx$

43. $\displaystyle\int \sqrt{x} \ln x \, dx$ 44. $\displaystyle\int (3x-x^3)(x^4-6x^2-6)^{-1} \, dx$

45. $\displaystyle\int (x-1)\cos(x^2-2x+1) \, dx$ 46. $\displaystyle\int (3-x)\sin 3x \, dx$

Evaluate the following improper integrals whenever they are convergent.

47. $\displaystyle\int_0^{\infty} e^{6-3x} \, dx$ 48. $\displaystyle\int_0^{\infty} x^{-2/3} \, dx$

49. $\displaystyle\int_{-\infty}^0 \frac{8}{(5-2x)^3} \, dx$ 50. $\displaystyle\int_0^{\infty} x^2 e^{-x^3} \, dx$

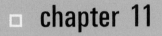

chapter 11

DIFFERENTIAL EQUATIONS

A *differential equation* is an equation in which derivatives of an unknown function $y = f(t)$ occur. Examples of such equations are

$$y' = 6t + 3$$

$$y' = 6y$$

$$y' + 3y + t = 0.$$

As we shall see, many physical processes can be described by differential equations. In this chapter we shall explore some topics in differential equations and use our resulting knowledge to study problems from many different fields, including business, genetics, and ecology.

11.1. Solutions of Differential Equations

A differential equation is an equation involving an unknown function y and one or more of the derivatives y', y'', y''', and so on. Suppose that y is a function of the variable t. A *solution* of a differential equation is any function $f(t)$ such that the differential equation becomes a true statement when y is replaced by $f(t)$, y' by $f'(t)$, y'' by $f''(t)$, and so forth.

EXAMPLE 1 Show that the function $f(t) = 5e^{-2t}$ is a solution of the differential equation

$$y' + 2y = 0. \tag{1}$$

Solution The differential equation (1) says that $y' + 2y$ equals zero for all values of t. We must show that this result holds if y is replaced by $5e^{-2t}$ and y' is replaced by $(5e^{-2t})' = -10e^{-2t}$. But

$$\overbrace{(5e^{-2t})'}^{y'} + 2\overbrace{(5e^{-2t})}^{y} = -10e^{-2t} + 10e^{-2t} = 0.$$

Therefore $y = 5e^{-2t}$ is a solution of the differential equation (1).

EXAMPLE 2 Show that the function $f(t) = \frac{1}{9}t + \sin 3t$ is a solution of the differential equation

$$y'' + 9y = t. \tag{2}$$

Solution If $f(t) = \frac{1}{9}t + \sin 3t$, then

$$f'(t) = \tfrac{1}{9} + 3 \cos 3t$$
$$f''(t) = -9 \sin 3t.$$

Substituting $f(t)$ for y and $f''(t)$ for y'' in the left side of (2), we obtain

$$\overbrace{-9 \sin 3t}^{y''} + \overbrace{9(\tfrac{1}{9}t + \sin 3t)}^{y} = -9 \sin 3t + t + 9 \sin 3t = t.$$

Therefore $y'' + 9y = t$ if $y = \frac{1}{9}t + \sin 3t$, and hence $y = \frac{1}{9}t + \sin 3t$ is a solution to $y'' + 9y = t$.

The process of determining all the functions that are solutions of a differential equation is called *solving the differential equation*. The process of antidifferentiation amounts to solving a simple type of differential equation. For example, a solution of the differential equation

$$y' = 3t^2 - 4 \tag{3}$$

is a function y whose derivative is $3t^2 - 4$. Thus solving (3) consists of finding all antiderivatives of $3t^2 - 4$. Clearly, y must be of the form $y = t^3 - 4t + C$,

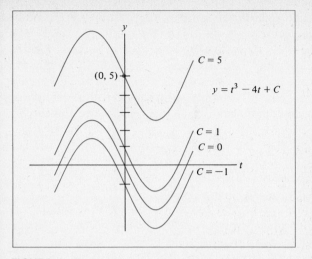

FIGURE 1 Typical solutions of $y' = 3t^2 - 4$.

where C is a constant. The solutions of (3) corresponding to several values of C are sketched in Fig. 1.

We encountered differential equations such as

$$y' = 2y \qquad (4)$$

in our discussion of exponential functions. Unlike (3), this equation does not give a specific formula for y' but instead describes a property of y', namely that y' is proportional to y (with 2 as the constant of proportionality). At the moment, the only way we can "solve" (4) is to simply know in advance what the solutions are. Recall from Chapter 4 that the solutions of (4) have the form $y = he^{2t}$ for any constant h. Some typical solutions of (4) are sketched in Fig. 2. In the next section we shall discuss a method for solving a class of differential equations that includes both (3) and (4) as special cases.

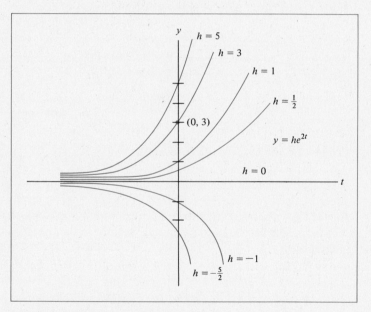

FIGURE 2 Typical solutions of $y' = 2y$.

Figures 1 and 2 illustrate two important differences between differential equations and algebraic equations (for instance, $ax^2 + bx + c = 0$). First, a solution of a differential equation is a *function* rather than a number. Second, a differential equation usually has infinitely many solutions.

Sometimes we want to find a particular solution that satisfies certain additional conditions called *initial conditions*. Initial conditions specify the values of a solution and a certain number of its derivatives at some specific value of t, often $t = 0$. If the solution to a differential equation is $y = f(t)$, we will often write $y(0)$ for $f(0)$, $y'(0)$ for $f'(0)$, and so on. Then some typical initial conditions are $y(0) = 5$, $y'(0) = -1$. The problem of determining a solution of a differential equation that satisfies given initial conditions is called an *initial value problem*.

EXAMPLE 3 (a) Solve the initial value problem $y' = 3t^2 - 4$, $y(0) = 5$.

(b) Solve the initial value problem $y' = 2y$, $y(0) = 3$.

Solution (a) We have already noted that the general solution of $y' = 3t^2 - 4$ is $f(t) = t^3 - 4t + C$. We want the particular solution that satisfies $f(0) = 5$. Geometrically, we are looking for the curve in Fig. 1 that passes through the point $(0, 5)$. Using the general formula for $f(t)$, we have

$$5 = f(0) = (0)^3 - 4(0) + C = C$$
$$C = 5.$$

Thus $f(t) = t^3 - 4t + 5$ is the desired solution.

(b) The general solution of $y' = 2y$ is $y = he^{2t}$. The condition $y(0) = 3$ means that y must be 3 when $t = 0$; that is, the point $(0, 3)$ must be on the graph of the solution to $y' = 2y$. (See Fig. 2.) We have

$$3 = y(0) = he^{2(0)} = h \cdot 1 = h$$
$$h = 3.$$

Thus $y = 3e^{2t}$ is the desired solution.

We conclude this section with an example of how a differential equation may be used to describe a physical process. (The equation can be solved by the method of the next section.) The example should be studied carefully, for it contains the key to understanding many similar problems that will appear in exercises.

EXAMPLE 4 (*Newton's Law of Cooling*) Suppose that a red-hot steel rod is plunged into a bath of cool water. Let $f(t)$ be the temperature of the rod at time t, and suppose that the water is maintained at a constant temperature M. According to Newton's law of cooling, the rate of change of $f(t)$ is proportional to the difference between the two temperatures M and $f(t)$. Find a differential equation that describes this physical law.

Solution The two key ideas are "rate of change" and "proportional." The rate of change of $f(t)$ is the derivative $f'(t)$. Since this is proportional to the difference $M - f(t)$, there exists a constant k such that

$$f'(t) = k[M - f(t)]. \tag{4}$$

The term "proportional" does not tell us whether k is positive or negative (or zero). We must decide this, if possible, from the context of the problem. In the present situation, the steel rod is hotter than the water, so $M - f(t)$ is negative. Also, $f(t)$ will decrease as time passes, so $f'(t)$ should be negative. Thus, to make $f'(t)$ negative in (4), k must be a positive number. From (4) we see that $y = f(t)$ satisfies a differential equation of the form

$$y' = k(M - y), \quad k \text{ a positive constant.}$$

Throughout this section we have assumed that our solution functions y are functions of the variable t, which in applications usually stands for time. In most of this chapter we will continue to use the variable t. However, if we use another variable occasionally, such as x, we will make the variable explicit by writing $\frac{dy}{dx}$ instead of y'.

PRACTICE PROBLEMS 1

1. Show that any function of the form $y = Ae^{t^3/3}$, where A is a constant, is a solution of the differential equation

$$y' - t^2 y = 0.$$

2. If the function $f(t)$ is a solution of the initial value problem

$$y' = (t + 2)y, \quad y(0) = 3,$$

find $f(0)$ and $f'(0)$.

EXERCISES 1

1. Show that the function $f(t) = \frac{3}{2}e^{t^2} - \frac{1}{2}$ is a solution of the differential equation $y' - 2ty = t$.

2. Show that the function $f(t) = t^2 - \frac{1}{2}$ is a solution of the differential equation $(y')^2 - 4y = 2$.

3. Show that the function $f(t) = (e^{-t} + 1)^{-1}$ satisfies $y' + y^2 = y$, $y(0) = \frac{1}{2}$.

4. Show that the function $f(t) = 5e^{2t}$ satisfies $y'' - 3y' + 2y = 0$, $y(0) = 5$, $y'(0) = 10$.

5. Solve $y' = \frac{1}{2}y$, $y(0) = 1$.

6. Solve $y' = e^{t/2}$, $y(0) = 1$.

7. Is the constant function $f(t) = 3$ a solution of the differential equation $y' = 6 - 2y$?

8. Is the constant function $f(t) = -4$ a solution of the differential equation $y' = t^2(y + 4)$?

9. If the function $f(t)$ is a solution of the initial value problem

$$y' = 2y - 3, \qquad y(0) = 4,$$

find $f(0)$ and $f'(0)$.

10. If the function $f(t)$ is a solution to the initial value problem $y' = e^t + y$, $y(0) = 0$, find $f(0)$ and $f'(0)$.

11. Let $f(t)$ denote the amount of capital invested by a certain business firm at time t. The rate of change of invested capital, $f'(t)$, is sometimes called the *rate of net investment*. Suppose that the management of the firm decides that the optimum level of investment should be C dollars and that, at any time, the rate of net investment should be proportional to the difference between C and the total capital invested. Construct a differential equation that describes this situation.

12. When a cool object is placed in a room, the rate at which it is warming up at any time is proportional to the difference between the room temperature and the temperature of the object. Letting $y = f(t)$ represent the temperature of the object at time t, give a differential equation that expresses this physical law.

SOLUTIONS TO PRACTICE PROBLEMS 1

1. If $y = Ae^{t^3/3}$, then

$$\overbrace{(Ae^{t^3/3})'}^{y'} - t^2(\overbrace{Ae^{t^3/3}}^{y}) = At^2 e^{t^3/3} - t^2 Ae^{t^3/3} = 0.$$

Therefore $y' - t^2 y = 0$ if $y = Ae^{t^3/3}$.

2. The initial condition $y(0) = 0$ says that $f(0) = 3$. Since $f(t)$ is a solution to $y' = (t + 2)y$,

$$f'(t) = (t + 2)f(t)$$

and hence

$$f'(0) = (0 + 2)f(0)$$

$$= 2 \cdot 3$$

$$= 6.$$

11.2. Separation of Variables

Here we describe a technique for solving an important class of differential equations—those of the form

$$y' = p(t)q(y),$$

where $p(t)$ is a function of t only and $q(y)$ is a function of y only. Two equations of this type are

$$y' = \frac{3t^2}{y^2} \qquad \left[p(t) = 3t^2, q(y) = \frac{1}{y^2} \right], \qquad (1)$$

$$y' = e^{-y}(2t + 1) \qquad [p(t) = 2t + 1, q(y) = e^{-y}]. \qquad (2)$$

The main feature of such equations is that we may *separate the variables*; that is, we may rewrite the equations so that y occurs only on one side of the equation and t on the other. For example, if we multiply both sides of equation (1) by y^2, then the equation becomes

$$y^2 y' = 3t^2;$$

if we multiply both sides of equation (2) by e^y, the equation becomes

$$e^y y' = 2t + 1.$$

It should be pointed out that the differential equation

$$y' = 3t^2 - 4$$

is of the preceding type. Here $p(t) = 3t^2 - 4$ and $q(y) = 1$. The variables are already separated, however. Similarly, the exponential differential equation

$$y' = 5y$$

is of the preceding type, with $p(t) = 5$, $q(y) = y$. We can separate the variables by writing the equation as

$$\frac{1}{y} y' = 5.$$

In the next example we present a procedure for solving differential equations in which the variables are separated.

EXAMPLE 1   Find all solutions of the differential equation $y^2 y' = 3t^2$.

Solution   (a) Write y' as $\dfrac{dy}{dt}$.

$$y^2 \frac{dy}{dt} = 3t^2.$$

(b) Integrate both sides with respect to t.

$$\int y^2 \frac{dy}{dt} \, dt = \int 3t^2 \, dt.$$

(c) Rewrite the left-hand side, "cancelling the dt."

$$\int y^2 \, dy = \int 3t^2 \, dt.$$

(d) Calculate the antiderivatives.

$$\tfrac{1}{3}y^3 + C_1 = t^3 + C_2.$$

(e) Solve for y in terms of t.

$$y^3 = 3(t^3 + C_2 - C_1)$$
$$y = \sqrt[3]{3t^3 + C}, \quad C \text{ a constant.}$$

We can check that this method works by showing that $y = \sqrt[3]{3t^3 + C}$ is a solution to $y^2 y' = 3t^2$: Since $y = (3t^3 + C)^{1/3}$, we have

$$y' = \tfrac{1}{3}(3t^3 + C)^{-2/3} \cdot 3 \cdot 3t^2 = 3t^2(3t^3 + C)^{-2/3}$$
$$y^2 y' = [(3t^3 + C)^{1/3}]^2 \cdot 3t^2(3t^3 + C)^{-2/3}$$
$$= 3t^2.$$

Discussion of step (c): Suppose that $y = f(t)$ is a solution of the differential equation $y^2 y' = 3t^2$. Then

$$[f(t)]^2 f'(t) = 3t^2.$$

Integrating, we have

$$\int [f(t)]^2 f'(t)\, dt = \int 3t^2\, dt.$$

Let us make the substitution $y = f(t)$, $dy = f'(t)\, dt$ in the left-hand side in order to get

$$\int y^2\, dy = \int 3t^2\, dt.$$

This is just the result of step (c). The process of "cancelling the dt" and integrating with respect to y is actually just equivalent to making the substitution $y = f(t)$, $dy = f'(t)\, dt$.

The technique used in Example 1 can be used for any differential equation with separated variables. Suppose that we are given such an equation:

$$h(y)y' = p(t).$$

where $h(y)$ is a function of y only and $p(t)$ is a function of t only. Our method of solution can be summarized as follows:

a. Write y' as $\dfrac{dy}{dt}$.

$$h(y)\frac{dy}{dt} = p(t).$$

b. Integrate both sides.

$$\int h(y)\frac{dy}{dt}\, dt = \int p(t)\, dt.$$

c. Rewrite the left-hand side by "cancelling the dt."

$$\int h(y)\, dy = \int p(t)\, dt.$$

d. Calculate the antiderivatives $H(y)$ for $h(y)$ and $P(t)$ for $p(t)$.

$$H(y) = P(t) + C.$$

e. Solve for y in terms of t.

$$y = \ldots.$$

Note: In step (d.) there is no need to write two constants of integration (as we did in Example 1), since they will be combined into one in step (e.).

EXAMPLE 2 Solve $e^y y' = 2t + 1$, $y(0) = 1$.

Solution (a) $e^y \dfrac{dy}{dt} = 2t + 1$.

(b) $\displaystyle\int e^y \frac{dy}{dt}\, dt = \int (2t + 1)\, dt$.

(c) $\displaystyle\int e^y\, dy = \int (2t + 1)\, dt$.

(d) $e^y = t^2 + t + C$.

(e) $y = \ln(t^2 + t + C)$.   [Take logarithms of both sides of the equation in step (d).]

If $y = \ln(t^2 + t + C)$ is to satisfy the initial condition $y(0) = 1$, then $1 = y(0) = \ln(0^2 + 0 + C) = \ln C$, so that $C = e$ and $y = \ln(t^2 + t + e)$.

EXAMPLE 3 Solve $y' = t^3 y^2 + y^2$.

Solution As the equation is given, the right-hand side is not in the form $p(t)q(y)$. However, we may rewrite the equation in the form $y' = (t^3 + 1)y^2$. Now we may separate the variables, dividing both sides by y^2, to get

$$\frac{1}{y^2} y' = t^3 + 1. \tag{3}$$

Then we apply our method of solution:

(a) $\dfrac{1}{y^2} \dfrac{dy}{dt} = t^3 + 1$.

(b) $\displaystyle\int \frac{1}{y^2} \frac{dy}{dt}\, dt = \int (t^3 + 1)\, dt$.

(c) $\int \frac{1}{y^2}\, dy = \int (t^3 + 1)\, dt.$

(d) $-\frac{1}{y} = \frac{1}{4}t^4 + t + C, \quad C$ a constant.

(e) $y = -\dfrac{1}{\frac{1}{4}t^4 + t + C}.$

The method that we have applied yields all the solutions of equation (3). However, we have ignored an important point. We wish to solve $y' = y^2(t^3 + 1)$ and not equation (3). Do the two equations have precisely the same solutions? We obtained equation (3) from the given equation by dividing by y^2. This is a permissible operation, provided that y is not equal to zero for all t. (Of course, if y is zero for some t, then the resulting differential equation is understood to hold only for some limited range of t.) Thus in dividing by y^2, we must assume that y is not the zero function. However, note that $y = 0$ is a solution of the original equation because

$$0 = (0)' = t^3 \cdot 0^2 + 0^2.$$

So when we divided by y^2, we "lost" the solution $y = 0$. Finally, we see that the solutions of the differential equation $y' = t^3 y^2 + y^2$ are

$$y = -\frac{1}{\frac{1}{4}t^4 + t + C}, \quad C \text{ a constant}$$

and

$$y = 0.$$

Warning: If the equation in Example 3 had been

$$y' = t^3 y^2 + 1,$$

we would not have been able to use the method of separation of variables because the expression $t^3 y^2 + 1$ cannot be written in the form $p(t)q(y)$.

EXAMPLE 4   Solve $y' = 5 - 2y$.

Solution   The algebra involved in solving an equation of the form $y' = a + by$ is always simplified by factoring out the coefficient of y. Thus we rewrite $y' = 5 - 2y$ as

$$y' = -2(y - \tfrac{5}{2}). \tag{4}$$

Clearly, the constant function $y = \frac{5}{2}$ is a solution of the differential equation because it makes both sides of (4) zero for all t. (The left-hand side is zero because the derivative of a constant function is zero.) Now, supposing $y \neq \frac{5}{2}$, we may divide by $y - \frac{5}{2}$, obtaining

$$\frac{1}{y - \frac{5}{2}} y' = -2$$

$$\int \frac{1}{y - \frac{5}{2}} \frac{dy}{dt}\, dt = \int -2\, dt$$

$$\int \frac{1}{y - \frac{5}{2}}\, dy = \int -2\, dt$$

$$\ln\left|y - \tfrac{5}{2}\right| = -2t + C, \quad C \text{ a constant}$$

$$\left|y - \tfrac{5}{2}\right| = e^{-2t+C} = e^C \cdot e^{-2t}. \tag{5}$$

In Section 4 we shall see that if $y \neq \frac{5}{2}$, then $y - \frac{5}{2}$ will be either positive for all t or negative for all t. Consequently, either

$$y - \tfrac{5}{2} = e^C \cdot e^{-2t} \qquad \text{or} \quad y - \tfrac{5}{2} = -e^C \cdot e^{-2t},$$

that is,

$$y = \tfrac{5}{2} + e^C \cdot e^{-2t} \quad \text{or} \qquad y = \tfrac{5}{2} - e^C \cdot e^{-2t}. \tag{6}$$

These two types of solutions and the constant solution $y = \frac{5}{2}$ may all be written in the form

$$y = \tfrac{5}{2} + Ae^{-2t}, \quad A \text{ any constant.}$$

The two types of solutions shown in (6) correspond to positive and negative values of A, respectively. The constant solution is given by setting $A = 0$.

Warning:

1. In Example 4 the "extra" solution $y = \frac{5}{2}$ corresponded to the value $A = 0$ of our final solution. Such is not always the case. In Example 3 the solution $y = 0$ does not correspond to any value of the constant C.

2. A common mistake in solving $y' = 5 - 2y$ is to *add* $2y$ to both sides to obtain $y' + 2y = 5$. In a sense, the variables are then separated but not in the proper way. For if we integrate both sides, we get

$$\int \left(\frac{dy}{dt} + 2y\right) dt = \int 5\, dt,$$

and there is no substitution that will simplify the left-hand side, because $\dfrac{dy}{dt}$ does not appear as a factor.

EXAMPLE 5 Solve $y' = te^t/y$, $y(0) = -5$.

Solution Separating the variables, we have

$$yy' = te^t$$

$$\int y\frac{dy}{dt}\, dt = \int te^t\, dt$$

$$\int y\, dy = \int te^t\, dt.$$

The integral $\int te^t\,dt$ may be computed by integration by parts:

$$\int te^t\,dt = te^t - \int 1 \cdot e^t\,dt = te^t - e^t + C.$$

Therefore

$$\tfrac{1}{2}y^2 = te^t - e^t + C$$
$$y^2 = 2te^t - 2e^t + C_1$$
$$y = \pm\sqrt{2te^t - 2e^t + C_1}.$$

Note that the $\pm$ appears, since there are two square roots of $2te^t - 2e^t + C_1$, differing from one another by a minus sign. Thus the solutions are of two sorts—namely,

$$y = +\sqrt{2te^t - 2e^t + C_1}$$
$$y = -\sqrt{2te^t - 2e^t + C_1}.$$

We must choose C_1 so that $y(0) = -5$. Since the values of y for the first solution are always positive, the given initial condition must correspond to the second solution, and we must have

$$-5 = y(0) = -\sqrt{2 \cdot 0 \cdot e^0 - 2e^0 + C_1} = -\sqrt{-2 + C_1}$$
$$-2 + C_1 = 25$$
$$C_1 = 27.$$

Hence the desired solution is

$$y = -\sqrt{2te^t - 2e^t + 27}.$$

When working the exercises below, it is a good practice to first find the constant solution(s), if any. A constant function $y = c$ is a solution of $y' = p(t)q(y)$ if and only if $q(c) = 0$. [For $y = c$ implies $y' = (c)' = 0$, and $p(t)q(y)$ will be zero for all t if and only if $q(y) = 0$—that is, $q(c) = 0$.] After listing the constant solutions, one may assume $q(y) \neq 0$ and go on to divide both sides of the equation $y' = p(t)q(y)$ by $q(y)$ to separate the variables.

PRACTICE PROBLEMS 2

1. Solve the initial value problem $y' = 5y$, $y(0) = 2$, by separation of variables.

2. Solve $y' = \sqrt{ty}$, $y(1) = 4$.

EXERCISES 2

Solve the following differential equations.

1. $\dfrac{dy}{dt} = \dfrac{5 - t}{y^2}$
2. $\dfrac{dy}{dt} = \left(\dfrac{e^t}{y}\right)^2$
3. $y' = y^2 - e^{3t}y^2$

4. $y' = e^{4y}t^3 - e^{4y}$ 5. $y' = te^{2y}$ 6. $y' = \dfrac{t^2y^2}{t^3 + 8}$

7. $y' = 4(y - 3)$ 8. $y' = -\tfrac{1}{2}(y - 4)$ 9. $y' = \dfrac{\ln t}{ty}$

10. $y' = 2 - y$ 11. $y' = 3y + 4$ 12. $yy' = t\cos(t^2 + 1)$

13. $\dfrac{dy}{dt} = \sqrt{\dfrac{y}{t}}$ 14. $\dfrac{dy}{dt} = e^{t-y}$ 15. $y' = y\cos t$

16. $y' = \dfrac{1}{ty + y}$ 17. $y' = (y - 3)^2 \ln t$ 18. $y' = \dfrac{2t(y - 1)}{t^2 + 1}$

Solve the following differential equations with given initial conditions.

19. $y' = t^2 e^{-3y}$, $y(0) = 2$ 20. $y' = \tfrac{1}{2}y - 3$, $y(0) = 4$

21. $y' = 3 - y$, $y(0) = 1$ 22. $y' = \dfrac{t^2}{y}$, $y(0) = -5$

23. $y^2 y' = t\sin t$, $y(0) = 2$ 24. $y' = 2te^{-2y} - e^{-2y}$, $y(0) = 3$

25. $y' = 5ty - 2t$, $y(0) = 1$ 26. $y' = [(1 + t)/(1 + y)]^2$, $y(0) = 2$

27. $\dfrac{dy}{dt} = \dfrac{t + 1}{ty}$, $t > 0$, $y(1) = -3$ 28. $\dfrac{dy}{dt} = -\dfrac{y}{t}$, $t > 0$, $y(1) = 4$

29. $\dfrac{dy}{dx} = \dfrac{\ln x}{\sqrt{xy}}$, $y(1) = 4$ 30. $\dfrac{dN}{dt} = -2tN^2$, $N(0) = 5$

31. A model that describes the relationship between the price and the weekly sales of a product is

$$\frac{dy}{dp} = -k\frac{y}{p + c},$$

where y is the volume of sales, p is the price per unit, and k and c are positive constants. That is, at any time, the rate of decrease of sales with respect to price is directly proportional to the sales level and inversely proportional to the sales price plus a constant. Solve this differential equation.

32. One problem in psychology is to determine the relation between some physical stimulus and the corresponding sensation or reaction produced in a subject. Suppose that, measured in appropriate units, the strength of a stimulus is s and the intensity of the corresponding sensation is some function of s, say $f(s)$. Some experimental data suggest that the rate of change of intensity of the sensation with respect to the stimulus is directly proportional to the intensity of the sensation and inversely proportional to the strength of the stimulus; that is, $f(s)$ satisfies the differential equation

$$\frac{dy}{ds} = k\frac{y}{s}$$

for some positive constant k. Solve this differential equation.

33. Let t represent the total number of hours that a truck driver spends during a year driving on a certain highway connecting two cities and let $p(t)$ represent the

probability that the driver will have at least one accident during these t hours. Then $0 \le p(t) \le 1$, and $1 - p(t)$ represents the probability of not having an accident. Under ordinary conditions the rate of increase in the probability of an accident (as a function of t) is proportional to the probability of not having an accident. Construct and solve a differential equation for this situation.

34. In certain learning situations there is a maximum amount of information that can be learned, and at any time the rate of learning is proportional to the amount yet to be learned. Let $y = f(t)$ be the amount of information learned up to time t. Construct and solve a differential equation that is satisfied by $f(t)$.

35. The Gompertz growth equation is

$$\frac{dy}{dt} = -ay \ln \frac{y}{b},$$

where a and b are positive constants. This equation is used in biology to describe the growth of certain populations. Find the general form of solutions to this differential equation.

36. When a certain liquid substance A is heated in a flask, it decomposes into a substance B at a rate (measured in units of A per hour) that at any time t is proportional to the square of the amount of substance A present. Let $y = f(t)$ be the amount of substance A present at time t. Construct and solve a differential equation that is satisfied by $f(t)$.

SOLUTIONS TO PRACTICE PROBLEMS 2

1. The constant function $y = 0$ is a solution of $y' = 5y$. If $y \ne 0$, we may divide by y and obtain

$$\frac{1}{y} y' = 5$$

$$\int \frac{1}{y} \frac{dy}{dt} \, dt = \int 5 \, dt$$

$$\int \frac{1}{y} \, dy = \int 5 \, dt$$

$$\ln |y| = 5t + C$$

$$|y| = e^{5t + C} = e^C \cdot e^{5t}$$

$$y = \pm e^C \cdot e^{5t}.$$

These two types of solutions and the constant solution may all be written in the form

$$y = Ae^{5t},$$

where A is an arbitrary constant (positive, negative, or zero). The initial condition $y(0) = 2$ implies that

$$2 = y(0) = Ae^{5(0)} = A.$$

Hence the solution of the initial-value problem is $y = 2e^{5t}$.

2. We rewrite $y' = \sqrt{ty}$ as $y' = \sqrt{t} \cdot \sqrt{y}$. The constant function $y = 0$ is one solution. To find the others, we suppose $y \neq 0$ and divide by $\sqrt{y}$ to obtain

$$\frac{1}{\sqrt{y}} y' = \sqrt{t}$$

$$\int y^{-1/2} \frac{dy}{dt} dt = \int t^{1/2} dt$$

$$\int y^{-1/2} dy = \int t^{1/2} dt$$

$$2y^{1/2} = \tfrac{2}{3}t^{3/2} + C$$

$$y^{1/2} = \tfrac{1}{3}t^{3/2} + C_1 \tag{7}$$

$$y = (\tfrac{1}{3}t^{3/2} + C_1)^2. \tag{8}$$

We must choose C_1 so that $y(1) = 4$. The quickest method is to use (7) instead of (8). We have $y = 4$ when $t = 1$, and so

$$4^{1/2} = \tfrac{1}{3}(1)^{3/2} + C_1$$

$$2 = \tfrac{1}{3} + C_1$$

$$C_1 = \tfrac{5}{3}.$$

Hence the desired solution is

$$y = (\tfrac{1}{3}t^{3/2} + \tfrac{5}{3})^2.$$

11.3. Numerical Solution of Differential Equations

Many differential equations that arise in real-life applications cannot be solved by *any* known method. However, approximate solutions may be obtained by several different numerical techniques. In this section we shall describe what is known as *Euler's method* for approximating solutions to initial-value problems of the form

$$y' = g(t, y), \qquad y(a) = y_0 \tag{1}$$

for values of t in some interval $a \leq t \leq b$. Here $g(t, y)$ is some reasonably well-behaved function of two variables. Equations of the form $y' = p(t)q(y)$ studied in the previous section are a special case of (1).

In the discussion below we assume that $f(t)$ is a solution of (1) for $a \leq t \leq b$. The basic idea on which Euler's method rests is the following: *If the graph of $y = f(t)$ passes through some given point (t, y), then the slope of the graph (that is, the value of y') at that point is just $g(t, y)$, because $y' = g(t, y)$.* Euler's method uses this observation to approximate the graph of $f(t)$ by a polygonal path, such as in Fig. 1.

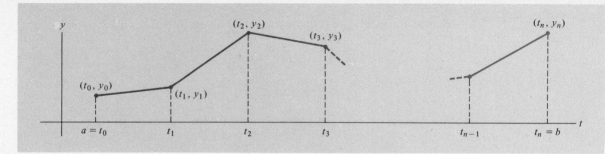

FIGURE 1

The t-axis from a to b is subdivided by the equally spaced points $t_0, t_1, \ldots, t_n$. Each of the n subintervals has length $h = (b - a)/n$. The initial condition $y(a) = y_0$ in (1) implies that the graph of the solution $f(t)$ passes through the point (t_0, y_0). As noted above, the slope of this graph at (t_0, y_0) must be $g(t_0, y_0)$. Thus on the first subinterval, Euler's method approximates the graph of $f(t)$ by the straight line

$$y = y_0 + g(t_0, y_0) \cdot (t - t_0)$$

which passes through (t_0, y_0) and has slope $g(t_0, y_0)$. When $t = t_1$, the y-coordinate on this line is

$$y_1 = y_0 + g(t_0, y_0) \cdot (t_1 - t_0)$$
$$= y_0 + g(t_0, y_0) \cdot h.$$

Since the graph of $f(t)$ is close to the point (t_1, y_1) on the line, the slope of the graph of $f(t)$ when $t = t_1$ will be close to $g(t_1, y_1)$. So we draw the straight line

$$y = y_1 + g(t_1, y_1) \cdot (t - t_1) \tag{2}$$

through (t_1, y_1) with slope $g(t_1, y_1)$ (Fig. 2), and we use this line to approximate $f(t)$ on the second subinterval. From (2) we determine an estimate y_2 for the

FIGURE 2

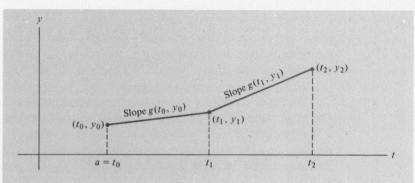

value of $f(t)$ at $t = t_2$:

$$y_2 = y_1 + g(t_1, y_1) \cdot h.$$

The slope of the graph of $f(t)$ at t_2 is now estimated by $g(t_2, y_2)$, and so on. Let us summarize this procedure:

Euler's Method The endpoints $(t_0, y_0), \ldots, (t_n, y_n)$ of the line segments approximating the solution of (1) on the interval $a \leq t \leq b$ are given by the following formulas, where $h = (b - a)/n$:

$$t_0 = a \quad \text{(given)}, \qquad y_0 \quad \text{(given)},$$
$$t_1 = t_0 + h, \qquad y_1 = y_0 + g(t_0, y_0) \cdot h,$$
$$t_2 = t_1 + h, \qquad y_2 = y_1 + g(t_1, y_1) \cdot h,$$
$$\vdots \qquad\qquad \vdots$$
$$t_n = t_{n-1} + h. \qquad y_n = y_{n-1} + g(t_{n-1}, y_{n-1}) \cdot h.$$

EXAMPLE 1 Use Euler's method with $n = 4$ to approximate the solution $f(t)$ to $y' = 2t - 3y$, $y(0) = 4$, for t in the interval $0 \leq t \leq 2$. In particular, estimate $f(2)$.

Solution Here $g(t, y) = 2t - 3y$, $a = 0$, $b = 2$, $y_0 = 4$, and $h = (2 - 0)/4 = \frac{1}{2}$. Starting with $(t_0, y_0) = (0, 4)$, we find that $g(0, 4) = -12$. Thus

$$t_1 = \tfrac{1}{2}, \qquad y_1 = 4 + (-12) \cdot \tfrac{1}{2} = -2.$$

Next, $g(\frac{1}{2}, -2) = 7$, and so

$$t_2 = 1, \qquad y_2 = -2 + 7 \cdot \tfrac{1}{2} = \tfrac{3}{2}.$$

Next, $g(1, \frac{3}{2}) = -\frac{5}{2}$, and so

$$t_3 = \tfrac{3}{2}, \qquad y_3 = \tfrac{3}{2} + (-\tfrac{5}{2}) \cdot \tfrac{1}{2} = \tfrac{1}{4}.$$

Finally, $g(\frac{3}{2}, \frac{1}{4}) = \frac{9}{4}$, and so

$$t_4 = 2, \qquad y_4 = \tfrac{1}{4} + \tfrac{9}{4} \cdot \tfrac{1}{2} = \tfrac{11}{8}.$$

Thus the approximation to the solution $f(t)$ is given by the polygonal path shown in Fig. 3. The last point $(2, \frac{11}{8})$ is close to the graph of $f(t)$ at $t = 2$, and so $f(2) \approx \frac{11}{8}$.

Actually, this polygonal path is somewhat misleading. The accuracy can be improved dramatically by increasing the value of n. Figure 4 shows the Euler approximation for $n = 8$ and $n = 20$. The graph of the exact solution is shown for comparison.

For many purposes, satisfactory graphs can be obtained by running Euler's method on a computer with large values of n. There is a limit to the accuracy obtainable, however, because each computer calculation involves a slight "round-

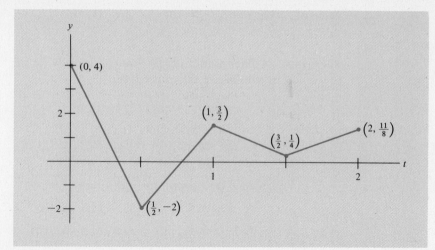

FIGURE 3

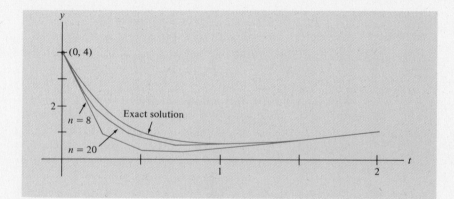

FIGURE 4

off" error. When n is very large, the cumulative round-off error can become significant. Several more sophisticated methods for approximating the solutions of differential equations are discussed in Chapter 8 of *Elementary Differential Equations*, 3d ed., by William Boyce and Richard DiPrima (New York: John Wiley & Sons, 1977).

PRACTICE PROBLEMS 3

Let $f(t)$ be the solution of $y' = \sqrt{ty}$, $y(1) = 4$.

1. Use Euler's method with $n = 2$ on the interval $1 \le t \le 2$ to estimate $f(2)$. (Note: $\sqrt{\frac{15}{2}} \approx 2.7386$.)

2. Draw the polygonal path corresponding to the application of Euler's method in Problem 1.

EXERCISES 3

1. Suppose $f(t)$ is a solution of the differential equation $y' = ty - 5$, and the graph of $f(t)$ passes through the point $(2, 4)$. What is the slope of the graph at this point?

2. Suppose $f(t)$ is a solution of $y' = t^2 - y^2$ and the graph of $f(t)$ passes through the point $(2, 3)$. Find the slope of the graph when $t = 2$.

3. Suppose $f(t)$ satisfies the initial-value problem $y' = y^2 + ty - 7$, $y(0) = 3$. Is $f(t)$ increasing or decreasing at $t = 0$?

4. Suppose $f(t)$ satisfies the initial-value problem $y' = y^2 + ty - 7$, $y(0) = 2$. Is the graph of $f(t)$ increasing or decreasing when $t = 0$?

5. Use Euler's method with $n = 2$ on the interval $0 \leq t \leq 1$ to approximate the solution $f(t)$ to $y' = t^2 y$, $y(0) = -2$. In particular, estimate $f(1)$.

6. Use Euler's method with $n = 2$ on the interval $2 \leq t \leq 3$ to approximate the solution $f(t)$ to $y' = t - 2y$, $y(2) = 3$. Estimate $f(3)$.

7. Use Euler's method with $n = 4$ to approximate the solution $f(t)$ to $y' = 2t - y + 1$, $y(0) = 5$ for $0 \leq t \leq 2$. Estimate $f(2)$.

8. Let $f(t)$ be the solution of $y' = y(2t - 1)$, $y(0) = 8$. Use Euler's method with $n = 4$ to estimate $f(1)$.

9. Let $f(t)$ be the solution of $y' = -(t + 1)y^2$, $y(0) = 1$. Use Euler's method with $n = 5$ to estimate $f(1)$. Then solve the differential equation, find an explicit formula for $f(t)$, and compute $f(1)$. How accurate is the estimated value of $f(1)$?

10. Let $f(t)$ be the solution of $y' = 10 - y$, $y(0) = 1$. Use Euler's method with $n = 5$ to estimate $f(1)$. Then solve the differential equation and find the exact value of $f(1)$.

11. Suppose the Consumer Products Safety Commission issues new regulations that affect the toy manufacturing industry. Every toy manufacturer will have to make certain changes in its manufacturing process. Let $f(t)$ be the fraction of manufacturers that have complied with the regulations within t months. Note that $0 \leq f(t) \leq 1$. Suppose that the rate at which new companies comply with the regulations is proportional to the fraction of companies who have not yet complied.

   (a) Construct a differential equation satisfied by $f(t)$.

   (b) Use Euler's method with $n = 3$ to estimate the fraction of companies who comply with the regulations within the first three months. (The answer will involve an unknown constant of proportionality, k.)

   (c) Solve the differential equation in (a) and compute $f(3)$.

   (d) Compare the answers in (b) and (c) when $k = .1$.

12. The Los Angeles Zoo plans to transport a California sea lion to the San Diego Zoo. The animal will be wrapped in a wet blanket during the trip. At any time t the blanket will lose water (owing to evaporation) at a rate proportional to the amount $f(t)$ of water in the blanket. Initially the blanket will contain two gallons of sea water.

(a) Set up the differential equation satisfied by $f(t)$.

(b) Use Euler's method with $n = 2$ to estimate the amount of moisture in the blanket after one hour. (The answer will involve an unknown constant of proportionality, k.)

(c) Solve the differential equation in (a) and compute $f(1)$.

(d) Compare the answers in (b) and (c) when $k = -.3$.

SOLUTIONS TO PRACTICE PROBLEMS 3

1. Here $g(t, y) = \sqrt{ty}$, $a = 1$, $b = 2$, $y_0 = 4$, and $h = (2 - 1)/2 = \frac{1}{2}$. We have

$$t_0 = 1, \qquad y_0 = 4, \qquad\qquad g(1, 4) = \sqrt{1 \cdot 4} = 2,$$
$$t_1 = \tfrac{3}{2}, \qquad y_1 = 4 + 2(\tfrac{1}{2}) = 5, \qquad g(\tfrac{3}{2}, 5) = \sqrt{\tfrac{3}{2} \cdot 5} \approx 2.7386,$$
$$t_2 = 2, \qquad y_2 = 5 + (2.7386)(\tfrac{1}{2}) = 6.3693.$$

Hence $f(2) \approx y_2 = 6.3693$.

[In Practice Problems 2 we found the solution of $y' = \sqrt{ty}$, $y(1) = 4$, to be $f(t) = (\tfrac{1}{3}t^{3/2} + \tfrac{5}{3})^2$. Using a calculator, we find that $f(2) = 6.8094$ (to four decimal places). The error of $6.8094 - 6.3693 = .4401$ in the approximation above is about 6.5%.]

2.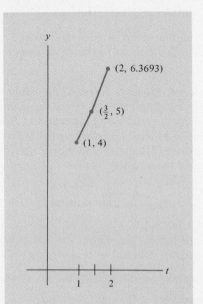

To find the polygonal path, plot the points (t_0, y_0), (t_1, y_1), and (t_2, y_2) and join them by line segments.

11.4. Qualitative Theory of Differential Equations

In this section we present a technique for sketching solutions to differential equations of the form $y' = g(y)$ *without having to solve the differential equation.* This technique is valuable for three reasons. First, there are many differential equations for which explicit solutions cannot be written down. Second, even when an explicit solution is available, we still face the problem of determining its behavior. For example, does the solution increase or decrease? If it increases, does it approach an asymptote or does it grow arbitrarily large? Third, and probably most significant, in many applications the explicit formula for a solution is unnecessary; only a general knowledge of the behavior of the solution is needed. That is, a qualitative understanding of the solution is sufficient.

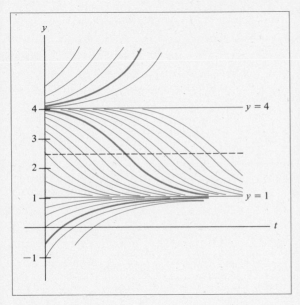

The theory introduced in this section is part of what is called the *qualitative theory of differential equations.* We shall limit our attention to differential equations of the form $y' = g(y)$. Such differential equations are called *autonomous.* The term "autonomous" here means "independent of time" and refers to the fact that the right-hand side of $y' = g(y)$ depends only on y and not on t. All of the applications studied in the next section will involve autonomous differential equations.

Throughout this section we shall consider the values of each solution $y = f(t)$ only for $t \geq 0$. To introduce the qualitative theory, let us examine the graphs of the various typical solutions of the differential equation $y' = \frac{1}{2}(1 - y)(4 - y)$. The solution curves in Fig. 1 illustrate the following properties.

FIGURE 1 Solutions of $y' = \frac{1}{2}(1 - y)(4 - y)$.

Property I Corresponding to each zero of $g(y)$ there is a constant solution of the differential equation. Specifically, if $g(c) = 0$, then the constant function $y = c$ is a solution. (The constant solutions in Fig. 1 are $y = 1$ and $y = 4$.)

Property II The constant solutions divide the t-y plane into horizontal strips. Each nonconstant solution lies completely in one strip.

Property III Each nonconstant solution is either strictly increasing or decreasing.

Property IV Each nonconstant solution either is asymptotic to a constant solution or else increases or decreases without bound.

Properties I–IV are valid for the solutions of any autonomous differential equation $y' = g(y)$ provided $g(y)$ is a "sufficiently well-behaved" function. We shall assume these properties in this chapter.

Using Properties I–IV, we can sketch the general shape of any solution curve by looking at the graph of the function $g(y)$ and the behavior of that graph near $y(0)$. The procedure for doing this is illustrated in the following example.

EXAMPLE 1 Sketch the solution to $y' = e^{-y} - 1$ that satisfies $y(0) = -2$.

Solution Here $g(y) = e^{-y} - 1$. On a y-z coordinate system we draw the graph of the function $z = g(y) = e^{-y} - 1$ [Fig. 2(a)]. The function $g(y) = e^{-y} - 1$ has a zero when $y = 0$. Therefore the equation $y' = e^{-y} - 1$ has the constant solution $y = 0$. We indicate this constant solution on a t-y coordinate system in Fig. 2(b). To begin the sketch of the solution satisfying $y(0) = -2$, we locate this initial value of y on the (horizontal) y-axis in Fig. 2(a) and on the (vertical) y-axis in Fig. 2(b).

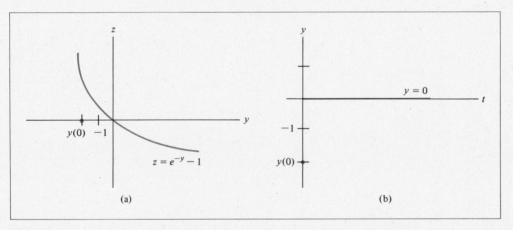

(a)   (b)

FIGURE 2

To determine whether the solution increases or decreases when it leaves the initial point $y(0)$ on the t-y graph, we look at the y-z graph and note that $z = g(y)$ is positive at $y = -2$ [Fig. 3(a)]. Consequently, since $y' = g(y)$, the derivative of the solution is positive, which implies that the solution is increasing. We indicate this by an arrow at the initial point in Fig. 3(b). Moreover, the solution y will increase asymptotically to the constant solution $y = 0$ by Properties III and **IV** of autonomous differential equations.

Next, we place an arrow in Fig. 4(a) to remind us that y will move from $y = -2$ toward $y = 0$. As y moves to the right toward $y = 0$ in Fig. 4(a), the z-coordinate of points on the graph of $g(y)$ becomes less positive; that is, $g(y)$ becomes less positive. Consequently, since $y' = g(y)$, the slope of the solution curve becomes less positive. Thus the solution curve is concave down, as we have shown in Fig. 4(b).

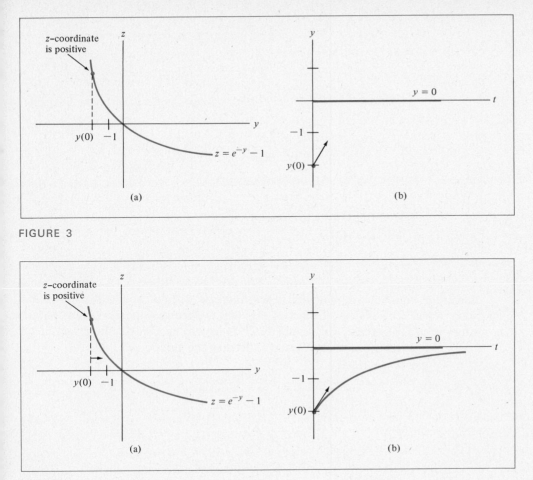

FIGURE 3

FIGURE 4

An important point to remember when sketching solutions is that z-coordinates on the y-z graph are values of $g(y)$, and since $y' = g(y)$, a z-coordinate gives the *slope* of the solution curve at the corresponding point on the t-y graph.

EXAMPLE 2   Sketch the graphs of the solutions to $y' = y + 2$ satisfying (a) $y(0) = 1$ and (b) $y(0) = -3$.

Solution   Here $g(y) = y + 2$. The graph of $z = g(y)$ is a straight line of slope 1 and z-intercept 2. [See Fig. 5(a).] This line crosses the y-axis only where $y = -2$. Thus the differential equation $y' = y + 2$ has one constant solution, $y = -2$. [See Fig. 5(b).]

(a) We locate the initial value $y(0) = 1$ on the y-axes of both graphs in Fig. 5. The corresponding z-coordinate on the y-z graph is positive; therefore the solution on the t-y graph has positive slope and is increasing as it leaves the initial

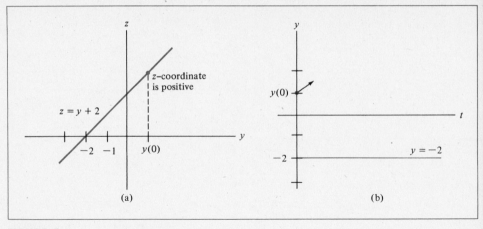

FIGURE 5

point. We indicate this by an arrow in Fig. 5(b). Now, Property IV of autonomous differential equations implies that y will increase without bound from its initial value. As we let y increase from 1 in Fig. 6(a), we see that the z-coordinates [i.e., values of $g(y)$] increase. Consequently, y' is increasing, and so the graph of the solution must be concave up. We have sketched the solution in Fig. 6(b).

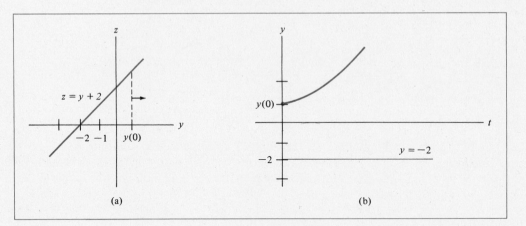

FIGURE 6

(b) Next we graph the solution for which $y(0) = -3$. From the graph of $z = y + 2$, we see that z is negative when $y = -3$. This implies that the solution is decreasing as it leaves the initial point. (See Fig. 7.) It follows that the values of y will continue to decrease without bound and become more and more negative. This means that on the y-z graph y must move to the *left* [Fig. 8(a)]. We now examine what happens to $g(y)$ as y moves to the left. (This is the opposite of the ordinary way to read a graph.) The z-coordinate becomes more negative, and

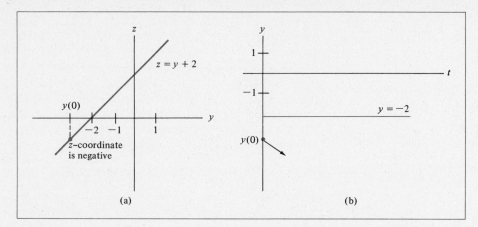

FIGURE 7

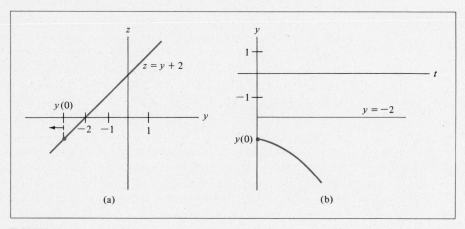

FIGURE 8

hence the slopes on the solution curve will become more negative. Thus the solution curve must be concave down, as in Fig. 8(b).

From the preceding examples we can state a few rules for sketching a solution to $y' = g(y)$ with $y(0)$ given:

1. Sketch the graph of $z = g(y)$ on a y-z coordinate system. Find and label the zeros of $g(y)$.

2. For each zero c of $g(y)$ draw the constant solution $y = c$ on the t-y coordinate system.

3. Plot $y(0)$ on the y-axes of the two coordinate systems.

4. Determine whether the value of $g(y)$ is positive or negative when $y = y(0)$. This tells us whether the solution is increasing or decreasing. On the t-y graph, indicate the direction of the solution through $y(0)$.

5. On the y-z graph indicate which direction y should move. (Note: If y is moving *down* on the t-y graph, then y moves to the *left* on the y-z graph.) As y moves in the proper direction on the y-z graph, determine whether g(y) becomes more positive, less positive, more negative, or less negative. This tells us the concavity of the solution.

6. Beginning at y(0) on the t-y graph, sketch the solution, being guided by the principle that the solution will grow (positively or negatively) without bound unless it encounters a constant solution. In this case, it will approach the constant solution asymptotically.

EXAMPLE 3   Sketch the solutions to $y' = y^2 - 4y$ satisfying $y(0) = 4.5$ and $y(0) = 3$.

Solution   Refer to Fig. 9. Since $g(y) = y^2 - 4y = y(y - 4)$, the zeros of $g(y)$ are 0 and 4, and hence the constant solutions are $y = 0$ and $y = 4$. The solution satisfying $y(0) = 4.5$ is increasing, because the z-coordinate is positive when $y = 4.5$ on the y-z graph. This solution continues to increase without bound. The solution satisfying $y(0) = 3$ is decreasing because the z-coordinate is negative when $y = 3$ on the y-z graph. This solution will decrease and approach asymptotically the constant solution $y = 0$.

An additional piece of information about the solution satisfying $y(0) = 3$ may be obtained from the graph of $z = g(y)$. We know that y decreases from 3 and approaches 0. From the graph of $z = g(y)$ in Fig. 9 it appears that at first the z-coordinates become more negative until y reaches 2 and then become less negative as y moves on toward 0. Since these z-coordinates are slopes on the solution curve, we conclude that as the solution moves downward from its initial point on the t-y coordinate system, its slope becomes more negative until the y-coordinate is 2 and then the slope becomes less negative as the y-coordinate approaches 0. Hence the solution is concave down until $y = 2$ and then is concave up. Thus there is an inflection point at $y = 2$, where the concavity changes.

FIGURE 9

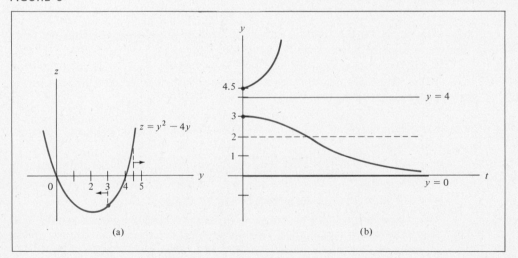

(a)   (b)

We saw in Example 3 that the inflection point at $y = 2$ was produced by the fact that $g(y)$ had a minimum at $y = 2$. A generalization (see below) of the argument in Example 3 shows that inflection points of solution curves occur at each value of y where $g(y)$ has a nonzero maximum or minimum point. Thus we may formulate an additional rule for sketching a solution of $y' = g(y)$.

7. On the t-y coordinate system draw dotted horizontal lines at all values of y at which $g(y)$ has a *nonzero* maximum or minimum point. A solution curve will have an inflection point whenever it crosses such a dotted line.

It is useful to note that when $g(y)$ is a quadratic function, as in Example 3, its maximum or minimum point occurs at a value of y halfway between the zeros of $g(y)$. This is because the graph of a quadratic function is a parabola, which is symmetric about a vertical line through its vertex.

EXAMPLE 4   Sketch a solution to $y' = e^{-y}$ with $y(0) > 0$.

Solution   Refer to Fig. 10. Since $g(y) = e^{-y}$ is always positive, there are no constant solutions to the differential equation and every solution will increase without bound. When drawing solutions that asymptotically approach a horizontal line, we have no choice as to whether to draw it concave up or concave down. This decision will be obvious from its increasing or decreasing nature and from knowledge of inflection points. However, for solutions that grow without bound, we must look at $g(y)$ in order to determine concavity. In this example, as t increases, the values of y increase. As y increases, $g(y)$ becomes less positive. Since $g(y) = y'$, we deduce that the slope of the solution curve becomes less positive; therefore the solution curve is concave down.

FIGURE 10

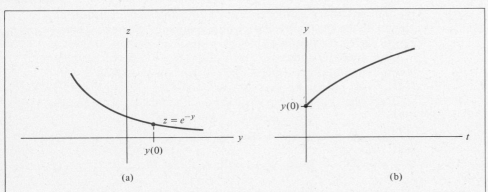

(a)   (b)

Consider the differential equation $y' = g(y)$, where $g(y)$ is the function whose graph is drawn below.

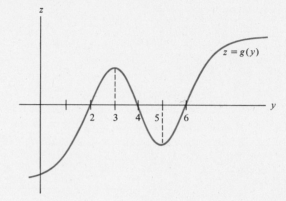

1. How many constant solutions are there to the differential equation $y' = g(y)$?

2. For what initial values, $y(0)$, will the corresponding solution of the differential equation be an increasing function?

3. Is it true that if the initial value $y(0)$ is near 4, then the corresponding solution will be asymptotic to the constant solution $y = 4$?

4. For what initial values, $y(0)$, will the corresponding solution of the differential equation have an inflection point?

EXERCISES 4

One or more initial conditions are given for each differential equation below. Use the qualitative theory of autonomous differential equations to sketch the graphs of the corresponding solutions. Include a y-z graph if one is not already provided. Always indicate the constant solutions on the t-y graph whether they are mentioned or not.

1. $y' = 2y - 6$, $y(0) = 1$, $y(0) = 4$. [The graph of $z = g(y)$ is drawn in Fig. 11.]

2. $y' = 5 - 2y$, $y(0) = 1$, $y(0) = 4$. (See Fig. 12.)

3. $y' = 4 - y^2$, $y(0) = -3$, $y(0) = -1$, $y(0) = 3$. (See Fig. 13.)

4. $y' = y^2 - 5$, $y(0) = -4$, $y(0) = 2$, $y(0) = 3$. (See Fig. 14.)

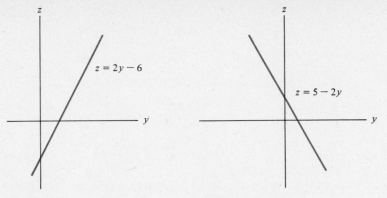

FIGURE 11

FIGURE 12

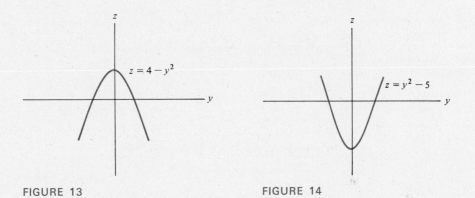

FIGURE 13

FIGURE 14

5. $y' = y^2 - 6y + 5$ or $y' = (y - 1)(y - 5)$, $y(0) = -1$, $y(0) = 2$, $y(0) = 4$, $y(0) = 6$. (See Fig. 15.)

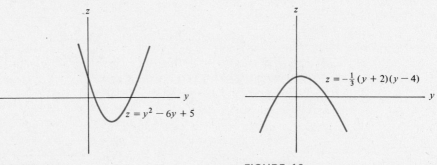

FIGURE 15

FIGURE 16

6. $y' = -\frac{1}{3}(y + 2)(y - 4)$, $y(0) = -3$, $y(0) = -1$, $y(0) = 6$. (See Fig. 16.)

7. $y' = y^3 - 9y$ or $y' = y(y^2 - 9)$, $y(0) = -4$, $y(0) = -1$, $y(0) = 2$, $y(0) = 4$. (See Fig. 17.)

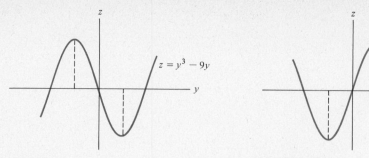

FIGURE 17   **FIGURE 18**

8. $y' = 9y - y^3$, $y(0) = -4$, $y(0) = -1$, $y(0) = 2$, $y(0) = 4$. (See Fig. 18.)

9. Use the graph in Fig. 19 to sketch the solutions to the Gompertz growth equation.

$$\frac{dy}{dt} = -\frac{1}{10} y \ln \frac{y}{100}$$

satisfying $y(0) = 10$ and $y(0) = 150$.

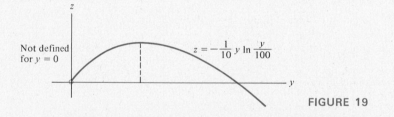

FIGURE 19

10. The graph of $z = -\frac{1}{2} y \ln(y/30)$, has the same general shape as the graph in Fig. 19 with maximum point at $y \approx 11.0364$ and y-intercept at $y = 30$. Sketch the solutions to the Gompertz growth equation

$$\frac{dy}{dt} = -\frac{1}{2} y \ln \frac{y}{30}$$

satisfying $y(0) = 1$, $y(0) = 20$, and $y(0) = 40$.

11. $y' = g(y)$, $y(0) = -.5$, $y(0) = .5$, where $g(y)$ is the function whose graph is given in Fig. 20.

FIGURE 20   **FIGURE 21**

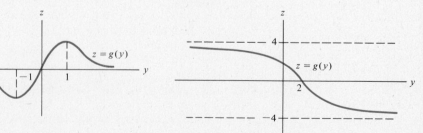

12. $y' = g(y)$, $y(0) = 0$, $y(0) = 4$, where the graph of $g(y)$ is given in Fig. 21.

13. $y' = g(y)$, $y(0) = 0$, $y(0) = 1.2$, $y(0) = 5$, $y(0) = 7$, where the graph of $g(y)$ is given in Fig. 22.

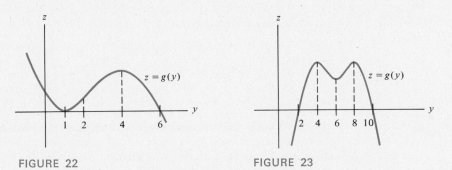

FIGURE 22   FIGURE 23

14. $y' = g(y)$, $y(0) = 1$, $y(0) = 3$, $y(0) = 11$, where the graph of $g(y)$ is given in Fig. 23.

15. $y' = 3 - \frac{1}{2}y$, $y(0) = 2$, $y(0) = 8$.

16. $y' = 3y$, $y(0) = -2$, $y(0) = 2$.

17. $y' = 5y - y^2$, $y(0) = 1$, $y(0) = 7$.

18. $y' = -y^2 + 10y - 21$, $y(0) = 1$, $y(0) = 4$.

19. $y' = y^2 - 3y - 4$, $y(0) = 0$, $y(0) = 3$.

20. $y' = \frac{1}{2}y^2 - 3y$, $y(0) = 3$, $y(0) = 6$, $y(0) = 9$.

21. $y' = y^2 + 2$, $y(0) = -1$, $y(0) = 1$.

22. $y' = y - \frac{1}{4}y^2$, $y(0) = -1$, $y(0) = 1$.

23. $y' = \sin y$, $y(0) = -\pi/6$, $y(0) = \pi/6$, $y(0) = 7\pi/4$.

24. $y' = 1 + \sin y$, $y(0) = 0$, $y(0) = \pi$.

25. $y' = 1/y$, $y(0) = -1$, $y(0) = 1$.

26. $y' = y^3$, $y(0) = -1$, $y(0) = 1$.

27. $y' = ky^2$, where k is a negative constant, $y(0) = -2$, $y(0) = 2$.

28. $y' = ky(M - y)$, where $k > 0$, $M > 10$, and $y(0) = 1$.

29. $y' = ky - A$, where k and A are positive constants. Sketch solutions where $0 < y(0) < A/k$ and $y(0) > A/k$.

30. $y' = k(y - A)$, where $k < 0$ and $A > 0$. Sketch solutions where $y(0) < A$ and $y(0) > A$.

31. Suppose that once a sunflower plant has started growing, the rate of growth at any time is proportional to the product of its height and the difference between its height at maturity and its current height. Give a differential equation that is satisfied by $f(t)$, the height at time t, and sketch the solution.

1. Three. The function $g(y)$ has zeros when y is 2, 4, and 6. Therefore, $y' = g(y)$ has the constant functions $y = 2$, $y = 4$, and $y = 6$ as solutions.

2. For $2 < y(0) < 4$ and $y(0) > 6$. Since nonconstant solutions are either strictly increasing or strictly decreasing, a solution is an increasing function provided that it is increasing at time $t = 0$. This is the case when the first derivative is positive at $t = 0$. When $t = 0$, $y' = g(y(0))$. Therefore, the solution corresponding to $y(0)$ is increasing whenever $g(y(0))$ is positive.

3. Yes. If $y(0)$ is slightly to the right of 4, then $g(y(0))$ is negative, and so the corresponding solution will be a decreasing function with values moving to the left closer and closer to 4. If $y(0)$ is slightly to the left of 4, then $g(y(0))$ is positive, and so the corresponding solution will be an increasing function with values moving to the right closer and closer to 4. (The constant solution $y = 4$ is referred to as a *stable* constant solution. The solution with initial value 4 stays at 4, and solutions with initial values near 4 move toward 4. The constant solution $y = 2$ is *unstable*. Solutions with initial values near 2 move away from 2.)

4. For $2 < y(0) < 3$ and $5 < y(0) < 6$. Inflection points of solutions correspond to maximum and minimum points of the function $g(y)$. If $2 < y(0) < 3$, the corresponding solution will be an increasing function. The values of y will move to the right (toward 4) and therefore will cross 3, a place at which $g(y)$ has a maximum. Similarly, if $5 < y(0) < 6$, the corresponding solution will be decreasing. The values of y on the y-z graph will move to the left and cross 5.

11.5. Applications of Differential Equations

Equations describing conditions in a physical process are often referred to as *mathematical models*. In this section we study real-life situations that may be modeled by an autonomous differential equation $y' = g(y)$. Here y will represent some quantity that is changing with time, and the equation $y' = g(y)$ will be obtained from a description of the rate of change of y.

We have already encountered two situations where the rate of change of y is *proportional* to some quantity:

1. $y' = ky$: "the rate of change of y is proportional to y" (exponential growth or decay);

2. $y' = k(M - y)$: "the rate of change of y is proportional to the difference between M and y" (Newton's law of cooling, for example).

The following example is of the same general type. It concerns the rate at which a technological innovation may spread through an industry, a subject of concern to both sociologists and economists.

EXAMPLE 1   The byproduct coke oven was first introduced into the iron and steel industry in 1894. It took about thirty years before all the major steel producers had adopted this innovation. Let $f(t)$ be the percentage of the producers that had installed the new coke ovens by time t. Then a reasonable model* for the way $f(t)$ increased is given by the assumption that the rate of change of $f(t)$ at time t was proportional to the product of $f(t)$ and the percentage of firms who had not yet installed the new coke ovens at time t. Write a differential equation that is satisfied by $f(t)$.

Solution   Since $f(t)$ is the *percentage* of firms that have the new coke oven, $100 - f(t)$ is the percentage of firms that still have not installed any new coke ovens. We are told that the rate of change of $f(t)$ is proportional to the product of $f(t)$ and $100 - f(t)$. Hence there is a constant of proportionality k such that

$$f'(t) = kf(t)[100 - f(t)].$$

Replacing $f(t)$ by y and $f'(t)$ by y', we obtain the desired differential equation,

$$y' = ky(100 - y).$$

Note that both y and $100 - y$ are nonnegative quantities. Clearly y' must be positive, because $y = f(t)$ is an increasing function. Hence the constant k must be positive.

The differential equation obtained in Example 1 is a special case of the *logistic differential equation*,

$$y' = ky(a - y), \tag{1}$$

where k and a are positive constants. This equation is used as a simple mathematical model of a wide variety of physical phenomena. In Section 6.3 we described applications of the logistic equation to restricted population growth and to the spread of an epidemic. Let us use the qualitative theory of differential equations to gain more insight into this important equation.

The first step in sketching solutions of (1) is to draw the y-z graph. Rewriting the equation $z = ky(a - y)$ in the form

$$z = -ky^2 + kay,$$

we see that the equation is quadratic in y and hence its graph will be a parabola. The parabola is concave down because the coefficient of y^2 is negative (since k is a positive constant). The zeros of the quadratic expression $ky(a - y)$ occur where $y = 0$ and $y = a$. Since a represents some positive constant, we select an arbitrary point on the positive y-axis and label it "a". With this information we

---

* Cf. E. Mansfield, "Technical Change and the Rate of Imitation." *Econometrica,* **29** (1961), 741–766.

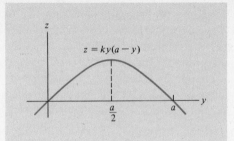

FIGURE 1

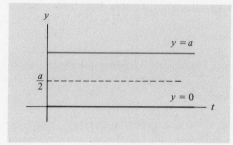

FIGURE 2

can sketch a representative graph, as in Fig. 1. Note that the vertex of the parabola occurs at $y = a/2$, halfway between the y-intercepts. (The reader should review how we obtained this graph, given only that k and a are positive constants. Similar situations will arise in the exercises.)

We begin the t-y graph as in Fig. 2, showing the constant solutions and placing a dotted line at $y = a/2$ where certain solution curves will have an inflection point. On either side of the constant solutions we choose initial values for y—say, y_1, y_2, y_3, y_4. Then we use the y-z graph to sketch the corresponding solution curves, as in Fig. 3.

The solution in Fig. 3(b) beginning at y_2 has the general shape usually referred to as a logistic curve. This is the type of solution that would model the situation described in Example 1. The solution in Fig. 3(b) beginning at y_1 usually has no physical significance. The other solutions shown in Fig. 3(b) can occur in practice, particularly in the study of population growth.

In ecology, the growth of a population is often described by a logistic equation written in the form

$$\frac{dN}{dt} = rN\left(\frac{K - N}{K}\right) \qquad (2)$$

or, equivalently,

$$\frac{dN}{dt} = \frac{r}{K}N(K - N),$$

where N is used instead of y to denote the size of the population at time t. Typical solutions of this equation are sketched in Fig. 4. The constant K is called the *carrying capacity* of the environment. When the initial population is close to zero,

FIGURE 3

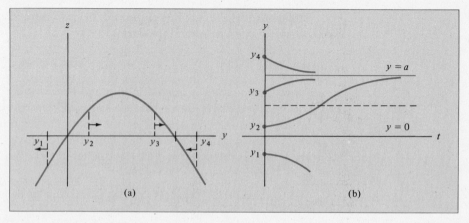

(a) (b)

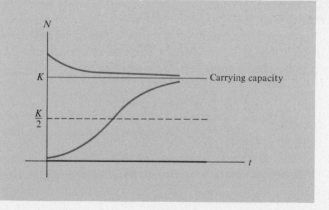

FIGURE 4

the population curve has the typical S-shaped appearance, and N approaches the carrying capacity asymptotically. When the initial population is greater than K, the population decreases in size, again asymptotically approaching the carrying capacity.

The quantity $(K - N)/K$ in (2) is a fraction between 0 and 1. It reflects the limiting effect of the environment upon the population and is close to 1 when N is close to 0. If this fraction were replaced by the constant 1, then (2) would become

$$\frac{dN}{dt} = rN.$$

This is the equation for ordinary exponential growth, where r is the growth rate. For this reason the parameter r in (2) is called the *intrinsic rate of growth* of the population. It expresses how the population would grow if the environment were to permit unrestricted exponential growth.

We now turn to applications that involve a different sort of autonomous differential equation. The main idea is illustrated in the following example.

EXAMPLE 2   A savings account earns 6% interest per year, compounded continuously, and continuous withdrawals are made from the account at the rate of $900 per year. Set up a differential equation that is satisfied by the amount $f(t)$ of money in the account at time t. Sketch typical solutions of the differential equation.

Solution   At first, let us ignore the withdrawals from the account. In Section 6.2 we discussed continuous compounding of interest and showed that if no deposits or withdrawals are made, then $f(t)$ satisfies the equation

$$y' = .06y.$$

That is, the savings account grows at a rate proportional to the size of the account. Since this growth comes from the interest, we conclude that *interest is being added to the account at a rate proportional to the amount in the account.*

Now suppose that continuous withdrawals are made from this same account at the rate of $900 per year. Then there are two influences on the way the amount of money in the account changes—the rate at which interest is added and the rate at which money is withdrawn. The rate of change of $f(t)$ is the *net effect* of these two influences. That is, $f(t)$ now satisfies the equation

$$y' \qquad = \qquad .06y \qquad - \qquad 900,$$

$$\begin{bmatrix} \text{rate of change} \\ \text{of } y \end{bmatrix} = \begin{bmatrix} \text{rate at which} \\ \text{interest is added} \end{bmatrix} - \begin{bmatrix} \text{rate at which} \\ \text{money is withdrawn} \end{bmatrix}.$$

The qualitative sketches for this differential equation are given in Fig. 5. The constant solution is found by solving $.06y - 900 = 0$, which gives $y = 900/.06 = 15,000$. If the initial amount $y(0)$ in the account is $15,000, then the balance in the account will always be $15,000. If the initial amount is greater than $15,000, the savings account will accumulate money without bound. If the initial amount is less than $15,000, the account balance will decrease. Presumably, the bank will stop withdrawals when the account balance reaches zero.

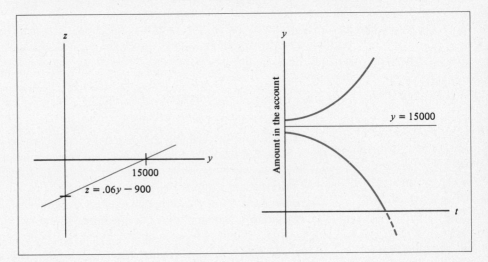

FIGURE 5

We may think of the savings account in Example 2 as a compartment or container into which money (interest) is being steadily added and also from which money is being steadily withdrawn. (See Fig. 6.)

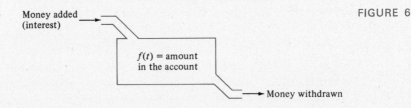

FIGURE 6

A similar situation arises frequently in physiology in what are called "one-compartment problems."* Typical examples of a compartment are a person's lungs, the digestive system, and the cardiovascular system. A common problem is to study the rate at which some substance in the compartment is changing, when two or more processes are increasing or decreasing the substance in the compartment. In many important cases, each of these processes changes the substance either at a constant rate or at a rate proportional to the amount in the compartment.

An earlier example of such a one-compartment problem, discussed in Section 6.3, concerned the continuous infusion of glucose into a patient's bloodstream. A similar situation is discussed in the next example.

EXAMPLE 3 (*A One-Compartment Mixing Process*) Consider a flask that contains 3 liters of salt water. Suppose water containing 25 grams per liter of salt is pumped into the flask at the rate of 2 liters per hour, and the mixture, being steadily stirred, is pumped out of the flask at the same rate. Find a differential equation satisfied by the amount of salt $f(t)$ in the flask at time t.

Solution Let $f(t)$ be the amount of salt measured in grams. Since the volume of the mixture in the flask is being held constant at 3 liters, the concentration of salt in the flask at time t is

$$[\text{concentration}] = \frac{[\text{amount of salt}]}{[\text{volume of mixture}]} = \frac{f(t) \text{ grams}}{3 \text{ liters}} = \frac{1}{3}f(t) \ \frac{\text{grams}}{\text{liter}}.$$

Next we compute the rates at which salt is entering and leaving the flask at time t:

$$[\text{rate of salt entering}] = [\text{concentration}] \times [\text{flow rate}]$$

$$= \left[25 \, \frac{\text{grams}}{\text{liter}}\right] \times \left[2 \, \frac{\text{liters}}{\text{hour}}\right]$$

$$= 50 \, \frac{\text{grams}}{\text{hour}}.$$

$$[\text{rate of salt leaving}] = [\text{concentration}] \times [\text{flow rate}]$$

$$= \left[\frac{1}{3}f(t) \, \frac{\text{grams}}{\text{liter}}\right] \times \left[2 \, \frac{\text{liters}}{\text{hour}}\right]$$

$$= \frac{2}{3}f(t) \, \frac{\text{grams}}{\text{hour}}.$$

The *net* rate of change of salt (in grams per hour) at time t is $f'(t) = 50 - \frac{2}{3}f(t)$. Hence the desired differential equation is

$$y' = 50 - \tfrac{2}{3}y.$$

_____

* Cf. William Simon, *Mathematical Techniques for Physiology and Medicine* (New York: Academic Press, 1972), Chap. V.

Differential Equations in Population Genetics In population genetics, hereditary phenomena are studied on a populational level rather than on an individual level. Consider a particular hereditary feature of an animal, such as the length of the hair. Suppose that basically there are two types of hair for a certain animal—long hair and short hair. Also suppose that long hair is the dominant type. Let A denote the gene responsible for long hair and a the gene responsible for short hair. Each animal has a pair of these genes—either AA ("dominant" individuals), or aa ("recessive" individuals), or Aa ("hybrid" individuals). If there are N animals in the population, then there are $2N$ genes in the population controlling hair length. Each Aa individual has one a gene, and each aa individual has two a genes. The total number of a genes in the population divided by $2N$ gives the fraction of a genes. This fraction is called the *gene frequency of a* in the population. Similarly, the fraction of A genes is called the gene frequency of A. Note that

$$\begin{bmatrix} \text{gene} \\ \text{frequency} \\ \text{of } a \end{bmatrix} + \begin{bmatrix} \text{gene} \\ \text{frequency} \\ \text{of } A \end{bmatrix} = \frac{[\text{number of } a \text{ genes}]}{2N} + \frac{[\text{number of } A \text{ genes}]}{2N}$$

$$= \frac{2N}{2N} = 1. \tag{3}$$

We shall denote the gene frequency of a by q. From (3) it follows that the gene frequency of A is $1 - q$.

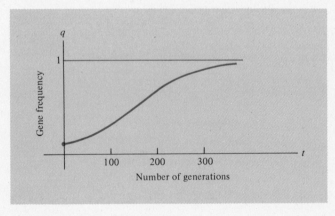

FIGURE 7

An important problem in population genetics involves the way the gene frequency q changes as the animals in the population reproduce. If each unit on the time axis represents one "generation," then we can consider q as a function of time t (Fig. 7). In general, many hundreds or thousands of generations are studied, and so the time for one generation is small when compared with the overall time period. For many purposes, q is considered as a differentiable function of t. In what follows we shall assume that the population mates at random and that the distribution of a and A genes is the same for males and females. In this case one can show from elementary probability theory that the gene frequency is essentially constant from one generation to the next when no "disturbing factors" are present, such as mutation or external influences on the population. We shall discuss differential equations that show the effect of such disturbing factors.*

---

    * Cf. C. C. Li, *Population Genetics* (Chicago: University of Chicago Press, 1955), pp. 240–263 and 283–286.

Suppose that in every generation a fraction v of the a genes mutate and become A genes. Then the rate of change of the gene frequency q due to this mutation is

$$\frac{dq}{dt} = -vq. \tag{4}$$

To understand this equation, think of q as a measure of the number of a genes, and think of the a genes as a population that is losing members at a constant percentage rate of $100v\%$ per generation (i.e., per unit time). This is an exponential decay process, and so q satisfies the exponential decay equation (4). Now suppose, instead, that in every generation a fraction μ of the A genes mutate into a genes. Since the gene frequency of A is $1 - q$, the decrease in the gene frequency of A due to mutation will be $\mu(1 - q)$ per generation. But this must match the increase in the gene frequency of a, since from (3) the sum of the two gene frequencies is constant. Thus the rate of change of the gene frequency of a per generation, due to the mutation from A to a, is given by

$$\frac{dq}{dt} = \mu(1 - q).$$

When mutation occurs, it frequently happens that in each generation a fraction μ of A mutate to a and at the same time a fraction v of a mutate to A. The *net effect* of these two influences on the gene frequency q is described by the equation

$$\frac{dq}{dt} = \mu(1 - q) - vq. \tag{5}$$

(The situation here is analogous to the one-compartment problems discussed earlier.)

Let us make a qualitative analysis of equation (5). To be specific, we take $\mu = .00003$ and $v = .00001$. Then

$$\frac{dq}{dt} = .00003(1 - q) - .00001q$$

$$= .00003 - .00004q,$$

or

$$\frac{dq}{dt} = -.00004(q - .75). \tag{6}$$

Figure 8(a) shows the graph of $z = -.00004(q - .75)$ with the z-axis scale greatly enlarged. Typical solution curves are sketched in Fig. 8(b). We see that the gene frequency $q = .75$ is an equilibrium value. If the initial value of q is smaller than .75, the value of q will rise under the effect of the mutations; after many generations it will be approximately .75. If the initial value of q is between .75 and 1.00, then q will eventually decrease to .75. The equilibrium value is completely determined by the magnitudes of the two opposing rates of mutation μ and v. From (6) we see that the rate of change of gene frequency is proportional to the difference between q and the equilibrium value .75.

In the study of how a population adapts to an environment over a long period, geneticists assume that some hereditary types have an advantage over

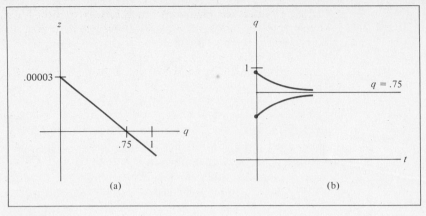

FIGURE 8

others in survival and reproduction. Suppose that the adaptive ability of the hybrid (Aa) individuals is slightly greater than that of both the dominant (AA) and the recessive (aa) individuals. In this case it turns out that the rate of change of gene frequency due to this "selection pressure" is

$$\frac{dq}{dt} = q(1 - q)(c - dq),\qquad (7)$$

where c and d are positive constants with $c < d$. On the other hand, if the adaptive ability of the hybrid individuals is slightly less than that of both the dominant and the recessive individuals, then it can be shown that

$$\frac{dq}{dt} = kq(1 - q)(2q - 1),\qquad (8)$$

where k is a constant between 0 and 1, called the *coefficient of selection against hybrids*.

It is possible to consider the joint effects of mutation and natural selection. Suppose that mutations from A to a occur at a rate μ per generation and from a to A at a rate ν per generation. Suppose also that selection is against recessive individuals (that is, recessives do not adapt as well as the rest of the population). Then the net rate of change in gene frequency turns out to be

$$\frac{dq}{dt} = \mu(1 - q) - \nu q - kq^2(1 - q).$$

Here $\mu(1 - q)$ represents the gain in a genes from mutations $A \to a$, the term νq is the loss in a genes from mutations $a \to A$, and the term $kq^2(1 - q)$ represents the loss in a genes due to natural selection pressures.

PRACTICE PROBLEMS 5

1. Refer to Example 3, involving the flow of salt water through a flask. Will $f(t)$ be an increasing or a decreasing function?

2. (Rate of litter accumulation) In a certain tropical forest, "litter" (mainly dead vegetation such as leaves and vines) forms on the ground at the rate of 10 grams per square centimeter per year. At the same time, however, the litter is decomposing at the rate of 80% per year. Let $f(t)$ be the amount of litter (in grams per square centimeter) present at time t. Find a differential equation satisfied by $f(t)$.

EXERCISES 5

1. (Social Diffusion) For information being spread by mass media, rather than through individual contact, the rate of spread of the information at any time is proportional to the percent of the population not having the information at that time. Give the differential equation that is satisfied by $y = f(t)$, the percentage of the population having the information at time t. Assume that $f(0) = 1$. Sketch the solution.

2. (Gravity) At one point in his study of a falling body starting from rest, Galileo conjectured that its velocity at any time is proportional to the distance it has dropped. Using this hypothesis, set up the differential equation whose solution is $y = f(t)$, the distance fallen by time t. By making use of the initial value, show why Galileo's original conjecture is invalid.

3. (Autocatalytic Reaction) In an autocatalytic reaction, one substance is converted into a second substance in such a way that the second substance catalyzes its own formation. This is the process by which trypsinogen is converted into the enzyme trypsin. The reaction starts only in the presence of some trypsin, and each molecule of trypsinogen yields one molecule of trypsin. The rate of formation of trypsin is proportional to the product of the amounts of the two substances present. Set up the differential equation that is satisfied by $y = f(t)$, the amount (number of molecules) of trypsin present at time t. Sketch the solution. For what value of y is the reaction proceeding the fastest? [Note: Letting M be the total amount of the two substances, the amount of trypsinogen present at time t is $M - f(t)$.]

4. (Drying) A porous material dries outdoors at a rate that is proportional to the moisture content. Set up the differential equation whose solution is $y = f(t)$, the amount of water at time t in a towel on a clothesline. Sketch the solution.

5. (Movement of Solutes Through a Cell Membrane) Let c be the concentration of a solute outside a cell that we assume to be constant throughout the process—that is, unaffected by the small influx of the solute across the membrane due to a difference in concentration. The rate of change of the concentration of the solute inside the cell at any time t is proportional to the difference between the outside concentration and the inside concentration. Set up the differential equation whose solution is $y = f(t)$, the concentration of the solute inside the cell at time t. Sketch a solution.

6. An experimenter reports that a certain strain of bacteria grows at a rate proportional to the square of the size of the population. Set up a differential equation which describes the growth of the population. Sketch a solution.

7. (Chemical Reaction) Suppose that substance A is converted into substance B at a rate that, at any time t, is proportional to the square of the amount of A. This situation occurs, for instance, when it is necessary for two molecules of A

to collide in order to create one molecule of B. Set up the differential equation that is satisfied by $y = f(t)$, the amount of substance A at time t. Sketch a solution.

8. (War Fever) L. F. Richardson proposed the following model to describe the spread of war fever.* If $y = f(t)$ is the percent of the population advocating war at time t, then the rate of change of $f(t)$ at any time is proportional to the product of the percent of the population advocating war and the percent not advocating war. Set up a differential equation that is satisfied by $y = f(t)$ and sketch a solution.

9. (Capital Investment Model) In economic theory, the following model is used to describe a possible capital investment policy. Let $f(t)$ represent the total invested capital of a company at time t. Additional capital is invested whenever $f(t)$ is below a certain equilibrium value E, and capital is withdrawn whenever $f(t)$ exceeds E. The rate of investment is proportional to the difference between $f(t)$ and E. Construct a differential equation whose solution is $f(t)$ and sketch two or three typical solution curves.

10. (Evans Price Adjustment Model) Consider a certain commodity that is produced by many companies and purchased by many other firms. Over a relatively short period there tends to be an equilibrium price p_0 per unit of the commodity that balances the supply and the demand. Suppose that, for some reason, the price is different from the equilibrium price. The Evans price adjustment model says that the rate of change of price with respect to time is proportional to the difference between the actual market price p and the equilibrium price. Write a differential equation that expresses this relation.

11. (Continuous Annuity) A *continuous annuity* is a steady stream of money that is paid to some person. Such an annuity may be established, for example, by making an initial deposit in a savings account and then making steady withdrawals to pay the continuous annuity. Suppose an initial deposit of $5400 is made into a savings account that earns $5\frac{1}{2}\%$ interest compounded continuously, and immediately continuous withdrawals are begun at the rate of $300 per year. Set up the differential equation that is satisfied by the amount $f(t)$ of money in the account at time t. Sketch the solution.

12. An initial deposit of $10,000 is made into a savings account that earns 6% interest compounded continuously. Six months later continuous withdrawals are begun at the rate of $600 per year. Set up the differential equation that is satisfied by the amount $f(t)$ of money in the account at time t for $t \geq \frac{1}{2}$. Sketch the solution.

13. A company wishes to set aside funds for future expansion and so arranges to make continuous *deposits* into a savings account at the rate of $10,000 per year. The savings account earns 5% interest compounded continuously.
    (a) Set up the differential equation that is satisfied by the amount $f(t)$ of money in the account at time t.
    (b) Solve the differential equation in (a), assuming that $f(0) = 0$, and determine how much money will be in the account at the end of five years.

14. A company arranges to make continuous deposits into a savings account at the rate of P dollars per year. The savings account earns 5% interest compounded

---

* See L. F. Richardson, "War moods I," *Psychometrica*, 1948, p. 13.

continuously. Find the approximate value of P that will make the savings account balance amount to $50,000 in four years.

15. The air in a crowded room full of people contains 0.25% carbon dioxide (CO_2). An air conditioner is turned on that blows fresh air into the room at the rate of 500 cubic feet per minute. The fresh air mixes with the stale air and the mixture leaves the room at the rate of 500 cubic feet per minute. The fresh air contains 0.01% CO_2, and the room has a volume of 2500 cubic feet.

    (a) Find a differential equation satisfied by the amount $f(t)$ of CO_2 in the room at time t.

    (b) The model developed in (a) ignores the CO_2 produced by the respiration of the people in the room. Suppose the people generate .08 cubic feet of CO_2 per minute. Modify the differential equation in (a) to take into account this additional source of CO_2.

16. A certain drug is administered intravenously to a patient at the continuous rate of 5 milligrams per hour. The patient's body removes the drug from the bloodstream at a rate proportional to the amount of the drug in the blood. Write a differential equation that is satisfied by the amount $f(t)$ of the drug in the blood at time t. Sketch a typical solution.

17. A single dose of iodine is injected intravenously into a patient. Suppose that the iodine mixes thoroughly in the blood before any is lost owing to metabolic processes (and ignore the time required for this mixing process). Iodine will leave the blood and enter the thyroid gland at a rate proportional to the amount of iodine in the blood. Also, iodine will leave the blood and pass into the urine at a (different) rate proportional to the amount of iodine in the blood. Suppose that the iodine enters the thyroid at the rate of 4% per hour, and the iodine enters the urine at the rate of 10% per hour. Let $f(t)$ denote the amount of iodine in the blood at time t. Write a differential equation satisfied by $f(t)$.

18. Show that the mathematical model in Problem 2 of Practice Problems 5 predicts that the amount of litter in the forest will eventually stabilize. What is the "equilibrium level" of litter in that problem? (Note: Today most forests are close to their equilibrium levels. This was not so during the Carboniferous Period when the great coal deposits were formed.)

19. In the study of the effect of natural selection on a population, one encounters the differential equation

$$\frac{dq}{dt} = -.0001q^2(1 - q),$$

where q is the frequency of a gene a and the selection pressure is against the recessive genotype aa. Sketch a solution of this equation when $q(0)$ is close to but slightly less than 1.

20. Typical values of c and d in equation (7) are $c = .15$, $d = .50$, and a typical value of k in equation (8) is $k = .05$. Sketch representative solutions for the equations

    (a) $\dfrac{dq}{dt} = q(1 - q)(.15 - .50q)$ (selection favoring hybrids).

(b) $\dfrac{dq}{dt} = .05q(1 - q)(2q - 1)$ (selection against hybrids).

Consider various initial conditions with $q(0)$ between 0 and 1. Discuss possible genetic interpretations of these curves; that is, describe the effect of "selection" on the gene frequency q in terms of the various initial conditions.

SOLUTIONS TO PRACTICE PROBLEMS 5

1. The nature of the function $f(t)$ depends on the initial amount of salt water in the flask. Figure 9 contains solutions for three different initial amounts, $y(0)$. If the initial amount is less than 75 grams, the amount of salt in the flask will increase asymptotically to 75. If the initial concentration is greater than 75 grams, the amount of salt in the flask will decrease asymptotically to 75. Of course, if the initial concentration is exactly 75 grams, the amount of salt in the flask will remain constant.

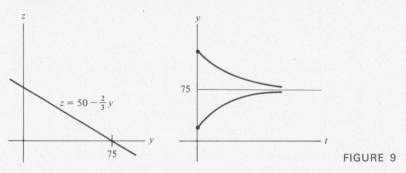

FIGURE 9

2. This problem resembles a one-compartment problem, where the forest floor is the compartment. We have

$$\begin{bmatrix} \text{rate of change of} \\ \text{litter} \end{bmatrix} = \begin{bmatrix} \text{rate of litter} \\ \text{formation} \end{bmatrix} - \begin{bmatrix} \text{rate of litter} \\ \text{decomposition} \end{bmatrix}.$$

If $f(t)$ is the amount of litter (in grams per square centimeter) at time t, then the 80% decomposition rate means that at time t the litter is decaying at the rate of $.80f(t)$ grams per square centimeter per year. Thus the net rate of change of litter is $f'(t) = 10 - .80f(t)$. The desired differential equation is

$$y' = 10 - .80y.$$

11.6. The Lotka-Volterra Model for the Competition Between Two Species

In each application so far, we have obtained information about a single unknown function by setting up and then analyzing a differential equation. Some applications involve two unknown functions that bear a close relationship to each other and whose behavior can be described by a *system of differential equations.* Our purpose here is to discuss an ecological situation in which a system of two differential equations will be analyzed in order to gain insight into the behavior of two unknown functions.

Consider the problem of determining the interaction of two species, one of which preys on the other. Let $N(t)$ denote the number of prey at time t and $P(t)$ the number of predators at time t. Let us assume that the prey die primarily because of attacks by the predator, and, in the absence of the predator, there are no other constraints on their multiplication. On the other hand, assume that the predators depend primarily on the prey for food, and, in the absence of the prey, they will die from hunger—the rate of death at any time being proportional to the number of predators. In effect, what we are assuming is that, in the absence of interaction with each other, $N(t)$ grows according to the organic growth law $N'(t) = aN(t), a > 0$, and $P(t)$ decreases according to the organic death law, $P'(t) = -cP(t), c > 0$.

How should we describe the interactions of predators and prey? Suppose that the predators and prey wander randomly about in a given area and that the number of kills per prey per unit time is proportional to the number of contacts between predator and prey. In turn, it seems reasonable to assume that the number of contacts per unit time at time t is proportional to $N(t) \cdot P(t)$. Thus, at time t, the prey are being born at a rate $aN(t)$ and being killed off at a rate $bN(t)P(t)$, for some positive constant b. So, for $N(t)$, we have the differential equation $N'(t) = aN(t) - bN(t)P(t)$.

Similarly, at time t, the predators are dying off at a rate $cP(t)$. It seems reasonable to assume that they are thriving and increasing (owing to being fed) at a rate that is proportional to the number of kills of prey. Therefore $P(t)$ is described by the differential equation $P'(t) = -cP(t) + dN(t)P(t)$, where $d > 0$.

$N(t)$ and $P(t)$ are thereby described by the system of differential equations

$$N'(t) = aN(t) - bN(t)P(t) = bN(t)\left[\frac{a}{b} - P(t)\right],$$

$$P'(t) = -cP(t) + dN(t)P(t) = dP(t)\left[N(t) - \frac{c}{d}\right],$$

where a, b, c, and d are positive constants. This system of differential equations was discovered independently by Lotka and Volterra, hence the name *Lotka-Volterra* equations. This model has been shown to predict fairly accurately the changes in population of elk and wolves or of rabbits and foxes when these species live in isolated environments. The goal of this section is to derive information about the behavior of $N(t)$ and $P(t)$ from the Lotka-Volterra system of equations. We will find that the populations fluctuate about the values c/d and a/b as time progresses.

From the physical problem we see that we can restrict ourselves to solutions for which $N(t)$ and $P(t)$ are positive for all t.

By inspection, we obtain the unique constant solution $N(t) = c/d$, $P(t) = a/b$. This solution tells us that if $N(0) = c/d$ and $P(0) = a/b$, then the two populations remain at these levels.

We would like to know when $N(t)$ is increasing and when it is decreasing.

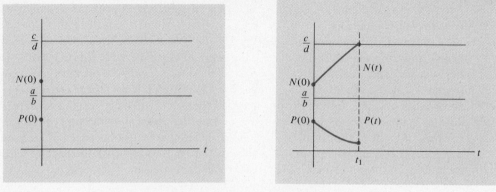

FIGURE 1

FIGURE 2

Since $N(t) > 0$, the first differential equation implies that $N'(t) > 0$ when $P(t) < a/b$ and that $N'(t) < 0$ when $P(t) > a/b$. We conclude that

1. When $P(t) < a/b$, $N(t)$ is increasing.
2. When $P(t) > a/b$, $N(t)$ is decreasing.

In the same manner, we obtain analogous results about $P(t)$.

3. When $N(t) > c/d$, $P(t)$ is increasing.
4. When $N(t) < c/d$, $P(t)$ is decreasing.

FIGURE 3

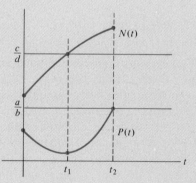

Statements 1 to 4 provide the key to sketching a general solution of the Lotka-Volterra equations. We begin by drawing in the lines $y = c/d$ and $y = a/b$ corresponding to the constant solution (Fig. 1). The initial values of $N(t)$ and $P(t)$ could be any positive numbers. We have selected two such numbers for the purpose of illustration.

At time $t = 0$, $P(t) < a/b$ and $N(t) < c/d$. Therefore statement 1 tells us that $N(t)$ is increasing and statement 4 tells us that $P(t)$ is decreasing. This behavior continues until that time, call it t_1, when $N(t)$ reaches the line $y = c/d$ (Fig. 2).

Since $P(t_1) < a/b$, we conclude from statement 1 that $N(t)$ continues to increase. Consequently, $N(t) > c/d$ for $t > t_1$. So, by statement 3, $P(t)$ is increasing for $t > t_1$. This behavior continues until that time, call it t_2, when $P(t)$ reaches the line $y = a/b$ (Fig. 3).

Continuing the above analysis, we obtain the graph shown in Fig. 4.

At time t_5, $N(t_5) = c/d = N(t_1)$. It can also be shown that $P(t_5) = P(t_1)$. Actually, there is a time, call it T, between t_4 and t_5 at which $N(t)$ and $P(t)$ achieve their initial values. The graphs merely repeat themselves from that point on. That is, $N(T + t) = N(t)$ and $P(T + t) = P(t)$ for all $t \geq 0$. We say that $N(t)$ and $P(t)$ are *periodic functions* with period T. (See the Appendix for proof.)

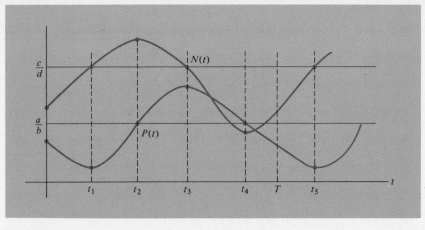

FIGURE 4

Let us compute the average number of predators during one period. Since $[\ln N(t)]' = N'(t)/N(t) = a - bP(t)$, we have

$$P(t) = \frac{1}{b}\left(a - [\ln N(t)]'\right).$$

The average value of $P(t)$ over the time interval from $t = 0$ to $t = T$ is

$$\frac{1}{T - 0}\int_0^T P(t)\, dt = \frac{1}{T}\cdot\frac{1}{b}\int_0^T \left(a - [\ln N(t)]'\right) dt$$

$$= \frac{1}{Tb}\left[at - \ln N(t)\right]\Big|_0^T$$

$$= \frac{1}{Tb}\left[aT - \ln N(T) + \ln N(0)\right]$$

$$= \frac{1}{Tb}\cdot aT$$

$$= \frac{a}{b}.$$

In the next to the last step we used the fact that $N(T) = N(0)$. Similarly, we can show that the average number of prey during one period is c/d. Thus we discover a remarkable fact. No matter how large the initial populations of the two species and no matter how extreme the fluctuations, the average populations are always the same. This property can be regarded as a law of conservation of nature.

APPENDIX Proof of the Periodicity of *N(t)* and *P(t)*

Referring to Fig. 4, we have chosen t_5 such that $N(t_5) = N(t_1)$ and $P(t_5)$, $P(t_1) < a/b$. To show periodicity, it is sufficient to show that $P(t_5) = P(t_1)$.

Before considering the system of differential equations, let us explore the function $F(x) = a \ln x - bx$, where $x, a, b > 0$. Looking at the derivative

$F'(x) = a/x - b$, we see that $F(x)$ is increasing for all $x < a/b$. (If $x < a/b$, then $xb < a$ or $0 < a - bx$. Dividing by x, $0 < a/x - b$.) So if $x_1, x_2 < a/b$ and $F(x_1) = F(x_2)$, then $x_1 = x_2$. In particular, with $x_1 = P(t_5)$ and $x_2 = P(t_1)$, if $F(P(t_5)) = F(P(t_1))$, then $P(t_5) = P(t_1)$.

The Lotka-Volterra equations give $a - bP(t) = N'(t)/N(t)$ and $P'(t)/P(t) = dN(t) - c$. Multiplying the left and right sides of these equations, we obtain

$$a\frac{P'(t)}{P(t)} - bP'(t) = dN'(t) - c\frac{N'(t)}{N(t)}.$$

An antiderivative of the left side is $a \ln P(t) - bP(t)$, and an antiderivative of the right side is $dN(t) - c \ln N(t)$. Since two antiderivatives of the same function differ by a constant, we have

$$a \ln P(t) - bP(t) - dN(t) + c \ln N(t) = k, \quad \text{a constant.}$$

Therefore, evaluating at $t = t_5$ and $t = t_1$,

$a \ln P(t_5) - bP(t_5) - dN(t_5) + c \ln N(t_5)$
$$= k = a \ln P(t_1) - bP(t_1) - dN(t_1) + c \ln N(t_1).$$

Since $N(t_5) = N(t_1)$, $a \ln P(t_5) - bP(t_5) = a \ln P(t_1) - bP(t_1)$.

Referring to the function $F(x)$ above, we have shown that $F(P(t_5)) = F(P(t_1))$ and hence $P(t_5) = P(t_1)$.

Chapter 11 : CHECKLIST

☐ Solution of a differential equation
☐ Separation of variables
☐ Euler's method
☐ Autonomous differential equation $y' = g(y)$
☐ Qualitative theory of differential equations
☐ Logistic differential equation $y' = ky(a - y)$

Chapter 11 : SUPPLEMENTARY EXERCISES

Solve the differential equations in Exercises 1 to 8.

1. $y^2 y' = 4t^3 - 3t^2 + 2$
2. $y'/(t + 1) = y + 1$
3. $y' = y/t - 3y, t > 0$
4. $(y')^2 = t$
5. $y = 7y' + ty', y(0) = 3$
6. $y' = te^{t+y}, y(0) = 0$
7. $yy' + t = 6t^2, y(0) = 7$
8. $y' = 5 - 8y, y(0) = 1$
9. Let $f(t)$ be the solution to $y' = 2e^{2t-y}$, $y(0) = 0$. Use Euler's method with $n = 4$ on $0 \le t \le 2$ to estimate $f(2)$. Then show that Euler's method gives the exact value of $f(2)$ by solving the differential equation.
10. Let $f(t)$ be the solution to $y' = (t + 1)/y$, $y(0) = 1$. Use Euler's method with $n = 3$ on $0 \le t \le 1$ to estimate $f(1)$. Then show that Euler's method gives the exact value of $f(1)$ by solving the differential equation.

11. Suppose $f(t)$ is a solution of $y' = (2 - y)e^{-y}$. Is $f(t)$ increasing or decreasing at some value of t where $f(t) = 3$?

12. Solve the initial value problem $y' = e^{y^2}(\cos y)(1 - e^{y-1})$, $y(0) = 1$.

13. Use Euler's method with $n = 6$ on the interval $0 \le t \le 3$ to approximate the solution $f(t)$ to $y' = .1y(20 - y)$, $y(0) = 2$.

14. Use Euler's method with $n = 5$ on the interval $0 \le t \le 1$ to approximate the solution $f(t)$ to $y' = \frac{1}{2}y(y - 10)$, $y(0) = 9$.

   Sketch the solutions of the differential equations in Exercises 15 to 24. In each case, also indicate the constant solutions.

15. $y' = 2\cos y$, $y(0) = 0$

16. $y' = 5 + 4y - y^2$, $y(0) = 1$

17. $y' = y^2 + y$, $y(0) = -\frac{1}{3}$

18. $y' = y^2 - 2y + 1$, $y(0) = -1$

19. $y' = \ln y$, $y(0) = 2$

20. $y' = 1 + \cos y$, $y(0) = -\frac{3}{4}$

21. $y' = 1/(y^2 + 1)$, $y(0) = -1$

22. $y' = 3/(y + 3)$, $y(0) = 2$

23. $y' = .4y^2(1 - y)$, $y(0) = -1$, $y(0) = .1$, $y(0) = 2$

24. $y' = y^3 - 6y^2 + 9y$, $y(0) = -\frac{1}{4}$, $y(0) = \frac{1}{4}$, $y(0) = 4$

25. The birth rate in a certain city is 3.5% per year and the death rate is 2% per year. Also, there is a net movement of population out of the city at a steady rate of 3000 people per year. Let $N = f(t)$ be the city's population at time t.

    (a) Write a differential equation satisfied by N.

    (b) Use a qualitative analysis of the equation to determine if there is a size at which the population would remain constant. Is it likely that a city would have such a constant population?

26. Suppose that in a chemical reaction, each gram of substance A combines with three grams of substance B to form four grams of substance C. Suppose the reaction begins with ten grams of A, fifteen grams of B, and zero grams of C present. Let $y = f(t)$ be the amount of C present at time t. Suppose that the rate at which substance C is formed is proportional to the product of the unreacted amounts of A and B present. That is, suppose $f(t)$ satisfies the differential equation

$$y' = k(10 - \tfrac{1}{4}y)(15 - \tfrac{3}{4}y), \quad y(0) = 0,$$

where k is a constant.

    (a) What do the quantities $10 - \frac{1}{4}f(t)$ and $15 - \frac{3}{4}f(t)$ represent?

    (b) Should the constant k be positive or negative?

    (c) Make a qualitative sketch of the solution of the above differential equation.

27. A bank account has $20,000 earning 5% interest compounded continuously. A pensioner uses the account to pay himself an annuity, drawing continuously at a $2000 annual rate. How long will it take for the balance in the account to drop to zero?

28. A continuous annuity of $12,000 per year is to be funded by steady withdrawals from a savings account that earns 6% interest compounded continuously.

    (a) What is the smallest initial amount in the account that will fund such an annuity forever?

    (b) What initial amount will fund such an annuity for exactly twenty years (at which time the savings account balance will be zero)?

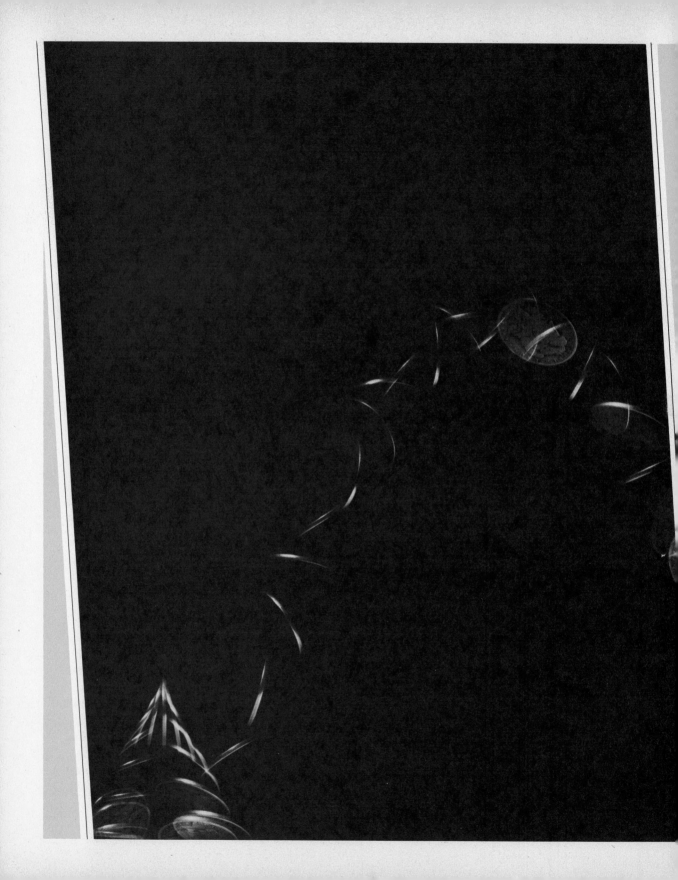

chapter 12

PROBABILITY AND CALCULUS

In this chapter we shall survey a few applications of calculus to the theory of probability. Since we do not intend this chapter to be a self-contained course in probability, we shall select only a few salient ideas to present a taste of probability theory and provide a starting point for further study.

12.1. Discrete Random Variables

We will motivate the concepts of mean, variance, standard deviation, and random variable by analyzing examination grades.

Suppose that the grades on an exam taken by 10 people are 50, 60, 60, 70, 70, 90, 100, 100, 100, 100. This information is displayed in a frequency table in Fig. 1.

Grade	50	60	70	90	100
Frequency	1	2	2	1	4

FIGURE 1

One of the first things we do when looking over the results of an exam is to compute the *mean* or *average* of the grades. We do this by totaling the grades and dividing by the number of people:

$$[\text{mean}] = \frac{50 \cdot 1 + 60 \cdot 2 + 70 \cdot 2 + 90 \cdot 1 + 100 \cdot 4}{10} = \frac{800}{10} = 80.$$

In order to get an idea of how spread out the grades are, we can compute the difference between each grade and the average grade. We have tabulated these differences in Fig. 2. For example, if a person received a 50, then [grade] − [mean] is $50 - 80 = -30$. As a measure of the spread of the grades, statisticians compute the average of the squares of these differences and call it the *variance* of the grade distribution. We have

$$[\text{variance}] = \frac{(-30)^2 \cdot 1 + (-20)^2 \cdot 2 + (-10)^2 \cdot 2 + (10)^2 \cdot 1 + (20)^2 \cdot 4}{10}$$

$$= \frac{900 + 800 + 200 + 100 + 1600}{10} = \frac{3600}{10} = 360.$$

[Grade] − [Mean]	−30	−20	−10	10	20
Frequency	1	2	2	1	4

FIGURE 2

The square root of the variance is called the *standard deviation* of the grade distribution. In this case, we have

$$[\text{standard deviation}] = \sqrt{360} \approx 18.97.$$

There is another way of looking at the grade distribution and its mean and variance. This new point of view is useful because it can be generalized to other situations. We begin by converting the frequency table to a relative frequency table. (See Fig. 3.) Below each grade we list the fraction of the class receiving that grade. The grade of 50 occurred $\frac{1}{10}$ of the time, the grade of 60 occurred $\frac{2}{10}$ of the time, and so on. Note that the relative frequencies add up to 1, because they represent the various fractions of the class grouped by test scores.

Grade	50	60	70	90	100
Relative frequency	$\frac{1}{10}$	$\frac{2}{10}$	$\frac{2}{10}$	$\frac{1}{10}$	$\frac{4}{10}$

FIGURE 3

It is sometimes helpful to display the data in the relative frequency table by constructing a *relative frequency histogram* as in Fig. 4. Over each grade we place a rectangle whose height equals the relative frequency of that grade.

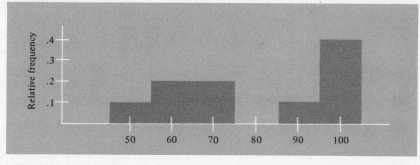

FIGURE 4

An alternate way to compute the mean grade is

$$[\text{mean}] = \frac{50 \cdot 1 + 60 \cdot 2 + 70 \cdot 2 + 90 \cdot 1 + 100 \cdot 4}{10}$$

$$= 50 \cdot \tfrac{1}{10} + 60 \cdot \tfrac{2}{10} + 70 \cdot \tfrac{2}{10} + 90 \cdot \tfrac{1}{10} + 100 \cdot \tfrac{4}{10}$$

$$= 5 + 12 + 14 + 9 + 40 = 80.$$

Looking at the second line of this computation, we see that the mean is a sum of the various grades times their relative frequencies. We say that the mean is the *weighted sum* of the grades. (Grades are weighted by their relative frequencies.)

In a similar manner we see that the variance is also a weighted sum.

$$[\text{variance}] = [(50 - 80)^2 \cdot 1 + (60 - 80)^2 \cdot 2 + (70 - 80)^2 \cdot 2$$
$$+ (90 - 80)^2 \cdot 1 + (100 - 80)^2 \cdot 4] \tfrac{1}{10}$$

$$= (50 - 80)^2 \cdot \tfrac{1}{10} + (60 - 80)^2 \cdot \tfrac{2}{10} + (70 - 80)^2 \cdot \tfrac{2}{10}$$

$$+ (90 - 80)^2 \cdot \tfrac{1}{10} + (100 - 80)^2 \cdot \tfrac{4}{10}$$

$$= 90 + 80 + 20 + 10 + 160 = 360.$$

A relative frequency table such as is given in Fig. 3 is also called a *probability table*. The reason for this terminology is as follows. Suppose we perform an *experiment* which consists of picking an exam paper at random from among the ten papers. If the experiment is repeated many times, we expect the grade of 50 to occur about one-tenth of the time, the grade of 60 about two-tenths of the time, and so on. We say that the *probability* of the grade of 50 being chosen is $\tfrac{1}{10}$, the probability of the grade of 60 being chosen is $\tfrac{2}{10}$, and so on. In other words, the probability associated with a given grade measures the likelihood that an exam having that grade is chosen.

In this section we will consider various experiments described by probability tables similar to the one in Fig. 3. The results of these experiments will be numbers (such as the exam scores above) called the *outcomes* of the experiment. We will also be given the probability of each outcome, indicating the relative frequency with which the given outcome is expected to occur, if the experiment is repeated very often. If the outcomes of an experiment are $a_1, a_2, \ldots, a_n$, with

Outcome	a_1	a_2	a_3	$\cdots$	a_n
Probability	p_1	p_2	p_3	$\cdots$	p_n

FIGURE 5

respective probabilities $p_1, p_2, \ldots, p_n$, then we describe the experiment by a probability table as in Fig. 5. Since the probabilities indicate relative frequencies, we see that

$$0 \leq p_i \leq 1$$

and

$$p_1 + p_2 + \ldots + p_n = 1.$$

This last equation indicates that the outcomes $a_1, \ldots, a_n$ comprise all possible results of the experiment. We will usually list the outcomes of our experiments in ascending order, so that $a_1 < a_2 < \ldots < a_n$.

We may display the data of a probability table in a histogram which has a rectangle of height p_i over the outcome a_i. (See Fig. 6.)

Let us define the *expected value* (or *mean*) of the probability table of Fig. 5 to be the weighted sum of the outcomes $a_1, \ldots, a_n$, each outcome weighted by the probability of its occurrence. That is,

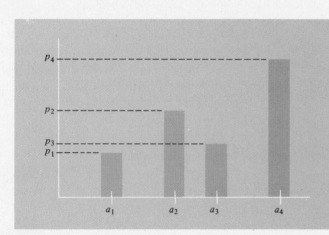

$$[\text{expected value}] = a_1 p_1 + a_2 p_2 + \ldots + a_n p_n.$$

FIGURE 6

Similarly, let us define the variance of the probability table to be the weighted sum of the squares of the differences between each outcome and the expected value. That is, if m denotes the expected value, then

$$[\text{variance}] = (a_1 - m)^2 p_1 + (a_2 - m)^2 p_2 + \ldots + (a_n - m)^2 p_n.$$

To keep from writing the word "outcome" so many times, we shall abbreviate by X the outcome of our experiment. That is, X is a variable which takes on the values $a_1, a_2, \ldots, a_n$ with respective probabilities $p_1, p_2, \ldots, p_n$. We will assume that our experiment is performed many times, being repeated in an unbiased (or random) way. Then X is a variable whose value depends on chance, and for this reason we say that X is a *random variable*. Instead of speaking of the expected value (mean) and the variance of a probability table, let us speak of the *expected value* and the *variance of the random variable* X which is associated with the probability table. We shall denote the expected value of X by $E(X)$ and the variance of X by $\text{Var}(X)$. The *standard deviation* of X is defined to be $\sqrt{\text{Var}(X)}$.

EXAMPLE 1 One bet in roulette is to wager \$1 on "red." The two possible outcomes are: "lose \$1" and "win \$1." These outcomes and their probabilities are given in Fig. 7. (Note: A roulette wheel in Las Vegas has 18 red numbers, 18 black numbers, and two green numbers.) Compute the expected value and the variance of the amount won.

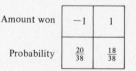

Amount won	-1	1
Probability	$\frac{20}{38}$	$\frac{18}{38}$

FIGURE 7

Solution Let X be the random variable "amount won." Then

$$E(X) = -1 \cdot \tfrac{20}{38} + 1 \cdot \tfrac{18}{38} = -\tfrac{2}{38} \approx -.0526,$$

$$\text{Var}(X) = [-1 - (-\tfrac{2}{38})]^2 \cdot \tfrac{20}{38} + [1 - (-\tfrac{2}{38})]^2 \cdot \tfrac{18}{38}$$

$$= (-\tfrac{36}{38})^2 \cdot \tfrac{20}{38} + (\tfrac{40}{38})^2 \cdot \tfrac{18}{38} \approx .997.$$

The expected value of the amount won is approximately $-5\tfrac{1}{4}$ ¢. In other words, sometimes we will win \$1 and sometimes we will lose \$1, but in the long run we can expect to lose an average of about $5\tfrac{1}{4}$ ¢ for each time we bet.

EXAMPLE 2 An experiment consists of selecting a number at random from the set of integers $\{1, 2, 3\}$. The probabilities are given by the table in Fig. 8. Let X designate the outcome. Find the expected value and the variance of X.

Number	1	2	3
Probability	$\frac{1}{3}$	$\frac{1}{3}$	$\frac{1}{3}$

FIGURE 8

Solution

$$E(X) = 1 \cdot \tfrac{1}{3} + 2 \cdot \tfrac{1}{3} + 3 \cdot \tfrac{1}{3} = 2,$$

$$\text{Var}(X) = (1 - 2)^2 \cdot \tfrac{1}{3} + (2 - 2)^2 \cdot \tfrac{1}{3} + (3 - 2)^2 \cdot \tfrac{1}{3}$$

$$= (-1)^2 \cdot \tfrac{1}{3} + 0 + (1)^2 \cdot \tfrac{1}{3} = \tfrac{2}{3}.$$

EXAMPLE 3 A cement company plans to bid on a contract for constructing the foundations of new homes in a housing development. The company is considering two bids: a high bid that will produce \$75,000 profit (if the bid is accepted), and a low bid that will produce \$40,000 profit. From past experience the company estimates that the high bid has a 30% chance of acceptance and the low bid a 50% chance. Which bid should the company make?

Solution The standard method of decision is to choose the bid that has the higher expected value. Let X be the amount the company makes if it submits the high bid, and let

Y be the amount it makes if it submits the low bid. Then the company must analyze the following probability tables.

	High Bid			Low Bid	
	Accepted	*Rejected*		*Accepted*	*Rejected*
Value of X	75,000	0	*Value of Y*	40,000	0
Probability	.30	.70	*Probability*	.50	.50

The expected values are

$$E(X) = (75{,}000)(.30) + 0(.70) = 22{,}500,$$

$$E(Y) = (40{,}000)(.50) + 0(.50) = 20{,}000.$$

If the cement company has many opportunities to bid on similar contracts, then a "high" bid each time will be accepted sufficiently often to produce an average profit of \$22,500 per bid. A consistently "low" bid will produce an average profit of \$20,000 per bid. Thus the company should submit the high bid.

When a probability table contains a large number of possible outcomes of an experiment, the associated histogram for the random variable X becomes a valuable aid for "visualizing" the data in the table. Look at Fig. 9, for example. Since the rectangles that make up the histogram all have the same width, their areas are in the same ratios as their heights. By an appropriate change of scale on the y-axis, we may assume that the *area* (instead of the height) of each rectangle gives the associated probability of X. Such a histogram is sometimes referred to as a *probability density histogram*.

A histogram that displays probabilities as areas is useful when one wishes to visualize the probability that X has a value between two specified numbers. For example, in Fig. 9 suppose that the probabilities associated with $X = 5$, $X = 6$,

FIGURE 9   Probabilities displayed as areas.

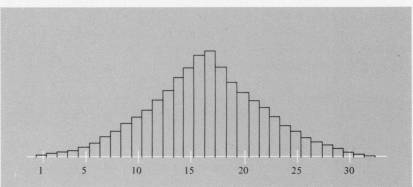

..., $X = 10$ are $p_5, p_6, \ldots, p_{10}$, respectively. Then the probability that X lies between 5 and 10 inclusive is $p_5 + p_6 + \ldots + p_{10}$. In terms of areas, this probability is just the total area of those rectangles over the values $5, 6, \ldots, 10$. (See Fig. 10.) We will consider analogous situations in the next section.

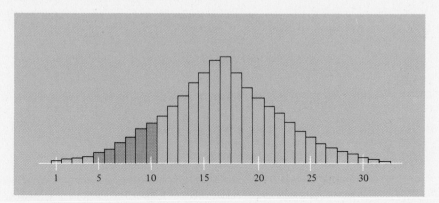

FIGURE 10   Probability that $5 \le X \le 10$.

PRACTICE PROBLEMS 1

1. Compute the expected value and the variance of the random variable X with the following probability table.

Value of X	−1	0	1	2
Probability	$\frac{1}{8}$	$\frac{1}{8}$	$\frac{3}{8}$	$\frac{3}{8}$

2. The production department at a radio factory sends CB radios to the inspection department in lots of 100. There an inspector examines three radios at random from each lot. If at least one of the three radios is defective and needs adjustment, the entire lot is sent back to the production department. Records of the inspection department show that the number X of defective radios in a sample of three radios has the following probability table.

Defectives	0	1	2	3
Probability	.7265	.2477	.0251	.0007

(a) What percentage of the lots does the inspection department reject?

(b) Find the mean number of defective radios in the samples of three radios.

(c) Based on the evidence in (b), estimate the average number of defective radios in each lot of 100 radios.

1. Table 1 is the probability table for a random variable X. Find $E(X)$, $Var(X)$, and the standard deviation of X.

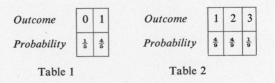

Outcome	0	1
Probability	$\frac{1}{5}$	$\frac{4}{5}$

Table 1

Outcome	1	2	3
Probability	$\frac{4}{9}$	$\frac{4}{9}$	$\frac{1}{9}$

Table 2

2. Find $E(X)$, $Var(X)$ and the standard deviation of X, where X is the random variable whose probability table is given in Table 2.

3. Compute the variances of the three random variables whose probability tables are given in Table 3(a), (b), and (c). Relate the sizes of the variances to the "spread" of the values of the random variable.

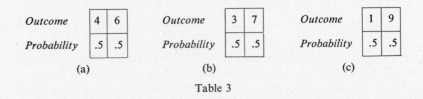

Outcome	4	6
Probability	.5	.5

(a)

Outcome	3	7
Probability	.5	.5

(b)

Outcome	1	9
Probability	.5	.5

(c)

Table 3

4. Compute the variances of the two random variables whose probability tables are given in Table 4(a), (b). Relate the sizes of the variances to the "spread" of the values of the random variables.

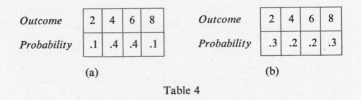

Outcome	2	4	6	8
Probability	.1	.4	.4	.1

(a)

Outcome	2	4	6	8
Probability	.3	.2	.2	.3

(b)

Table 4

5. The number of accidents per week at a busy intersection was recorded for a year. There were 11 weeks with no accidents, 26 weeks with one accident, 13 weeks with two accidents, and 2 weeks with three accidents. A week is to be selected at random and the number of accidents noted. Let X be the outcome. Then X is a random variable taking on the values 0, 1, 2, and 3.

(a) Write out a probability table for X.

(b) Compute $E(X)$.

(c) Interpret $E(X)$.

6. The number of phone calls coming into a telephone switchboard during each minute was recorded during an entire hour. During 30 of the one-minute intervals there were no calls, during 20 intervals there was one call, and during 10 intervals there were two calls. A one-minute interval is to be selected at random and the number of calls noted. Let X be the outcome. Then X is a random variable taking on the values 0, 1, and 2.

   (a) Write out a probability table for X.

   (b) Compute $E(X)$.

   (c) Interpret $E(X)$.

7. Consider a circle with radius 1.

   (a) What percentage of the points lie within $\frac{1}{2}$ unit of the center?

   (b) Let c be a constant with $0 < c < 1$. What percentage of the points lie within c units of the center?

8. Consider a circle with circumference 1. An arrow (that is, a spinner) is attached at the center so that, when flicked, it spins freely. Upon stopping, it points to a particular point on the circumference of the circle. Determine the likelihood that the point is:

   (a) On the top half of the circumference.

   (b) On the top quarter of the circumference.

   (c) On the top one-hundredth of the circumference.

   (d) Exactly at the top of the circumference.

9. A citrus grower anticipates a profit of $100,000 this year if the nightly temperatures remain mild. Unfortunately, the weather forecast indicates a 25% chance that the temperatures will drop below freezing during the next week. Such freezing weather will destroy 40% of the crop and reduce the profit to $60,000. However, the grower can protect the citrus fruit against the possible freezing (using smudge pots, electric fans, and so on) at a cost of $5000. Should the grower spend the $5000 and thereby reduce the profit to $95,000? [Hint: Compute $E(X)$, where X is the profit the grower will get if he does nothing to protect the fruit.]

10. Suppose the weather forecast in Exercise 9 indicates a 10% chance that cold weather will reduce the citrus grower's profit from $100,000 to $85,000, and a 10% chance that cold weather will reduce the profit to $75,000. Should the grower spend $5000 to protect the citrus fruit against the possible bad weather?

SOLUTIONS TO PRACTICE PROBLEMS 1

1.
$$E(X) = (-1) \cdot \tfrac{1}{8} + 0 \cdot \tfrac{1}{8} + 1 \cdot \tfrac{3}{8} + 2 \cdot \tfrac{3}{8} = 1,$$
$$\text{Var}(X) = (-1 - 1)^2 \cdot \tfrac{1}{8} + (0 - 1)^2 \cdot \tfrac{1}{8} + (1 - 1)^2 \cdot \tfrac{3}{8} + (2 - 1)^2 \cdot \tfrac{3}{8}$$
$$= 4 \cdot \tfrac{1}{8} + 1 \cdot \tfrac{1}{8} + 0 + 1 \cdot \tfrac{3}{8} = 1.$$

2. (a) There are three cases where a lot will be rejected: $X = 1, 2,$ or 3. Adding the corresponding probabilities, we find that the probability of rejecting a lot is .2477 + .0251 + .0007 = .2735, or 27.35%. [An alternate method of solution uses the fact that the sum of the probabilities for *all* possible cases must be 1. From the table we see that the probability of accepting a lot is .7265, so the probability of rejecting a lot is 1 − .7265 = .2735.]

(b) $E(X) = 0(.7265) + 1(.2477) + 2(.0251) + 3(.0007) = .3000.$

(c) In (b) we found that an average of .3 radios in every sample of three radios is defective. Thus about 10% of the radios in the sample are defective. Since the samples are chosen at random, we may assume that about 10% of *all* the radios are defective. Thus we estimate that on the average 10 out of each lot of 100 radios will be defective.

12.2. Continuous Random Variables

Consider a cell population that is growing vigorously. Suppose that when a cell is T days old it divides and forms two new "daughter" cells. If the population is sufficiently large, it will contain cells of many different ages between 0 and T. It turns out that the proportion of cells of various ages remains constant. That is, if a and b are any two numbers between 0 and T, with $a < b$, then the proportion of cells whose ages lie between a and b is essentially constant from one moment to the next, even though individual cells are aging and new cells are being formed all the time. In fact, biologists have found that under the ideal circumstances described, the proportion of cells whose ages are between a and b is given by the area under the graph of the function $f(x) = 2ke^{-kx}$ from $x = a$ to $x = b$, where $k = (\ln 2)/T$.* (See Fig. 1.)

Now consider an experiment where we select a cell at random from the population and observe its age, X. Then the probability that X lies between a and b is given by the area under the graph of $f(x) = 2ke^{-kx}$ from a to b, as in Fig. 1. Let us denote this probability by $\Pr(a \leq X \leq b)$. Using the fact that the area under the graph of $f(x)$ is given by a definite integral, we have

$$\Pr(a \leq X \leq b) = \int_a^b f(x)\, dx = \int_a^b 2ke^{-kx}\, dx. \tag{1}$$

Since X can assume any one of the (infinitely many) numbers in the continuous interval from 0 to T, we say that X is a *continuous random variable*. The function $f(x)$ that determines the probability in (1) for each a and b is called the (*probability*) *density function* of X (or of the experiment whose outcome is X).

Suppose that we consider an experiment whose outcome may be any value between A and B. The outcome of the experiment, denoted X, is called a *con-*

---

* See J. R. Cook and T. W. James, "Age Distribution of Cells in Logarithmically Growing Cell Populations," in *Synchrony in Cell Division and Growth*, Erik Zeuthen, ed. (New York: John Wiley and Sons, 1964), pp. 485–495.

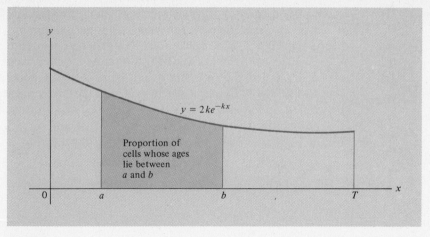

Proportion of cells whose ages lie between a and b

$y = 2ke^{-kx}$

FIGURE 1

tinuous random variable. For the cell population described above, $A = 0$ and $B = T$. Another typical experiment might consist of choosing a number X at random between $A = 5$ and $B = 10$. Or one could observe the duration X of a random telephone call passing through a given telephone switchboard. If we have no way of knowing how long a call might last, then X might be any non-negative number. In this case it is convenient to say that X lies between 0 and ∞ and to take $A = 0$ and $B = \infty$. On the other hand, if the possible values of X for some experiment include rather large negative numbers, then one sometimes takes $A = -\infty$.

Given an experiment whose outcome is a continuous random variable X, the probability $\Pr(a \leq X \leq b)$ is a measure of the likelihood that an outcome of the experiment will lie between a and b. If the experiment is repeated many times, then the proportion of times X has a value between a and b should be close to $\Pr(a \leq X \leq b)$. In experiments of practical interest involving a continuous random variable X it is usually possible to find a function $f(x)$ such that

$$\Pr(a \leq X \leq b) = \int_a^b f(x)\, dx \tag{2}$$

for all a and b in the range of possible values of X. Such a function $f(x)$ is called a *probability density function* and satisfies the following properties:

I. $f(x) \geq 0, \qquad A \leq x \leq B.$

II. $\int_A^B f(x)\, dx = 1.$

Indeed, property I means that for x between A and B, the graph of $f(x)$ must lie on or above the x-axis. Property II simply says that there is probability 1 that X has a value between A and B. (Of course, if $B = \infty$ and/or $A = -\infty$, then the integral in property II is an improper integral.) Properties I and II characterize probability density functions, in the sense that any function $f(x)$ satisfying I and II is the probability density function for some continuous random variable X. Moreover, $\Pr(a \le X \le b)$ can then be calculated using Equation (2).

Unlike a probability table for a discrete random variable, a density function $f(x)$ does *not* give the probability that X has a certain value. Instead, $f(x)$ gives the probability that X is *near* a specific value in the following sense. If x_0 is a number between A and B and if Δx is the width of a small interval centered at x_0, then the probability that X is between $x_0 - \frac{1}{2}\Delta x$ and $x_0 + \frac{1}{2}\Delta x$ is approximately $f(x_0)\,\Delta x$—that is, the area of the rectangle shown in Fig. 2.

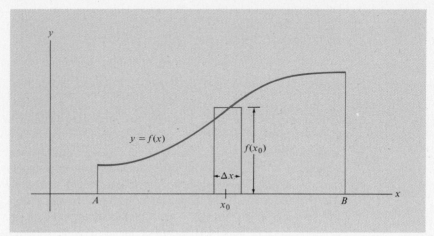

FIGURE 2

EXAMPLE 1   Consider the cell population described earlier. Let $f(x) = 2ke^{-kx}$, where $k = (\ln 2)/T$. Show that $f(x)$ is indeed a probability density function on $0 \le x \le T$.

Solution   Clearly $f(x) \ge 0$, since $\ln 2$ is positive and the exponential function is never negative. Thus property I is satisfied. For property II we check that

$$\int_0^T f(x)\,dx = \int_0^T 2ke^{-kx}\,dx = -2e^{-kx}\Big|_0^T = -2e^{-kT} + 2e^0$$
$$= -2e^{-[(\ln 2)/T]T} + 2 = -2e^{-\ln 2} + 2$$
$$= -2e^{\ln(1/2)} + 2 = -2(\tfrac{1}{2}) + 2 = 1.$$

EXAMPLE 2   Let $f(x) = kx^2$. (a) Find the value of k that makes $f(x)$ a probability density function on $0 \le x \le 4$. (b) Let X be a continuous random variable whose density function is $f(x)$. Compute $\Pr(1 \le X \le 2)$.

Solution   (a) We must have $k \ge 0$ so that property I is satisfied. For property II, we calculate

$$\int_0^4 f(x)\,dx = \int_0^4 kx^2\,dx = \tfrac{1}{3}kx^3\Big|_0^4 = \tfrac{1}{3}k(4)^3 - 0$$
$$= \tfrac{64}{3}k.$$

To satisfy property II we must have $\frac{64}{3}k = 1$, or $k = \frac{3}{64}$. Thus $f(x) = \frac{3}{64}x^2$.

(b) $$\Pr(1 \le X \le 2) = \int_1^2 f(x)\, dx = \int_1^2 \tfrac{3}{64}x^2\, dx$$

$$= \tfrac{1}{64}x^3 \Big|_1^2 = \tfrac{8}{64} - \tfrac{1}{64} = \tfrac{7}{64}.$$

The area corresponding to this probability is shown in Fig. 3.

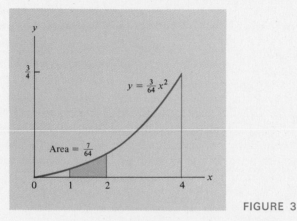

FIGURE 3

The density function in the next example is a special case of what statisticians sometimes call a *beta* probability density.

EXAMPLE 3 The parent corporation for a franchised chain of fast food restaurants claims that the proportion of their new restaurants that make a profit during their first year of operation has the probability density

$$f(x) = 12x(1 - x)^2, \quad 0 \le x \le 1.$$

(a) What is the probability that less than 40% of the restaurants opened this year will make a profit during their first year of operation?

(b) What is the probability that more than 50% of the restaurants will make a profit during their first year of operation?

Solution Let X be the proportion of new restaurants opened this year that make a profit during their first year of operation. Then the possible values of X range between 0 and 1.

(a) The probability that X is less than .4 equals the probability that X is between 0 and .4. We note that $f(x) = 12x(1 - 2x + x^2) = 12x - 24x^2 + 12x^3$, and, therefore

$$\Pr(0 \le X \le .4) = \int_0^{.4} f(x)\, dx = \int_0^{.4} (12x - 24x^2 + 12x^3)\, dx$$

$$= (6x^2 - 8x^3 + 3x^4)\Big|_0^{.4} = .5248.$$

(b) The probability that X is greater than .5 equals the probability that X is between .5 and 1. Thus,

$$\Pr(.5 \le X \le 1) = \int_{.5}^1 (12x - 24x^2 + 12x^3)\, dx$$

$$= (6x^2 - 8x^3 + 3x^4)\Big|_{.5}^1 = .3125.$$

Each probability density function is closely related to another important function called a cumulative distribution function. To describe this relationship, let us consider an experiment whose outcome is a continuous random variable X, with values between A and B, and let $f(x)$ be the associated density function. For each number x between A and B let $F(x)$ be the probability that X is less than or equal to the number x. Sometimes we write $F(x) = \Pr(X \le x)$; however, since X is never less than A, we also may write

$$F(x) = \Pr(A \le X \le x) \tag{3}$$

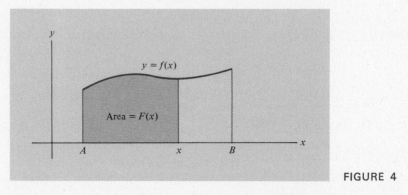

FIGURE 4

Graphically, $F(x)$ is the area under the graph of the density function $f(x)$ from A to x. (See Fig. 4.) The function $F(x)$ is called the *cumulative distribution function* of the random variable X (or of the experiment whose outcome is X). Note that $F(x)$ also has the properties

$$F(A) = \Pr(A \le X \le A) = 0, \tag{4}$$

$$F(B) = \Pr(A \le X \le B) = 1. \tag{5}$$

Since $F(x)$ is an "area function" that gives the area under the graph of $f(x)$ from A to x, we know from Section 7.2 that $F(x)$ is an antiderivative of $f(x)$. That is,

$$F'(x) = f(x), \quad A \le x \le B. \tag{6}$$

It follows that one may use $F(x)$ to compute probabilities, since

$$\Pr(a \le X \le b) = \int_a^b f(x)\,dx = F(b) - F(a), \tag{7}$$

for any a and b between A and B.

The relation (6) between $F(x)$ and $f(x)$ makes it possible to find one of these functions when the other is known, as we see in the following two examples.

EXAMPLE 4 Let X be the age of a cell selected at random from the cell population described earlier. The density function for X is $f(x) = 2ke^{-kx}$, where $k = (\ln 2)/T$. Find the cumulative distribution function $F(x)$ for X.

Solution Since $F(x)$ is an antiderivative of $f(x) = 2ke^{-kx}$, we have $F(x) = -2e^{-kx} + C$ for some constant C. Now $F(x)$ is defined for $0 \le x \le T$. Thus (4) implies that $F(0) = 0$. Setting $F(0) = -2e^0 + C = 0$, we find that $C = 2$, and so

$$F(x) = -2e^{-kx} + 2.$$

EXAMPLE 5 Let X be the random variable associated with the experiment which consists of selecting a point at random from a circle of radius 1 and observing its distance from the center. Find the probability density function $f(x)$ and cumulative distribution function $F(x)$ of X.

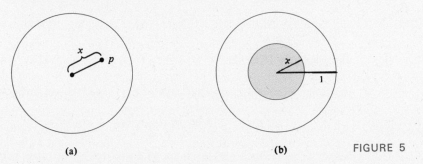

(a)   (b)   FIGURE 5

Solution The distance of a point from the center of the unit circle is a number between 0 and 1. Suppose that $0 \le x \le 1$. Let us first compute the cumulative distribution function $F(x) = \Pr(0 \le X \le x)$. That is, let us find the probability that a point selected at random lies within x units of the center of the circle—in other words, lies inside the circle of radius x. See the shaded region in Fig. 5(b). Since the area of this shaded region is πx^2 and the area of the entire unit circle is $\pi \cdot 1^2 = \pi$, the proportion of points inside the shaded region is $\pi x^2/\pi = x^2$. Thus the probability is x^2 that a point selected at random will be in this shaded region. Hence

$$F(x) = x^2.$$

Differentiating, we find that the probability density function for X is

$$f(x) = F'(x) = 2x.$$

Our final example involves a continuous random variable X whose possible values lie between $A = 1$ and $B = +\infty$; that is, X is any number greater than or equal to 1.

EXAMPLE 6 Let $f(x) = 3x^{-4}$, $x \ge 1$.

(a) Show that $f(x)$ is the probability density function of some random variable X.

(b) Find the cumulative distribution function $F(x)$ of X.

(c) Compute $\Pr(X \le 4)$, $\Pr(4 \le X \le 5)$, and $\Pr(4 \le X)$.

Solution (a) It is clear that $f(x) \ge 0$ for $x \ge 1$. Thus property I holds. In order to check property II, we must compute

$$\int_1^\infty 3x^{-4}\, dx.$$

But

$$\int_1^b 3x^{-4}\, dx = -x^{-3}\bigg|_1^b = -b^{-3} + 1 \to 1$$

as $b \to \infty$. Thus

$$\int_1^\infty 3x^{-4}\, dx = 1,$$

and property II holds.

(b) Since $F(x)$ is an antiderivative of $f(x) = 3x^{-4}$, we have

$$F(x) = \int 3x^{-4}\, dx = -x^{-3} + C.$$

Since X has values greater than or equal to 1, we must have $F(1) = 0$. Setting $F(1) = -1 + C = 0$, we find that $C = 1$, and so

$$F(x) = 1 - x^{-3}.$$

(c) $\Pr(X \le 4) = F(4) = 1 - 4^{-3} = 1 - \frac{1}{64} = \frac{63}{64}.$

Since we know $F(x)$, we may use it to compute $\Pr(4 \le X \le 5)$, as follows:

$$\Pr(4 \le X \le 5) = F(5) - F(4) = (1 - 5^{-3}) - (1 - 4^{-3})$$

$$= \frac{1}{4^3} - \frac{1}{5^3} \approx .0076.$$

We may compute $\Pr(4 \le X)$ directly by evaluating the improper integral

$$\int_4^\infty 3x^{-4}\, dx.$$

However, there is a simpler method. We know that

$$\int_1^4 3x^{-4}\, dx + \int_4^\infty 3x^{-4}\, dx = \int_1^\infty 3x^{-4}\, dx = 1. \tag{8}$$

In terms of probabilities, (8) may be written as

$$\Pr(X \le 4) + \Pr(4 \le X) = 1.$$

Hence,

$$\Pr(4 \le X) = 1 - \Pr(X \le 4) = 1 - \tfrac{63}{64} = \tfrac{1}{64}.$$

PRACTICE PROBLEMS 2

1. Suppose that in a certain farming region, and in a certain year, the number of bushels of wheat produced per acre is a random variable X with a density function

$$f(x) = \frac{x - 30}{50}, \quad 30 \le x \le 40.$$

   (a) What is the probability that an acre selected at random produced less than 35 bushels of wheat?

   (b) If the farming region had 20,000 acres of wheat, how many acres produced less than 35 bushels of wheat?

2. The density function for a continuous random variable X on the interval $1 \le x \le 2$ is $f(x) = 8/3x^3$. Find the corresponding cumulative distribution function for X.

EXERCISES 2

Verify that each of the following functions is a probability density function.

1. $f(x) = \frac{1}{18}x,\ 0 \le x \le 6$ 2. $f(x) = 2(x - 1),\ 1 \le x \le 2$

3. $f(x) = \frac{1}{4},\ 1 \le x \le 5$ 4. $f(x) = \frac{8}{9}x,\ 0 \le x \le \frac{3}{2}$

5. $f(x) = 5x^4, 0 \le x \le 1$ 6. $f(x) = \frac{3}{2}x - \frac{3}{4}x^2, 0 \le x \le 2$

In Exercises 7–12 find the value of k that makes the given function a probability density function on the specified interval.

7. $f(x) = kx, 1 \le x \le 3$ 8. $f(x) = kx^2, 0 \le x \le 2$

9. $f(x) = k, 5 \le x \le 20$ 10. $f(x) = k/\sqrt{x}, 1 \le x \le 4$

11. $f(x) = kx^2(1 - x), 0 \le x \le 1$ 12. $f(x) = k(3x - x^2), 0 \le x \le 3$

13. The density function of a continuous random variable X is $f(x) = \frac{1}{8}x, 0 \le x \le 4$. Sketch the graph of $f(x)$ and shade in the areas corresponding to (a) $\Pr(X \le 1)$, (b) $\Pr(2 \le X \le 2.5)$, (c) $\Pr(3.5 \le X)$.

14. The density function of a continuous random variable X is $f(x) = 3x^2, 0 \le x \le 1$. Sketch the graph of $f(x)$ and shade in the areas corresponding to (a) $\Pr(X \le .3)$, (b) $\Pr(.5 \le X \le .7)$, (c) $\Pr(.8 \le X)$.

15. Find $\Pr(1 \le X \le 2)$ when X is a random variable whose density function is given in Exercise 1.

16. Find $\Pr(1.5 \le X \le 1.7)$ when X is a random variable whose density function is given in Exercise 2.

17. Find $\Pr(X \le 3)$ when X is a random variable whose density function is given in Exercise 3.

18. Find $\Pr(1 \le X)$ when X is a random variable whose density function is given in Exercise 4.

19. Suppose that the lifetime X (in hours) of a certain type of flashlight battery is a random variable on the interval $30 \le x \le 50$ with density function $f(x) = \frac{1}{20}$, $30 \le x \le 50$. Find the probability that a battery selected at random will last at least 35 hours.

20. Suppose that at a certain supermarket the amount of time one must wait at the express lane is a random variable with density function $f(x) = 11/[10(x + 1)^2]$, $0 \le x \le 10$. Find the probability of having to wait less than four minutes at the express lane.

21. The cumulative distribution function for a random variable X on the interval $1 \le x \le 2$ is $F(x) = \frac{4}{3} - (4/3x^2)$. Find the corresponding density function.

22. The cumulative distribution function for a random variable X on the interval $1 \le x \le 5$ is $F(x) = \frac{1}{2}\sqrt{x - 1}$. Find the corresponding density function.

23. Compute the cumulative distribution function corresponding to the density function $f(x) = \frac{1}{5}, 2 \le x \le 7$.

24. Compute the cumulative distribution function corresponding to the density function $f(x) = \frac{1}{2}(3 - x), 1 \le x \le 3$.

25. The time (in minutes) required to complete a certain subassembly is a random variable X with density function $f(x) = \frac{1}{21}x^2, 1 \le x \le 4$.

(a) Use $f(x)$ to compute $\Pr(2 \le X \le 3)$.

(b) Find the corresponding cumulative distribution function $F(x)$.

(c) Use $F(x)$ to compute $\Pr(2 \le X \le 3)$.

26. The density function for a continuous random variable X on the interval $1 \leq x \leq 4$ is $f(x) = \frac{4}{9}x - \frac{1}{9}x^2$.

(a) Use $f(x)$ to compute $\Pr(3 \leq X \leq 4)$.

(b) Find the corresponding cumulative distribution function $F(x)$.

(c) Use $F(x)$ to compute $\Pr(3 \leq X \leq 4)$.

An experiment consists of selecting a point at random from the square in Fig. 6(a). Let X be the maximum of the coordinates of the point.

27. Show that the cumulative distribution function of X is $F(x) = x^2/4$, $0 \leq x \leq 2$.

28. Find the corresponding density function of X.

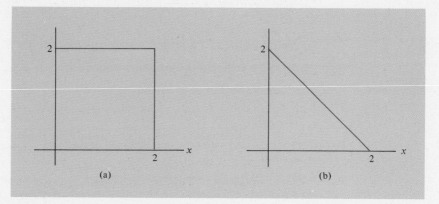

(a)   (b)

FIGURE 6

An experiment consists of selecting a point at random from the triangle in Fig. 6(b). Let X be the sum of the coordinates of the point.

29. Show that the cumulative distribution function of X is $F(x) = x^2/4$, $0 \leq x \leq 2$.

30. Find the corresponding density function of X.

Suppose that in a certain cell population, cells divide every 10 days and the age of a cell selected at random is a random variable X with the density function $f(x) = 2ke^{-kx}$, $0 \leq x \leq 10$, $k = (\ln 2)/10$.

31. Find the probability that a cell is at most five days old.

32. Upon examination of a slide, 10% of the cells are found to be undergoing mitosis (a change in the cell leading to division). Compute the length of time required for mitosis; that is, find the number M such that

$$\int_{10-M}^{10} 2ke^{-kx}\, dx = .10.$$

33. A random variable X has a density function $f(x) = \frac{1}{3}$, $0 \leq x \leq 3$. Find b such that $\Pr(0 \leq X \leq b) = .6$.

34. A random variable X has a density function $f(x) = \frac{2}{3}x$ on $1 \leq x \leq 2$. Find a such that $\Pr(a \leq X) = \frac{1}{3}$.

35. A random variable X has a cumulative distribution function $F(x) = \frac{1}{4}x^2$ on $0 \le x \le 2$. Find b such that $\Pr(X \le b) = .09$.

36. A random variable X has a cumulative distribution function $F(x) = (x - 1)^2$ on $1 \le x \le 2$. Find b such that $\Pr(X \le b) = \frac{1}{4}$.

37. Let X be a continuous random variable with values between $A = 1$ and $B = \infty$, and with the density function $f(x) = 4x^{-5}$.

    (a) Verify that $f(x)$ is a probability density function for $x \ge 1$.

    (b) Find the corresponding cumulative distribution function $F(x)$.

    (c) Use $F(x)$ to compute $\Pr(1 \le X \le 2)$ and $\Pr(2 \le X)$.

38. Let X be a continuous random variable with the density function $f(x) = 2(x + 1)^{-3}$, $x \ge 0$.

    (a) Verify that $f(x)$ is a probability density function for $x \ge 0$.

    (b) Find the cumulative distribution function for X.

    (c) Compute $\Pr(1 \le X \le 2)$ and $\Pr(3 \le X)$.

SOLUTIONS TO PRACTICE PROBLEMS 2

1. (a) $\displaystyle \Pr(X \le 35) = \int_{30}^{35} \frac{x - 30}{50}\, dx = \left. \frac{(x - 30)^2}{100} \right|_{30}^{35}$

   $$= \frac{5^2}{100} - 0 = .25.$$

   (b) Using part (a), we see that 25% of the 20,000 acres, or 5000 acres, produced less than 35 bushels of wheat.

2. The cumulative distribution function $F(x)$ is an antiderivative of $f(x) = 8/3x^3 = \frac{8}{3}x^{-3}$. Thus $F(x) = -\frac{4}{3}x^{-2} + C$ for some constant C. Since X varies over the interval $1 \le x \le 2$, we must have $F(1) = 0$; that is, $-\frac{4}{3}(1)^{-2} + C = 0$. Thus $C = \frac{4}{3}$, and

   $$F(x) = \tfrac{4}{3} - \tfrac{4}{3}x^{-2}.$$

12.3. Expected Value and Variance

When studying the cell population described in Section 2, one might reasonably ask for the average age of the cells. In general, if one is given an experiment described by a random variable X and a probability density function $f(x)$, it is often important to know the "average" outcome of the experiment, and the degree to which the experimental outcomes are spread out around the average. To provide this information in Section 1 we introduced the concepts of expected value and variance of a discrete random variable. Let us now examine the analogous definitions for a continuous random variable.

Definition Let X be a continuous random variable whose possible values lie between A and B, and let $f(x)$ be the probability density function for X. Then the *expected value* (or *mean*) of X is the number $E(X)$ defined by

$$E(X) = \int_A^B xf(x)\,dx. \qquad (1)$$

The *variance* of X is the number $Var(X)$ defined by

$$Var(X) = \int_A^B [x - E(X)]^2 f(x)\,dx. \qquad (2)$$

The expected value of X has the same interpretation as in the discrete case—namely, if the experiment whose outcome is X is performed many times, then the average of all the outcomes will approximately equal $E(X)$. As in the case of a discrete random variable, the variance of X is a quantitative measure of the likely spread of the values of X about the mean $E(X)$ when the experiment is performed many times.

To explain why the definition (1) of $E(X)$ is analogous to the definition in Section 1, let us approximate the integral in (1) by a Riemann sum of the form

$$x_1 f(x_1)\,\Delta x + x_2 f(x_2)\,\Delta x + \ldots + x_n f(x_n)\,\Delta x. \qquad (3)$$

Here $x_1, \ldots, x_n$ are the midpoints of subintervals of the interval from A to B, each subinterval of width $\Delta x = (B - A)/n$. (See Fig. 1.) Now recall from Section 2 that, for $i = 1, \ldots, n$, the quantity $f(x_i)\,\Delta x$ is approximately the probability that X is close to x_i—that is, the probability that X lies in the subinterval centered at x_i. If we write $\Pr(X \approx x_i)$ for this probability, then (3) is nearly the same as

$$x_1 \cdot \Pr(X \approx x_1) + x_2 \cdot \Pr(X \approx x_2) + \ldots + x_n \cdot \Pr(X \approx x_n). \qquad (4)$$

As the number of subintervals increases, the sum becomes closer and closer to the integral in (1) defining $E(X)$. Furthermore, each approximating sum in (4) resembles the sum in the definition of the expected value of a discrete random variable, where one computes the weighted sum over all possible outcomes, with each outcome weighted by the probability of its occurrence.

FIGURE 1

A similar analysis would show that the definition (2) of variance is analogous to the definition for the discrete case.

EXAMPLE 1   Let us consider the experiment of selecting a number at random from among the numbers between 0 and B. Let X denote the associated random variable. Determine the cumulative distribution function of X, the density function of X, and the mean and variance of X.

Solution
$$F(x) = \frac{[\text{length of the interval from 0 to } x]}{[\text{length of the interval from 0 to } B]}$$

$$= \frac{x}{B}.$$

Since $f(x) = F'(x)$, we see that $f(x) = 1/B$. Thus, we have

$$\mathrm{E}(X) = \int_0^B x \cdot \frac{1}{B}\, dx = \frac{1}{B}\int_0^B x\, dx = \frac{1}{B} \cdot \frac{B^2}{2} = \frac{B}{2},$$

$$\mathrm{Var}(X) = \int_0^B \left(x - \frac{B}{2}\right)^2 \cdot \frac{1}{B}\, dx$$

$$= \frac{1}{B}\int_0^B \left(x - \frac{B}{2}\right)^2 dx$$

$$= \frac{1}{B} \cdot \frac{1}{3}\left(x - \frac{B}{2}\right)^3 \Big|_0^B$$

$$= \frac{1}{3B}\left[\left(\frac{B}{2}\right)^3 - \left(-\frac{B}{2}\right)^3\right]$$

$$= \frac{B^2}{12}.$$

The graph of the density function $f(x)$ is shown in Fig. 2. Since the density function has a flat graph, the random variable X is called the *uniform random variable* on the interval from 0 to B.

FIGURE 2

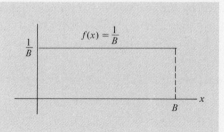

EXAMPLE 2   Let X be the age of a cell chosen at random from the population described in Section 2, where the density function for X was given as

$$f(x) = 2ke^{-kx}, \quad 0 \le x \le T,$$

and $k = (\ln 2)/T$. Find the average age, $E(X)$, of the cell population.

Solution   By definition

$$E(X) = \int_0^T x \cdot 2ke^{-kx} \, dx.$$

To calculate this integral we need integration by parts, with $f(x) = 2x$, $g(x) = ke^{-kx}$, $f'(x) = 2$, and $G(x) = -e^{-kx}$. We have

$$\int_0^T 2xke^{-kx} \, dx = -2xe^{-kx} \Big|_0^T - \int_0^T -2e^{-kx} \, dx$$

$$= -2Te^{-kT} - \left(\frac{2}{k} e^{-kx} \right) \Big|_0^T$$

$$= -2Te^{-kT} - \frac{2}{k} e^{-kT} + \frac{2}{k}.$$

This formula for $E(X)$ may be simplified by noting that $e^{-kT} = e^{-\ln 2} = \frac{1}{2}$. Thus

$$E(X) = -2T \left(\frac{1}{2} \right) - \frac{2}{k} \left(\frac{1}{2} \right) + \frac{2}{k} = \frac{1}{k} - T$$

$$= \frac{T}{\ln 2} - T = \left(\frac{1}{\ln 2} - 1 \right) T$$

$$\approx .4427T.$$

EXAMPLE 3   Consider the experiment of selecting a point at random in a circle of radius 1, and let X be the distance from this point to the center. Compute the expected value and variance of the random variable X.

Solution   We showed in Example 5 of Section 2 that the density function for X is given by $f(x) = 2x$, $0 \le x \le 1$. Therefore, we see that

$$E(X) = \int_0^1 x \cdot 2x \, dx = \int_0^1 2x^2 \, dx$$

$$= \frac{2x^3}{3} \Big|_0^1 = \tfrac{2}{3},$$

and

$$\text{Var}(X) = \int_0^1 (x - \tfrac{2}{3})^2 \cdot 2x \, dx \qquad [\text{since } \text{E}(X) = \tfrac{2}{3}]$$

$$= \int_0^1 (x^2 - \tfrac{4}{3}x + \tfrac{4}{9}) \cdot 2x \, dx$$

$$= \int_0^1 (2x^3 - \tfrac{8}{3}x^2 + \tfrac{8}{9}x) \, dx$$

$$= (\tfrac{1}{2}x^4 - \tfrac{8}{9}x^3 + \tfrac{4}{9}x^2) \Big|_0^1$$

$$= \tfrac{1}{2} - \tfrac{8}{9} + \tfrac{4}{9} = \tfrac{1}{18}.$$

From our first calculation, we see that if a large number of points are chosen randomly from a circle of radius 1, then their average distance to the center should be about $\tfrac{2}{3}$.

The following alternate formula for the variance of a random variable is usually easier to use than the actual definition of $\text{Var}(X)$.

Let X be a continuous random variable whose values lie between A and B, and let $f(x)$ be the density function for X. Then

$$\text{Var}(X) = \int_A^B x^2 f(x) \, dx - \text{E}(X)^2. \tag{5}$$

To prove (5), we let $m = \text{E}(X) = \int_A^B xf(x) \, dx$. Then

$$\text{Var}(X) = \int_A^B (x - m)^2 f(x) \, dx = \int_A^B (x^2 - 2xm + m^2)f(x) \, dx$$

$$= \int_A^B x^2 f(x) \, dx - 2m \int_A^B xf(x) \, dx + m^2 \int_A^B f(x) \, dx$$

$$= \int_A^B x^2 f(x) \, dx - 2m \cdot m + m^2 \cdot 1 \quad (\text{by property II})$$

$$= \int_A^B x^2 f(x) \, dx - m^2.$$

EXAMPLE 4   A college library has found that, in any given month during a school year, the proportion of students who make some use of the library is a random variable X with the cumulative distribution function

$$F(x) = 4x^3 - 3x^4, \quad 0 \le x \le 1.$$

(a) Compute $E(X)$ and give an interpretation of this quantity.

(b) Compute $\text{Var}(X)$.

Solution (a) To compute $E(X)$ we first find the probability density function $f(x)$. From Section 2 we know that

$$f(x) = F'(x) = 12x^2 - 12x^3.$$

Hence,

$$E(X) = \int_0^1 xf(x)\,dx = \int_0^1 (12x^3 - 12x^4)\,dx$$

$$= (3x^4 - \tfrac{12}{5}x^5)\Big|_0^1 = 3 - \tfrac{12}{5} = \tfrac{3}{5}.$$

The meaning of $E(X)$ in this example is that over a period of many months (during school years) the average proportion of students each month who make some use of the library should be close to $\tfrac{3}{5}$.

(b) We first compute

$$\int_0^1 x^2 f(x)\,dx = \int_0^1 (12x^4 - 12x^5)\,dx = (\tfrac{12}{5}x^5 - 2x^6)\Big|_0^1$$

$$= \tfrac{12}{5} - 2 = \tfrac{2}{5}.$$

Then, from the alternate formula (5) for the variance, we find that

$$\text{Var}(X) = \tfrac{2}{5} - E(X)^2 = \tfrac{2}{5} - (\tfrac{3}{5})^2 = \tfrac{1}{25}.$$

PRACTICE PROBLEMS 3

1. Find the expected value and variance of the random variable X whose density function is $f(x) = 1/2\sqrt{x}$, $1 \leq x \leq 4$.

2. An insurance company finds that the proportion X of their salesmen who sell more than \$25,000 worth of insurance in a given week is a random variable with the beta probability density function

$$f(x) = 60x^3(1 - x)^2, \quad 0 \leq x \leq 1.$$

(a) Compute $E(X)$ and give an interpretation of this quantity.

(b) Compute $\text{Var}(X)$.

EXERCISES 3

Find the expected value and variance for each random variable whose probability density function is given below. When computing the variance, use formula (5).

1. $f(x) = \tfrac{1}{18}x$, $0 \leq x \leq 6$

2. $f(x) = 2(x - 1)$, $1 \leq x \leq 2$

3. $f(x) = \tfrac{1}{4}$, $1 \leq x \leq 5$

4. $f(x) = \tfrac{8}{9}x$, $0 \leq x \leq \tfrac{3}{2}$

5. $f(x) = 5x^4, 0 \le x \le 1$

6. $f(x) = \frac{3}{2}x - \frac{3}{4}x^2, 0 \le x \le 2$

7. $f(x) = 12x(1 - x)^2, 0 \le x \le 1$

8. $f(x) = 3\sqrt{x}/16, 0 \le x \le 4$

9. A newspaper publisher estimates that the proportion X of space devoted to advertising on a given day is a random variable with the beta probability density $f(x) = 30x^2(1 - x)^2, 0 \le x \le 1$.

   (a) Find the cumulative distribution function for X.

   (b) Find the probability that less than 25% of the newspaper's space on a given day contains advertising.

   (c) Find $E(X)$ and give an interpretation of this quantity.

   (d) Compute $Var(X)$.

10. Let X be the proportion of new restaurants in a given year that make a profit during their first year of operation, and suppose the density function for X is $f(x) = 20x^3(1 - x), 0 \le x \le 1$.

    (a) Find $E(X)$ and give an interpretation of this quantity.

    (b) Compute $Var(X)$.

11. The useful life (in hundreds of hours) of a certain machine component is a random variable X with the cumulative distribution function $F(x) = \frac{1}{9}x^2, 0 \le x \le 3$.

    (a) Find $E(X)$ and give an interpretation of this quantity.

    (b) Compute $Var(X)$.

12. The time (in minutes) required to complete an assembly on a production line is a random variable X with the cumulative distribution function $F(x) = \frac{1}{125}x^3$, $0 \le x \le 5$.

    (a) Find $E(X)$ and give an interpretation of this quantity.

    (b) Compute $Var(X)$.

13. Suppose that the amount of time (in minutes) a person spends reading the editorial page of the newspaper is a random variable with the density function $f(x) = \frac{1}{72}x$, $0 \le x \le 12$. Find the average time spent reading the editorial page.

14. Suppose that at a certain bus stop the time between buses is a random variable X with the density function $f(x) = 6x(10 - x)/1000, 0 \le x \le 10$. Find the average time between buses.

15. When preparing a bid on a large construction project, a contractor analyzes how long each phase of the construction will take. Suppose the contractor estimates that the time required for the electrical work will be X hundred man-hours, where X is a random variable with the density function $f(x) = x(6 - x)/18, 3 \le x \le 6$.

    (a) Find the cumulative distribution function $F(x)$.

    (b) What is the likelihood that the electrical work will take less than 500 man-hours?

    (c) Find the mean time to complete the electrical work.

    (d) Find $Var(X)$.

16. The amount of milk (in thousands of gallons) a dairy sells each week is a random variable X with the density function $f(x) = 4(x - 1)^3$, $1 \leq x \leq 2$.

    (a) What is the likelihood that the dairy will sell more than 1500 gallons?

    (b) What is the average amount of milk the dairy sells each week?

17. Let X be a continuous random variable with values between $A = 1$ and $B = \infty$, and with the density function $f(x) = 4x^{-5}$. Compute $E(X)$ and $\text{Var}(X)$.

18. Let X be a continuous random variable with density function $f(x) = 3x^{-4}$, $x \geq 1$. Compute $E(X)$ and $\text{Var}(X)$.

    If X is a random variable with a density function $f(x)$ on $A \leq x \leq B$, then the *median* of X is that number M such that $\int_A^M f(x)\,dx = \frac{1}{2}$. In other words, $\Pr(X \leq M) = \frac{1}{2}$.

19. Find the median of the random variable whose density function is $f(x) = \frac{1}{18}x$, $0 \leq x \leq 6$.

20. Find the median of the random variable whose density function is $f(x) = 2(x - 1)$, $1 \leq x \leq 2$.

21. A machine component described in Exercise 11 has a 50% chance of lasting at least how long?

22. In Exercise 12, find the length of time T such that half of the assemblies are completed in T minutes or less.

23. In Exercise 20 of Section 2, find the length of time T such that about half of the time one waits only T minutes or less in the express lane at the supermarket.

24. Find the number M such that half of the time the dairy in Exercise 16 sells M thousand gallons of milk or less.

25. Show that $E(X) = B - \int_A^B F(x)\,dx$, where $F(x)$ is the cumulative distribution function for X on $A \leq x \leq B$.

26. Use the formula in Exercise 25 to compute $E(X)$ for the random variable X in Exercise 12.

SOLUTIONS TO PRACTICE PROBLEMS 3

1. $E(X) = \int_1^4 x \cdot \frac{1}{2\sqrt{x}}\,dx = \int_1^4 \frac{1}{2} x^{1/2}\,dx = \frac{1}{3} x^{3/2} \Big|_1^4$

   $= \frac{1}{3}(4)^{3/2} - \frac{1}{3} = \frac{8}{3} - \frac{1}{3} = \frac{7}{3}.$

   To find $\text{Var}(X)$, we first compute

   $$\int_1^4 x^2 \cdot \frac{1}{2\sqrt{x}}\,dx = \int_1^4 \frac{1}{2} x^{3/2}\,dx = \frac{1}{5} x^{5/2} \Big|_1^4$$

   $$= \frac{1}{5}(4)^{5/2} - \frac{1}{5} = \frac{32}{5} - \frac{1}{5} = \frac{31}{5}.$$

Then, from formula (5),

$$\text{Var}(X) = \tfrac{31}{5} - (\tfrac{7}{3})^2 = \tfrac{34}{45}.$$

2. (a) First note that $f(x) = 60x^3(1 - x)^2 = 60x^3(1 - 2x + x^2) = 60x^3 - 120x^4 + 60x^5$. Then

$$E(X) = \int_0^1 xf(x)\,dx = \int_0^1 (60x^4 - 120x^5 + 60x^6)\,dx$$

$$= (12x^5 - 20x^6 + \tfrac{60}{7}x^7)\Big|_0^1 = 12 - 20 + \tfrac{60}{7} = \tfrac{4}{7}.$$

Thus, in an "average" week, about four-sevenths of the salesmen sell more than $25,000 worth of insurance. More precisely, over a period of many weeks, we expect an average of four-sevenths of the salesmen each week to sell more than $25,000 worth of insurance.

(b) $\displaystyle\int_0^1 x^2 f(x)\,dx = \int_0^1 (60x^5 - 120x^6 + 60x^7)\,dx$

$$= (10x^6 - \tfrac{120}{7}x^7 + \tfrac{60}{8}x^8)\Big|_0^1$$

$$= 10 - \tfrac{120}{7} + \tfrac{60}{8} = \tfrac{5}{14}.$$

Hence

$$\text{Var}(X) = \tfrac{5}{14} - (\tfrac{4}{7})^2 = \tfrac{3}{98}.$$

12.4. Exponential and Normal Random Variables

This section is devoted to the two most important types of probability density functions—the exponential and normal density functions. These functions are associated with random variables that arise in a wide variety of applications. We will describe some typical examples.

A. Exponential Density Functions Let k be a positive constant. Then the function

$$f(x) = ke^{-kx}, \quad x \geq 0,$$

is called an *exponential density function*. (See Fig. 1.) This function is indeed a probability density function. First, $f(x)$ is clearly greater than or equal to 0. Second,

$$\int_0^b ke^{-kx}\,dx = -e^{-kx}\Big|_0^b = 1 - e^{-kb} \to 1 \quad \text{as } b \to \infty,$$

so that

$$\int_0^\infty ke^{-kx}\,dx = 1.$$

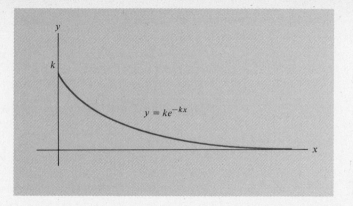

FIGURE 1 An exponential density function.

A random variable X with an exponential density function is called an *exponential random variable*, and the values of X are said to be *exponentially distributed*. Exponential random variables are used in reliability calculations to represent the lifetime (or "time to failure") of electronic components such as tubes and transistors. They are used to describe the length of time between two successive random events, such as the interarrival times between successive telephone calls at a switchboard. Also, exponential random variables can arise in the study of service times, e.g., the length of time a person spends in a doctor's office or at a gas station.

Let us compute the expected value of an exponential random variable X, namely,

$$E(X) = \int_0^\infty xf(x)\, dx = \int_0^\infty xke^{-kx}\, dx.$$

We may approximate this improper integral by a definite integral and use integration by parts to find that

$$\int_0^b xke^{-kx}\, dx = \left. -xe^{-kx} \right|_0^b - \int_0^b -e^{-kx}\, dx$$

$$= (-be^{-kb} + 0) - \frac{1}{k} e^{-kx} \Big|_0^b$$

$$= -be^{-kb} - \frac{1}{k} e^{-kb} + \frac{1}{k}. \tag{1}$$

As $b \to \infty$, this quantity approaches $1/k$, because the numbers $-be^{-kb}$ and $-(1/k)e^{-kb}$ both approach 0. (See Exercise 22 of Section 13.3.) Thus

$$E(X) = \int_0^\infty xke^{-kx}\, dx = \frac{1}{k}.$$

Let us now compute the variance of X. From the alternate formula for $\text{Var}(X)$ given in Section 3, we have

$$\text{Var}(X) = \int_0^\infty x^2 f(x)\, dx - E(X)^2$$
$$= \int_0^\infty x^2 k e^{-kx}\, dx - \frac{1}{k^2}. \tag{2}$$

Using integration by parts, we obtain

$$\int_0^b x^2 k e^{-kx}\, dx = x^2(-e^{-kx})\Big|_0^b - \int_0^b 2x(-e^{-kx})\, dx$$
$$= (-b^2 e^{-kb} + 0) + 2\int_0^b x e^{-kx}\, dx$$
$$= -b^2 e^{-kb} + \frac{2}{k}\int_0^b x k e^{-kx}\, dx. \tag{3}$$

Now let $b \to \infty$. We know from our calculation (1) of $E(X)$ that the integral in the second term of (3) approaches $1/k$; also, it can be shown that $-b^2 e^{-kb}$ approaches 0 (see Exercise 24 of Section 13.3). Therefore,

$$\int_0^\infty x^2 k e^{-kx}\, dx = \frac{2}{k} \cdot \frac{1}{k} = \frac{2}{k^2}.$$

And by equation (2), we have

$$\text{Var}(X) = \frac{2}{k^2} - \frac{1}{k^2} = \frac{1}{k^2}.$$

Let us summarize our results:

> Let X be a random variable with an exponential density function $f(x) = ke^{-kx}$ ($x \geq 0$). Then
> $$E(X) = \frac{1}{k} \quad \text{and} \quad \text{Var}(X) = \frac{1}{k^2}.$$

EXAMPLE 1 Suppose that the number of days of continuous use provided by a certain brand of light bulb is an exponential random variable X with expected value 100 days.

(a) Find the density function of X.

(b) Find the probability that a randomly-chosen bulb will last between 80 and 90 days.

(c) Find the probability that a randomly-chosen bulb will last for more than 40 days.

(d) Suppose we simultaneously turn on 10,000 new light bulbs. How many days will elapse before 2% of the bulbs have burned out?

Solution   (a) Since X is an exponential random variable, its density function must be of the form $f(x) = ke^{-kx}$ for some $k > 0$. Since the expected value of such a density function is $1/k$ and is equal to 100 in this case, we see that

$$\frac{1}{k} = 100,$$

$$k = \frac{1}{100} = .01.$$

Thus,

$$f(x) = .01e^{-.01x}.$$

(b)
$$\Pr(80 \leq X \leq 90) = \int_{80}^{90} .01e^{-.01x}\, dx$$

$$= -e^{-.01x}\Big|_{80}^{90}$$

$$= -e^{-.9} + e^{-.8} \approx .04276.$$

(c)
$$\Pr(X \geq 40) = \int_{40}^{\infty} .01e^{-.01x}\, dx = 1 - \int_{0}^{40} .01e^{-.01x}\, dx$$

(since $\int_{0}^{\infty} f(x)\, dx = 1$), so that

$$\Pr(X \geq 40) = 1 + (e^{-.01x})\Big|_{0}^{40}$$

$$= 1 + (e^{-.4} - 1)$$

$$= e^{-.4} \approx .67032.$$

(d) Let b denote the number of days before 2% of the bulbs burn out. Then $\Pr(X \leq b) = .02$, so that

$$.02 = \int_{0}^{b} .01e^{-.01x}\, dx = -e^{-.01x}\Big|_{0}^{b} = -e^{-.01b} + 1.$$

Thus, we see that $e^{-.01b} = .98$. Therefore,

$$-.01b = \ln .98$$

$$b = \frac{\ln .98}{-.01} \approx 2.02027.$$

Thus, in our batch of 10,000 bulbs, it will take approximately 2 days for 200 bulbs to burn out.

EXAMPLE 2 During a certain part of the day, the interarrival time between successive phone calls at a central telephone exchange is an exponential random variable X with expected value $\frac{1}{3}$ seconds.

(a) Find the density function of X.

(b) Find the probability that between $\frac{1}{3}$ and $\frac{2}{3}$ seconds elapse between consecutive phone calls.

(c) Find the probability that the time between successive phone calls is more than 2 seconds.

Solution (a) Since X is an exponential random variable, its density function is $f(x) = ke^{-kx}$ for some $k > 0$. Since the expected value of X is $1/k = \frac{1}{3}$, we have $k = 3$ and $f(x) = 3e^{-3x}$.

(b)
$$\Pr(\tfrac{1}{3} \le X \le \tfrac{2}{3}) = \int_{1/3}^{2/3} 3e^{-3x} = -e^{-3x} \Big|_{1/3}^{2/3} = -e^{-2} + e^{-1}$$

$$\approx .23254.$$

(c)
$$\Pr(X \ge 2) = \int_{2}^{\infty} 3e^{-3x} = 1 - \int_{0}^{2} 3e^{-3x}$$

$$= 1 + (e^{-3x}) \Big|_{0}^{2}$$

$$= e^{-6} \approx .00248.$$

In other words, about .25% of the time, the waiting time between consecutive calls is at least 2 seconds.

B. Normal Density Functions Let μ, σ be given numbers, with $\sigma > 0$. Then the function

$$f(x) = \frac{1}{\sigma\sqrt{2\pi}} e^{-(1/2)[(x-\mu)/\sigma]^2} \tag{4}$$

is called a *normal density function*. A random variable X whose density function has this form is called a *normal random variable*, and the values of X are said to be *normally distributed*. Many random variables in applications are approximately normal. For example, errors which occur in physical measurements and various manufacturing processes, as well as many human physical and mental characteristics are all conveniently modeled by normal random variables.

The graph of the density function in (4) is called a *normal curve* (Fig. 2). A normal curve is symmetric about the line $x = \mu$ and has inflection points at $\mu - \sigma$ and $\mu + \sigma$. Figure 3 shows three normal curves corresponding to different values of σ. The parameters μ and σ determine the shape of the curve. The value of μ

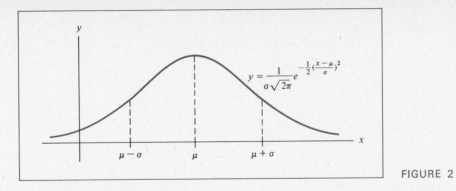

FIGURE 2

determines the point where the curve reaches its maximum height, and the value of σ determines how sharp a peak the curve has.

It can be shown that the constant $1/\sigma\sqrt{2\pi}$ in the definition (4) of a normal density function $f(x)$ is needed in order to make the area under the normal curve equal to 1, i.e., to make $f(x)$ a probability density function. The theoretical values of a normal random variable X include all positive and negative numbers, but the normal curve approaches the horizontal axis so rapidly beyond the inflection points that the probabilities associated with intervals on the x-axis far to the left or right of $x = \mu$ are negligible.

Using techniques outside the scope of this book, one can verify the following basic facts about a normal random variable.

Let X be a random variable with the normal density function

$$f(x) = \frac{1}{\sigma\sqrt{2\pi}}\, e^{-(1/2)[(x-\mu)/\sigma]^2}.$$

Then the expected value (mean), variance, and standard deviation of X are given by

$$E(X) = \mu, \quad \operatorname{Var}(X) = \sigma^2, \quad \text{and} \quad \sqrt{\operatorname{Var}(X)} = \sigma.$$

A normal random variable with expected value $\mu = 0$ and standard deviation $\sigma = 1$ is called a *standard normal random variable* and is often denoted by the letter Z. Using these values for μ and σ in (4), and writing z in place of the variable x, we see that the density function for Z is

$$f(z) = \frac{1}{\sqrt{2\pi}}\, e^{-(1/2)z^2}.$$

The graph of this function is called the *standard normal curve*. See Fig. 4.

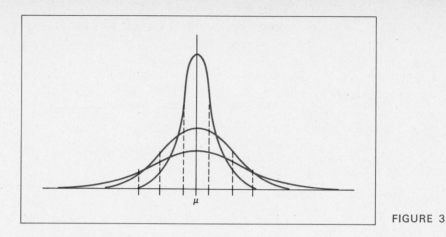

FIGURE 3

Probabilities involving a standard normal random variable Z may be written in the form

$$\Pr(a \le Z \le b) = \int_a^b \frac{1}{\sqrt{2\pi}} e^{-(1/2)z^2} \, dz.$$

Such an integral cannot be evaluated directly because the density function for Z cannot be antidifferentiated in terms of elementary functions. However, tables of such probabilities have been compiled, using numerical approximations to the definite integrals. For $z \ge 0$, let $A(z) = \Pr(0 \le Z \le z)$ and $A(-z) = \Pr(-z \le Z \le 0)$. That is, let $A(z)$ and $A(-z)$ be the areas of the regions shown

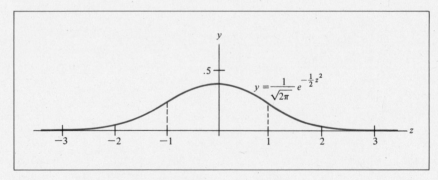

FIGURE 4   The standard normal curve.

FIGURE 5   Areas under the standard normal curve.

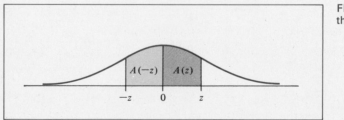

in Fig. 5. From the symmetry of the standard normal curve, it is clear that $A(-z) = A(z)$. The values of $A(z)$ for $z \geq 0$ are listed in Table 4 on page A 8.

EXAMPLE 3 Let Z be a standard normal random variable. Use Table 4 to compute the following probabilities:

(a) $\Pr(0 \leq Z \leq 1.84)$ (b) $\Pr(-1.65 \leq Z \leq 0)$ (c) $\Pr(.7 \leq Z)$

(d) $\Pr(.5 \leq Z \leq 2)$ (e) $\Pr(-.75 \leq Z \leq 1.46)$

Solution (a) $\Pr(0 \leq Z \leq 1.84) = A(1.84)$. In Table 4 page A8, we move down the column under "z" until we reach 1.8; then we move to the right in the same row to the column with the heading ".04." There we find that $A(1.84) = .4671$.

(b) $\Pr(-1.65 \leq Z \leq 0) = A(-1.65) = A(1.65) = .4505$ (from Table 4).

(c) Since the area under the normal curve is 1, the symmetry of the curve implies that the area to the right of the y-axis is .5. Now $\Pr(.7 \leq Z)$ is the area under the curve to the right of .7, and so we can find this area by subtracting from .5 the area between 0 and .7. (See Fig. 6(a).) Thus

$$\begin{aligned} \Pr(.7 \leq Z) &= .5 - \Pr(0 \leq Z \leq .7) \\ &= .5 - A(.7) = .5 - .2580 \quad \text{(from Table 4)} \\ &= .2420. \end{aligned}$$

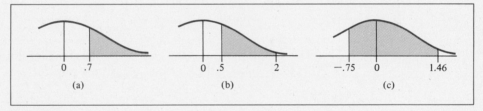

FIGURE 6

(d) The area under the standard normal curve from .5 to 2 equals the area from 0 to 2 minus the area from 0 to .5. See Fig. 6(b). Thus we have

$$\begin{aligned} \Pr(.5 \leq Z \leq 2) &= A(2) - A(.5) \\ &= .4772 - .1915 = .2857. \end{aligned}$$

(e) The area under the standard normal curve from $-.75$ to 1.46 equals the area from $-.75$ to 0 plus the area from 0 to 1.46. See Fig. 6(c). Thus

$$\begin{aligned} \Pr(-.75 \leq Z \leq 1.46) &= A(-.75) + A(1.46) \\ &= A(.75) + A(1.46) \\ &= .2734 + .4279 = .7013. \end{aligned}$$

When X is an *arbitrary* normal random variable, with mean μ and standard deviation σ, one may compute a probability such as $\Pr(a \le X \le b)$ by making the change of variable $z = (x - \mu)/\sigma$. This converts the integral for $\Pr(a \le X \le b)$ into an integral involving the standard normal density function. The following example illustrates this procedure.

EXAMPLE 4 A metal flange on a truck must be between 92.1 and 94 mm long in order to fit properly. Suppose the lengths of the flanges supplied to the truck manufacturer are normally distributed with mean $\mu = 93$ mm and standard deviation $\sigma = .4$ mm.

(a) What percentage of the flanges have an acceptable length?

(b) What percentage of the flanges are too long?

Solution Let X be the length of a metal flange selected at random from the supply of flanges.

(a) We have

$$\Pr(92.1 \le X \le 94) = \int_{92.1}^{94} \frac{1}{(.4)\sqrt{2\pi}} e^{-(1/2)[(x-93)/.4]^2}\, dx.$$

Using the substitution $z = (x - 93)/.4$, $dz = (1/.4)\, dx$, we note that if $x = 92.1$ then $z = (92.1 - 93)/.4 = -.9/.4 = -2.25$, and if $x = 94$ then $z = (94 - 93)/.4 = 1/.4 = 2.5$. Hence,

$$\Pr(92.1 \le X \le 94) = \int_{-2.25}^{2.5} \frac{1}{\sqrt{2\pi}} e^{-(1/2)z^2}\, dz.$$

The value of this integral is the area under the standard normal curve from -2.25 to 2.5, which equals the area from -2.25 to 0 plus the area from 0 to 2.5. Thus,

$$\Pr(92.1 \le X \le 94) = A(-2.25) + A(2.5)$$

$$= A(2.25) + A(2.5)$$

$$= .4878 + .4938 = .9816.$$

From this probability we conclude that about 98% of the flanges will have an acceptable length.

(b) $$\Pr(94 \le X) = \int_{94}^{\infty} \frac{1}{(.4)\sqrt{2\pi}} e^{-(1/2)[(x-93)/.4]^2}\, dx.$$

This integral is approximated by an integral from $x = 94$ to $x = b$ where b is large. If we substitute $z = (x - 93)/.4$, we find that

$$\int_{94}^{b} \frac{1}{(.4)\sqrt{2\pi}} e^{-(1/2)[(x-93)/.4]^2}\, dx = \int_{2.5}^{(b-93)/.4} \frac{1}{\sqrt{2\pi}} e^{-(1/2)z^2}\, dz. \tag{5}$$

Now when $b \to \infty$, the quantity $(b - 93)/.4$ also becomes arbitrarily large. Since the left integral in (5) approaches $\Pr(94 \leq X)$, we conclude that

$$\Pr(94 \leq X) = \int_{2.5}^{\infty} \frac{1}{\sqrt{2\pi}} e^{-(1/2)z^2} \, dz.$$

To calculate this integral we use the method of Example 3(c). The area under the standard normal curve to the right of 2.5 equals the area to the right of 0 minus the area from 0 to 2.5. That is,

$$\Pr(94 \leq X) = .5 - A(2.5)$$

$$= .5 - .4938 = .0062.$$

Approximately .6% of the flanges exceed the maximum acceptable length.

PRACTICE PROBLEMS 4

1. The emergency flasher on an automobile is warranted for the first 12,000 miles that the car is driven. During that period a defective flasher will be replaced free. Suppose the time before failure of the emergency flasher (measured in thousands of miles) is an exponential random variable X with mean 50 (thousand miles). What percentage of the flashers will have to be replaced during the warranty period?

2. Suppose the lead time between ordering furniture from a certain company and receiving delivery is a normal random variable with $\mu = 18$ weeks and $\sigma = 5$ weeks. Find the likelihood that a customer will have to wait more than 16 weeks.

EXERCISES 4

Find (by inspection) the expected values and variances of the exponential random variables with the density functions given in Exercises 1–4.

1. $3e^{-3x}$ 2. $\frac{1}{4}e^{-x/4}$ 3. $.2e^{-.2x}$ 4. $1.5e^{-1.5x}$

Suppose that in a large factory there is an average of two accidents per day and the time between accidents has an exponential density function with expected value of $\frac{1}{2}$ day.

5. Find the probability that the time between two accidents will be more than $\frac{1}{2}$ day and less than one day.

6. Find the probability that the time between accidents will be less than 8 hours (that is, $\frac{1}{3}$ day).

Suppose that the amount of time required to serve a customer at a bank has an exponential density function with mean 3 minutes.

7. Find the probability that a customer is served in less than 2 minutes.

8. Find the probability that a customer will require more than 5 minutes.

During a certain part of the day, the time between arrivals of automobiles at the tollgate on a turnpike is an exponential random variable with expected value 20 seconds.

9. Find the probability that the time between successive arrivals is more than 60 seconds.

10. Find the probability that the time between successive arrivals is greater than 10 seconds and less than 30 seconds.

Upon studying the vacancies occurring in the U.S. Supreme Court, it has been determined that the time elapsed between successive resignations is an exponential random variable with expected value 2 years.*

11. A new president takes office at the same time a justice retires. Find the probability that the next vacancy will take place during his four-year term.

12. Find the probability that the composition of the U.S. Supreme Court will remain unchanged for a period of five years or more.

13. Suppose the average lifespan of an electronic component is 72 months, and the lifespans are exponentially distributed.

(a) Find the probability that a component lasts for more than 24 months.

(b) The *reliability function* $r(t)$ gives the probability that a component will last for more than t months. Compute $r(t)$ in this case.

14. Consider a group of patients that have been treated for an acute disease such as cancer, and let X be the number of years a person lives after receiving the treatment (the "survival time"). Under suitable conditions, the density function for X will be $f(x) = ke^{-kx}$ for some constant k.

(a) The *survival function* $S(x)$ is the probability that a person chosen at random from the group of patients survives until at least time x. Explain why $S(x) = 1 - F(x)$, where $F(x)$ is the cumulative distribution function for X, and compute $S(x)$.

(b) Suppose that the probability is .90 that a patient will survive at least 5 years, i.e., $S(5) = .90$. Find the constant k in the exponential density function $f(x)$.

Find the expected values and standard deviations (by inspection) of the normal random variables with the density functions given in Exercises 15–18.

15. $\dfrac{1}{\sqrt{2\pi}} e^{-(1/2)(x-4)^2}$ 16. $\dfrac{1}{\sqrt{2\pi}} e^{-(1/2)(x+5)^2}$

17. $\dfrac{1}{3\sqrt{2\pi}} e^{-(1/18)x^2}$ 18. $\dfrac{1}{5\sqrt{2\pi}} e^{-(1/2)[(x-3)/5]^2}$

19. Show that the function $f(x) = e^{-x^2/2}$ has a maximum at $x = 0$.

---

* See W. A. Wallis, "The Poisson Distribution and the Supreme Court," *J. Am. Statistical Assoc.* **31** (1936), 376–380.

20. Show that the function $f(x) = e^{-[(x-\mu)/\sigma]^2}$ has a maximum at $x = \mu$.

21. Show that the function $f(x) = e^{-x^2/2}$ has inflection points at $x = \pm 1$.

22. Show that the function $f(x) = e^{-(1/2)[(x-\mu)/\sigma]^2}$ has inflection points at $x = \mu \pm \sigma$.

23. Let Z be a standard normal random variable. Calculate

   (a) $\Pr(-1.3 \le Z \le 0)$,

   (b) $\Pr(.25 \le Z)$,

   (c) $\Pr(-1 \le Z \le 2.5)$,

   (d) $\Pr(Z \le 2)$.

24. Calculate the area under the standard normal curve for values of z

   (a) between .5 and 1.5,

   (b) between $-.75$ and .75,

   (c) to the left of $-.3$,

   (d) to the right of -1.

25. The gestation period (length of pregnancy) of pregnant females of a certain species is approximately normally distributed with a mean of 6 months and standard deviation of $\frac{1}{2}$ month.

   (a) Find the percentage of births that occur after a gestation period of between 6 and 7 months.

   (b) Find the percentage of births that occur after a gestation period of between 5 and 6 months.

26. Suppose the lifespan of a certain automobile tire is normally distributed with $\mu = 25,000$ miles and $\sigma = 2000$ miles.

   (a) Find the probability that a tire will last between 28,000 and 30,000 miles.

   (b) Find the probability that a tire will last more than 29,000 miles.

27. Suppose that the amount of milk in a gallon container is a normal random variable with $\mu = 128.2$ ounces and $\sigma = .2$ ounces. Find the probability that a random bottle of milk contains less than 128 ounces.

28. The amount of weight required to break a certain brand of twine has a normal density function with $\mu = 43$ kilograms and $\sigma = 1.5$ kilograms. Find the probability that the breaking weight of a piece of the twine is less than 40 kilograms.

29. A student with an eight o'clock class at the University of Maryland commutes to school. She has discovered that along each of two possible routes her traveling time to school (including the time to get to class) is approximately a normal random variable. If she uses the Capitol Beltway for most of her trip, then $\mu = 25$ minutes and $\sigma = 5$ minutes. If she drives a longer route over local city streets, then $\mu = 28$ minutes and $\sigma = 3$ minutes. Which route should the student take if she leaves home at 7:30 A.M.? (Assume that the best route is one that minimizes the probability of being late to class.)

30. Which route should the student in Exercise 29 take if she leaves home at 7:26 A.M.?

31. A certain type of bolt must fit through a 20 mm test hole, or else it is discarded. If the diameters of the bolts are normally distributed with $\mu = 18.2$ mm and $\sigma = .8$ mm, what percentage of the bolts will be discarded?

32. The Math SAT scores of a recent freshman class at a university were normally distributed with $\mu = 535$ and $\sigma = 100$.

    (a) What percent of the scores were between 500 and 600?

    (b) Find the minimum score needed to be in the top 10% of the class.

33. Let X be the time to failure (in years) of a transistor, and suppose the transistor has been operating properly for a years. Then it can be shown that the probability that the transistor will fail within the next b years is

$$\frac{\Pr(a \leq X \leq a + b)}{\Pr(a \leq X)}. \qquad (6)$$

Compute this probability for the case when X is an exponential random variable with density function $f(x) = ke^{-kx}$, and show that this probability equals $\Pr(0 \leq X \leq b)$. This means that the probability given by (6) does not depend on how long the transistor has already been operating. Due to this fact, exponential random variables are said to be *memoryless*.

34. The *median* of an exponential density function is that number M such that $\Pr(X \leq M) = \frac{1}{2}$. Show that $M = (\ln 2)/k$. (We see that the median is less than the mean.)

SOLUTIONS TO PRACTICE PROBLEMS 4

1. The density function for X is $f(x) = ke^{-kx}$, where $1/k = 50$ (thousand miles), and $k = 1/50 = .02$. Then

$$\Pr(X \leq 12) = \int_0^{12} .02e^{-.02x}\, dx = -e^{-.02x}\Big|_0^{12}$$
$$= 1 - e^{-.24} \approx .21337.$$

About 21% of the flashers will have to be replaced during the warranty period.

2. Let X be the time between ordering and receiving the furniture. Since $\mu = 18$ and $\sigma = 5$, we have

$$\Pr(16 \leq X) = \int_{16}^{\infty} \frac{1}{5\sqrt{2\pi}} e^{-(1/2)[(x-18)/5]^2}\, dx.$$

If we substitute $z = (x - 18)/5$, then $dz = \frac{1}{5}\, dx$, and $z = -.4$ when $x = 16$.

$$\Pr(16 \leq X) = \int_{-.4}^{\infty} \frac{1}{\sqrt{2\pi}} e^{-(1/2)z^2}\, dz.$$

(A similar substitution was made in Example 4(b).) The above integral gives the area under the standard normal curve to the right of $-.4$. Since the area between $-.4$ and 0 is $A(-.4) = A(.4)$, and the area to the right of 0 is .5, we have

$$\Pr(16 \leq X) = A(.4) + .5 = .1554 + .5 = .6554.$$

Chapter 12 : CHECKLIST

☐ Relative frequency histogram

☐ Probability table

☐ Discrete random variable

☐ Expected value (mean) of a discrete random variable

☐ Variance of a discrete random variable

☐ Standard deviation

☐ Probability density histogram

☐ Continuous random variable

☐ Probability density function

☐ Cumulative distribution function

☐ Expected value (mean) of a continuous random variable

☐ Variance of a continuous random variable

☐ Alternate formula for variance

☐ Exponential density function

☐ Expected value and variance of an exponential random variable

☐ Normal density function

☐ Expected value and variance of a normal random variable

☐ Standard normal density function

Chapter 12 : SUPPLEMENTARY EXERCISES

1. Let X be a continuous random variable on $0 \le x \le 2$, with the density function $f(x) = \frac{3}{8}x^2$.

   (a) Calculate $\Pr(X \le 1)$ and $\Pr(1 \le X \le 1.5)$.

   (b) Find $E(X)$ and $\text{Var}(X)$.

2. Let X be a continuous random variable on $3 \le x \le 4$, with the density function $f(x) = 2(x - 3)$.

   (a) Calculate $\Pr(3.2 \le X)$ and $\Pr(3 \le X)$.

   (b) Find $E(X)$ and $\text{Var}(X)$.

3. Verify that for any number $A, f(x) = e^{A-x}, x \ge A$, is a density function. Compute the associated cumulative distribution function for X.

4. Verify that for any positive constants k and A, the function $f(x) = kA^k/x^{k+1}$, $x \geq A$, is a density function. The associated cumulative distribution function $F(x)$ is called a *Pareto distribution*. Compute $F(x)$.

5. For any positive integer n, the function $f_n(x) = c_n x^{(n-2)/2}e^{-x/2}$, $x \geq 0$, where c_n is an appropriate constant, is called the *Chi-Square density function* with n degrees of freedom. Find c_2 and c_4 such that $f_2(x)$ and $f_4(x)$ are probability density functions.

6. Verify that for any positive number k, $f(x) = (1/2k^3)x^2e^{-x/k}$, $x \geq 0$, is a density function.

7. A medical laboratory tests many blood samples for a certain disease that occurs in about 5% of the samples. The lab collects samples from 10 persons and mixes together some blood from each sample. If a test on the mixture is positive, an additional 10 tests must be run, one on each individual sample. But if the test on the mixture is negative, no other tests are needed. It can be shown that the test of the mixture will be negative with probability $(.95)^{10} = .599$, because each of the 10 samples has a 95% chance of being free of the disease. If X is the total number of tests required, then X has the following probability table.

| | Test of mixture | |
	Negative	Positive
Total tests	1	11
Probability	.599	.401

(a) Find $E(X)$.

(b) If the laboratory uses the above procedure on 200 blood samples, about how many tests can it expect to run?

8. Suppose the laboratory in Exercise 7 uses batches of 5 instead of 10 samples. The probability of a negative test on the mixture of 5 samples is $(.95)^5 = .774$. Thus the following table gives the probabilities for the number X of tests required.

| | Test of mixture | |
	Negative	Positive
Total tests	1	6
Probability	.774	.226

(a) Find $E(X)$.

(b) If the laboratory uses this procedure on 200 blood samples, about how many tests can it expect to run?

9. A certain gas station sells X thousand gallons of gas each week. Suppose that the cumulative distribution function for X is $F(x) = 1 - \frac{1}{4}(2 - x)^2$, $0 \leq x \leq 2$.

(a) If the tank contains 1.6 thousand gallons at the beginning of the week, find the probability that the gas station will have enough gas for its customers throughout the week.

(b) How much gas must be in the tank at the beginning of the week in order to have a probability of .99 that there will be enough gasoline for the week?

(c) Compute the density function for X.

10. A service contract on a new electric typewriter costs $100 per year. The contract covers all necessary maintenance and repairs on the typewriter. Suppose that the actual cost to the manufacturer for providing this service is a random variable X (measured in hundreds of dollars) whose probability density function is $f(x) = (x - 5)^4/625$, $0 \leq x \leq 5$. Compute $E(X)$ and determine how much money the manufacturer expects to make on each service contract on the average.

11. A random variable X has a uniform density function $f(x) = \frac{1}{5}$ on $20 \leq X \leq 25$.

(a) Find $E(X)$ and $Var(X)$.

(b) Find b such that $Pr(X \leq b) = .3$.

12. A random variable X has a cumulative distribution function $F(x) = (x^2 - 9)/16$ on $3 \leq x \leq 5$.

(a) Find the density function for X.

(b) Find a such that $Pr(a \leq X) = \frac{1}{4}$.

13. The annual income of the households in a certain community ranges between 5 and 25 thousand dollars. Let X represent the annual income (in thousands of dollars) of a household chosen at random in this community, and suppose the probability density function for X is $f(x) = kx$, $5 \leq x \leq 25$.

(a) Find the value of k that makes $f(x)$ a density function.

(b) Find the fraction of households whose income exceeds $20,000.

(c) Find the mean annual income of the households in the community.

14. For each positive integer k, the function $f(x) = k(k + 1)x^{k-1}(1 - x)$ is a probability density function on $0 \leq x \leq 1$.

(a) Find the associated cumulative distribution function.

(b) Show that $E(X) = k/(k + 2)$.

15. A point is selected at random from the rectangle of Fig. 7(a); call its coordinates (θ, y). Find the probability that $y \leq \sin \theta$.

FIGURE 7

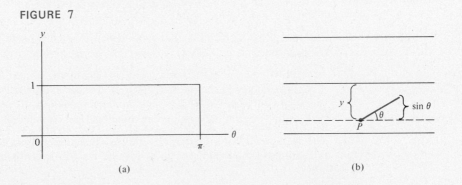

(a) (b)

16. (Buffon Needle Problem) A needle of length one unit is dropped on a floor which is ruled with parallel lines, one unit apart. See Fig. 7(b). Let P be the lowest point of the needle, y the distance of P from the above ruled line, and θ the angle that the needle makes with a line parallel to the ruled lines. Show that the needle touches a ruled line if and only if $y \leq \sin \theta$. Conclude that the probability of the needle touching a ruled line is the probability found in Exercise 15.

17. The lifetime of a certain TV picture tube is an exponential random variable with an expected value of 5 years. Suppose the tube manufacturer sells the tube for $100 but will give a complete refund if the tube burns out within 3 years. Then the revenue the manufacturer receives on each tube is a discrete random variable Y with values 100 and 0. Determine the expected revenue per tube.

18. The condenser motor in an air conditioner costs $300 to replace, but a home air conditioning service will guarantee to replace it free when it burns out if you will pay an annual insurance premium of $25. Suppose that the lifespan of the motor is an exponential random variable with an expected life of 10 years. Should you take out the insurance for the first year? [Hint: Consider the random variable Y such that $Y = 300$ if the motor burns out during the year and $Y = 0$ otherwise. Compare $E(Y)$ with the cost of one year's insurance.]

19. An exponential random variable X has been used to model the relief times (in minutes) of arthritic patients who have taken an analgesic for their pain. Suppose the density function for X is $f(x) = ke^{-kx}$ and suppose that a certain analgesic provides relief within 4 minutes for 75% of a large group of patients. Then one may estimate that $\Pr(X \leq 4) = .75$. Use this estimate to find an approximate value for k. [Hint: First show that $\Pr(X \leq 4) = 1 - e^{-4k}$.]

20. A piece of new equipment has a useful life of X thousand hours, where X is a random variable with the gamma density function $f(x) = .01xe^{-x/10}$, $x \geq 0$. A manufacturer expects the machine to generate $5000 of additional income for every thousand hours of use, but the machine costs $60,000. Should the manufacturer purchase the new equipment? [Hint: Compute the expected value of the additional earnings generated by the machine.]

21. Extensive records are kept of the lifespans (in months) of a certain product, and a relative frequency histogram is constructed from the data, using areas to represent relative frequencies (as in Fig. 4 in Section 1). It turns out that the upper boundary of the relative frequency histogram is approximated closely by the graph of the function

$$f(x) = \frac{1}{8\sqrt{2\pi}} e^{-(1/2)[(x-50)/8]^2}.$$

Determine the probability that the lifespan of such a product is between 30 and 50 months.

22. A certain machine part has a nominal length of 80 mm, with a tolerance of $\pm.05$ mm. Suppose the actual length of the parts supplied is a normal random variable with mean 79.99 mm and standard deviation .02 mm. How many parts in a lot of 1000 should you expect to lie outside the tolerance limits?

23. The men hired by a certain city police department must be at least 69 inches tall. Suppose the heights of adult men in the city are normally distributed with

$\mu = 70$ inches and $\sigma = 2$ inches. What percentage of the men are tall enough to be eligible for recruitment by the police department?

24. Suppose the police force in Exercise 23 maintains the same height requirements for women as men, and suppose the heights of women in the city are normally distributed, with $\mu = 65$ inches and $\sigma = 1.6$ inches. What percentage of the women are eligible for recruitment?

25. Let Z be a standard normal random variable. Find the number a such that $\Pr(a \leq Z) = .40$.

26. Scores on a school's entrance exam are normally distributed, with $\mu = 500$ and $\sigma = 100$. If the school wishes to admit only the students in the top 40%, what should be the cutoff grade?

27. It is useful in some applications to know that about 68% of the area under the standard normal curve lies between -1 and 1.

    (a) Verify this statement.

    (b) Let X be a normal random variable with expected value μ and variance σ^2. Compute $\Pr(\mu - \sigma \leq X \leq \mu + \sigma)$.

28. (a) Show that about 95% of the area under the standard normal curve lies between -2 and 2.

    (b) Let X be a normal random variable with expected value μ and variance σ^2. Compute $\Pr(\mu - 2\sigma \leq X \leq \mu + 2\sigma)$.

29. The Chebyshev Inequality says that for any random variable X with expected value μ and standard deviation σ,

$$\Pr(\mu - n\sigma \leq X \leq \mu + n\sigma) \geq 1 - \frac{1}{n^2}.$$

    (a) Take $n = 2$. Apply the Chebyshev Inequality to an exponential random variable.

    (b) By integrating, find the exact value of the probability in (a).

30. Do the same as in Exercise 29 with a normal random variable.

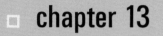

chapter 13

TAYLOR POLYNOMIALS AND INFINITE SERIES

In earlier chapters, we introduced the functions e^x, $\ln x$, $\sin x$, $\cos x$, $\tan x$. Whenever we needed the value of one of these functions for a particular value of x, such as $e^{.023}$, $\ln 5.8$, $\sin .25$, we had to use either a table or a scientific calculator. Now we shall take up the problem of numerically computing the values of such functions for particular choices of the variable x. The computational methods developed have many applications—for example, to differential equations and probability theory.

13.1. Taylor Polynomials

In this section we take up the problem of approximating the value $f(a)$ of a function at $x = a$. At first sight this might seem a frivolous problem. Why not just evaluate the function directly? To appreciate the difficulties involved, consider the familiar function $f(x) = e^x$. Suppose that we wish to evaluate it for $x = \pm.3$, $\pm.2$, $\pm.1$, 0. We could, of course, consult a table or use a scientific calculator to determine the desired values. However, using a table or calculator merely begs the question. After all, how is the table compiled? How does the calculator arrive at its result? What could we do if neither were available? To answer these questions, let us use some calculus.

In Fig. 1 we have drawn the graph of $f(x) = e^x$ for values of x near 0. We have drawn the tangent line at $x = 0$. The slope of the tangent line is $f'(0) = 1$. Moreover, the line passes through the point $(0, f(0)) = (0, 1)$. So the equation of the tangent line is

$$y - f(0) = f'(0)(x - 0)$$

$$y = f(0) + f'(0)x$$

$$y = 1 + x.$$

From our discussion of the derivative, we know that the tangent line at $x = 0$ closely approximates the graph for values of x near 0. This suggests that we approximate $f(x) = e^x$ by the function $p_1(x) = 1 + x$. In Table 1 we have listed some exact values of e^x (to five decimal places) and have tabulated the corresponding values of $p_1(x)$. Note the reasonable agreement between e^x and $p_1(x)$.

The discussion above suggests that, in general, for values of x near 0, a given function $f(x)$ may be approximated by the function

$$p_1(x) = f(0) + f'(0)x,$$

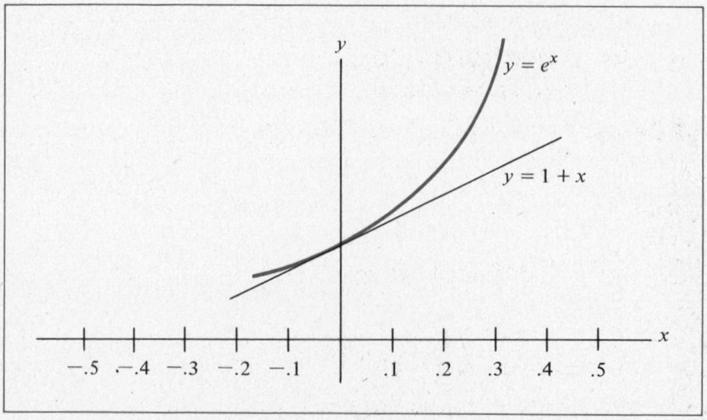

FIGURE 1

TABLE 1

x	e^x	$p_1(x)$
.3	1.34986	1.3
.2	1.22140	1.2
.1	1.10517	1.1
0	1.00000	1.0
−.1	.90484	.9
−.2	.81873	.8
−.3	.74082	.7

which is called the *first Taylor polynomial of* $f(x)$ *at* $x = 0$. The graph of $p_1(x)$ is just the tangent line to $y = f(x)$ at $x = 0$.

The first Taylor polynomial "resembles" $f(x)$ near $x = 0$ in the following sense. If we compute $p_1(0)$ and $p_1'(0)$, we find that

$$p_1(0) = f(0),$$

$$p_1'(0) = f'(0).$$

In other words, $p_1(x)$ coincides with $f(x)$ in both its value at $x = 0$ and the value of its derivative at $x = 0$. This suggests that to approximate $f(x)$ even more closely at $x = 0$, we look for a polynomial which coincides with $f(x)$ in its value at $x = 0$ and in the values of its first *and second* derivatives at $x = 0$. Yet a further approximation can be obtained by going out to third derivatives, and so forth.

Let us now show how to construct such approximating polynomials. Recall that a *polynomial* is a function of the form

$$p(x) = a_0 + a_1 x + \ldots + a_n x^n,$$

where $a_0, a_1, \ldots, a_n$ are given numbers.

If we are given values for $p(0), p'(0), \ldots, p^{(n)}(0)$, then it is always possible to find a polynomial $p(x)$ having these derivatives at $x = 0$.

EXAMPLE 1   Find a polynomial $p(x)$ of degree 3 such that $p(0) = 1$, $p'(0) = -2$, $p''(0) = 7$, $p'''(0) = -5$.

Solution   Since $p(x)$ is to have degree 3, we must have

$$p(x) = a_0 + a_1 x + a_2 x^2 + a_3 x^3.$$

Then

$$p'(x) = a_1 + 2a_2 x + 3a_3 x^2,$$

$$p''(x) = 2a_2 + 3 \cdot 2a_3 x,$$

$$p'''(x) = 3 \cdot 2a_3.$$

Setting $x = 0$ in each of the above four expressions gives

$$1 = p(0) = a_0,$$
$$-2 = p'(0) = a_1,$$
$$7 = p''(0) = 2a_2,$$
$$-5 = p'''(0) = 3 \cdot 2a_3.$$

Therefore,

$$a_0 = 1, \qquad a_1 = -2, \qquad a_2 = \frac{7}{2}, \qquad a_3 = \frac{-5}{3 \cdot 2},$$

and, rewriting the coefficients slightly, we have

$$p(x) = 1 + \frac{(-2)}{1} x + \frac{7}{1 \cdot 2} x^2 + \frac{(-5)}{1 \cdot 2 \cdot 3} x^3.$$

The form in which we have written $p(x)$ clearly exhibits the values $1, -2, 7, -5$ of the derivatives which were given.

The procedure of Example 1 can be generalized to yield the following result.

Suppose that $p(x)$ is a polynomial of degree at most n and that $p(0) = b_0$, $p'(0) = b_1$, $p''(0) = b_2, \ldots, p^{(n)}(0) = b_n$, where $p^{(n)}(0)$ denotes the value of the nth derivative of $p(x)$ at $x = 0$. Then $p(x)$ is given by the formula

$$p(x) = b_0 + \frac{b_1}{1} x + \frac{b_2}{1 \cdot 2} x^2 + \ldots + \frac{b_n}{1 \cdot 2 \cdot \ldots \cdot n} x^n.$$

EXAMPLE 2   Find a polynomial $p(x)$ of degree 5 such that $p(0) = 1$, $p'(0) = -1$, $p''(0) = 1$, $p'''(0) = -1$, $p^{(4)}(0) = 1$, $p^{(5)}(0) = -1$.

Solution   Here $b_0 = b_2 = b_4 = 1$, $b_1 = b_3 = b_5 = -1$. Then

$$p(x) = 1 + \frac{(-1)}{1} x + \frac{1}{1 \cdot 2} x^2 + \frac{(-1)}{1 \cdot 2 \cdot 3} x^3 + \frac{1}{1 \cdot 2 \cdot 3 \cdot 4} x^4 + \frac{(-1)}{1 \cdot 2 \cdot 3 \cdot 4 \cdot 5} x^5$$

$$= 1 - x + \frac{1}{2} x^2 - \frac{1}{6} x^3 + \frac{1}{24} x^4 - \frac{1}{120} x^5.$$

Another form for polynomials is useful when we are concerned with values near $x = a$. A polynomial written in the form

$$p(x) = a_0 + a_1(x - a) + a_2(x - a)^2 + \ldots + a_n(x - a)^n$$

is called a *polynomial in* $x - a$. Polynomials in $x - a$ can be specified by giving the values of their derivatives at $x = a$, just as polynomials in x are specified by giving the values of their derivatives at $x = 0$.

Suppose $p(x)$ is a polynomial of degree at most n such that

$$p(a) = b_0, \quad p'(a) = b_1, \quad p''(a) = b_2, \ldots, \quad p^{(n)}(a) = b_n.$$

Then $p(x)$ is given by the formula

$$p(x) = b_0 + \frac{b_1}{1}(x - a) + \frac{b_2}{1 \cdot 2}(x - a)^2 + \ldots + \frac{b_n}{1 \cdot 2 \cdot \ldots \cdot n}(x - a)^n. \quad (1)$$

Let us now turn to the central topic of this section, the approximation of functions by polynomials. Suppose that we are given a function $f(x)$ and a value of x, say $x = a$. We wish to approximate $f(x)$ by a polynomial so that the behavior of $f(x)$ at $x = a$ is closely mirrored by the polynomial. As we have seen, the behavior of $f(x)$ near $x = a$ is determined by the values of the derivatives of $f(x)$ at $x = a$. This suggests that we approximate $f(x)$ by a polynomial $p(x)$ so that the values of the derivatives of $p(x)$ at $x = a$ are the same as those of $f(x)$. More specifically, suppose that we wish to approximate $f(x)$ by a polynomial $p(x)$ of degree at most n. Then we should choose $p(x)$ so that

$$p(a) = f(a),$$
$$p'(a) = f'(a),$$
$$p''(a) = f''(a),$$
$$\vdots$$
$$p^{(n)}(a) = f^{(n)}(a).$$

Applying formula (1), we obtain the following result:

Let $f(x)$ be a given function and let $p_n(x)$ be the polynomial defined by

$$p_n(x) = f(a) + \frac{f'(a)}{1}(x - a) + \frac{f''(a)}{1 \cdot 2}(x - a)^2 + \ldots$$
$$+ \frac{f^{(n)}(a)}{1 \cdot 2 \cdot \ldots \cdot n}(x - a)^n. \quad (2)$$

Then

$$p_n(a) = f(a), \quad p_n'(a) = f'(a), \quad \ldots, \quad p_n^{(n)}(a) = f^{(n)}(a).$$

The polynomial $p_n(x)$ of (2) is called the nth *Taylor polynomial* of $f(x)$ at $x = a$. This polynomial closely mirrors the behavior of $f(x)$ near $x = a$.

EXAMPLE 3   Calculate the second Taylor polynomial of $f(x) = \sqrt{x}$ at $x = 1$.

Solution Here $a = 1$. Since we want the second Taylor polynomial, we must calculate the values of $f(x)$ and of its first two derivatives at $x = 1$.

$$f(x) = x^{1/2}, \quad f'(x) = \tfrac{1}{2}x^{-1/2}, \quad f''(x) = -\tfrac{1}{4}x^{-3/2},$$

$$f(1) = 1, \quad f'(1) = \tfrac{1}{2}, \quad f''(1) = -\tfrac{1}{4}.$$

Therefore, the desired Taylor polynomial is

$$p_2(x) = 1 + \frac{\tfrac{1}{2}}{1}(x - 1) + \frac{(-\tfrac{1}{4})}{1 \cdot 2}(x - 1)^2$$

$$= 1 + \tfrac{1}{2}(x - 1) - \tfrac{1}{8}(x - 1)^2.$$

EXAMPLE 4 Use the result of Example 3 to estimate $\sqrt{1.02}$.

Solution Since 1.02 is close to 1, $p_2(1.02)$ gives a good approximation to $f(1.02) = \sqrt{1.02}$.

$$p_2(1.02) = 1 + \tfrac{1}{2}(1.02 - 1) - \tfrac{1}{8}(1.02 - 1)^2$$

$$= 1 + \tfrac{1}{2}(.02) - \tfrac{1}{8}(.02)^2$$

$$= 1 + .01 - .00005$$

$$= 1.00995.$$

This approximation is exceptionally good, since to nine decimal places $\sqrt{1.02} = 1.009950494$.

EXAMPLE 5 Determine the first three Taylor polynomials of $f(x) = e^x$ at $x = 0$ and sketch their graphs.

Solution Since all derivatives of e^x are e^x, we see that

$$f(0) = f'(0) = f''(0) = f'''(0) = e^0 = 1.$$

Thus, the desired Taylor polynomials are

$$p_1(x) = 1 + \frac{1}{1} \cdot x = 1 + x,$$

$$p_2(x) = 1 + \frac{1}{1} \cdot x + \frac{1}{1 \cdot 2}x^2 = 1 + x + \frac{1}{2}x^2,$$

$$p_3(x) = 1 + \frac{1}{1} \cdot x + \frac{1}{1 \cdot 2}x^2 + \frac{1}{1 \cdot 2 \cdot 3}x^3 = 1 + x + \frac{1}{2}x^2 + \frac{1}{6}x^3.$$

The graphs of these polynomials can be sketched by the curve-sketching techniques of Chapter 2. (See Fig. 2.) Notice that successive Taylor polynomials provide successively better approximations to $f(x)$ near $x = 0$.

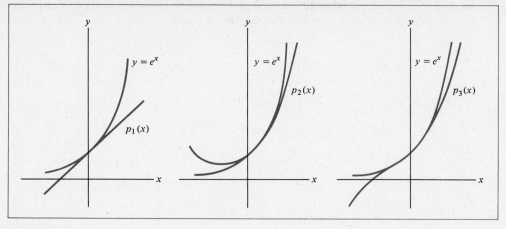

FIGURE 2

EXAMPLE 6   Determine the nth Taylor polynomial for $f(x) = 1/(1 - x)$ about $x = 0$.

Solution

$$f(x) = (1 - x)^{-1}, \qquad\qquad f(0) = 1,$$

$$f'(x) = (1 - x)^{-2}, \qquad\qquad f'(0) = 1,$$

$$f''(x) = 2(1 - x)^{-3}, \qquad\qquad f''(0) = 2,$$

$$f'''(x) = 2 \cdot 3(1 - x)^{-4}, \qquad\qquad f'''(0) = 2 \cdot 3,$$

$$f^{(4)}(x) = 2 \cdot 3 \cdot 4(1 - x)^{-5}, \qquad\qquad f^{(4)}(0) = 2 \cdot 3 \cdot 4,$$

$$\vdots \qquad\qquad\qquad\qquad \vdots$$

$$f^{(n)}(x) = 2 \cdot 3 \cdot 4 \cdot \ldots \cdot n(1 - x)^{-n-1}, \qquad f^{(n)}(0) = 2 \cdot 3 \cdot \ldots \cdot n,$$

Therefore,

$$p_n(x) = 1 + \frac{1}{1} x + \frac{2}{1 \cdot 2} x^2 + \frac{2 \cdot 3}{1 \cdot 2 \cdot 3} x^3 + \frac{2 \cdot 3 \cdot 4}{1 \cdot 2 \cdot 3 \cdot 4} x^4 + \ldots$$

$$+ \frac{2 \cdot 3 \cdot 4 \cdot \ldots \cdot n}{1 \cdot 2 \cdot 3 \cdot \ldots \cdot n} x^n$$

$$= 1 + x + x^2 + x^3 + x^4 + \ldots + x^n.$$

PRACTICE PROBLEMS 1

1. (a) Determine the third Taylor polynomial of $f(x) = \cos x$ about $x = 0$.
   (b) Use the result of (a) to estimate $\cos(.12)$.

2. Determine all Taylor polynomials of $f(x) = x^3 - 3x - 17$ about $x = 3$.

EXERCISES 1

In Exercises 1–6 determine a polynomial $p(x)$ having the given properties.

1. $p(0) = -1, p'(0) = 3, p''(0) = -2$

2. $p(0) = \frac{1}{2}, p'(0) = -1, p''(0) = -4$

3. $p(1) = 0, p'(1) = -2, p''(1) = 0, p'''(1) = -3$

4. $p(-2) = 4, p'(-2) = 0, p''(-2) = 1, p'''(-2) = 2$

5. $p(0) = 1, p'(0) = 2, p''(0) = 3, p'''(0) = 4, p^{(4)}(0) = 5$

6. $p(-5) = p'(-5) = p''(-5) = p'''(-5) = p^{(4)}(-5) = 12$

In Exercises 7–16 determine the third Taylor polynomial of the given function about the given value of a.

7. $f(x) = \ln x, a = 1$

8. $f(x) = \sin x, a = 0$

9. $f(x) = \cos x^2, a = 0$

10. $f(x) = 1/x, a = 2$

11. $f(x) = x^4, a = 0$

12. $f(x) = xe^x, a = 0$

13. $f(x) = x^2, a = 3$

14. $f(x) = 1/(2 - x), a = 1$

15. $f(x) = e^{x^2}, a = 0$

16. $f(x) = \sqrt{1 - x}, a = 0$

17. Determine the fourth Taylor polynomial of $f(x) = e^x$ at $x = 0$ and use it to estimate $e^{.01}$.

18. Determine the fourth Taylor polynomial of $f(x) = \ln(1 - x)$ at $x = 0$ and use it to estimate $\ln(.9)$.

19. Determine the second Taylor polynomial of $f(x) = \sqrt{x}$ at $x = 9$ and use it to estimate $\sqrt{8.8}$.

20. Determine the second Taylor polynomial of $f(x) = \tan x$ at $x = 0$ and use it to estimate $\tan(.1)$.

21. Sketch the graph of $f(x) = 1/(1 - x)$ and its first three Taylor polynomials at $x = 0$.

22. Sketch the graph of $f(x) = \cos x$ and its first three Taylor polynomials at $x = 0$.

23. Determine the nth Taylor polynomial for $f(x) = e^x$ at $x = 0$.

24. Determine the nth Taylor polynomial for $f(x) = 1/x$ at $x = 1$.

25. Determine the nth Taylor polynomial for $f(x) = \ln(1 - x)$ at $x = 0$.

26. Determine all Taylor polynomials for $f(x) = x^2 + 2x + 1$ at $x = 0$.

SOLUTIONS TO PRACTICE PROBLEMS 1

1. (a) $f(x) = \cos x,$ $\qquad f(0) = 1,$

   $f'(x) = -\sin x,$ $\qquad f'(0) = 0,$

   $f''(x) = -\cos x,$ $\qquad f''(0) = -1,$

   $f'''(x) = \sin x,$ $\qquad f'''(0) = 0.$

   Therefore,

   $$p_3(x) = 1 + \frac{0}{1} \cdot x + \frac{-1}{1 \cdot 2} x^2 + \frac{0}{1 \cdot 2 \cdot 3} x^3$$

   $$= 1 - \frac{1}{2} x^2.$$

   [Notice that here the third Taylor polynomial is actually a polynomial of degree 2. The important thing about $p_3(x)$ is not its degree but rather the fact that it agrees with $f(x)$ at $x = 0$ up to its third derivative.]

   (b) By (a), $\cos x \approx 1 - \frac{1}{2}x^2$ when x is near 0. Therefore,

   $$\cos(.12) \approx 1 - \frac{1}{2}(.12)^2 = .9928.$$

   (Note: To 10 decimal places, $\cos .12 = .9928086359.$)

2. $f(x) = x^3 - 3x - 17,$ $\qquad f(3) = 1,$

   $f'(x) = 3x^2 - 3,$ $\qquad f'(3) = 24,$

   $f''(x) = 6x,$ $\qquad f''(3) = 18,$

   $f'''(x) = 6,$ $\qquad f'''(3) = 6,$

   $f^{(4)}(x) = 0,$ $\qquad f^{(4)}(3) = 0.$

   Clearly, $f^{(n)}(3) = 0$ for all $n \geq 4$. Therefore,

   $$p_1(x) = 1 + \frac{24}{1}(x - 3) = 1 + 24(x - 3),$$

   $$p_2(x) = 1 + \frac{24}{1}(x - 3) + \frac{18}{1 \cdot 2}(x - 3)^2 = 1 + 24(x - 3) + 9(x - 3)^2,$$

   $$p_3(x) = 1 + \frac{24}{1}(x - 3) + \frac{18}{1 \cdot 2}(x - 3)^2 + \frac{6}{1 \cdot 2 \cdot 3}(x - 3)^3$$

   $$= 1 + 24(x - 3) + 9(x - 3)^2 + (x - 3)^3.$$

   Since $f^{(n)}(3) = 0$ for all $n \geq 4$, we see that

   $$p_n(x) = 1 + 24(x - 3) + 9(x - 3)^2 + (x - 3)^3, \quad \text{for all } n \geq 4.$$

13.2. The Newton-Raphson Algorithm

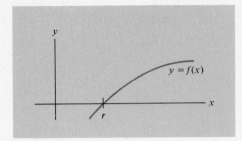

FIGURE 1

Many applications of mathematics involve the solution of equations. Often one has a function $f(x)$ and must find a value of x, say $x = r$, such that $f(r) = 0$. Such a value of x is called a *zero* of the function or, equivalently, a *root* of the equation $f(x) = 0$. Graphically, a zero of $f(x)$ is a value of x where the graph of $y = f(x)$ crosses the x-axis (see Fig. 1). When $f(x)$ is a polynomial, it is sometimes possible to factor $f(x)$ and quickly discover the zeros of $f(x)$. Unfortunately, in most realistic applications there is no simple way to locate zeros. However, there are several methods for finding an approximate value of a zero to any desired degree of accuracy. We shall describe one such method—the Newton-Raphson algorithm.

Suppose that we know that a zero of $f(x)$ is approximately x_0. The idea of the Newton-Raphson algorithm is to obtain an even better approximation to the zero by replacing $f(x)$ by its first Taylor polynomial at x_0—that is, by

$$p(x) = f(x_0) + \frac{f'(x_0)}{1}(x - x_0).$$

Since $p(x)$ closely resembles $f(x)$ near $x = x_0$, the zero of $f(x)$ should be close to the zero of $p(x)$. But solving the equation $p(x) = 0$ for x gives

$$f(x_0) + f'(x_0)(x - x_0) = 0$$
$$xf'(x_0) = f'(x_0)x_0 - f(x_0)$$
$$x = x_0 - \frac{f(x_0)}{f'(x_0)}.$$

That is, if x_0 is an approximation to the zero r, then the number

$$x_1 = x_0 - \frac{f(x_0)}{f'(x_0)} \tag{1}$$

FIGURE 2

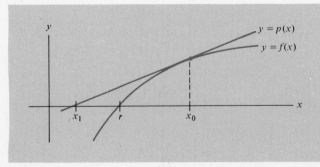

generally provides an improved approximation.

We may visualize the situation geometrically as in Fig. 2. The first Taylor polynomial $p(x)$ at x_0 has as its graph the tangent line to $y = f(x)$ at the point $(x_0, f(x_0))$. The value of x for which $p(x) = 0$—that is, $x = x_1$—corresponds to the point where the tangent line crosses the x-axis.

Now let us use x_1 in place of x_0 as an approximation to the zero r. We obtain a new approximation x_2 from x_1 in the same way we obtained x_1 from x_0—namely,

$$x_2 = x_1 - \frac{f(x_1)}{f'(x_1)}.$$

We may repeat this process over and over. At each stage a new approximation x_{new} is obtained from the old approximation x_{old} by the formula

$$x_{\text{new}} = x_{\text{old}} - \frac{f(x_{\text{old}})}{f'(x_{\text{old}})}.$$

In this way we obtain a sequence of approximations $x_0, x_1, x_2, \ldots$, which usually approach as close to r as desired. (See Fig. 3.)

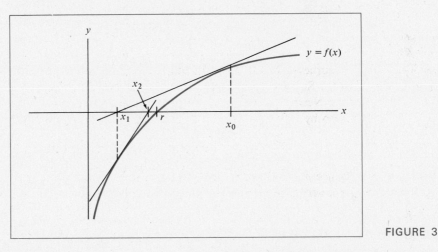

FIGURE 3

EXAMPLE 1 The polynomial $f(x) = x^3 - x - 2$ has a zero between 1 and 2. Let $x_0 = 1$ and find the next three approximations of the zero of $f(x)$ using the Newton-Raphson algorithm.

Solution Since $f'(x) = 3x^2 - 1$, formula (1) becomes

$$x_1 = x_0 - \frac{x_0^3 - x_0 - 2}{3x_0^2 - 1}.$$

With $x_0 = 1$, we have

$$x_1 = 1 - \frac{1^3 - 1 - 2}{3(1)^2 - 1} = 1 - \frac{-2}{2} = 2,$$

$$x_2 = 2 - \frac{2^3 - 2 - 2}{3(2)^2 - 1} = 2 - \frac{4}{11} = \frac{18}{11},$$

$$x_3 = \frac{18}{11} - \frac{(\frac{18}{11})^3 - \frac{18}{11} - 2}{3(\frac{18}{11})^2 - 1} \approx 1.530.$$

The actual value of r to three decimal places is 1.521.

EXAMPLE 2 Use four repetitions of the Newton-Raphson algorithm to approximate $\sqrt{2}$.

Solution $\sqrt{2}$ is a zero of the function $f(x) = x^2 - 2$. Since $\sqrt{2}$ clearly lies between 1 and 2, let us take our initial approximation as $x_0 = 1$. ($x_0 = 2$ would do just as well.) Since $f'(x) = 2x$, we have

$$x_1 = x_0 - \frac{x_0^2 - 2}{2x_0}$$

$$= 1 - \frac{1^2 - 2}{2(1)} = 1 - \left(-\frac{1}{2}\right) = 1.5,$$

$$x_2 = 1.5 - \frac{(1.5)^2 - 2}{2(1.5)} \approx 1.4167,$$

$$x_3 = 1.4167 - \frac{(1.4167)^2 - 2}{2(1.4167)} \approx 1.41422,$$

$$x_4 = 1.41422 - \frac{(1.41422)^2 - 2}{2(1.41422)} \approx 1.41421.$$

This approximation to $\sqrt{2}$ is correct to five decimal places.

EXAMPLE 3 Approximate the zeros of the polynomial $x^3 + x + 3$.

Solution By applying our curve-sketching techniques, we can make a rough sketch of the graph of $y = x^3 + x + 3$, as in Fig. 4. The graph crosses the x-axis between $x = -2$ and $x = -1$. So the polynomial has one zero lying between -2 and -1.

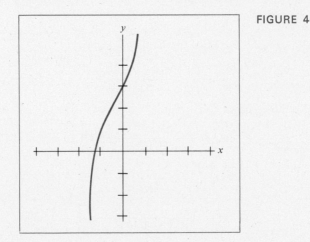

FIGURE 4

Let us therefore set $x_0 = -1$. Since $f'(x) = 3x^2 + 1$, we have

$$x_1 = x_0 - \frac{x_0^3 + x_0 + 3}{3x_0^2 + 1}$$

$$= -1 - \frac{(-1)^3 + (-1) + 3}{3(-1)^2 + 1} = -1.25,$$

$$x_2 = -1.25 - \frac{(-1.25)^3 + (-1.25) + 3}{3(-1.25)^2 + 1} \approx -1.21429,$$

$$x_3 \approx -1.21341,$$

$$x_4 \approx -1.21342.$$

Therefore the zero of the given polynomial is approximately -1.21342.

EXAMPLE 4 Approximate the positive solution of $e^x - 4 = x$.

Solution The rough sketches of the two graphs in Fig. 5 indicate that the solution lies near 2. Let $f(x) = e^x - 4 - x$. Then the solution of the original equation will be a zero of $f(x)$. Let's apply the Newton-Raphson algorithm to $f(x)$ with $x_0 = 2$.

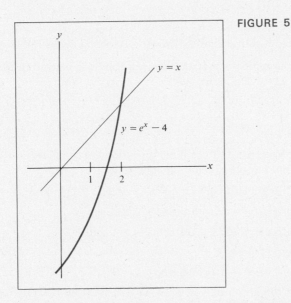

FIGURE 5

Since $f'(x) = e^x - 1$,

$$x_1 = x_0 - \frac{e^{x_0} - 4 - x_0}{e^{x_0} - 1}$$

$$= 2 - \frac{e^2 - 4 - 2}{e^2 - 1} = 2 - \frac{1.38906}{6.38906} \approx 1.78,$$

$$x_2 = 1.78 - \frac{e^{1.78} - 4 - (1.78)}{e^{1.78} - 1} = 1.78 - \frac{.14986}{4.92986} \approx 1.75,$$

$$x_3 = 1.75 - \frac{e^{1.75} - 4 - (1.75)}{e^{1.75} - 1} = 1.75 - \frac{.0046}{4.7546} \approx 1.749.$$

Therefore, an approximate solution is $x = 1.749$.

Comments

1. The values of successive approximations will depend on the extent of roundoff used during the calculation. It is best to use a computer or hand calculator, since they carry out numbers to eight or more places of accuracy.

2. The Newton-Raphson algorithm is an excellent computational tool. However, in some cases it will not work. For instance, if $f'(x_n) = 0$ for some approximation x_n, then there is no way to compute the next approximation. Other instances in which the algorithm fails are presented in Exercises 15 and 16.

3. It can be shown that if $f(x)$, $f'(x)$, and $f''(x)$ are continuous near r [a zero of $f(x)$] and $f'(r) \neq 0$, then the Newton-Raphson algorithm will definitely work provided that the initial approximation x_0 is not too far away.

PRACTICE PROBLEMS 2

1. Use three repetitions of the Newton-Raphson algorithm to estimate $\sqrt[3]{7}$.

2. Use three repetitions of the Newton-Raphson algorithm to estimate the zeros of $f(x) = 2x^3 + 3x^2 + 6x - 3$.

EXERCISES 2

In Exercises 1–8 use three repetitions of the Newton-Raphson algorithm to approximate the following:

1. $\sqrt{5}$ 2. $\sqrt{7}$ 3. $\sqrt[3]{6}$ 4. $\sqrt[3]{11}$

5. The zero of $x^2 - x - 5$ between 2 and 3.

6. The zero of $x^2 + 3x - 11$ between -5 and -6.

7. The zero of $\sin x + x^2 - 1$ near $x_0 = 0$.

8. The zero of $e^x + 10x - 3$ near $x_0 = 0$.

9. Sketch the graph of $y = x^3 + 2x + 2$ and use the Newton-Raphson algorithm (three repetitions) to approximate all x-intercepts.

10. Sketch the graph of $y = x^3 + x - 1$ and use the Newton-Raphson algorithm (three repetitions) to approximate all x-intercepts.

11. Use the Newton-Raphson algorithm to find an approximate solution to $e^{-x} = x^2$.

12. Use the Newton-Raphson algorithm to find an approximate solution to $e^{5-x} = 10 - x$.

13. What special occurrence takes place when the Newton-Raphson algorithm is applied to a linear function, $f(x) = mx + b$ with $m \neq 0$?

14. What happens when the first approximation, x_0, is actually a zero of $f(x)$?

Exercises 15 and 16 present two examples in which successive repetitions of the Newton-Raphson algorithm do not approach a root.

15. Apply the Newton-Raphson algorithm to the function $f(x) = x^{1/3}$ whose graph is drawn in Fig. 6(a). Let $x_0 = 1$.

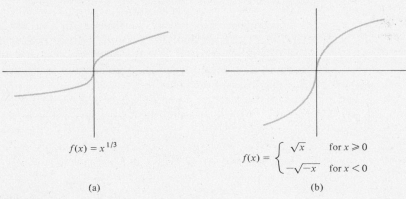

$$f(x) = x^{1/3}$$

(a)

$$f(x) = \begin{cases} \sqrt{x} & \text{for } x \geqslant 0 \\ -\sqrt{-x} & \text{for } x < 0 \end{cases}$$

(b)

FIGURE 6

16. Apply the Newton-Raphson algorithm to the function whose graph is drawn in Fig. 6(b). Use $x_0 = 1$.

SOLUTIONS TO PRACTICE PROBLEMS 2

1. We wish to approximate a zero of $f(x) = x^3 - 7$. Since $f(1) = -6 < 0$ and $f(2) = 1 > 0$, the graph of $f(x)$ crosses the x-axis somewhere between $x = 1$ and $x = 2$. Take $x_0 = 2$ as the initial approximation to the zero. Since $f'(x) = 3x^2$, we have

$$x_1 = x_0 - \frac{x_0^3 - 7}{3x_0^2}$$

$$= 2 - \frac{2^3 - 7}{3(2)^2} = \frac{23}{12} \approx 1.9167,$$

$$x_2 = 1.9167 - \frac{(1.9167)^3 - 7}{3(1.9167)^2} \approx 1.91294,$$

$$x_3 = 1.91294 - \frac{(1.91294)^3 - 7}{3(1.91294)^2} \approx 1.91293.$$

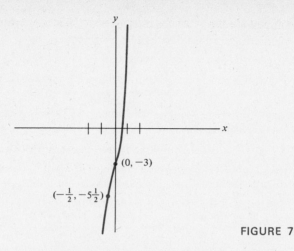

(0, −3)

$(-\frac{1}{2}, -5\frac{1}{2})$

FIGURE 7

2. As a preliminary step, we use the methods of Chapter 2 to sketch the graph of $f(x)$ (Fig. 7). We see that $f(x)$ has a zero which occurs for a positive value of x. Since $f(0) = -3$ and $f(1) = 8$, the graph crosses the x-axis between 0 and 1. Let us choose $x_0 = 0$ as our initial approximation to the zero of $f(x)$. Since $f'(x) = 6x^2 + 6x + 6$. We then have

$$x_1 = x_0 - \frac{2x_0^3 + 3x_0^2 + 6x_0 - 3}{6x_0^2 + 6x_0 + 6}$$

$$= 0 - \frac{-3}{6} = \frac{1}{2},$$

$$x_2 = \frac{1}{2} - \frac{1}{\frac{21}{2}} = \frac{1}{2} - \frac{2}{21} = \frac{17}{42} \approx .40476.$$

Continuing, we find $x_3 \approx .39916$.

13.3. Infinite Series

An *infinite series* is an infinite addition of numbers

$$a_1 + a_2 + a_3 + a_4 + \dots.$$

Here are some examples:

$$1 + \tfrac{1}{2} + \tfrac{1}{4} + \tfrac{1}{8} + \tfrac{1}{16} + \dots, \tag{1}$$

$$1 + 1 + 1 + 1 + \dots, \tag{2}$$

$$1 - 1 + 1 - 1 + \dots. \tag{3}$$

To certain infinite series it is possible to associate a "sum." To illustrate how this is done, let us consider the infinite series (1). If we add up the first two, three, four, five, and six terms of the infinite series (1), we obtain

$$1 + \tfrac{1}{2} = 1\tfrac{1}{2},$$
$$1 + \tfrac{1}{2} + \tfrac{1}{4} = 1\tfrac{3}{4},$$
$$1 + \tfrac{1}{2} + \tfrac{1}{4} + \tfrac{1}{8} = 1\tfrac{7}{8},$$
$$1 + \tfrac{1}{2} + \tfrac{1}{4} + \tfrac{1}{8} + \tfrac{1}{16} = 1\tfrac{15}{16},$$
$$1 + \tfrac{1}{2} + \tfrac{1}{4} + \tfrac{1}{8} + \tfrac{1}{16} + \tfrac{1}{32} = 1\tfrac{31}{32}.$$

It appears from the above calculations that by increasing the number of terms, we bring the sum arbitrarily close to 2. Indeed, this is supported by further calculation. For example,

$$\underbrace{1 + \frac{1}{2} + \frac{1}{4} + \ldots + \frac{1}{2^9}}_{10 \text{ terms}} = 2 - \frac{1}{2^9} \approx 1.998047,$$

$$\underbrace{1 + \frac{1}{2} + \frac{1}{4} + \ldots + \frac{1}{2^{19}}}_{20 \text{ terms}} = 2 - \frac{1}{2^{19}} \approx 1.999998,$$

$$\underbrace{1 + \frac{1}{2} + \frac{1}{4} + \ldots + \frac{1}{2^{n-1}}}_{n \text{ terms}} = 2 - \frac{1}{2^{n-1}}.$$

Therefore, it seems reasonable to assign the infinite series (1) the "sum" 2:

$$1 + \tfrac{1}{2} + \tfrac{1}{4} + \tfrac{1}{8} + \tfrac{1}{16} + \ldots = 2. \tag{4}$$

The sum of the first n terms of an infinite series is called its nth *partial sum* and is denoted S_n. In series (1), however, we were very fortunate that the partial sums approached a limiting value, 2. This is not always the case. For example, consider the infinite series (2). If we form the first few partial sums, we get

$$S_2 = 1 + 1 \qquad\qquad = 2,$$
$$S_3 = 1 + 1 + 1 \qquad\quad = 3,$$
$$S_4 = 1 + 1 + 1 + 1 = 4.$$

And we see that these sums do not approach any limit. Rather, they become larger and larger, eventually exceeding any specified number.

The partial sums need not grow without bound in order that an infinite series not have a sum. For example, consider the infinite series (3). Here the sums of initial terms are

$$S_2 = 1 - 1 \qquad\qquad\qquad = 0,$$
$$S_3 = 1 - 1 + 1 \qquad\qquad\quad = 1,$$
$$S_4 = 1 - 1 + 1 - 1 \qquad\quad = 0,$$
$$S_5 = 1 - 1 + 1 - 1 + 1 = 1,$$

and so forth. The sums oscillate between 0 and 1 and do not approach a limit. So the infinite series (3) has no sum.

An infinite series whose partial sums approach a limit is called *convergent*. The limit is then called the *sum* of the infinite series. An infinite series whose partial sums do not approach a limit is called *divergent*. From our discussion above, we know that the infinite series (1) is convergent, whereas (2) and (3) are divergent.

It is often an extremely difficult task to determine whether or not a given infinite series is convergent. And intuition is not always an accurate guide. For example, one might at first suspect that the infinite series

$$1 + \tfrac{1}{2} + \tfrac{1}{3} + \tfrac{1}{4} + \tfrac{1}{5} + \ldots$$

(the so-called *harmonic series*) is convergent. However, it is not. Ths sums of its initial terms increase without bound, although they do so very slowly. For example, it takes about 22,000 terms before the sum exceeds 10 and about 2.7×10^{43} terms before the sum exceeds 100. Nevertheless, the sum eventually exceeds any prescribed number. (See Exercise 26).

In the remainder of this section we shall exhibit a number of interesting series which do converge. In each case we shall assume the convergence and use it to draw some interesting conclusions.

Let r be a given number. A series of the form

$$1 + r + r^2 + r^3 + r^4 + \ldots$$

is called a *geometric series with ratio r*. (The "ratio" of consecutive terms is r.) Geometric series occur in many parts of mathematics. Their use in probability theory will be discussed in Section 4. The convergence properties of the geometric series can be summarized as follows (see Exercise 25 for a proof):

> *The Geometric Series*   The infinite series
>
> $$1 + r + r^2 + r^3 + r^4 + \ldots$$
>
> converges if $|r| < 1$. In this case the sum is $1/(1 - r)$.

For example, if $r = \tfrac{1}{2}$, then we obtain the infinite series (1). In this case $1/(1 - \tfrac{1}{2}) = 1/\tfrac{1}{2} = 2$, in agreement with our previous observation.

EXAMPLE 1   Calculate the sum of these infinite series:

(a) $1 + \dfrac{1}{5} + \dfrac{1}{5^2} + \dfrac{1}{5^3} + \dfrac{1}{5^4} + \ldots$.

(b) $\dfrac{1}{4^2} + \dfrac{1}{4^3} + \dfrac{1}{4^4} + \dfrac{1}{4^5} + \dfrac{1}{4^6} + \ldots$.

Solution   (a) This is a geometric series with $r = \frac{1}{5}$. Its sum is

$$\frac{1}{1-r} = \frac{1}{1-\frac{1}{5}} = \frac{1}{\frac{4}{5}} = \frac{5}{4}.$$

(b) If we add $1 + \frac{1}{4}$ to this series, we get the geometric series with $r = \frac{1}{4}$, whose sum is

$$\frac{1}{1-r} = \frac{1}{1-\frac{1}{4}} = \frac{1}{\frac{3}{4}} = \frac{4}{3}.$$

Therefore, the sum of the given series is

$$\tfrac{4}{3} - (1 + \tfrac{1}{4}) = \tfrac{4}{3} - \tfrac{5}{4} = \tfrac{1}{12}.$$

EXAMPLE 2   Calculate the sums of the following infinite series:

(a) $1 + \dfrac{1}{3^2} + \dfrac{1}{3^4} + \dfrac{1}{3^6} + \dfrac{1}{3^8} + \cdots.$

(b) $\dfrac{2}{3^2} + \dfrac{2}{3^4} + \dfrac{2}{3^6} + \dfrac{2}{3^8} + \dfrac{2}{3^{10}} + \cdots.$

Solution   (a) Rewrite the infinite series as

$$1 + \frac{1}{9} + \frac{1}{9^2} + \frac{1}{9^3} + \frac{1}{9^4} + \cdots$$

(since $3^2 = 9$, $3^4 = 9^2$, $3^6 = 9^3$, $3^8 = 9^4$ and so on). This is a geometric series with $r = \frac{1}{9}$ and has sum

$$\frac{1}{1-r} = \frac{1}{1-\frac{1}{9}} = \frac{9}{8}.$$

(b) Write the infinite series as

$$\frac{2}{3^2}\left(1 + \frac{1}{3^2} + \frac{1}{3^4} + \frac{1}{3^6} + \frac{1}{3^8} + \cdots\right).$$

By part (a) the series in parentheses has sum $\frac{9}{8}$. So the given series has sum

$$\frac{2}{3^2}\left(\frac{9}{8}\right) = \frac{1}{4}.$$

The next example gives an application of infinite series to economics.

EXAMPLE 3   (*The Multiplier Effect*)   The Federal Government enacts a wage tax cut of $10 billion. Assume that each wage earner spends 93% of his take-home pay. Estimate the effect of the tax cut on economic activity, neglecting all other factors except savings.

Solution   Express all amounts of money in billions of dollars. Of the increase in take-home pay created by the tax cut, $(.93)(10)$ billion dollars will be spent. Of those dollars,

93% will be spent again and 7% saved, so an additional spending of (.93)(.93)(10) billion dollars is created. The recipients of those dollars will spend 93% of them, creating yet additional spending of

$$(.93)(.93)(.93)(10) = 10(.93)^3$$

billion dollars. And so on. The total amount of new spending created by the tax cut is thus given by the infinite series

$$10(.93) + 10(.93)^2 + 10(.93)^3 + \ldots.$$

The sum of this series is

$$10(.93)(1 + .93 + .93^2 + \ldots) = 9.3\left(\frac{1}{1 - .93}\right)$$

$$= 9.3\left(\frac{1}{.07}\right)$$

$$\approx 132.86.$$

Thus, a \$10 billion tax cut creates additional spending of about \$132.86 billion. This demonstrates a particular instance of what economists call the *multiplier effect*.

Taylor Series In Section 1 we defined the nth Taylor polynomial of the function $f(x)$ at $x = 0$ to be the polynomial function

$$f(0) + \frac{f'(0)}{1} x + \frac{f''(0)}{1 \cdot 2} x^2 + \ldots + \frac{f^{(n)}(0)}{1 \cdot 2 \cdot \ldots \cdot n} x^n.$$

It gives an approximation to the original function for values of x near 0.

Now that we have the concept of infinite series, we can define the *Taylor series* of $f(x)$ at $x = 0$ to be the function defined by the infinite series

$$f(0) + \frac{f'(0)}{1} x + \frac{f''(0)}{1 \cdot 2} x^2 + \frac{f'''(0)}{1 \cdot 2 \cdot 3} x^3 + \ldots.$$

By restricting x to those values for which the series converges, the series defines a function. The startling result is that for "suitable functions" the Taylor series provides more than just an approximation to $f(x)$. Rather, it actually has the same values as $f(x)$. We may summarize this result as follows:

If $f(x)$ is a suitable function, then

$$f(x) = f(0) + \frac{f'(0)}{1} x + \frac{f''(0)}{1 \cdot 2} x^2 + \frac{f'''(0)}{1 \cdot 2 \cdot 3} x^3 + \ldots \tag{5}$$

for all values of x for which the infinite series converges.

Note: A precise definition for the term "suitable" is complicated to formulate. All of the functions that we will consider in this section of the text are suitable.

EXAMPLE 4   Determine the Taylor series for $f(x) = 1/(1 - x)$.

Solution   As we saw in Section 1,

$$f(0) = 1, \quad f'(0) = 1, \quad f''(0) = 1 \cdot 2, \quad f'''(0) = 1 \cdot 2 \cdot 3, \quad \ldots.$$

Therefore, by (5) we obtain

$$f(x) = f(0) + \frac{f'(0)}{1} x + \frac{f''(0)}{1 \cdot 2} x^2 + \frac{f'''(0)}{1 \cdot 2 \cdot 3} x^3 + \ldots,$$

$$\frac{1}{1 - x} = 1 + \frac{1}{1} x + \frac{1 \cdot 2}{1 \cdot 2} x^2 + \frac{1 \cdot 2 \cdot 3}{1 \cdot 2 \cdot 3} x^3 + \ldots$$

$$= 1 + x + x^2 + x^3 + \ldots.$$

In this case the Taylor series expansion is the familiar geometric series which converges for all values of x with $|x| < 1$.

EXAMPLE 5   Determine the Taylor series for $f(x) = e^x$.

Solution   Since $f^{(n)}(x) = e^x$ for all n, $f^{(n)}(0) = e^0 = 1$ for all n. Therefore, by (5) we obtain

$$f(x) = f(0) + \frac{f'(0)}{1} x + \frac{f''(0)}{1 \cdot 2} x^2 + \frac{f'''(0)}{1 \cdot 2 \cdot 3} x^3 + \ldots$$

$$e^x = 1 + x + \frac{1}{1 \cdot 2} x^2 + \frac{1}{1 \cdot 2 \cdot 3} x^3 + \ldots.$$

It can be shown that this infinite series converges for every value of x.

Taylor series are computed just like Taylor polynomials except that you don't stop. However, there are some other methods of obtaining Taylor series. For instance, we can differentiate or integrate both sides of a Taylor series expansion. If we differentiate each term in the Taylor series expansion of a function $f(x)$, we obtain the Taylor series expansion for $f'(x)$. The analogous result holds for integration.

EXAMPLE 6   Differentiate both sides of the Taylor series expansion of $f(x) = 1/(1 - x)$ ($|x| < 1$).

Solution   We begin with

$$\frac{1}{1 - x} = 1 + x + x^2 + x^3 + x^4 + x^5 + \ldots \quad \text{for } |x| < 1.$$

Since

$$\frac{d}{dx} \left(\frac{1}{1 - x} \right) = \frac{1}{(1 - x)^2},$$

we obtain

$$\frac{1}{(1-x)^2} = 1 + 2x + 3x^2 + 4x^3 + 5x^4 + \ldots \quad \text{for } |x| < 1. \quad (6)$$

EXAMPLE 7   Use the result of Example 6 to determine

$$1 + \frac{2}{7} + \frac{3}{7^2} + \frac{4}{7^3} + \frac{5}{7^4} + \ldots.$$

Solution   Set $x = \frac{1}{7}$ in the Taylor series (6). Then

$$\frac{1}{(1-\frac{1}{7})^2} = 1 + 2\left(\frac{1}{7}\right) + 3\left(\frac{1}{7}\right)^2 + 4\left(\frac{1}{7}\right)^3 + 5\left(\frac{1}{7}\right)^4 + \ldots.$$

Now,

$$\frac{1}{(1-\frac{1}{7})^2} = \frac{1}{(\frac{6}{7})^2} = \frac{1}{\frac{36}{49}} = \frac{49}{36}.$$

Therefore,

$$\frac{49}{36} = 1 + \frac{2}{7} + \frac{3}{7^2} + \frac{4}{7^3} + \frac{5}{7^4} + \ldots.$$

EXAMPLE 8   Integrate both sides of the Taylor series expansion of $f(x) = 1/(1 - x)$ $(|x| < 1)$.

Solution   We begin with

$$\frac{1}{1-x} = 1 + x + x^2 + x^3 + x^4 + \ldots \quad (|x| < 1).$$

Now,

$$\int \frac{1}{1-x}\,dx = \int (1 + x + x^2 + x^3 + x^4 + \ldots)\,dx,$$

$$-\ln(1-x) + C = x + \tfrac{1}{2}x^2 + \tfrac{1}{3}x^3 + \tfrac{1}{4}x^4 + \tfrac{1}{5}x^5 + \ldots \quad (|x| < 1).$$

where C is the constant of integration. If we set $x = 0$ in both sides, we obtain

$$0 + C = 0 + \tfrac{1}{2}\cdot 0^2 + \tfrac{1}{3}\cdot 0^3 + \tfrac{1}{4}\cdot 0^4 + \tfrac{1}{5}\cdot 0^5 + \ldots = 0$$

so $C = 0$. Thus,

$$-\ln(1-x) = x + \tfrac{1}{2}x^2 + \tfrac{1}{3}x^3 + \tfrac{1}{4}x^4 + \tfrac{1}{5}x^5 + \ldots \quad (|x| < 1).$$

EXAMPLE 9   Use the result of Example 8 to compute ln 1.1.

Solution   Let $x = -.1$ in the Taylor series for $-\ln(1 - x)$. Then

$$-\ln(1 - (-.1)) = (-.1) + \tfrac{1}{2}(-.1)^2 + \tfrac{1}{3}(-.1)^3 + \tfrac{1}{4}(-.1)^4 + \tfrac{1}{5}(-.1)^5 + \ldots$$

$$-\ln 1.1 = -.1 + \frac{.01}{2} - \frac{.001}{3} + \frac{.0001}{4} - \frac{.00001}{5} + \ldots$$

$$\ln 1.1 = .1 - \frac{.01}{2} + \frac{.001}{3} - \frac{.0001}{4} + \frac{.00001}{5} - \ldots$$

This infinite series can be used to compute ln 1.1 to any degree of accuracy required. For instance, summing the first five terms gives ln 1.1 ≈ .09531 (correct to five decimal places).

PRACTICE PROBLEMS 3

1. Determine the sum of the infinite series

$$1 - \tfrac{1}{3} + \tfrac{1}{9} - \tfrac{1}{27} + \tfrac{1}{81} - \ldots.$$

2. Find the rational number whose decimal expansion is .45454545.... That is, calculate the sum of the infinite series

$$\tfrac{45}{100} + \tfrac{45}{10000} + \tfrac{45}{1000000} + \tfrac{45}{100000000} + \ldots.$$

3. Determine the function whose Taylor series is

$$1 \cdot 2 + 2 \cdot 3x + 3 \cdot 4x^2 + 4 \cdot 5x^3 + 5 \cdot 6x^4 + \ldots, \quad |x| < 1.$$

[Hint: Refer to Example 6.]

EXERCISES 3

Determine the sums of the following infinite series.

1. $1 + \dfrac{1}{6} + \dfrac{1}{6^2} + \dfrac{1}{6^3} + \dfrac{1}{6^4} + \ldots.$

2. $\dfrac{1}{3^2} + \dfrac{1}{3^3} + \dfrac{1}{3^4} + \dfrac{1}{3^5} + \dfrac{1}{3^6} + \ldots.$

3. $\dfrac{1}{5} + \dfrac{1}{5^4} + \dfrac{1}{5^7} + \dfrac{1}{5^{10}} + \dfrac{1}{5^{13}} + \ldots.$

4. $1 + \dfrac{1}{6^2} + \dfrac{1}{6^4} + \dfrac{1}{6^6} + \dfrac{1}{6^8} + \ldots.$

5. $1 - .2 + .04 - .008 + .0016 - \ldots.$

6. $1 - \dfrac{2}{3} + \dfrac{4}{9} - \dfrac{8}{27} + \dfrac{16}{81} - \ldots.$

Sum an appropriate infinite series to find the rational number whose decimal expansion is given.

7. .23232323...

8. .173173173173...

9. .22222...

10. .15151515...

11. .012012012012...

12. .44444...

13. Determine the Taylor series for sin x and differentiate to obtain the Taylor series for cos x.

14. Determine the Taylor series for $\sqrt{1 + x}$ and differentiate to obtain the Taylor series for $1/\sqrt{1 + x}$.

15. As we saw in Example 6,

$$\frac{1}{(1 - x)^2} = 1 + 2x + 3x^2 + 4x^3 + 5x^4 + \dots \quad (|x| < 1).$$

We can multiply by x to obtain the Taylor series,

$$\frac{x}{(1 - x)^2} = x + 2x^2 + 3x^3 + 4x^4 + 5x^5 + \dots \quad (|x| < 1).$$

Find the function whose Taylor series is

$$1 + 4x + 9x^2 + 16x^3 + 25x^4 + \dots.$$

16. Substitute x^2 for x in the Taylor series of e^x to obtain the Taylor series of e^{x^2}. Find an infinite series that converges to $\int_0^1 e^{x^2} \, dx$.

17. The Taylor series for $\sin x^2$ is

$$x^2 - \frac{x^6}{1 \cdot 2 \cdot 3} + \frac{x^{10}}{1 \cdot 2 \cdot 3 \cdot 4 \cdot 5} - \frac{x^{14}}{1 \cdot 2 \cdot \ldots \cdot 7} + \frac{x^{18}}{1 \cdot 2 \cdot \ldots \cdot 9} - \dots$$

Find an infinite series that converges to $\int_0^1 \sin x^2 \, dx$.

18. Differentiate the Taylor series for e^x to verify that $\dfrac{d}{dx}(e^x) = e^x$.

19. Determine the Taylor series for e^{-x}.

20. Determine the Taylor series for $1/(1 + x^2)$.

21. (a) Use the Taylor series for e^x at $x = 0$ to show that $e^x > x^2/2$ for $x > 0$.

    (b) Deduce that $e^{-x} < 2/x^2$ for $x > 0$.

    (c) Show that xe^{-x} approaches 0 as $x \to \infty$.

22. Let k be a positive constant.

    (a) Show that $e^{kx} > \dfrac{k^2 x^2}{2}$, for $x > 0$.

    (b) Deduce that $e^{-kx} < \dfrac{2}{k^2 x^2}$, for $x > 0$.

    (c) Show that xe^{-kx} approaches 0 as $x \to \infty$.

23. Show that $e^x > x^3/6$ for $x > 0$, and from this deduce that $x^2 e^{-x}$ approaches 0 as $x \to \infty$.

24. If k is a positive constant, show that $x^2 e^{-kx}$ approaches 0 as $x \to \infty$.

25. Let r be any number other than 1.
    (a) Show that $(1 - r)(1 + r + r^2 + \dots + r^n) = 1 - r^{n+1}$.
    (b) Divide both sides of the result of (a) by $1 - r$ and explain why the geometric series converges when $|r| < 1$.

26. Show that the infinite series

$$1 + \tfrac{1}{2} + \tfrac{1}{3} + \tfrac{1}{4} + \tfrac{1}{5} + \dots$$

diverges. [Hint: $\tfrac{1}{3} + \tfrac{1}{4} > \tfrac{1}{2}$; $\tfrac{1}{5} + \tfrac{1}{6} + \tfrac{1}{7} + \tfrac{1}{8} > \tfrac{1}{2}$; $\tfrac{1}{9} + \dots + \tfrac{1}{16} > \tfrac{1}{2}$; etc.]

SOLUTIONS TO PRACTICE PROBLEMS 3

1. This infinite series is a geometric series with $r = -\frac{1}{3}$. Therefore, its sum is

$$\frac{1}{1 - (-\frac{1}{3})} = \frac{1}{1 + \frac{1}{3}} = \frac{1}{\frac{4}{3}} = \frac{3}{4}.$$

2. This infinite series can be rewritten as

$$\frac{45}{100}\left(1 + \frac{1}{100} + \left(\frac{1}{100}\right)^2 + \left(\frac{1}{100}\right)^3 + \cdots\right).$$

Therefore its sum is

$$\frac{45}{100}\left(\frac{1}{1 - \frac{1}{100}}\right) = \frac{45}{100} \cdot \frac{1}{\frac{99}{100}} = \frac{45}{100} \cdot \frac{100}{99} = \frac{45}{99} = \frac{5}{11}.$$

3. Start with the geometric series

$$\frac{1}{1 - x} = 1 + x + x^2 + x^3 + x^4 + x^5 + x^6 + \ldots, \quad |x| < 1,$$

and differentiate it twice:

$$\frac{1}{(1 - x)^2} = 1 + 2x + 3x^2 + 4x^3 + 5x^4 + 6x^5 + \ldots,$$

$$\frac{2}{(1 - x)^3} = 1 \cdot 2 + 2 \cdot 3x + 3 \cdot 4x + 4 \cdot 5x + 5 \cdot 6x^4 + \ldots.$$

13.4. Infinite Series and Probability Theory

One of the most important applications of infinite series is to the field of probability. A complete description of the connection between the two subjects is beyond the scope of this book. In this section we will highlight a few of its most salient features.

In Chapter 12, we considered one class of experiments having an infinite number of possible outcomes. By use of infinite series we can obtain results about another important class of such experiments—namely, those whose possible outcomes are the numbers 0, 1, 2, 3, ... Such experiments abound. Here are a few examples.

1. Flip a coin until a head appears and observe the number of consecutive tails which occurred. This number may be 0, 1, 2, 3, ...

2. Observe the number of telephone calls arriving at a switchboard during a given minute. This number may be 0, 1, 2, 3, ...

3. Observe the number of fire insurance claims filed (with a particular insurance company) during a given month. The number of claims is one of the numbers 0, 1, 2, 3, ...

Suppose that we consider an experiment whose possible outcomes are the numbers 0, 1, 2, 3, ... As in Chapter 12, we denote the outcome of the experiment by X and call X a *random variable*. To each possible value of X, there is an associated probability of occurrence, p_n. That is,

$$\Pr(X = 0) = p_0,$$
$$\Pr(X = 1) = p_1,$$
$$\vdots$$
$$\Pr(X = n) = p_n,$$
$$\vdots$$

Note that since $p_0, p_1, \ldots, p_n, \ldots$ are probabilities, each lies between 0 and 1. Furthermore, the sum of all these probabilities is 1. (One of the outcomes 0, 1, 2, 3, ... always occurs.) That is,

$$p_0 + p_1 + \ldots + p_n + \ldots = 1.$$

In analogy with the case of experiments having a finite number of possible outcomes, we may define the *expected value* (or average value) of the random variable X (or of the experiment whose outcome is X) to be the number $E(X)$ given by the following formula:

$$E(X) = 0 \cdot p_0 + 1 \cdot p_1 + 2 \cdot p_2 + 3 \cdot p_3 + \ldots.$$

(provided the infinite series converges). That is, the expected value $E(X)$ is formed by adding the products of the possible outcomes by their respective probabilities of occurrence.

In a similar fashion, letting μ denote $E(X)$, we define the *variance* of X by

$$\text{Var}(X) = (0 - \mu)^2 \cdot p_0 + (1 - \mu)^2 \cdot p_1 + (2 - \mu)^2 \cdot p_2 + (3 - \mu)^2 \cdot p_3 + \ldots.$$

EXAMPLE 1 Suppose we toss a coin until a head appears and observe the number X of consecutive tails preceding it.

(a) Determine the probability p_n that exactly n consecutive tails occur.

(b) Determine the probability that an odd number of consecutive tails occur.

(c) Determine the average number of consecutive tails which occur.

(d) Write down the infinite series that gives the variance for the number of consecutive tails.

Solution   (a) The probability p_0 corresponds to tossing a head on the first throw. And since a head is just as likely as a tail, we have $p_0 = \frac{1}{2}$. To calculate p_1, notice that there are four possible results of tossing a coin twice:

heads heads;   tails tails;   heads tails;   tails heads.

Each of these results is equally likely, so each has probability $\frac{1}{4}$. The probability p_1 corresponds to the result "tails heads," and so $p_1 = \frac{1}{4} = 1/2^2$. Similarly, to compute p_2 we note that there are eight possible results for tossing a coin three times, of which only one,

tails tails heads

relates to p_2. So $p_2 = \frac{1}{8} = 1/2^3$. In a similar fashion, we find that $p_n = 1/2^{n+1}$ for $n = 0, 1, 2, 3, \ldots$.

(b) The probability that an odd number of consecutive tails occurs is

$$
\begin{aligned}
p_1 + p_3 + p_5 + p_7 + \ldots &= \frac{1}{2^2} + \frac{1}{2^4} + \frac{1}{2^6} + \frac{1}{2^8} + \ldots \\
&= \frac{1}{2^2}\left(1 + \frac{1}{2^2} + \frac{1}{2^4} + \frac{1}{2^6} + \ldots\right) \\
&= \frac{1}{4}\left(1 + \frac{1}{4} + \frac{1}{4^2} + \frac{1}{4^3} + \ldots\right) \\
&= \frac{1}{4} \cdot \frac{1}{1 - \frac{1}{4}} = \frac{1}{4} \cdot \frac{4}{3} \\
&= \frac{1}{3}.
\end{aligned}
$$

(c) The average number of consecutive tails is

$$
\begin{aligned}
E(X) &= 0 \cdot p_0 + 1 \cdot p_1 + 2 \cdot p_2 + 3 \cdot p_3 + \ldots \\
&= 0 \cdot \frac{1}{2} + 1 \cdot \frac{1}{2^2} + 2 \cdot \frac{1}{2^3} + 3 \cdot \frac{1}{2^4} + \ldots \\
&= \frac{1}{2^2} + \frac{2}{2^3} + \frac{3}{2^4} + \ldots \\
&= \frac{1}{4}\left(1 + 2 \cdot \frac{1}{2} + 3 \cdot \frac{1}{2^2} + \ldots\right).
\end{aligned}
$$

Recall that we showed in the preceding section that for $|x| < 1$ we have

$$
1 + 2x + 3x^2 + \ldots = \frac{1}{(1 - x)^2}.
$$

Setting $x = \frac{1}{2}$ in the formula gives

$$
1 + 2 \cdot \frac{1}{2} + 3 \cdot \frac{1}{2^2} + \ldots = \frac{1}{(1 - \frac{1}{2})^2} = 4.
$$

Therefore,

$$E(X) = \frac{1}{4} \cdot 4 = 1.$$

So the average number of consecutive tails is 1, in agreement with our intuition.

(d) The variance for the number of consecutive tails is

$$\text{Var}(X) = (0 - 1)^2 \cdot \frac{1}{2} + (1 - 1)^2 \cdot \frac{1}{2^2} + (2 - 1)^2 \cdot \frac{1}{2^3}$$

$$+ (3 - 1)^2 \cdot \frac{1}{2^4} + (4 - 1)^2 \cdot \frac{1}{2^5} + \dots$$

$$= \frac{1}{2} + 1^2 \cdot \frac{1}{2^3} + 2^2 \cdot \frac{1}{2^4} + 3^2 \cdot \frac{1}{2^5} + \dots.$$

(It can be shown that this infinite series converges to 2. See Exercise 22.)

EXAMPLE 2   An electronics plant manufactures electronic calculators. After manufacture, each calculator is inspected for defects. The probability that exactly n nondefective calculators are observed before a defective one is observed is $p_n = \frac{1}{50}(.98)^n$. What is the average number of nondefective calculators between consecutive defective ones?

Solution   Our experiment consists of observing the number X of nondefective calculators in a row. We are given that $p_n = \frac{1}{50}(.98)^n$ and are asked for the expected value $E(X)$ of the experiment. But

$$E(X) = 0 \cdot \frac{1}{50}(.98)^0 + 1 \cdot \frac{1}{50}(.98)^1 + 2 \cdot \frac{1}{50}(.98)^2 + 3 \cdot \frac{1}{50}(.98)^3 + \dots$$

$$= \frac{.98}{50}[1 + 2 \cdot (.98)^1 + 3 \cdot (.98)^2 + \dots].$$

As we observed in the preceding example, the series in parentheses is the series

$$1 + 2x + 3x^2 + 4x^3 + \dots = \frac{1}{(1 - x)^2},$$

evaluated at $x = .98$. Thus,

$$E(X) = \frac{.98}{50}\frac{1}{(1 - .98)^2} = \frac{.98}{50}\frac{1}{(.02)^2} = 49.$$

In other words, on the average 49 nondefective calculators will occur between consecutive defective ones.

Poisson Experiments In many experiments occurring in applications the probability p_n is given by a formula of the following type:

$$p_0 = e^{-\lambda},$$

$$p_1 = \frac{\lambda}{1} e^{-\lambda},$$

$$p_2 = \frac{\lambda^2}{1 \cdot 2} e^{-\lambda},$$

$$p_3 = \frac{\lambda^3}{1 \cdot 2 \cdot 3} e^{-\lambda},$$

$$\vdots$$

$$p_n = \frac{\lambda^n}{1 \cdot 2 \cdot \ldots \cdot n} e^{-\lambda}, \tag{1}$$

where λ is a constant depending on the particular experiment under consideration. An experiment for which equation (1) holds is called a *Poisson experiment*, and the outcome X is called a *Poisson random variable*. The probabilities for X given by equation (1) are said to be *Poisson distributed* (with parameter λ).

Poisson random variables constitute one of the most important classes of random variables studied in probability theory. This is because of the wide variety of physical phenomena which can be accurately modeled as Poisson experiments. For example, experiments 2 and 3 given at the beginning of this section are Poisson experiments. Here are some others:

4. Observe the annual number of deaths due to a particular disease.

5. Observe the monthly number of breakdowns of a particular type of machine in a given factory.

6. Observe the number of typographical errors per page in a given book.

7. Observe the number of persons arriving during five-minute intervals at a super-market checkout counter.

8. Observe the number of protozoa in various drop-size samples of water drawn from a pond.

Note that for a Poisson experiment the requirement that
$$p_0 + p_1 + p_2 + p_3 + \ldots = 1$$
can be expressed by the equation

$$e^{-\lambda} + \frac{\lambda}{1} e^{-\lambda} + \frac{\lambda^2}{1 \cdot 2} e^{-\lambda} + \frac{\lambda^3}{1 \cdot 2 \cdot 3} e^{-\lambda} + \ldots = 1.$$

To verify this, we rewrite the left side in the form

$$e^{-\lambda}\left(1 + \frac{\lambda}{1} + \frac{\lambda^2}{1 \cdot 2} + \frac{\lambda^3}{1 \cdot 2 \cdot 3} + \cdots\right),$$

and we note that the quantity inside the parentheses is the Taylor series for e^{λ}. Thus

$$e^{-\lambda}\left(1 + \frac{\lambda}{1} + \frac{\lambda^2}{1 \cdot 2} + \frac{\lambda^3}{1 \cdot 2 \cdot 3} + \ldots\right) = e^{-\lambda}e^{\lambda} = e^0 = 1.$$

The following facts about Poisson experiments provide an interpretation for the parameter λ.

Let X be a random variable whose probabilities are Poisson distributed with parameter λ, that is,

$$p_0 = e^{-\lambda},$$

$$p_n = \frac{\lambda^n}{1 \cdot 2 \cdots \cdots n} e^{-\lambda} \quad (n = 1, 2, \ldots).$$

Then the expected value and variance of X are given by

$$\mathrm{E}(X) = \lambda, \qquad \mathrm{Var}(X) = \lambda.$$

We shall verify only the statement about $\mathrm{E}(X)$. The argument uses the Taylor series for e^{λ}. We have

$$\mathrm{E}(X) = 0 \cdot p_0 + 1 \cdot p_1 + 2 \cdot p_2 + 3 \cdot p_3 + 4 \cdot p_4 + \ldots$$

$$= 0 \cdot e^{-\lambda} + 1 \cdot \frac{\lambda}{1} e^{-\lambda} + 2 \cdot \frac{\lambda^2}{1 \cdot 2} e^{-\lambda}$$

$$+ 3 \cdot \frac{\lambda^3}{1 \cdot 2 \cdot 3} e^{-\lambda} + 4 \cdot \frac{\lambda^4}{1 \cdot 2 \cdot 3 \cdot 4} e^{-\lambda} + \ldots$$

$$= \lambda e^{-\lambda} + \frac{\lambda^2}{1} e^{-\lambda} + \frac{\lambda^3}{1 \cdot 2} e^{-\lambda} + \frac{\lambda^4}{1 \cdot 2 \cdot 3} e^{-\lambda} + \ldots$$

$$= \lambda e^{-\lambda}\left(1 + \frac{\lambda}{1} + \frac{\lambda^2}{1 \cdot 2} + \frac{\lambda^3}{1 \cdot 2 \cdot 3} + \ldots\right)$$

$$= \lambda e^{-\lambda} \cdot e^{\lambda}$$

$$= \lambda.$$

The next two examples illustrate some applications of Poisson experiments.

EXAMPLE 3 Suppose that we observe the number X of calls received by a telephone switchboard during a one-minute interval. Experience suggests that X is Poisson distributed with $\lambda = 3$.

(a) Determine the probability that zero, one, or two calls arrive during a particular minute.

(b) Determine the probability that three or more calls arrive during a particular minute.

(c) Determine the average number of calls received per minute.

Solution (a) The probability that zero, one, or two calls arrive during a given minute is $p_0 + p_1 + p_2$. Moreover

$$p_0 = e^{-\lambda} = e^{-3} \approx .04979,$$

$$p_1 = \frac{\lambda}{1} e^{-\lambda} = 3e^{-3} \approx .14937,$$

$$p_2 = \frac{\lambda^2}{1 \cdot 2} e^{-\lambda} = \frac{9}{2} e^{-3} \approx .22406.$$

Thus $p_0 + p_1 + p_2 \approx .42322$. That is, during approximately 42% of the minutes, either zero, one, or two calls are received.

(b) The probability of receiving three or more calls is the same as the probability of *not* receiving zero, one, or two calls and so is equal to

$$1 - (p_0 + p_1 + p_2) = 1 - .42322$$

$$= .57678.$$

(c) The average number of calls received per minute is equal to λ. That is, on the average the switchboard receives three calls per minute.

EXAMPLE 4 Drop-size water samples are drawn from a New England pond. The number of protozoa in each sample is estimated and the average number is found to be 10. Suppose a sample is chosen at random. What is the probability that it contains four or fewer protozoa?

Solution The probability that a sample contains exactly n protozoa is given by the Poisson distribution

$$p_n = \frac{\lambda^n}{1 \cdot 2 \cdot \ldots \cdot n} e^{-\lambda},$$

where $\lambda = 10$, the expected value. The probability that there are 4 or fewer protozoa equals

$$p_0 + p_1 + p_2 + p_3 + p_4.$$

To calculate this probability we use a calculator with an e^x key and find that it equals .02925. In other words, the probability that a random sample contains 4 or fewer protozoa is .02925.

PRACTICE PROBLEMS 4

1. Suppose we toss a coin until a tail appears and observe the number of consecutive heads preceding it. What is the probability that this number is divisible by 3?

2. A Public Health Officer is tracking down the source of a bacterial infection in a certain city. She analyzes the reported incidence of the infection in each city block and finds an average of three cases per block. A certain block is found to have seven cases. What is the probability that a randomly chosen block has seven or more cases, assuming that the number of cases per block is Poisson distributed?

EXERCISES 4

Suppose we toss a coin until a tail appears and observe the number of consecutive heads preceding the tail.

1. What is the probability that exactly three heads occur?

2. What is the probability that exactly three or five heads occur?

3. What is the probability that the number of consecutive heads is at least three?

4. What is the probability that the number of consecutive heads is even?

5. What is the probability that the number of consecutive heads is odd and at least three?

6. What is the probability that the number of consecutive heads is divisible by five?

   Suppose that in a certain town there are two competing taxi companies, Red Cab and Blue Cab. The taxis mix with downtown traffic in a random manner. The Red fleet is twice the size of the Blue fleet. Suppose that we stand on a downtown street corner and watch the taxis passing by, observing the number of consecutive Red taxis before a Blue taxi appears.

7. Determine the formula for p_n, the probability of exactly n consecutive Red taxis.

8. Determine the average number of consecutive Red taxis.

9. What is the probability that the number of consecutive Red taxis is divisible by three?

   The number of typographical errors per page of a certain newspaper is approximately Poisson distributed with $\lambda = 5$.

10. What is the probability that a given page is error-free?

11. What is the probability that a given page has exactly two errors?

12. What is the probability that a given page has either two or three errors?

13. What is the probability that a given page has at most four errors?

14. What is the average number of errors per page?

15. What is the probability that a given page has at least two errors?

   The monthly number of fire insurance claims filed with the Firebug Insurance Company is Poisson distributed with $\lambda = 10$.

16. What is the probability that in a given month no claims are filed?

17. What is the probability that in a given month at least three claims are filed?

18. What is the average number of claims filed per month?

19. What is the probability that the number of claims is either one, two, or three?

20. A person shooting at a target has five successive hits and then a miss. If x is the probability of success on each shot, then the probability of having five successive hits followed by a miss is $p = x^5(1 - x)$. Take first and second derivatives to determine the value of x for which p has its maximum value.

21. In a production process a box of 100 fuses is examined and found to contain two defective fuses. Suppose that the probability of having two defective fuses in a box of 100 is $p = (\lambda^2/2)e^{-\lambda}$ for some λ. Take first and second derivatives to determine the value of λ for which p has its maximum value.

22. Use the result of Exercise 15 of Section 3 to show that the variance in Example 1 is 2.

23. When raisins are mixed into batter that is made into cookies, the number of raisins a particular cookie will have is Poisson distributed. If 100 cookies are being made, how many raisins should be used so that the probability of a cookie's having no raisins is .01?

SOLUTIONS TO PRACTICE PROBLEMS 4

1. If p_n is the probability that exactly n consecutive heads occur, then we know from Example 1 that $p_n = 1/2^{n+1}$. The probability that the number of consecutive heads is divisible by 3 equals

$$
\begin{aligned}
p_0 + p_3 + p_6 + p_9 + p_{12} + \ldots &= \frac{1}{2} + \frac{1}{2^4} + \frac{1}{2^7} + \frac{1}{2^{10}} + \frac{1}{2^{13}} + \ldots \\
&= \frac{1}{2}\left(1 + \frac{1}{2^3} + \frac{1}{2^6} + \frac{1}{2^9} + \frac{1}{2^{12}}\right) + \ldots \\
&= \frac{1}{2}\left[\frac{1}{1 - (1/2^3)}\right] \\
&= \frac{1}{2} \cdot \frac{8}{7} \\
&= \frac{4}{7}.
\end{aligned}
$$

2. The number of cases per block is Poisson distributed with $\lambda = 3$. So the probability of having seven or more cases in a given block is

$$p_7 + p_8 + p_9 + \ldots = 1 - (p_0 + p_1 + p_2 + p_3 + p_4 + p_5 + p_6).$$

However,

$$p_n = \frac{3^n}{1 \cdot 2 \cdot \ldots \cdot n} e^{-3},$$

so that

$$p_0 = .04979, \quad p_1 = .14937,$$
$$p_2 = .22406, \quad p_3 = .22406,$$
$$p_4 = .16804, \quad p_5 = .10082,$$
$$p_6 = .05041.$$

Therefore, the probability of seven or more cases in a given block equals

$$1 - (.04979 + .14937 + .22406 + .22406 + .16804 + .10082 + .05041) = .03345.$$

Chapter 13: CHECKLIST

☐ Taylor polynomial of $f(x)$ at $x = a$

☐ Newton-Raphson algorithm

☐ Infinite series

☐ Partial sum of an infinite series

☐ Sum of an infinite series

☐ Convergent infinite series

☐ Divergent infinite series

☐ Geometric series

☐ Taylor series of $f(x)$ at $x = 0$

☐ Poisson random variable

Chapter 13: SUPPLEMENTARY EXERCISES

1. Determine a polynomial $p(x)$ for which $p(3) = 2$, $p'(3) = 0$, $p''(3) = -1$, and $p'''(3) = 3$.

2. Determine a polynomial $p(x)$ for which $p(-2) = 1$, $p'(-2) = 6$, $p''(-2) = 2$, $p'''(-2) = 11$, $p^{(4)}(-2) = 5$.

3. Determine the fourth Taylor polynomial of $f(x) = 3x^2 - 2x$ at $x = 3$.

4. Determine the second Taylor polynomial of $f(x) = \sqrt{\cos x}$ at $x = 0$.

5. Use the second Taylor polynomial of $f(x) = x^{3/2}$ at $x = 4$ to estimate $(4.06)^{3/2}$.

6. Use the second Taylor polynomial of $f(x) = \ln x$ at $x = 1$ to estimate $\ln(1.2)$.

7. Use the Newton-Raphson algorithm (two repetitions) to approximate the zero of $x^2 - 3x - 2$ near $x_0 = 4$.

8. Sketch the graph of $y = x^3 - 6x^2 + 12x - 4$ and use the Newton-Raphson algorithm (two repetitions) to approximate all x-intercepts.

9. Use the Newton-Raphson algorithm to approximate the solution to the equation $e^{2x} = 1 + e^{-x}$.

10. Use the Newton-Raphson algorithm (three repetitions) to approximate $\sqrt{11}$.

11. Determine the sum of the infinite series,

$$1 - \tfrac{3}{4} + \tfrac{9}{16} - \tfrac{27}{64} + \tfrac{81}{256} - \ldots.$$

12. Sum an appropriate infinite series to find the rational number whose decimal expansion is $.13131313\ldots$.

13. Determine the Taylor series for $\cos 2x$ at $x = 0$ and integrate to derive the Taylor series for $\sin 2x$.

14. Determine the Taylor series for $\ln(1 + x)$ at $x = 0$ and differentiate to obtain the Taylor series for $1/(1 + x)$.

15. Determine the Taylor series for e^{3x} at $x = 0$. Use the Taylor series to demonstrate that

$$\frac{d}{dx} e^{3x} = 3e^{3x}.$$

16. Find an infinite series that converges to

$$\int_0^{1/2} \frac{e^x - 1}{x}\, dx.$$

(Note: The Taylor series for $(e^x - 1)/x$ can be obtained from the Taylor series of e^x by subtracting 1 and dividing by x.)

A pair of dice is rolled until a "seven" or an "eleven" appears, and the number of rolls preceding the final roll is observed. The probability of rolling "seven or eleven" is $\tfrac{2}{9}$.

17. Determine the formula for p_n, the probability of exactly n consecutive rolls preceding the final roll.

18. What is the probability that an odd number of consecutive rolls precede the final roll?

19. Determine the average number of consecutive rolls preceding the final roll.

A small volume of blood is to be selected, examined under a microscope, and the number of white blood cells counted. Suppose that for healthy people the number of white blood cells in such a specimen is Poisson distributed with $\lambda = 4$.

20. What is the probability that a specimen from a healthy person has exactly four white blood cells?

21. What is the probability that a specimen from a healthy person has eight or more white blood cells?

22. What is the average number of white blood cells per specimen from a healthy person?

APPENDIX
TABLES

TABLE 1   The Exponential Function

x	e^x	e^{-x}	x	e^x	e^{-x}	x	e^x	e^{-x}
.00	1.00000	1.00000	.40	1.49182	.67032	.80	2.22554	.44933
.01	1.01005	.99005	.41	1.50682	.66365	.81	2.24791	.44486
.02	1.02020	.98020	.42	1.52196	.65705	.82	2.27050	.44043
.03	1.03045	.97045	.43	1.53726	.65051	.83	2.29332	.43605
.04	1.04081	.96079	.44	1.55271	.64404	.84	2.31637	.43171
.05	1.05127	.95123	.45	1.56831	.63763	.85	2.33965	.42741
.06	1.06184	.94176	.46	1.58407	.63128	.86	2.36316	.42316
.07	1.07251	.93239	.47	1.59999	.62500	.87	2.38691	.41895
.08	1.08329	.92312	.48	1.61607	.61878	.88	2.41090	.41478
.09	1.09417	.91393	.49	1.63232	.61263	.89	2.43513	.41066
.10	1.10517	.90484	.50	1.64872	.60653	.90	2.45960	.40657
.11	1.11628	.89583	.51	1.66529	.60050	.91	2.48432	.40252
.12	1.12750	.88692	.52	1.68203	.59452	.92	2.50929	.39852
.13	1.13883	.87810	.53	1.69893	.58860	.93	2.53451	.39455
.14	1.15027	.86936	.54	1.71601	.58275	.94	2.55998	.39063
.15	1.16183	.86071	.55	1.73325	.57695	.95	2.58571	.38674
.16	1.17351	.85214	.56	1.75067	.57121	.96	2.61170	.38289
.17	1.18530	.84366	.57	1.76827	.56553	.97	2.63794	.37908
.18	1.19722	.83527	.58	1.78604	.55990	.98	2.66446	.37531
.19	1.20925	.82696	.59	1.80399	.55433	.99	2.69123	.37158
.20	1.22140	.81873	.60	1.82212	.54881	1.00	2.71828	.36788
.21	1.23368	.81058	.61	1.84043	.54335	1.01	2.74560	.36422
.22	1.24608	.80252	.62	1.85893	.53794	1.02	2.77319	.36059
.23	1.25860	.79453	.63	1.87761	.53259	1.03	2.80107	.35701
.24	1.27125	.78663	.64	1.89648	.52729	1.04	2.82922	.35345
.25	1.28403	.77880	.65	1.91554	.52205	1.05	2.85765	.34994
.26	1.29693	.77105	.66	1.93479	.51685	1.06	2.88637	.34646
.27	1.30996	.76338	.67	1.95424	.51171	1.07	2.91538	.34301
.28	1.32313	.75578	.68	1.97388	.50662	1.08	2.94468	.33960
.29	1.33643	.74826	.69	1.99372	.50158	1.09	2.97427	.33622
.30	1.34986	.74082	.70	2.01375	.49659	1.10	3.00417	.33287
.31	1.36343	.73345	.71	2.03399	.49164	1.11	3.03436	.32956
.32	1.37713	.72615	.72	2.05443	.48675	1.12	3.06485	.32628
.33	1.39097	.71892	.73	2.07508	.48191	1.13	3.09566	.32303
.34	1.40495	.71177	.74	2.09594	.47711	1.14	3.12677	.31982
.35	1.41907	.70469	.75	2.11700	.47237	1.15	3.15819	.31664
.36	1.43333	.69768	.76	2.13828	.46767	1.16	3.18993	.31349
.37	1.44773	.69073	.77	2.15977	.46301	1.17	3.22199	.31037
.38	1.46228	.68386	.78	2.18147	.45841	1.18	3.25437	.30728
.39	1.47698	.67706	.79	2.20340	.45384	1.19	3.28708	.30422

x	e^x	e^{-x}	x	e^x	e^{-x}	x	e^x	e^{-x}
1.20	3.32012	.30119	**1.60**	4.95303	.20190	**4.0**	54.598	.01832
1.21	3.35348	.29820	1.61	5.00281	.19989	4.1	60.340	.01657
1.22	3.38719	.29523	1.62	5.05309	.19790	4.2	66.686	.01500
1.23	3.42123	.29229	1.63	5.10387	.19593	4.3	73.700	.01357
1.24	3.45561	.28938	1.64	5.15517	.19398	4.4	81.451	.01228
1.25	3.49034	.28650	1.65	5.20698	.19205	4.5	90.017	.01111
1.26	3.52542	.28365	1.66	5.25931	.19014	4.6	99.484	.01005
1.27	3.56085	.28083	1.67	5.31217	.18825	4.7	109.947	.00910
1.28	3.59664	.27804	1.68	5.36556	.18637	4.8	121.510	.00823
1.29	3.63279	.27527	1.69	5.41948	.18452	4.9	134.290	.00745
1.30	3.66930	.27253	**1.70**	5.47395	.18268	**5.0**	148.41	.00674
1.31	3.70617	.26982	1.71	5.52896	.18087	5.1	164.02	.00610
1.32	3.74342	.26714	1.72	5.58453	.17907	5.2	181.27	.00552
1.33	3.78104	.26448	1.73	5.64065	.17728	5.3	200.34	.00499
1.34	3.81904	.26185	1.74	5.69734	.17552	5.4	221.41	.00452
1.35	3.85743	.25924	1.75	5.75460	.17377	5.5	244.69	.00409
1.36	3.89619	.25666	1.80	6.04965	.16530	5.6	270.43	.00370
1.37	3.93535	.25411	1.85	6.35982	.15724	5.7	298.87	.00335
1.38	3.97490	.25158	1.90	6.68589	.14957	5.8	330.30	.00303
1.39	4.01485	.24908	1.95	7.02869	.14227	5.9	365.04	.00274
1.40	4.05520	.24660	**2.0**	7.3891	.13534	**6.0**	403.43	.00248
1.41	4.09596	.24414	2.1	8.1662	.12246	6.1	445.86	.00224
1.42	4.13712	.24171	2.2	9.0250	.11080	6.2	492.75	.00203
1.43	4.17870	.23931	2.3	9.9742	.10026	6.3	544.57	.00184
1.44	4.22070	.23693	2.4	11.0232	.09072	6.4	601.85	.00166
1.45	4.26311	.23457	2.5	12.1825	.08208	6.5	665.14	.00150
1.46	4.30596	.23224	2.6	13.4637	.07427	6.6	735.10	.00136
1.47	4.34924	.22993	2.7	14.8797	.06721	6.7	812.41	.00123
1.48	4.39295	.22764	2.8	16.4446	.06081	6.8	897.85	.00111
1.49	4.43710	.22537	2.9	18.1741	.05502	6.9	992.27	.00101
1.50	4.48169	.22313	**3.0**	20.086	.04979	**7.0**	1096.6	.00091
1.51	4.52673	.22091	3.1	22.198	.04505	7.5	1808.0	.00055
1.52	4.57223	.21871	3.2	24.533	.04076	8.0	2981.0	.00034
1.53	4.61818	.21654	3.3	27.113	.03688	8.5	4914.8	.00020
1.54	4.66459	.21438	3.4	29.964	.03337	9.0	8103.1	.00012
1.55	4.71147	.21225	3.5	33.115	.03020	9.5	13360	.00007
1.56	4.75882	.21014	3.6	36.598	.02732	10.0	22026	.00005
1.57	4.80665	.20805	3.7	40.447	.02472	10.5	36316	.00003
1.58	4.85496	.20598	3.8	44.701	.02237	11.0	59874	.00002
1.59	4.90375	.20393	3.9	49.402	.02024	11.5	98716	.00001

TABLE 2 The Natural Logarithm Function

x	ln x	x	ln x	x	ln x	x	ln x
		0.40	−0.91629	0.80	−0.22314	1.20	0.18232
0.01	−4.60517	0.41	−0.89160	0.81	−0.21072	1.21	0.19062
0.02	−3.91202	0.42	−0.86750	0.82	−0.19845	1.22	0.19885
0.03	−3.50656	0.43	−0.84397	0.83	−0.18633	1.23	0.20701
0.04	−3.21888	0.44	−0.82098	0.84	−0.17435	1.24	0.21511
0.05	−2.99573	0.45	−0.79851	0.85	−0.16252	1.25	0.22314
0.06	−2.81341	0.46	−0.77653	0.86	−0.15082	1.26	0.23111
0.07	−2.65926	0.47	−0.75502	0.87	−0.13926	1.27	0.23902
0.08	−2.52573	0.48	−0.73397	0.88	−0.12783	1.28	0.24686
0.09	−2.40795	0.49	−0.71335	0.89	−0.11653	1.29	0.25464
0.10	−2.30259	0.50	−0.69315	0.90	−0.10536	1.30	0.26236
0.11	−2.20727	0.51	−0.67334	0.91	−0.09431	1.31	0.27003
0.12	−2.12026	0.52	−0.65393	0.92	−0.08338	1.32	0.27763
0.13	−2.04022	0.53	−0.63488	0.93	−0.07257	1.33	0.28518
0.14	−1.96611	0.54	−0.61619	0.94	−0.06188	1.34	0.29267
0.15	−1.89712	0.55	−0.59784	0.95	−0.05129	1.35	0.30010
0.16	−1.83258	0.56	−0.57982	0.96	−0.04082	1.36	0.30748
0.17	−1.77196	0.57	−0.56212	0.97	−0.03046	1.37	0.31481
0.18	−1.71480	0.58	−0.54473	0.98	−0.02020	1.38	0.32208
0.19	−1.66073	0.59	−0.52763	0.99	−0.01005	1.39	0.32930
0.20	−1.60944	0.60	−0.51083	1.00	0.00000	1.40	0.33647
0.21	−1.56065	0.61	−0.49430	1.01	0.00995	1.41	0.34359
0.22	−1.51413	0.62	−0.47804	1.02	0.01980	1.42	0.35066
0.23	−1.46968	0.63	−0.46204	1.03	0.02956	1.43	0.35767
0.24	−1.42712	0.64	−0.44629	1.04	0.03922	1.44	0.36464
0.25	−1.38629	0.65	−0.43078	1.05	0.04879	1.45	0.37156
0.26	−1.34707	0.66	−0.41552	1.06	0.05827	1.46	0.37844
0.27	−1.30933	0.67	−0.40048	1.07	0.06766	1.47	0.38526
0.28	−1.27297	0.68	−0.38566	1.08	0.07696	1.48	0.39204
0.29	−1.23787	0.69	−0.37106	1.09	0.08618	1.49	0.39878
0.30	−1.20397	0.70	−0.35667	1.10	0.09531	1.50	0.40547
0.31	−1.17118	0.71	−0.34249	1.11	0.10436	1.51	0.41211
0.32	−1.13943	0.72	−0.32850	1.12	0.11333	1.52	0.41871
0.33	−1.10866	0.73	−0.31471	1.13	0.12222	1.53	0.42527
0.34	−1.07881	0.74	−0.30111	1.14	0.13103	1.54	0.43178
0.35	−1.04982	0.75	−0.28768	1.15	0.13976	1.55	0.43825
0.36	−1.02165	0.76	−0.27444	1.16	0.14842	1.56	0.44469
0.37	−0.99425	0.77	−0.26136	1.17	0.15700	1.57	0.45108
0.38	−0.96758	0.78	−0.24846	1.18	0.16551	1.58	0.45742
0.39	−0.94161	0.79	−0.23572	1.19	0.17395	1.59	0.46373

TABLE 2   The Natural Logarithm Function *(continued)*

x	ln x	x	ln x	x	ln x	x	ln x
1.60	0.47000	2.00	0.69315	6.00	1.79176	10.0	2.30259
1.61	0.47623	2.10	0.74194	6.10	1.80829	11.0	2.39790
1.62	0.48243	2.20	0.78846	6.20	1.82455	12.0	2.48491
1.63	0.48858	2.30	0.83291	6.30	1.84055	13.0	2.56495
1.64	0.49470	2.40	0.87547	6.40	1.85630	14.0	2.63906
1.65	0.50078	2.50	0.91629	6.50	1.87180	15.0	2.70805
1.66	0.50682	2.60	0.95551	6.60	1.88707	16.0	2.77259
1.67	0.51282	2.70	0.99325	6.70	1.90211	17.0	2.83321
1.68	0.51879	2.80	1.02962	6.80	1.91692	18.0	2.89037
1.69	0.52473	2.90	1.06471	6.90	1.93152	19.0	2.94444
1.70	0.53063	3.00	1.09861	7.00	1.94591	20.0	2.99573
1.71	0.53649	3.10	1.13140	7.10	1.96009	21.0	3.04452
1.72	0.54232	3.20	1.16315	7.20	1.97408	22.0	3.09104
1.73	0.54812	3.30	1.19392	7.30	1.98787	23.0	3.13549
1.74	0.55389	3.40	1.22378	7.40	2.00148	24.0	3.17805
1.75	0.55962	3.50	1.25276	7.50	2.01490	25.0	3.21888
1.76	0.56531	3.60	1.28093	7.60	2.02815	26.0	3.25810
1.77	0.57098	3.70	1.30833	7.70	2.04122	27.0	3.29584
1.78	0.57661	3.80	1.33500	7.80	2.05412	28.0	3.33220
1.79	0.58222	3.90	1.36098	7.90	2.06686	29.0	3.36730
1.80	0.58779	4.00	1.38629	8.00	2.07944	30.0	3.40120
1.81	0.59333	4.10	1.41099	8.10	2.09186	31.0	3.43399
1.82	0.59884	4.20	1.43508	8.20	2.10413	32.0	3.46574
1.83	0.60432	4.30	1.45862	8.30	2.11626	33.0	3.49651
1.84	0.60977	4.40	1.48160	8.40	2.12823	34.0	3.52636
1.85	0.61519	4.50	1.50408	8.50	2.14007	35.0	3.55535
1.86	0.62058	4.60	1.52606	8.60	2.15176	36.0	3.58352
1.87	0.62594	4.70	1.54756	8.70	2.16332	37.0	3.61092
1.88	0.63127	4.80	1.56862	8.80	2.17475	38.0	3.63759
1.89	0.63658	4.90	1.58924	8.90	2.18605	39.0	3.66356
1.90	0.64185	5.00	1.60944	9.00	2.19722	40.0	3.68888
1.91	0.64710	5.10	1.62924	9.10	2.20827	41.0	3.71357
1.92	0.65233	5.20	1.64866	9.20	2.21920	42.0	3.73767
1.93	0.65752	5.30	1.66771	9.30	2.23001	43.0	3.76120
1.94	0.66269	5.40	1.68640	9.40	2.24071	44.0	3.78419
1.95	0.66783	5.50	1.70475	9.50	2.25129	45.0	3.80666
1.96	0.67294	5.60	1.72277	9.60	2.26176	46.0	3.82864
1.97	0.67803	5.70	1.74047	9.70	2.27213	47.0	3.85015
1.98	0.68310	5.80	1.75786	9.80	2.28238	48.0	3.87120
1.99	0.68813	5.90	1.77495	9.90	2.29253	49.0	3.89182

x	$\ln x$	x	$\ln x$	x	$\ln x$	x	$\ln x$
50.0	3.91202	90.0	4.49981	400.	5.99146	800.	6.68461
51.0	3.93183	91.0	4.51086	410.	6.01616	810.	6.69703
52.0	3.95124	92.0	4.52179	420.	6.04025	820.	6.70930
53.0	3.97029	93.0	4.53260	430.	6.06379	830.	6.72143
54.0	3.98898	94.0	4.54329	440.	6.08677	840.	6.73340
55.0	4.00733	95.0	4.55388	450.	6.10925	850.	6.74524
56.0	4.02535	96.0	4.56435	460.	6.13123	860.	6.75693
57.0	4.04305	97.0	4.57471	470.	6.15273	870.	6.76849
58.0	4.06044	98.0	4.58497	480.	6.17379	880.	6.77992
59.0	4.07754	99.0	4.59512	490.	6.19441	890.	6.79122
60.0	4.09434	100.	4.60517	500.	6.21461	900.	6.80239
61.0	4.11087	110.	4.70048	510.	6.23441	910.	6.81344
62.0	4.12713	120.	4.78749	520.	6.25383	920.	6.82437
63.0	4.14313	130.	4.86753	530.	6.27288	930.	6.83518
64.0	4.15888	140.	4.94164	540.	6.29157	940.	6.84588
65.0	4.17439	150.	5.01064	550.	6.30992	950.	6.85646
66.0	4.18965	160.	5.07517	560.	6.32794	960.	6.86693
67.0	4.20469	170.	5.13580	570.	6.34564	970.	6.87730
68.0	4.21951	180.	5.19296	580.	6.36303	980.	6.88755
69.0	4.23411	190.	5.24702	590.	6.38012	990.	6.89770
70.0	4.24850	200.	5.29832	600.	6.39693	1000.	6.90776
71.0	4.26268	210.	5.34711	610.	6.41346	—	—
72.0	4.27667	220.	5.39363	620.	6.42972	—	—
73.0	4.29046	230.	5.43808	630.	6.44572	—	—
74.0	4.30407	240.	5.48064	640.	6.46147	—	—
75.0	4.31749	250.	5.52146	650.	6.47697	—	—
76.0	4.33073	260.	5.56068	660.	6.49224	—	—
77.0	4.34381	270.	5.59842	670.	6.50728	—	—
78.0	4.35671	280.	5.63479	680.	6.52209	—	—
79.0	4.36945	290.	5.66988	690.	6.53669	—	—
80.0	4.38203	300.	5.70378	700.	6.55108	—	—
81.0	4.39445	310.	5.73657	710.	6.56526	—	—
82.0	4.40672	320.	5.76832	720.	6.57925	—	—
83.0	4.41884	330.	5.79909	730.	6.59304	—	—
84.0	4.43082	340.	5.82895	740.	6.60665	—	—
85.0	4.44265	350.	5.85793	750.	6.62007	—	—
86.0	4.45435	360.	5.88610	760.	6.63332	—	—
87.0	4.46591	370.	5.91350	770.	6.64639	—	—
88.0	4.47734	380.	5.94017	780.	6.65929	—	—
89.0	4.48864	390.	5.96615	790.	6.67203	—	—

TABLE 3 Trigonometric Functions in Radians

t	$\sin t$	$\cos t$	$\tan t$	t	$\sin t$	$\cos t$	$\tan t$
.0	.00000	1.00000	.00000	3.1	.04158	−.99914	−.04162
.1	.09983	.99500	.10033	π	.00000	−1.00000	.00000
.2	.19867	.98007	.20271	3.2	−.05837	−.99829	.05847
.3	.29552	.95534	.30934	3.3	−.15775	−.98748	.15975
.4	.38942	.92106	.42279	3.4	−.25554	−.96680	.26432
.5	.47943	.87758	.54630	3.5	−.35078	−.93646	.37459
.6	.56464	.82534	.68414	3.6	−.44252	−.89676	.49347
.7	.64422	.76484	.84229	3.7	−.52984	−.84810	.62473
.8	.71736	.69671	1.02964	3.8	−.61186	−.79097	.77356
.9	.78333	.62161	1.26016	3.9	−.68777	−.72593	.94742
1.0	.84147	.54030	1.55741	4.0	−.75680	−.65364	1.15782
1.1	.89121	.45360	1.96476	4.1	−.81828	−.57482	1.42353
1.2	.93204	.36236	2.57215	4.2	−.87158	−.49026	1.77778
1.3	.96356	.26750	3.60210	4.3	−.91617	−.40080	2.28585
1.4	.98545	.16997	5.79788	4.4	−.95160	−.30733	3.09632
1.5	.99749	0.7074	14.10142	4.5	−.97753	−.21080	4.63733
$\pi/2$	1.00000	.00000	********	4.6	−.99369	−.11215	8.86017
1.6	.99957	−.02920	−34.23253	4.7	−.99992	−.01239	80.71269
1.7	.99166	−.12884	−7.69660	$3\pi/2$	−1.00000	.00000	********
1.8	.97385	−.22720	−4.28626	4.8	−.99616	.08750	−11.38487
1.9	.94630	−.32329	−2.92710	4.9	−.98245	.18651	−5.26749
2.0	.90930	−.41615	−2.18504	5.0	−.95892	.28366	−3.38052
2.1	.86321	−.50485	−1.70985	5.1	−.92581	.37798	−2.44939
2.2	.80850	−.58850	−1.37382	5.2	−.88345	.46852	−1.88564
2.3	.74571	−.66628	−1.11921	5.3	−.83227	.55437	−1.50127
2.4	.67546	−.73739	−.91601	5.4	−.77276	.63469	−1.21754
2.5	.59847	−.80114	−.74702	5.5	−.70554	.70867	−.99558
2.6	.51550	−.85689	−.60160	5.6	−.63127	.77557	−.81394
2.7	.42738	−.90407	−.47273	5.7	−.55069	.83471	−.65973
2.8	.33499	−.94222	−.35553	5.8	−.46460	.88552	−.52467
2.9	.23925	−.97096	−.24641	5.9	−.37388	.92748	−.40311
3.0	.14112	−.98999	−.14255	6.0	−.27942	.96017	−.29101
—	—	—	—	6.1	−.18216	.98327	−.18526
—	—	—	—	6.2	−.08309	.99654	−.08338
—	—	—	—	2π	.00000	1.00000	.00000

TABLE 4   Areas Under the Standard Normal Curve

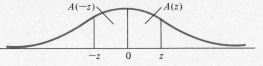

z	.00	.01	.02	.03	.04	.05	.06	.07	.08	.09
0.0	.0000	.0040	.0080	.0120	.0160	.0199	.0239	.0279	.0319	.0359
0.1	.0398	.0438	.0478	.0517	.0557	.0596	.0636	.0675	.0714	.0754
0.2	.0793	.0832	.0871	.0910	.0948	.0987	.1026	.1064	.1103	.1141
0.3	.1179	.1217	.1255	.1293	.1331	.1368	.1406	.1443	.1480	.1517
0.4	.1554	.1591	.1628	.1664	.1700	.1736	.1772	.1808	.1844	.1879
0.5	.1915	.1950	.1985	.2019	.2054	.2088	.2123	.2157	.2190	.2224
0.6	.2258	.2291	.2324	.2357	.2389	.2422	.2454	.2486	.2518	.2549
0.7	.2580	.2612	.2642	.2673	.2704	.2734	.2764	.2794	.2823	.2852
0.8	.2881	.2910	.2939	.2967	.2996	.3023	.3051	.3078	.3106	.3133
0.9	.3159	.3186	.3212	.3238	.3264	.3289	.3315	.3340	.3365	.3389
1.0	.3413	.3438	.3461	.3485	.3508	.3531	.3554	.3577	.3599	.3621
1.1	.3643	.3665	.3686	.3708	.3729	.3749	.3770	.3790	.3810	.3820
1.2	.3849	.3869	.3888	.3907	.3925	.3944	.3962	.3980	.3997	.4015
1.3	.4032	.4049	.4066	.4082	.4099	.4115	.4131	.4147	.4162	.4177
1.4	.4192	.4207	.4222	.4236	.4251	.4265	.4279	.4292	.4306	.4319
1.5	.4332	.4345	.4357	.4370	.4382	.4394	.4406	.4418	.4429	.4441
1.6	.4452	.4463	.4474	.4484	.4495	.4505	.4515	.4525	.4535	.4545
1.7	.4554	.4564	.4573	.4582	.4591	.4599	.4608	.4616	.4625	.4633
1.8	.4641	.4649	.4656	.4664	.4671	.4678	.4686	.4693	.4699	.4706
1.9	.4713	.4719	.4726	.4732	.4738	.4744	.4750	.4756	.4761	.4767
2.0	.4772	.4778	.4783	.4788	.4793	.4798	.4803	.4808	.4812	.4817
2.1	.4821	.4826	.4830	.4834	.4838	.4842	.4846	.4850	.4854	.4857
2.2	.4861	.4864	.4868	.4871	.4875	.4878	.4881	.4884	.4887	.4890
2.3	.4893	.4896	.4898	.4901	.4904	.4906	.4909	.4911	.4913	.4916
2.4	.4918	.4920	.4922	.4925	.4927	.4929	.4931	.4932	.4934	.4936
2.5	.4938	.4940	.4941	.4943	.4945	.4946	.4948	.4949	.4951	.4952
2.6	.4953	.4955	.4956	.4957	.4959	.4960	.4961	.4962	.4963	.4964
2.7	.4965	.4966	.4967	.4968	.4969	.4970	.4971	.4972	.4973	.4974
2.8	.4974	.4975	.4976	.4977	.4977	.4978	.4979	.4979	.4980	.4981
2.9	.4981	.4982	.4982	.4983	.4984	.4984	.4985	.4985	.4986	.4986
3.0	.4987	.4987	.4987	.4988	.4988	.4989	.4989	.4989	.4990	.4990
3.1	.4990	.4991	.4991	.4991	.4992	.4992	.4992	.4992	.4993	.4993
3.2	.4993	.4993	.4994	.4994	.4994	.4994	.4994	.4995	.4995	.4995
3.3	.4995	.4995	.4995	.4996	.4996	.4996	.4996	.4996	.4996	.4997
3.4	.4997	.4997	.4997	.4997	.4997	.4997	.4997	.4997	.4997	.4998
3.5	.4998	.4998	.4998	.4998	.4998	.4998	.4998	.4998	.4998	.4998

ANSWERS TO ODD-NUMBERED EXERCISES

1. $0, 10, 0, 70$ 3. $0, 0, -\frac{9}{8}, a^3 + a^2 - a - 1$ 5. $\frac{1}{3}, 3, \frac{a+1}{a+2}$

7. (a) 1980 sales, (b) 60   9. $x \neq 1, 2$ 11. $x < 3$ 13. Function

15. Not a function   17. Not a function   19. 1   21. 3

23. Positive   25. Positive   27. $-1, 5, 9$ 29. .03   31. .04

33. No   **35.** Yes   **37.**

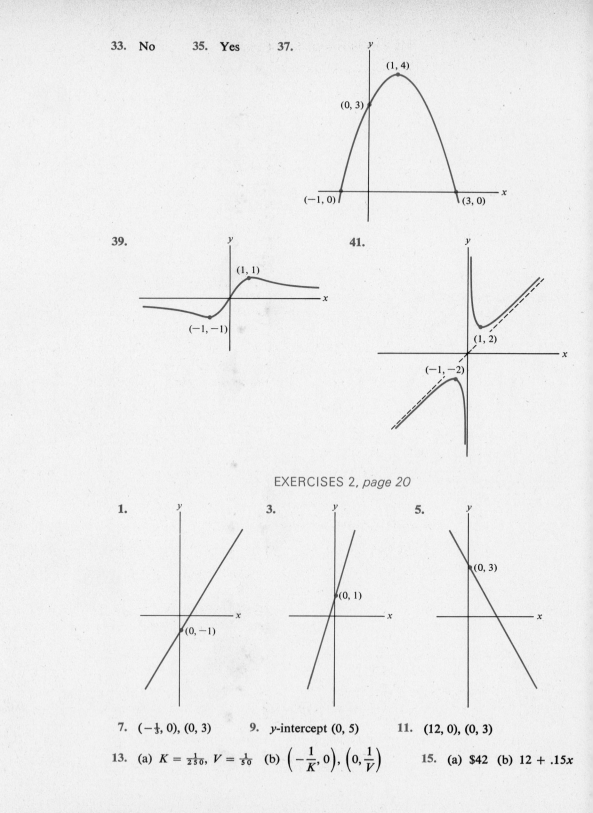

39.

41.

EXERCISES 2, *page 20*

1.

3.

5.

7. $\left(-\frac{1}{3}, 0\right)$, $(0, 3)$ **9.** *y*-intercept $(0, 5)$ **11.** $(12, 0)$, $(0, 3)$

13. (a) $K = \frac{1}{250}$, $V = \frac{1}{50}$ (b) $\left(-\frac{1}{K}, 0\right)$, $\left(0, \frac{1}{V}\right)$ **15.** (a) $\$42$ (b) $12 + .15x$

17. $150 + 135n$, $n =$ number

21. $a = -2, b = 3, c = 1$

27. 2   29. 4   31.

1. $x^2 + 9x + 1$ 3. $9x$

9. $\dfrac{4x}{x^2 - 12x + 32}$ 11.   1. $\frac{1}{3}$ 13. $-4, y - 4 = -4(x + 2)$

15. $\dfrac{-x^2 + 5x}{x^2 + 3x - 10}$ 17.   17. $y - 2.25 = 3(x - 1.5)$ 19. $(\frac{5}{6}, \frac{25}{36})$

19. $f(x)g(x) = x, f(x)/g(x) =$ 25. $\frac{3}{4}$ 27. $y + 1 = 3(x + 1)$

21. $f(x)g(x) = x^{25/12}, f(x)/g(x)$

EXERCISES 3, page 62

25. $\left(\dfrac{x}{1 - x}\right)^3 - 5\left(\dfrac{x}{1 - x}\right)^2 +$ $\frac{5}{2}x^{3/2}$ 7. $\frac{1}{3}x^{-2/3}$ 9. $-2x^{-3}$

31. $4h - h^2 - 2th$ 33.   15. $-3x^{-4}$ 17. -192 19. $-\frac{1}{9}$

25. 108 27. $\frac{1}{6}$ 29. $25, -10$

$\frac{8}{3}$ 35. $48, y - 64 = 48(x - 4)$ 37. $8x^7$

43. $\frac{1}{5}x^{-4/5}$

1. $2, \frac{3}{2}$ 3. $\frac{3}{2}$ 5.

EXERCISES 4, page 71

11. $2 + \sqrt{6}/3, 2 - \sqrt{6}/3$ 5. No limit   7. -5 9. 5

17. $3(x + 2)^2$ 19. $-2(x$ 15. 0   17. 3   19. -4

23. $-2x(x - \sqrt{3})(x + \sqrt{3})$ 25. No limit   27. $-\frac{2}{11}$ 29. 6   31. 3

29. $(0, 8), (2, 6)$ 31. $(0,$ 37. 0   39. 3   41. 0   43. 0

33. $-7, 3$ 35. $-2, 3$

CHAPTER 0: SU

EXERCISES 5, page 78

1. $2, 27\frac{1}{3}, -2, -2\frac{1}{8}, \dfrac{5\sqrt{2}}{2}$ 5. No   7. No   9. Yes   11. No

le   15. Not continuous, not differentiable

7. All x 9. Yes   ntiable   19. Continuous, differentiable

15. $-\frac{2}{5}, 1$ 17. $\left(\dfrac{5 + \sqrt{4}}{10}\right.$

EXERCISES 6, page 84

21. $x^{5/2} - 2x^{3/2}$ 23. $x^{3/2}$ $+ 3$ 5. $5x^4 - \dfrac{1}{x^2}$ 7. $4x^3 + 3x^2 + 1$

9. $6x$ 11. $3x^2 + 14x$ 13. $-\dfrac{8}{x^3}$ 15. $3 + \dfrac{1}{x^2}$

17. $x^2 - x + 1$ 19. $\dfrac{1}{x^6}$ 21. $\dfrac{-1}{2\sqrt{x}}$ 23. $30(3x + 1)^9$

25. $\dfrac{9x^2 + 1}{2\sqrt{3x^3 + x + 1}}$ 27. $6(4x - 1)(2x^2 - x + 4)^5$ 29. 4   31. 15

33. $f'(4) = 48,\ y = 48x - 191$

35. $f'(x) = 36x^3 + 18x^2 - 22x - 4 = 2(3x^2 + x - 2)(6x + 1)$

EXERCISES 7, *page 90*

1. $10t(t^2 + 1)^4$ 3. $(2t - 1)^{-1/2}$ 5. $\tfrac{2}{3}(T^3 + 5T)^{-1/3}(3T^2 + 5)$

7. $6P - \tfrac{1}{2}$ 9. $2a^2t + b^2$ 11. $f'(x) = x - 7,\ f''(x) = 1$

13. $y' = \tfrac{1}{2}x^{-1/2},\ y'' = -\tfrac{1}{4}x^{-3/2}$ 15. $f'(r) = 2\pi(hr + 1),\ f''(r) = 2\pi h$

17. $g'(x) = -5,\ g''(x) = 0$ 19. $f'(P) = 15(3P + 1)^4,\ f''(P) = 180(3P + 1)^3$

21. 20   23. 54   25. 34   27. $8k(2P - 1)^{-3}$

29. $f'(3) = -\tfrac{1}{2},\ f''(3) = -\tfrac{1}{8}$ 31. 20

33. (a) $f'''(x) = 60x^2 - 24x$ (b) $f'''(x) = \dfrac{15}{2\sqrt{x}}$

EXERCISES 8, *page 96*

1. 13   3. 63 units per hour   5. 1 gram per week

7. $-\tfrac{200}{27}$ units per day, $\tfrac{8}{5}$ units per day, increasing

9. (a) \$16   (b) extra cost $= \$16.10$ 11. (a) \$9.01   (b) $\approx \$9.10$

13. (a) 7 km/hr   (b) 16.5 km   15. (a) 160 ft/sec   (b) 96 ft/sec

     (c) -32 ft/sec^2 (d) $t = 10$ (e) -160 ft/sec   17. $t = 20$

19. $(2 \pm \sqrt{2})/2$

CHAPTER 1: SUPPLEMENTARY EXERCISES, *page 99*

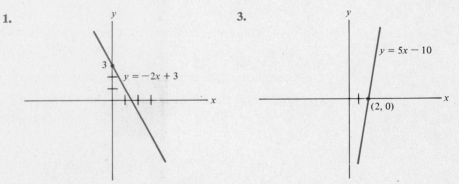

1.

3.

5.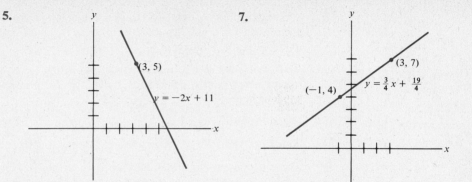

(3, 5)

$y = -2x + 11$

7.

(3, 7)

$y = \frac{3}{4}x + \frac{19}{4}$

(−1, 4)

9. $7x^6 + 3x^2$ **11.** $\dfrac{3}{\sqrt{x}}$ **13.** $-\dfrac{3}{x^2}$ **15.** $48x(3x^2 - 1)^7$

17. $-\dfrac{5}{(5x - 1)^2}$ **19.** $\dfrac{x}{\sqrt{x^2 + 1}}$ **21.** $-4x^{-5}$ **23.** 0

25. $10[x^5 - (x - 1)^5]^9[5x^4 - 5(x - 1)^4]$ **27.** $\frac{3}{2}t^{-1/2} + \frac{3}{2}t^{-3/2}$

29. $\dfrac{2(9t^2 - 1)}{(t - 3t^3)^2}$ **31.** $\frac{9}{4}x^{1/2} - 4x^{-1/3}$ **33.** 28 **35.** $14, 3$

37. $\frac{15}{2}$ **39.** 33 **41.** $4x^3 - 4x$ **43.** $-\frac{3}{2}(1 - 3P)^{-1/2}$ **45.** 29

47. $300(5x + 1)^2$ **49.** -2 **51.** $3x^{-1/2}$

53. slope -4; tangent $y = -4x + 6$

55.

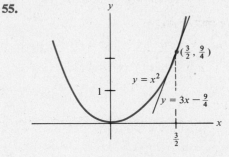

$\left(\frac{3}{2}, \frac{9}{4}\right)$

$y = x^2$

$y = 3x - \frac{9}{4}$

$\frac{3}{2}$

57. $y = 2$ **59.** 96 ft/sec **61.** 4 **63.** Does not exist

65. $-\frac{1}{50}$

CHAPTER 2
EXERCISES 1, *page 109*

1. (a), (e), (f) **3.** (b), (c), (d)

5. Decreasing for $x < -2$, min at $(-2, -2)$, increasing for $x > -2$, concave up, y-intercept $(0, 0)$, x-intercepts $(0, 0)$ and $(-3.6, 0)$.

7. Decreasing for $x < 0$, min at $(0, 2)$, increasing for $0 < x < 2$, max at $(2, 4)$, decreasing for $x > 2$, concave up for $x < 1$, concave down for $x > 1$, inflection point at $(1, 3)$, y-intercept $(0, 2)$, x-intercept $(3.4, 0)$.

9. Decreasing for $x < 2$, min at $(2, 3)$, increasing for $x > 2$, concave up for all x, no inflection points, defined for $x > 0$, the line $y = x$ is an asymptote, the y-axis is an asymptote.

11. Slope increases for all x.

13. Slope decreases for $x < 3$, increases for $x > 3$. Minimum slope occurs at $x = 3$.

15. Oxygen content decreases until time a, at which time it reaches a minimum. After a, oxygen content steadily increases. The rate of increase increases until b, and then decreases. Time b is the time when oxygen content is increasing fastest.

17.

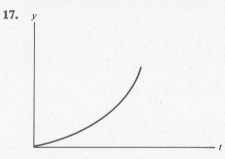

19. The parachutist's speed levels off to 15 feet per second.

21. (a) Yes (b) Yes

23.   **25.**

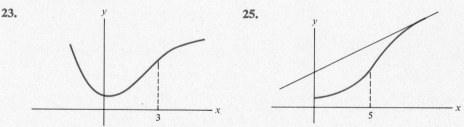

EXERCISES 2, *page 116*

1. (b), (c), (f)   **3.** (d), (e), (f)

5.   **7.**

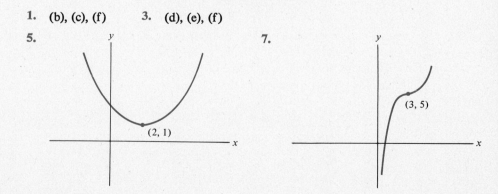

9.

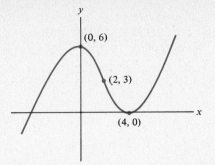

11. The second curve **13.** The second curve

15.

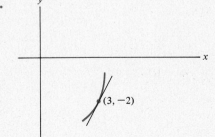

17.

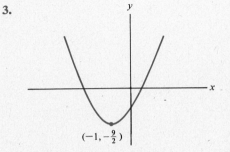

19.

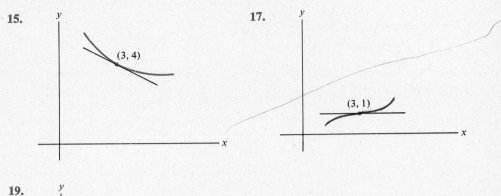

EXERCISES 3, *page 125*

1.

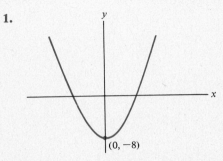

3.

5.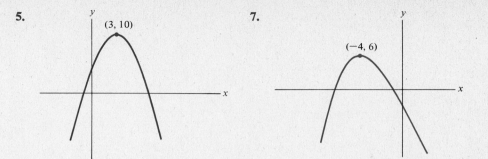

(3, 10)

7.

(−4, 6)

9.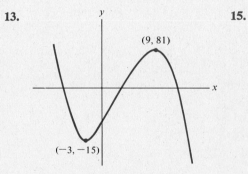

(−3, 0)

(−1, −4)

11.

(−2, 16)

(2, −16)

13.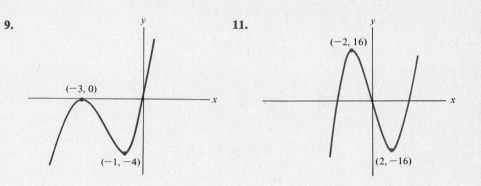

(9, 81)

(−3, −15)

15.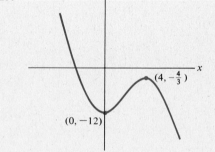

$(4, -\frac{4}{3})$

(0, −12)

17.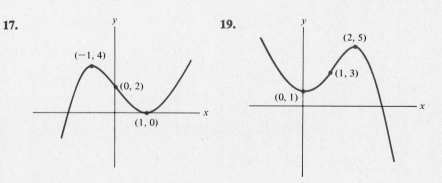

(−1, 4)

(0, 2)

(1, 0)

19.

(2, 5)

(1, 3)

(0, 1)

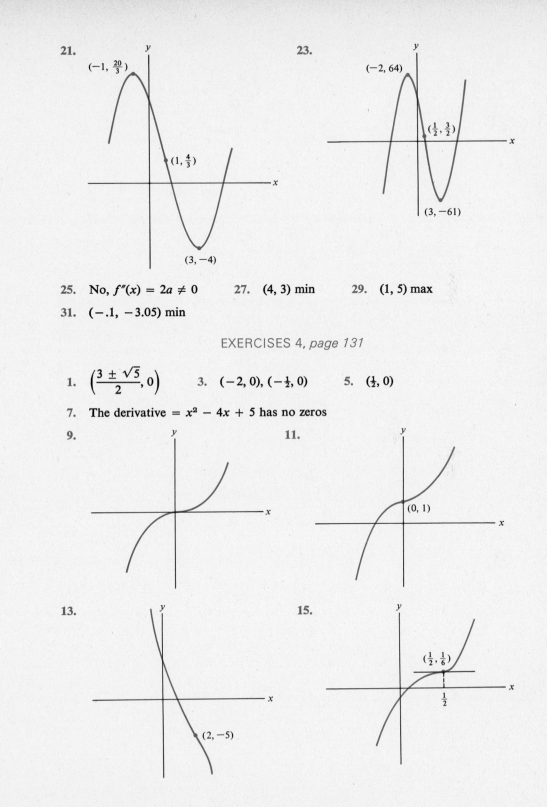

21.

$(-1, \frac{20}{3})$

$(1, \frac{4}{3})$

$(3, -4)$

23.

$(-2, 64)$

$(\frac{1}{2}, \frac{3}{2})$

$(3, -61)$

25. No, $f''(x) = 2a \neq 0$ **27.** (4, 3) min   **29.** (1, 5) max

31. $(-.1, -3.05)$ min

EXERCISES 4, *page 131*

1. $\left(\dfrac{3 \pm \sqrt{5}}{2}, 0\right)$ **3.** $(-2, 0), (-\frac{1}{2}, 0)$ **5.** $(\frac{1}{2}, 0)$

7. The derivative $= x^2 - 4x + 5$ has no zeros

9.

11.

$(0, 1)$

13.

$(2, -5)$

15.

$(\frac{1}{2}, \frac{1}{6})$

$\frac{1}{2}$

17.

19.

21.

23.

25.

27.

29.

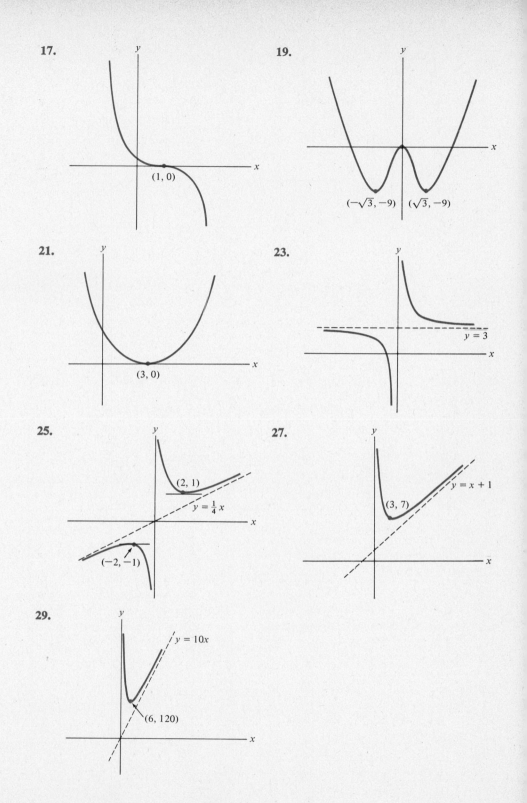

31.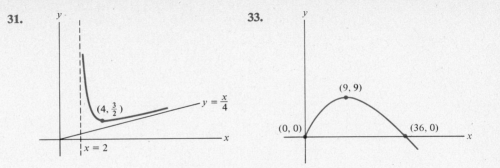

33.

EXERCISES 5, *page 139*

1. 36   3. $t = 4$, $f(4) = 8$ 5. 20   7. 150 ft   9. 250

11. $x = 50$, $h = 50$ 13. $x = 4$ in., $h = 2$ in.   15. $x = \dfrac{14}{4 + \pi}$ ft

17. $x = 10$ ft, $h = 7.5$ ft   19. 20 cm × 20 cm × 20 cm

21. 5 ft × 5 ft × 6 ft

EXERCISES 6, *page 146*

1. 10 m × 10 m   3. $x = \dfrac{220}{\pi}$ yd   5. $\frac{30}{4}$ in.   7. 100 ft × 120 ft

9. $x = 6$ meters, $h = 9$ meters   11. 150   13. $1.10

15. $x = 40{,}000$ 17. 400   19. $t = 0$ 21. $2\sqrt{3} \times 6$

EXERCISES 7, *page 159*

1. $1   3. 32   5. 5   7. $x = 20$ units, $p = 133.33

9. 2 million tons, $156 per ton   11. (a) $1.00   (b) $1.15

13. (a) $x = 15 \cdot 10^5$, $p = 45.   (b) No. Profit is maximized when price is increased to $50.

CHAPTER 2: SUPPLEMENTARY EXERCISES, *page 161*

1. Graph goes through (1, 2), increasing at $x = 1$

3. Increasing and concave up at $x = 3$

5. (10, 2) is a minimum point

7. Graph goes through (5, −1), decreasing at $x = 5$

9. $(-2, 0)$ is a maximum point

11.

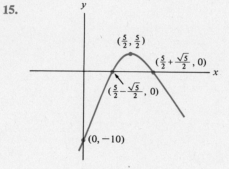

13.

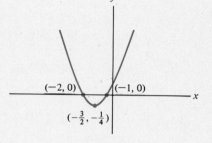

15.

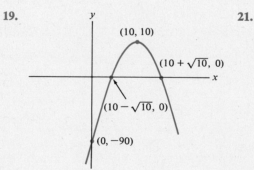

17.

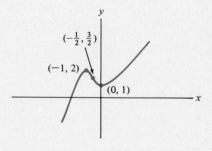

19.

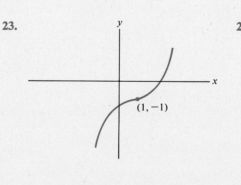

21.

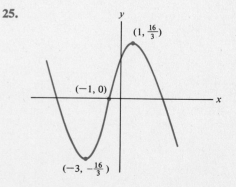

23.

25.

27.

29.

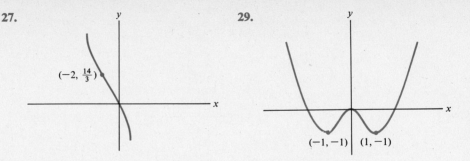

$(-2, \frac{14}{3})$

$(-1, -1)$ $(1, -1)$

31. $f'(x) = 3x(x^2 + 2)^{1/2}$, $f'(0) = 0$

33. $f''(x) = -2x(1 + x^2)^{-2}$, $f''(x)$ is positive for $x < 0$ and negative for $x > 0$

35. $1\frac{1}{2}$ ft wide $\times$ 3 ft long $\times$ 2 ft deep

37. $x = 25\sqrt{3}$, $y = 25\sqrt{6}$

39. 2000 cans

CHAPTER 3
EXERCISES 1, *page 170*

1. $4x^3 + 3x^2 + 10x + 7$ **3.** $24x^3 - 6x^2 + 2x + 1$

5. $15x^4 + 20x^3 + 3$ **7.** $3x^2 - 6x + 3 - \dfrac{2}{x^2}$ **9.** $(x - 1)^4(12x - 37)$

11. $2x(x^2 - 3)^9(11x^2 + 27)$ **13.** $(6x - 47)(2x + 9)^{-3}$

15. $(3x^2 - 1)^2(\frac{39}{2}x^{3/2} - \frac{1}{2}x^{-1/2})$ **17.** $(x - 4x^2)(x^2 - 1)^{-1/2} - 4(x^2 - 1)^{1/2}$

19. $\dfrac{4x}{(x^2 + 1)^2}$ **21.** $-\dfrac{10x + 2}{(5x^2 + 2x + 5)^2}$ **23.** $\dfrac{\dfrac{1}{2\sqrt{x}} - \dfrac{\sqrt{x}}{2}}{(x + 1)^2}$

25. $\dfrac{-3x^2 + 1}{(x^2 + 1)^3}$ **27.** $\dfrac{2}{\sqrt{(x - 2)(x + 2)^3}}$ **29.** $-\dfrac{5x + 6}{(4x + x^2)^{3/2}}$

31. $\dfrac{x(x^5 + 1)^2(33x^5 + 15x^3 - 12)}{(3x^2 + 1)^3}$

33. (a) $f(0) = 0$, $f(\frac{1}{2}) = \frac{1}{2}$, $f(1) = 1$ (d)

(c) Concave up for $0 < x < \frac{1}{2}$; concave down for $\frac{1}{2} < x < 1$; inflection point at $(\frac{1}{2}, \frac{1}{2})$

43. $f(x)$

EXERCISES 2, *page 176*

1. $\dfrac{1}{x^3 + 1}$ 3. $\sqrt{\dfrac{x^2}{x^2 - 1}}$ 5. x^4

7. $f(x) = \sqrt{x}, g(x) = x^2 + 3x - 1$ 9. $f(x) = \sqrt{x}, g(x) = \dfrac{x - 1}{x + 1}$

11. $f(x) = x^5, g(x) = x^3 + 9x - 2$ 13. $f(x) = \dfrac{x - 1}{x + 1}, g(x) = x^2$

15. $f(x) = \sqrt{x + 1}, g(x) = x^{1/3}$ 17. $f(x) = x^2 + 1, g(x) = x^3 + 1$

19. $f(x) = x(x^2 + 1)^2, g(x) = x^2$ 21. $30x(x^2 + 1)^{14}$

23. $2x^3(1 + x^4)^{-1/2}$ 25. $3\left[\dfrac{2}{x} + 5(3x - 1)^4\right]^2\left[-\dfrac{2}{x^2} + 60(3x - 1)^3\right]$

27. $8x(x^3 - 1)(x^2 + 1)^3 + 3x^2(x^2 + 1)^4$

29. $3(4 - x)^3(4 + x)^2 - 3(4 - x)^2(4 + x)^3$

31. $\dfrac{x}{2}(2 + \sqrt{1 + x^2})^{-1/2}(1 + x^2)^{-1/2}$

33. $-\dfrac{\sqrt{4x + x^2} + (3 - x)(2 + x)(4x + x^2)^{-1/2}}{4x + x^2}$ 35. $\dfrac{2(2x)^{-1/2}}{(1 - \sqrt{2x})^2}$

37.   41. $x^3 + 1$

CHAPTER 3: SUPPLEMENTARY EXERCISES, *page 178*

1. $\frac{7}{2}x^{5/2} + \frac{1}{2}x^{-1/2}$ 3. $\dfrac{1}{2\sqrt{x}(\sqrt{x} + 1)^2}$

5. $(x^3 - 2)^9(32x^4 + 155x^3 + 270x^2 - 4x - 10)$ 7. $\dfrac{-8x^2 + 16}{(x^2 + 2)^2}$

9. $(10 + 4x)(x + 2 + (x + 2)^2)$

11. $(\sqrt{x} - 2)(9(3x + 1)^2 + 6(3x + 1)) + \dfrac{(3x + 1)^3 + (3x + 1)^2 + 1}{2\sqrt{x}}$

13. $\dfrac{1}{2\sqrt{x + \sqrt{x + \sqrt{x}}}}\left(1 + \dfrac{1}{2\sqrt{x + \sqrt{x}}}\right)\left(1 + \dfrac{1}{2\sqrt{x}}\right)$

15. $\dfrac{1}{3\left(x + \sqrt{x + \sqrt{x}}\right)^{2/3}}\left(1 + \dfrac{1}{2\sqrt{x + \sqrt{x}}}\right)\left(1 + \dfrac{1}{2\sqrt{x}}\right)$ 17. $\frac{3}{2}$

CHAPTER 4
EXERCISES 1, *page 184*

1. $4, 27, \frac{1}{8}$ 3. $\frac{1}{16}, 512, \frac{2}{3}$ 5. $4, \frac{1}{9}$ 7. $9, 8$ 9. $32, 5$

11. (a) 8 (b) $\frac{1}{8}$ (c) 5.66 (d) 17.15 (e) 1.15 (f) 1.87 (g) $.18$ (h) $.07$

13. 1 15. 2 17. -1 19. $\frac{1}{5}$ 21. $\frac{5}{2}$ 23. 2^{15x}

25. $5^{3x} - 1$ 27. $3^{5x} + 1$

EXERCISES 2, *page 190*

1. $1.1612, 1.105, 1.10$ 3. $1.005, 1.002, 1$ 5. $10e^{10x}$ 7. $4e^{4x}$

9. $e^{2x} + e^{5x}$ 11. e^{1+x} 13. e^{2x} 15. 7.3891 17. $.60653$

19. $x = 4$ 21. $x = 4, -2$ 23. $xe^x + e^x$ 25. $\dfrac{e^x}{(1 + e^x)^2}$

27. $20e^x(1 + 5e^x)^3$

EXERCISES 3, *page 196*

1. $-e^{-x}$ 3. $5e^x$ 5. $2te^{t^2}$ 7. $\dfrac{e^x + e^{-x}}{2}$ 9. $-2(e^{-2x} + 1)$

11. $3(e^x + e^{-x})^2(e^x - e^{-x})$ 13. $-\frac{2}{3}e^{3 - 2x}$ 15. $3e^{3t}$

17. $\left(3x^2 + 1 + \dfrac{1}{x^2}\right)e^{x^3 + x - (1/x)}$ 19. $4(2x + 1 - e^{2x+1})^3(2 - 2e^{2x+1})$

21. $3x^2e^{x^2} + 2x^4e^{x^2}$ 23. $3xe^{x^3} - x^{-2}e^{x^3}$ 25. $-xe^{-x+2}$

27. $\left(-\dfrac{1}{x^2} + \dfrac{1}{x} + 3\right)e^x$ 29. $\dfrac{2e^x}{(e^x + 1)^2}$ 31. Max at $x = 1$

33. Min at $x = \frac{11}{2}$ 35. Max at $x = 3$ 37. Min at $x = 1$; max at $x = 3$

39. Max at $x = -6$; min at $x = -5$ 41. $.02e^{-2e^{-.01x}}e^{-.01x}$

43. $y = Ce^{-4x}$ 45. $y = e^{-.5x}$

1. 81   3. $\frac{1}{25}$ 5. 4   7. 9   9. e^{3x^2} 11. e^{2x}

13. $e^{11x} + 7e^x$ 15. $x = 4$ 17. $x = -5$ 19. $70e^{7x}$

21. $e^{x^2} + 2x^2 e^{x^2}$ 23. $e^x \cdot e^{e^x} = e^{x + e^x}$

25. $\dfrac{(2x - 1)(e^{3x} + 3) - 3e^{3x}(x^2 - x + 5)}{(e^{3x} + 3)^2}$ 27. $y = Ce^{-t}$

29. $y = 2e^{1.5t}$

31. 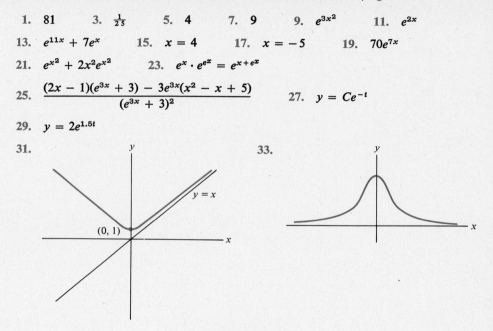   33.

CHAPTER 5
EXERCISES 1, *page 206*

1. -1 3. $-\ln 1.7$ 5. $e^{2.2}$ 7. 2   9. e 11. 1

13. $\frac{1}{2}\ln 5$ 15. $4 - e^{1/2}$ 17. $\pm e^3$ 19. $\dfrac{\ln(.5)}{-.00012}$ 21. $\frac{3}{5}$

23. $\dfrac{e}{2}$ 25. $3 \ln \frac{9}{2}$ 27. $5 \ln 6$ 29. $\frac{1}{5}\ln \frac{2}{5}$ 31. $-\ln \frac{3}{2}$

33. $(-\ln 3, 3 - 3\ln 3)$, minimum   35. $(\frac{1}{2}\ln \frac{3}{2}, \frac{1}{2})$, minimum

37. 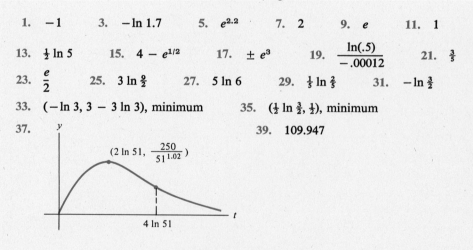   39. 109.947

EXERCISES 2, *page 210*

1. $\dfrac{1}{x}$ 3. $\dfrac{1}{x + 5}$ 5. $-\dfrac{\ln(x + 1)}{x^2} + \dfrac{1}{x(x + 1)}$ 7. $\left(\dfrac{1}{x} + 1\right)e^{\ln x + x}$

9. $\dfrac{1}{x}$ 11. $\dfrac{2\ln x}{x} + \dfrac{1}{x}$ 13. $\dfrac{1}{x}$ 15. $\dfrac{\ln x - 2}{(\ln x)^3}$

17. $2e^{2x} \ln x + \dfrac{e^{2x}}{x}$ 19. $\dfrac{5e^{5x}}{e^{5x} + 1}$ 21. $2(\ln 4)t$

23. $\dfrac{6 \ln t - 3(\ln t)^2}{t^2}$ 25. $y = 1$ 27. $\left(e^2, \dfrac{2}{e}\right)$, yes

29.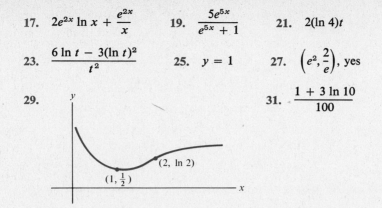

31. $\dfrac{1 + 3 \ln 10}{100}$

33. $P(x) = 300 \ln(x + 1) - 2x$ and $P'(149) = 0$. Since $P''(149) < 0$, the graph of $P(x)$ is concave down at $x = 149$. So $P(x)$ has a maximum there.

35. $\dfrac{1}{e}$

EXERCISES 3, *page 215*

1. $\ln 5x$ 3. $\ln 3$ 5. $\ln 2$ 7. x^2 9. $\ln \dfrac{x^5 z^3}{\sqrt{y}}$

11. $3 \ln x$ 13. $3 \ln 3$ 15. $\dfrac{1}{x + 5} + \dfrac{2}{2x - 1} - \dfrac{1}{4 - x}$

17. $\dfrac{1}{x + 1} + \dfrac{3}{3x - 2} - \dfrac{1}{x + 2}$ 19. $\dfrac{1}{2x} - \dfrac{2x}{x^2 + 1}$

21. $(x + 1)^3(4x - 1)^2 \left[\dfrac{3}{x + 1} + \dfrac{8}{4x - 1} \right]$

23. $(x - 2)^3(x - 3)^5(x + 2)^{-7} \left[\dfrac{3}{x - 2} + \dfrac{5}{x - 3} - \dfrac{7}{x + 2} \right]$ 25. $x^x[1 + \ln x]$

27. $e^x \sqrt{x^2 - 1} \left[1 + \dfrac{x}{x^2 - 1} \right]$ 29. $x^{\ln x} \cdot \dfrac{2 \ln x}{x}$

31. $\dfrac{\sqrt{x - 1}(x - 2)}{x^2 - 3} \cdot \left[\dfrac{1}{2} \cdot \dfrac{1}{x - 1} + \dfrac{1}{x - 2} - \dfrac{2x}{x^2 - 3} \right]$ 33. $y = cx^k$

35. $h = 3, k = \ln 2$

CHAPTER 5: SUPPLEMENTARY EXERCISES, *page 217*

1. $\tfrac{2}{3}$ 3. 1 5. $\sqrt{5}$ 7. $\tfrac{1}{2} \ln 5$ 9. $e^{5/2}$ 11. $e, \dfrac{1}{e}$

13. $\dfrac{5}{5x - 7}$ 15. $\dfrac{2 \ln x}{x}$ 17. $\dfrac{6x^5 + 12x^3}{x^6 + 3x^4 + 1}$ 19. $\dfrac{1}{x} + 1 - \dfrac{1}{2(1 + x)}$

21. $\dfrac{1}{x \ln x}$ 23. $\ln x$ 25. $\dfrac{e^x}{x} + e^x \ln x$

27. $(x^2 + 5)^6(x^3 + 7)^8(x^4 + 9)^{10}\left[\dfrac{12x}{x^2 + 5} + \dfrac{24x^2}{x^3 + 7} + \dfrac{40x^3}{x^4 + 9}\right]$

29. $10^x \ln 10$

31.   33.

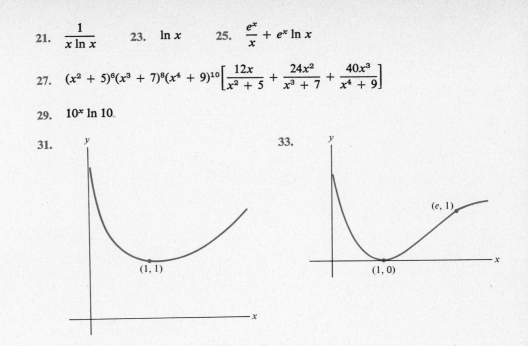

CHAPTER 6
EXERCISES 1, *page 227*

1. (a) $P(t) = 5000e^{.03t}$ (b) $f(x) = 30e^{2x}$ (c) $P(t) = 10^4 e^{-.2t}$

    (d) $f(x) = 5.3e^{-x/2}$ (e) $S(t) = 50,000e^{-.05t}$ (f) $f(x) = 4e^{2x/3}$

3. $f(x) = e^{2x}$

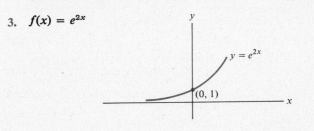

5. 5.3 years   7. $S(t) = S_0 e^{-.11t}$ 9. 9000 years   11. 180 years

13. 10,000   15. 20,000 years   17. (a) $P(t) = 500e^{-.23t}$ (b) 10 months

EXERCISES 2, *page 234*

1. $A = 10,000(1.02)^{12}$ 3. \$786.63   5. \$580.92   7. 13.86%

9. 1991   11. 16.324%

EXERCISES 3, *page 244*

1. (a) $f'(x) = 10e^{-2x} > 0$, $f(x)$ increasing; $f''(x) = -20e^{-2x} < 0$, $f(x)$ concave down   (b) As x becomes large, $e^{-2x} = \dfrac{1}{e^{2x}}$ approaches 0

(c)

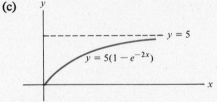

3. $y' = 2e^{-x} = 2 - (2 - 2e^{-x}) = 2 - y$

5. $y' = 30e^{-10x} = 30 - (30 - 30e^{-10x}) = 30 - 10y = 10(3 - y)$,
$f(0) = 3(1 - 1) = 0$ 7. 4.8 hours

CHAPTER 6: SUPPLEMENTARY EXERCISES, *page 246*

1. $f(x) = \frac{5}{2}e^{3x}$ 3. $y = -e^{.1x}$ 5. $y = 8e^{x-2}$ 7. $y = e^{-6x+18}$

9. $A = \frac{1}{2}$, $k = \ln 2$ 11. $.1e^{-1} \approx .04$ mg/cc   13. 51.3 mg

15. 11.3 hours   17. 18.31 years

CHAPTER 7
EXERCISES 1, *page 256*

1. $\frac{1}{2}x^2 + C$ 3. $\frac{1}{3}e^{3x} + C$ 5. $3x + C$ 7. $-\frac{1}{4}$ 9. $\frac{2}{3}$

11. -2 13. $-\frac{5}{2}$ 15. $\frac{1}{2}$ 17. -1 19. 1   21. $\frac{1}{15}$

23. $\dfrac{x^3}{3} - \dfrac{x^2}{2} - x + C$ 25. $4\sqrt{x} - 2x^{3/2} + C$ 27. $4t + e^{-5t} + \dfrac{e^{2t}}{6} + C$

29. $\frac{2}{5}t^{5/2} + C$ 31. C 33. $\dfrac{x^2}{2} + 3$ 35. $\frac{2}{3}x^{3/2} + x - \frac{28}{3}$

37. $2\ln|x| + 2$ 39. (a) $-16t^2 + 96t + 256$ (b) 8 seconds   (c) 400 ft

41. $P(t) = 60t + t^2 - \frac{1}{12}t^3$ 43. $20 - 25e^{-.4t}$ degrees Celsius

45. $-95 + 1.3x + .03x^2 - .0006x^3$ 47. $5875(e^{.016t} - 1)$

EXERCISES 2, *page 264*

1. 0   3. 5   5. 30   7. $\frac{4}{3}(1 - e^{-3})$ 9. 14

11. $5(e - 1)$ 13. $\ln 2$ 15. $1\frac{7}{9}$ 17. $\frac{1}{5}$ 19. $3\frac{3}{4}$

21. $\frac{115}{6} + \ln\frac{7}{9}$ 23. 323.2 km   25. 80.8   27. $1,185.75

29. The additional profit generated by raising the sales level from a to b units.

31. 10   33. $2(e^{1/2} - 1)$ 35. $6\frac{3}{5}$ 37. 15   39. 9

43. 10

1. $\frac{64}{3}$ 3. $\frac{52}{3}$ 5. 20   7. 18   9. $\frac{32}{3}$ 11. $\frac{32}{3}$

13. (a) $\frac{9}{2}$ (b) $\frac{19}{3}$ (c) $\frac{79}{6}$ 15. $\frac{3}{2}$ 17. $2 + 12\ln\left(\frac{3}{2}\right)$

19. 4.48 billion barrels saved

1. .21875, .2421875, actual .25   3. .63107, actual value .63212   5. $20

7. $50   9. $200   11. $25

13. Intersection (100, 10), consumers' surplus = $100, producers' surplus = $250

15. $3236.68   17. 4.52P dollars   19. 3   21. 0   23. 8

25. 55 degrees   27. $\approx$ 82 grams

29. The sum is closely approximated by $\int_0^3 (3-x)^2\,dx$, and the value of this integral is 9

31. The sum is closely approximated by $\int_0^1 (2x+x^3)\,dx$, and the value of this integral is $\frac{5}{4}$

33. 100.08; they estimate the amount of oil consumed in 1976, 1977, 1978, and 1979

35. $\frac{4}{3}\pi r^3$ 37. $\dfrac{31\pi}{5}$ 39. 8π 41. $\dfrac{\pi}{2}\left(1-\dfrac{1}{e^{2r}}\right)$

CHAPTER 7: SUPPLEMENTARY EXERCISES, *page 289*

1. $-2e^{-x/2} + C$ 3. $\frac{3}{5}x^5 - x^4 + C$ 5. $-\frac{2}{3}(4-x)^{3/2} + C$
7. $\frac{3}{4}$ 9. $\frac{1}{3}\ln 4$ 11. $\frac{5}{32}$ 13. 8
15. $\frac{1}{3}(x-5)^3 - 7$ 19. $.02x^2 + 150x + 500$ dollars
21. The total quantity of drug (in cubic centimeters) injected during the first four minutes
23. 68.5 quadrillion Btu   25. $\frac{40}{99}$, exact value: .40547
27. $433.33   29. 15   31. 11.48

33. $f(t) = Q - \dfrac{Q}{A}t$ (b) $Q/2$

35. (a) The area under the curve $y = 1/(1+t^2)$ from $t = 0$ to $t = 3$.
    (b) $1/(1+x^2)$

CHAPTER 8
EXERCISES 1, *page 298*

1. $f(1, 0) = 1$, $f(0, 1) = 8$, $f(3, 2) = 25$

3. $f(0, 1) = 0$, $f(3, 12) = 18$, $f(a, b) = 3\sqrt{ab}$

5. $f(2, 3, 4) = -2, f(7, 46, 44) = \frac{7}{2}$

11. $50. $50 invested at 5% continuously compounded interest will yield $100 in 13.8 years

EXERCISES 2, page 306

1. $5y, 5x$ 3. $4xe^y, 2x^2e^y$ 5. $-\dfrac{y^2}{x^2}, \dfrac{2y}{x}$

7. $4(2x - y + 5), -2(2x - y + 5)$ 9. $(2xe^{3x} + 3x^2e^{3x}) \ln y, x^2e^{3x}/y$

11. $\dfrac{2y}{(x + y)^2}, -\dfrac{2x}{(x + y)^2}$ 13. $\dfrac{3\sqrt{K}}{2\sqrt{L}}$ 15. $\dfrac{2xy}{z}, \dfrac{x^2}{z}, -\dfrac{1 + x^2y}{z^2}$

17. $ze^{yz}, xz^2e^{yz}, x(yz + 1)e^{yz}$ 19. 1, 3   21. -12

23. $\dfrac{\partial f}{\partial x} = 3x^2y + 2y^2, \dfrac{\partial^2 f}{\partial x^2} = 6xy, \dfrac{\partial f}{\partial y} = x^3 + 4xy, \dfrac{\partial^2 f}{\partial y^2} = 4x,$

$\dfrac{\partial^2 f}{\partial y\,\partial x} = \dfrac{\partial^2 f}{\partial x\,\partial y} = 3x^2 + 4y$

25. (a) Marginal productivity of labor = 12,000; of capital = 1000

    (b) 6000 fewer units produced

27. As the price of a bus ride increases, fewer people will ride the bus if the price of a train ticket remains constant. An increase in train ticket prices, coupled with constant bus fare, should cause more people to ride the bus.

29. $\dfrac{\partial f}{\partial r}, \dfrac{\partial f}{\partial m}, \dfrac{\partial f}{\partial s}$ 31. $\dfrac{\partial V}{\partial P}(20,300) = -.06, \dfrac{\partial V}{\partial T} = .004$

33. $\dfrac{\partial^2 f}{\partial x^2} = -\frac{4.5}{4}x^{-5/4}y^{1/4}$. Marginal productivity of labor is decreasing.

EXERCISES 3, page 316

1. $(-2, 1)$ 3. $(26, 11)$ 5. $(1, -3), (-1, -3)$

7. $(\sqrt{5}, 1)\,(\sqrt{5}, -1); (-\sqrt{5}, 1); (-\sqrt{5}, -1)$ 9. $(\frac{1}{3}, \frac{4}{3})$

11. $(0, 0)$ min   13. $(-1, -4)$ max   15. $(0, -1)$ min

17. $(-1, 2)$ max; $(1, 2)$ neither max nor min

19. $(\frac{1}{4}, 2)$ min; $(\frac{1}{4}, -2)$ neither max nor min

21. $(\frac{1}{2}, \frac{1}{6}, \frac{1}{4})$ 23. 14 in. × 14 in. × 28 in.   25. $x = 120, y = 80$

EXERCISES 4, page 327

1. 58 at $x = 6, y = 2, \lambda = 12$ 3. 13 at $x = 8, y = -3, \lambda = 13$

5. $x = \frac{1}{2}, y = 2$ 7. $x = 2, y = 3, z = 1$ 9. $x = 2, y = 2, z = 3$

11. $x = \frac{\sqrt{2}}{2}, y = \frac{\sqrt{2}}{2}$ 13. $x = 4, y = 4, z = 2$

15. (a) $x = 81, y = 16$ (b) $\lambda = 3$ 17. $x = 20, y = 60$

EXERCISES 5, page 334

1. 5.009   3. .99   5. 4.95   7. 6.175   9. 75.6

11. Decrease in profit = \$4300   13. 4400π cu mm   17. 1.5

EXERCISES 6, page 341

1. $y = -2x + \frac{7}{3}$ 3. $y = x + 2$ 5. $y = .4x + .9$

7. $y = .7x - .6$ 9. $y = .5875x + 1.85$

11. (a) $y = -4.2x + 22$,   (b) $8.6°$ C   13. 46

EXERCISES 7, page 347

1. $e^2 - 2e + 1$ 3. $2 - e^{-2} - e^2$ 5. $\frac{38}{3}$

7. $e^{-5} + e^{-2} - e^{-3} - e^{-4}$

CHAPTER 8: SUPPLEMENTARY EXERCISES, page 348

1. $2, \frac{5}{6}, 0$ 3. ≈ 20. Ten dollars increases to twenty dollars in 11.5 years

5. $6x + y, x + 10y$ 7. $\frac{1}{y} e^{x/y}, -\frac{x}{y^2} e^{x/y}$ 9. $3x^2, -z^2, -2yz$

11. $6, 1$ 13. $20x^3 - 12xy, 6y^2, -6x^2, -6x^2$

15. $-201, 5.5$. At the level $p = 25$, $t = 10,000$, an increase in price of \$1 will result in a loss in sales of approximately 201 calculators, and an increase in advertising of \$1 will result in the sales of approximately 5.5 additional calculators.

17. $(3, 2)$ 19. $(0, 1), (-2, 1)$ 21. Min at $(2, 3)$

23. Min at $(1, 4)$; neither max nor min at $(-1, 4)$

25. $20; x = 3, y = -1, \lambda = 8$ 27. $x = \frac{1}{2}, y = \frac{3}{2}, z = 2$

29. $x = 10, y = 20$ 31. 23.6   33. 5.85   35. $y = \frac{5}{2}x - \frac{5}{3}$

37. $y = -2x + 1$ 39. 170   41. 40

CHAPTER 9
EXERCISES 1, page 357

1. $\frac{\pi}{6}, \frac{2\pi}{3}, \frac{7\pi}{4}$ 3. $\frac{5\pi}{2}, -\frac{7\pi}{6}, -\frac{\pi}{2}$ 5. 4π 7. $\frac{7\pi}{2}$ 9. -3π

11. $\dfrac{2\pi}{3}$

13.

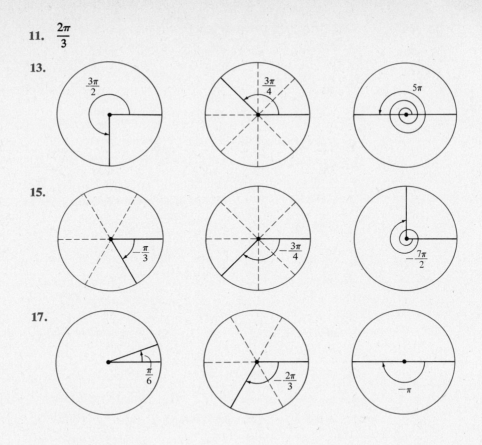

15.

17.

EXERCISES 2, *page 365*

1. $\sin t = \frac{1}{2}, \cos t = \dfrac{\sqrt{3}}{2}$ 3. $\sin t = \dfrac{2}{\sqrt{13}}, \cos t = \dfrac{3}{\sqrt{13}}$

5. $\sin t = \frac{1}{4}, \cos t = \dfrac{\sqrt{15}}{4}$ 7. $\sin t = \dfrac{1}{\sqrt{5}}, \cos t = -\dfrac{2}{\sqrt{5}}$

9. $\sin t = \dfrac{\sqrt{2}}{2}, \cos t = -\dfrac{\sqrt{2}}{2}$ 11. $\sin t = -.8, \cos t = -.6$

13. .4 radians   15. 3.59   17. 10.86   19. $b = 1.31, c = 2.73$

21. $\dfrac{\pi}{6}$ 23. $\dfrac{3\pi}{4}$ 25. $\dfrac{5\pi}{8}$ 27. $\dfrac{\pi}{4}$ 29. $\dfrac{\pi}{3}$ 31. $-\dfrac{\pi}{6}$

33. $\dfrac{\pi}{4}$ 35. $\cos t$ decreases from 1 to -1 37. 0, 0, 1, -1

EXERCISES 3, page 376

1. $4 \cos 4t$ 3. $4 \cos t$ 5. $-6 \sin 3t$ 7. $1 - \pi \sin \pi t$

9. $-\cos(\pi - t)$ 11. $-3 \cos^2 t \sin t$ 13. $\dfrac{\cos(\sqrt{x-1})}{2\sqrt{x-1}}$

15. $\dfrac{\cos(x-1)}{2\sqrt{\sin(x-1)}}$ 17. $8(1 + \cos t)^7 \cdot (-\sin t)$

19. $-6x^2 \cos x^3 \sin x^3$ 21. $e^x(\sin x + \cos x)$

23. $2 \cos(2x) \cos(3x) - 3 \sin(2x) \sin(3x)$ 25. $\cos^{-2} t$

27. $-\dfrac{\sin t}{\cos t}$ 29. $\dfrac{\cos(\ln t)}{t}$ 31. -3 33. $y = 2$

35. $\frac{1}{2} \sin 2x + C$ 37. $-\dfrac{\cos(4x+1)}{4} + C$

39. (a) max $= 120$ at $0, \dfrac{\pi}{3}$; min $= 80$ at $\dfrac{\pi}{6}, \dfrac{\pi}{2}$ (b) 57

EXERCISES 4, page 381

1. $\sec t = \dfrac{\text{hypotenuse}}{\text{adjacent}}$ 3. $\tan t = \frac{5}{12}, \sec t = \frac{13}{12}$

5. $\tan t = -\frac{1}{2}, \sec t = -\dfrac{\sqrt{5}}{2}$ 7. $\tan t = -1, \sec t = -\sqrt{2}$

9. $\tan t = \frac{4}{3}, \sec t = -\frac{5}{3}$ 11. $75 \tan(.7) \approx 63$ feet   13. $\tan t \sec t$

15. $-\csc^2 t$ 17. $4 \sec^2(4t)$ 19. $-3 \sec^2(\pi - x)$

21. $4(2x+1) \sec^2(x^2 + x + 3)$ 23. $\dfrac{\sec^2 \sqrt{x}}{2\sqrt{x}}$ 25. $\tan x + x \sec^2 x$

27. $2 \tan x \sec^2 x$ 29. $6[1 + \tan(2t)]^2 \sec^2(2t)$ 31. $\sec t$

CHAPTER 9: SUPPLEMENTARY EXERCISES, page 384

1. $\dfrac{3\pi}{2}$ 3. $-\dfrac{3\pi}{4}$ 5.

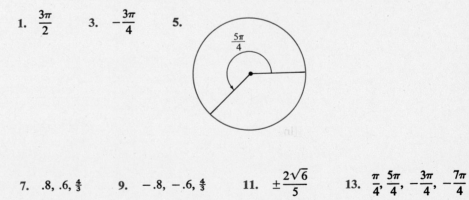

7. $.8, .6, \frac{4}{3}$ 9. $-.8, -.6, \frac{4}{3}$ 11. $\pm\dfrac{2\sqrt{6}}{5}$ 13. $\dfrac{\pi}{4}, \dfrac{5\pi}{4}, -\dfrac{3\pi}{4}, -\dfrac{7\pi}{4}$

15. Negative   **17.** 32.6 ft   **19.** $3 \cos t$ **21.** $(\cos \sqrt{t}) \cdot \frac{1}{2} t^{-1/2}$

23. $x^3 \cos x + 3x^2 \sin x$ **25.** $-\dfrac{2 \sin(3x) \sin(2x) + 3 \cos(2x) \cos(3x)}{\sin^2(3x)}$

27. $-12 \cos^2(4x) \sin(4x)$ **29.** $[\sec^2(x^4 + x^2)](4x^3 + 2x)$

31. $\cos(\tan x) \sec^2 x$ **33.** $\sin x \sec^2 x + \sin x$ **35.** $\dfrac{\cos x}{\sin x}$

37. $3e^{3x} \sin^4 x + 4e^{3x} \sin^3 x \cos x$ **39.** $\dfrac{\tan(3t) \cos t - 3 \sin t \sec^2(3t)}{\tan^2(3t)}$

41. $e^{\tan t} \sec^2 t$ **43.** $2(\cos^2 t - \sin^2 t)$

45. $\dfrac{\partial f}{\partial s} = \cos s \cos(2t), \dfrac{\partial f}{\partial t} = -2 \sin s \sin(2t)$

47. $\dfrac{\partial f}{\partial s} = t^2 \cos(st), \dfrac{\partial f}{\partial t} = \sin(st) + st \cos(st)$ **49.** $y - 1 = 2\left(t - \dfrac{\pi}{4}\right)$

51.

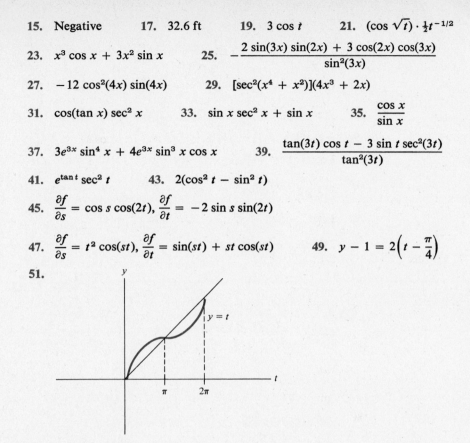

53. 4   **55.** $\dfrac{\pi^2}{2} - 2$

CHAPTER 10
EXERCISES 1, *page 393*

1. $\frac{1}{6}(x^2 + 4)^6 + C$ **3.** $\frac{1}{4}(x^2 - 5x)^4 + C$ **5.** $\ln|x^3 - 1| + C$

7. $\frac{1}{2}[\ln(2x)]^2 + C$ **9.** $-\frac{1}{4}\cos^4 x + C$ **11.** $\frac{1}{2}\sin^2 x + C$

13. $\sqrt{2x + 1} + C$ **15.** $\frac{2}{9}(x^3 - 1)^{3/2} + C$

17. $2e^{\sqrt{x+5}} + C$ **19.** $\frac{1}{3}[\sin(2x)]^{3/2} + C$

21. $-\frac{1}{2}[\ln x]^{-2} + C$ **23.** $\sqrt{x^2 + 1} + C$ **25.** $\frac{1}{10}[\ln(x^5 - 7)]^2 + C$

27. $\ln|\sin x| + C$ **29.** $-\frac{1}{2}e^{-x^2} + C$

31. $\frac{1}{3}\ln|x^3 - 3x^2 + 1| + C$ **33.** $\frac{1}{2}(x^4 + 4)^{1/2} + C$

35. $\frac{1}{2}\ln|e^{2x} + e^{-2x}| + C$

1. $\frac{1}{5}xe^{5x} - \frac{1}{25}e^{5x} + C$ 3. $\frac{x}{10}(2x + 1)^5 - \frac{1}{120}(2x + 1)^6 + C$

5. $-xe^{-x} - e^{-x} + C$ 7. $2x(x + 1)^{1/2} - \frac{4}{3}(x + 1)^{3/2} + C$

9. $-\frac{1}{15}x(1 - 3x)^5 - \frac{1}{270}(1 - 3x)^6 + C$ 11. $-3xe^{-x} - 3e^{-x} + C$

13. $\frac{2}{3}x(x + 1)^{3/2} - \frac{4}{15}(x + 1)^{5/2} + C$ 15. $\frac{2}{3}x^{3/2} \ln x - \frac{2}{9}x^{3/2} + C$

17. $x^2 \sin x + 2x \cos x - 2 \sin x + C$ 19. $\frac{x^2}{2} \ln 5x - \frac{1}{4}x^2 + C$

21. $\frac{1}{2}x^2 e^{x^2} - \frac{1}{2}e^{x^2} + C$ 23. $\frac{1}{6}[\ln x]^6 + C$

25. $2\sqrt{x} \ln x - 4\sqrt{x} + C$ 27. $-e^{-x}(x^2 + 2x + 2) + C$

1. $\frac{1}{18}$ 3. 0   5. $\frac{1}{2}$ 7. $\frac{64}{3}$ 9. $\frac{23}{6}$ 11. 9

13. $\frac{1}{3}(e^{27} - e)$ 15. 0   17. 4   19. $\frac{1}{\pi}$ 21. $\frac{\pi}{2}$ 23. $\frac{9\pi}{2}$

1. .34, exact value $\frac{1}{3}$ 3. .977, exact value 1

5. 4, 2, exact value $\frac{4}{3}$ 7. .3125, .265625, exact value .25

9. Trapezoidal rule 1, rectangle rule 1, exact value 1

11. Trapezoidal rule .634226, rectangle rule .63107, exact .63212

13. 25,750 sq ft   15. .386 miles

1. $6890   3. 290,635   5. 1,491,000

1. 0   3. No limit   5. $\frac{1}{4}$ 7. 2   9. 5

11. 6   13. 2   15. $\frac{1}{2}$ 17. 2   19. 1

21. Area under curve from 1 to b is $-4 + 4b^{1/4}$. This has no limit as $b \to \infty$.

23. $\frac{1}{2}$ 25. Divergent   27. $\frac{1}{8}$ 29. $\frac{2}{3}$

31. 1   33. $2e$ 35. $\frac{1}{4}$ 37. 2

CHAPTER 10: SUPPLEMENTARY EXERCISES, *page 424*

1. $-\frac{1}{6}\cos(3x^2) + C$ 3. $-\frac{1}{36}(1 - 3x^2)^6 + C$ 5. $\frac{1}{3}[\ln x]^3 + C$

7. $-\frac{1}{3}x^2(4 - x^2)^{3/2} - \frac{2}{15}(4 - x^2)^{5/2} + C$ 9. $-\frac{1}{3}e^{-x^3} + C$

11. $\frac{1}{3}x^2 \sin(3x) - \frac{2}{27}\sin(3x) + \frac{2}{9}x \cos(3x) + C$

13. $2(x \ln x - x) + C$ 15. $\frac{2}{3}x(3x - 1)^{1/2} - \frac{4}{27}(3x - 1)^{3/2} + C$

17. $\frac{1}{4}\frac{x}{(1 - x)^4} - \frac{1}{12}\frac{1}{(1 - x)^3} + C$ 19. $\frac{3}{8}$ 21. $1 - e^{-2}$

23. $\frac{3}{4}e^{-2} - \frac{5}{4}e^{-4}$ 25. $\frac{1}{4}[\ln 7]^4$ 27. $.79621$ 29. 1.11667

31. $f(x) = x, g(x) = e^{2x}$ 33. $u = \sqrt{x + 1}$ 35. $u = x^4 - x^2 + 4$

37. $f(x) = (3x - 1)^2, g(x) = e^{-x}$; then integrate by parts again

39. $f(x) = 500 - 4x, g(x) = e^{-x/2}$ 41. $u = x^2 + 6x$

43. $f(x) = \ln x, g(x) = \sqrt{x}$ 45. $u = x^2 - 2x + 1$

47. $\frac{1}{3}e^6$ 49. $\frac{2}{25}$

CHAPTER 11
EXERCISES 1, *page 431*

5. $y = e^{\frac{1}{2}t}$ 7. Yes   9. $f(0) = 4, f'(0) = 5$

11. $y' = k(C - y), \quad k > 0$

EXERCISES 2, *page 438*

1. $y = \sqrt[3]{15t - \frac{3}{2}t^2 + C}$ 3. $y = \dfrac{3}{e^{3t} - 3t + C}$ or $y = 0$

5. $y = -\frac{1}{2}\ln(-t^2 + C)$ 7. $y = 3 + Ce^{4t}$

9. $y = \pm\sqrt{(\ln t)^2 + C}$ 11. $y = Ce^{3t} - \frac{4}{3}$

13. $y = (\sqrt{t} + C)^2$ or $y = 0$ 15. $y = Ce^{\sin t}$

17. $y = \dfrac{1}{t - t \ln t + C} + 3$ or $y = 3$ 19. $y = \frac{1}{3}\ln(t^3 + e^6)$

21. $y = 3 - 2e^{-t}$ 23. $y = \sqrt[3]{3 \sin t - 3t \cos t + 8}$

25. $y = \frac{3}{5}e^{(5/2)t^2} + \frac{2}{5}$ 27. $y = -\sqrt{2t + 2 \ln t + 7}$

29. $y = (3x^{1/2} \ln x - 6x^{1/2} + 14)^{2/3}$

31. $y = B(p + c)^{-k}, \quad B$ any number

33. $y' = k(1 - y)$ where $y = p(t), y(0) = 0; y = 1 - e^{-kt}$

35. $y = be^{Ce^{-at}}, \quad C$ any number

1. 3   **3.** Increasing   **5.** $f(1) \approx -\frac{9}{4}$

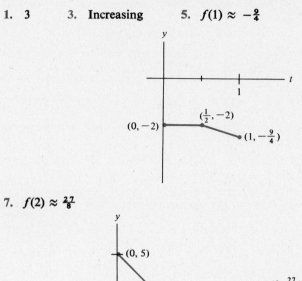

(0, −2) $(\frac{1}{2}, -2)$ $(1, -\frac{9}{4})$

7. $f(2) \approx \frac{27}{8}$

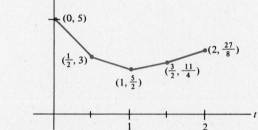

(0, 5) $(\frac{1}{2}, 3)$ $(1, \frac{5}{2})$ $(\frac{3}{2}, \frac{11}{4})$ $(2, \frac{27}{8})$

9. Euler's method yields $f(1) \approx .37011$; solution:

$$f(t) = 1/(\tfrac{1}{2}t^2 + t + 1); \quad f(1) = .4; \quad \text{error} = .02989$$

11. (a) $y' = k(1 - y)$, $y(0) = 0$, where $k > 0$

(b) $3k - 3k^2 + k^3$ (c) $y = 1 - e^{-kt}$, $y(3) = 1 - e^{-3k}$

(d) Euler's method gives .27100, exact solution gives .25918

1.   **3.**

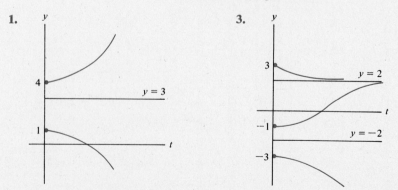

5.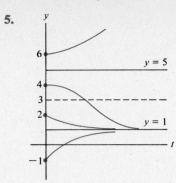

7.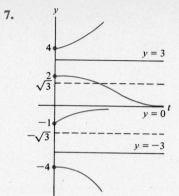

9.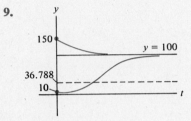

11.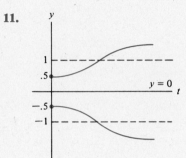

13.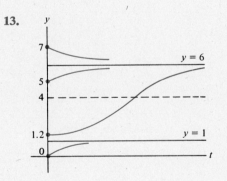

15.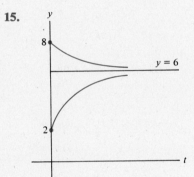

17.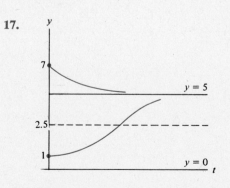

19.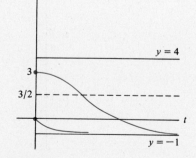

21.

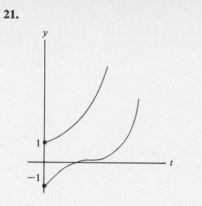

23.

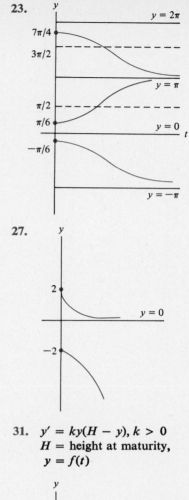

25.

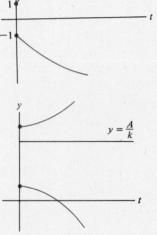

27.

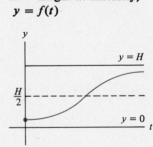

29.

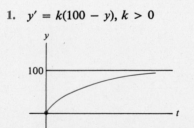

31. $y' = ky(H - y), k > 0$
    $H =$ **height at maturity,**
    $y = f(t)$

EXERCISES 5, *page 467*

1. $y' = k(100 - y), k > 0$

3. $y' = ky(M - y)$, $k > 0$

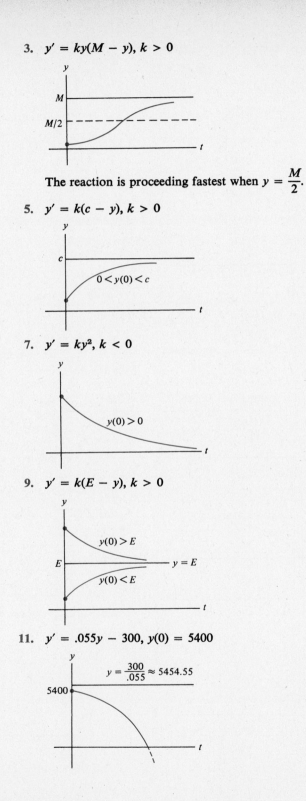

The reaction is proceeding fastest when $y = \dfrac{M}{2}$.

5. $y' = k(c - y)$, $k > 0$

7. $y' = ky^2$, $k < 0$

9. $y' = k(E - y)$, $k > 0$

11. $y' = .055y - 300$, $y(0) = 5400$

13. (a) $y' = .05y + 10,000, y(0) = 0$ (b) $y = 200,000(e^{.05t} - 1)$, \$56,806

15. (a) $y' = .05 - .2y, y(0) = 6.25$ (b) $y' = .13 - .2y, y(0) = 6.25$

17. $y' = -.14y$ 19.

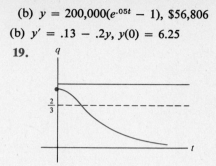

CHAPTER 11: SUPPLEMENTARY EXERCISES, *page 474*

1. $y = \sqrt[3]{3t^4 - 3t^3 + 6t + C}$ 3. $y = Ate^{-3t}$

5. $y = \frac{3}{7}t + 3$ 7. $y = \sqrt{4t^3 - t^2 + 49}$

9. $f(t) = 2t, f(2) = 4$ 11. **Decreasing**

13.

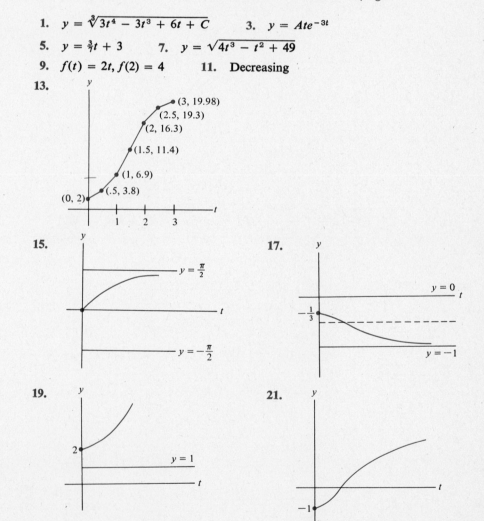

A42   *Answers to Odd-Numbered Exercises*

23.

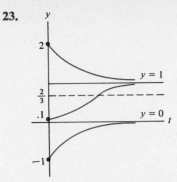

25. (a) $N' = .015N - 3000$

(b) There is a constant solution $N = 200,000$, but it is unstable. It is unlikely a city would have such a constant population.

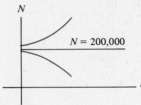

27. 13.863 years

CHAPTER 12
EXERCISES 1, *page 484*

1. $E(X) = \frac{4}{5}$, Var$(X) = .16$, standard deviation $= .4$

3. (a) Var$(X) = 1$ (b) Var$(X) = 4$ (c) Var$(X) = 16$

5.

(a)

Accidents	0	1	2	3
Probability	.21	.5	.25	.04

(b) $E(X) = 1.12$

(c) Average of 1.12 accidents per week during year.

7. (a) 25%   (b) $100c^2$%

9. $E(X) = 90,000$. The grower should spend the $5000.

EXERCISES 2, *page 493*

7. $\frac{1}{4}$ 9. $\frac{1}{15}$ 11. 12   13.

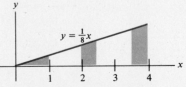

15. $\frac{1}{12}$ 17. $\frac{1}{2}$ 19. $\frac{3}{4}$ 21. $8/3x^3$ 23. $\frac{1}{5}x - \frac{2}{5}$

25. (a) $\frac{19}{63}$ (b) $(x^3 - 1)/63$ 31. .586   33. 1.8

35. .6   37. (b) $1 - x^{-4}$ (c) $\frac{15}{16}, \frac{1}{16}$

EXERCISES 3, *page 501*

1. $E(X) = 4$, $\text{Var}(X) = 2$ 3. $E(X) = 3$, $\text{Var}(X) = \frac{4}{3}$

5. $E(X) = \frac{5}{6}$, $\text{Var}(X) = \frac{5}{252}$ 7. $E(X) = \frac{2}{5}$, $\text{Var}(X) = \frac{1}{25}$

9. (a) $F(x) = 10x^3 - 15x^4 + 6x^5$ (b) $\frac{53}{512}$

    (c) $\frac{1}{2}$. About half of the time less than 25% of the newspaper's space is devoted to advertising.

    (d) $\frac{1}{28}$

11. (a) 2. The average useful life of the component is 200 hours.   (b) $\frac{1}{2}$

13. 8 minutes

15. (a) $F(x) = \frac{1}{6}x^2 - \frac{1}{54}x^3 - 1$,   (b) $\frac{23}{27}$ (c) 412.5 man-hours

    (d) .5344

17. $E(X) = \frac{4}{3}$, $\text{Var}(X) = \frac{2}{9}$ 19. $3\sqrt{2}$ 21. $3/\sqrt[3]{2}$ hundred hours

23. $\frac{5}{6}$ minute   25. Hint: Compute $\int_B^A xf(x)\, dx$ using integration by parts.

EXERCISES 4, *page 513*

1. $E(X) = \frac{1}{3}$, $\text{Var}(X) = \frac{1}{9}$ 3. $E(X) = 5$, $\text{Var}(X) = 25$

5. $e^{-1} - e^{-2}$ 7. $1 - e^{-2/3}$ 9. e^{-3}

11. $1 - e^{-2}$ 13. (a) $e^{-1/3}$ (b) $r(t) = e^{-t/72}$

15. $\mu = 4$, $\sigma = 1$ 17. $\mu = 0$, $\sigma = 3$

23. (a) .4032   (b) .4013   (c) .8351   (d) .9772

25. (a) .4772   (b) .4772   27. .1587   29. The Capitol Beltway

31. 1.22%   33. Hint: First show that $\Pr(a \le X \le b) = e^{-ka}(1 - e^{-kb})$.

CHAPTER 12: SUPPLEMENTARY EXERCISES, *page 517*

1. (a) .125, .2969   (b) $E(X) = 1.5$, $\text{Var}(X) = .15$

3. $1 - e^{A-x}$ 5. $c_2 = \frac{1}{2}$, $c_4 = \frac{1}{4}$ 7. (a) 5.01   (b) 100

9. (a) .96   (b) 1.8 thousand gallons   (c) $1 - x/2$

11. (a) $E(X) = 22.5$, $\text{Var}(X) = 2.0833$ (b) 21.5

13. (a) 1/300   (b) .375   (c) $17,222   15. $2/\pi$

17. $54.88   **19.** $k \approx .35$ **21.** .52   **23.** 69.15%

25. $a \approx .25$ **27.** (b) .6826

29. (a) $\Pr\left(-\frac{1}{k} \le X \le \frac{3}{k}\right) = \Pr\left(0 \le X \le \frac{3}{k}\right) \ge \frac{3}{4}$ (b) .95

CHAPTER 13
EXERCISES 1, *page 530*

1. $p(x) = -1 + 3x - x^2$ **3.** $p(x) = -2(x - 1) - \frac{1}{2}(x - 1)^3$

5. $p(x) = 1 + 2x + \frac{3}{2}x^2 + \frac{2}{3}x^3 + \frac{5}{24}x^4$

7. $p_3(x) = (x - 1) - \frac{1}{2}(x - 1)^2 + \frac{1}{3}(x - 1)^3$

9. $p_3(x) = 1$ **11.** $p_3(x) = 0$

13. $p_3(x) = 9 + 6(x - 3) + (x - 3)^2$ **15.** $p_3(x) = 1 + x^2$

17. $p_4(x) = 1 + x + \frac{1}{2}x^2 + \frac{1}{6}x^3 + \frac{1}{24}x^4$, $p_4(.01) \approx 1.01005$

19. $p_2(x) = 3 + \frac{1}{6}(x - 9) - \frac{1}{216}(x - 9)^2$, $p_2(8.8) = 2.96648$

21.

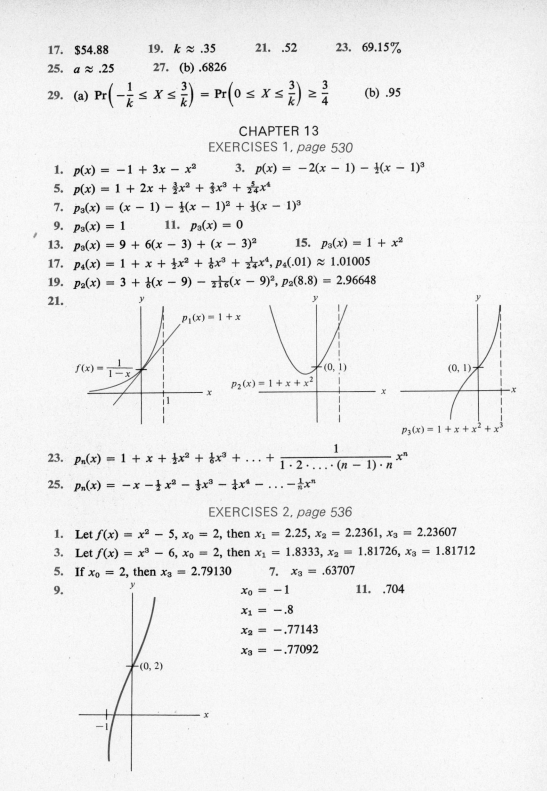

23. $p_n(x) = 1 + x + \frac{1}{2}x^2 + \frac{1}{6}x^3 + \ldots + \dfrac{1}{1 \cdot 2 \cdot \ldots \cdot (n - 1) \cdot n} x^n$

25. $p_n(x) = -x - \frac{1}{2}x^2 - \frac{1}{3}x^3 - \frac{1}{4}x^4 - \ldots - \frac{1}{n}x^n$

EXERCISES 2, *page 536*

1. Let $f(x) = x^2 - 5$, $x_0 = 2$, then $x_1 = 2.25$, $x_2 = 2.2361$, $x_3 = 2.23607$

3. Let $f(x) = x^3 - 6$, $x_0 = 2$, then $x_1 = 1.8333$, $x_2 = 1.81726$, $x_3 = 1.81712$

5. If $x_0 = 2$, then $x_3 = 2.79130$ **7.** $x_3 = .63707$

9.

$x_0 = -1$ **11.** .704

$x_1 = -.8$

$x_2 = -.77143$

$x_3 = -.77092$

13. x_1 will be the exact root   15. $x_0 = 1$, $x_1 = -2$, $x_2 = 4$, $x_3 = -8$

EXERCISES 3, *page 545*

1. $\frac{6}{5}$ 3. $\frac{25}{124}$ 5. $.8333$ 7. $\frac{23}{99}$ 9. $\frac{2}{9}$ 11. $\frac{12}{999}$

13. $\sin x = x - \frac{1}{6}x^3 + \frac{1}{120}x^5 - \frac{1}{5040}x^7 + \ldots$,

$\cos x = 1 - \frac{1}{2}x^2 + \frac{1}{24}x^4 - \frac{1}{720}x^6 + \ldots$ 15. $f(x) = \dfrac{1 - x^2}{(1 - x)^4}$

17. $\dfrac{1}{3} - \dfrac{1}{1 \cdot 2 \cdot 3 \cdot 7} + \dfrac{1}{1 \cdot 2 \cdot \ldots \cdot 5 \cdot 11} - \dfrac{1}{1 \cdot 2 \cdot \ldots \cdot 7 \cdot 15}$

$+ \dfrac{1}{1 \cdot 2 \cdot \ldots \cdot 9 \cdot 19} - \ldots$

19. $1 - x + \frac{1}{2}x^2 - \frac{1}{6}x^3 + \frac{1}{24}x^4 - \ldots$

EXERCISES 4, *page 554*

1. $\frac{1}{16}$ 3. $\frac{1}{8}$ 5. $\frac{1}{12}$ 7. $(\frac{1}{3})(\frac{2}{3})^n$ 9. $\frac{9}{19}$

11. $.08425$ 13. $.44063$ 15. $.95956$ 17. $.99723$

19. $.01029$ 21. $\lambda = 2$ 23. 461

CHAPTER 13: SUPPLEMENTARY EXERCISES, *page 556*

1. $p(x) = 2 - \frac{1}{2}(x - 3)^2 + \frac{1}{2}(x - 3)^3$

3. $p_4(x) = 21 + 16(x - 3) + 3(x - 3)^2$ 5. 8.18068

7. $x_0 = 4$, $x_1 = 3.6$, $x_2 = 3.5619$ 9. $.28248$ 11. $\frac{4}{7}$

13. $\cos 2x = 1 - 2x^2 + \frac{2}{3}x^4 - \frac{4}{45}x^6 + \ldots$, $\sin 2x = 2x - \frac{4}{3}x^3 + \frac{4}{15}x^5 - \frac{8}{315}x^7 + \ldots$

15. $e^{3x} = 1 + 3x + \frac{9}{2}x^2 + \frac{9}{2}x^3 + \frac{27}{8}x^4 + \ldots$

17. $(\frac{2}{9})(\frac{7}{9})^n$ 19. $\frac{7}{2}$ 21. $.05113$

INDEX

A

Absolute value function, 19
Acceleration, 95
Advertising, 226
Age distribution of cells, 303
Air pressure in the lungs, 145
Allometric equation, 215
Annuity, continuous, 468
Antiderivative, 250
  rules, 252–53
Antidifferentiation, 249
Approximation:
  of average value of a function, 283
  to change of a function, 303, 331
  of compound interest, 232

of definite integrals, 277, 408
of tangent line by secant lines, 59
using a total differential, 331
Architectural design, 296, 326
Area:
  between two curves, 267ff
  under a curve, 261ff
Asymptote, 107
Autocatalytic reaction, 467
Autonomous differential equation, 447
Average cost, 17, 168
Average rate of change, 92
Average value of a function, 283, 346
Average velocity, 94

B

Bacteria growth, 220–21
Basal metabolic rate, 376
Beer, consumption, 307

Body surface area, 308
Buffon needle problem, 520

C

Calcium metabolism, 93
Cancer death rates, 339
Carbon dating, 225–26
Carbon-14, 224–26
Cardiac output, 332
Carrying capacity, 460
Chain rule, 174
Change of limits rule, 401
Chebyshev inequality, 521
Chi-square density function, 518
Cigarette consumption, 339
Cobb-Douglas production functions, 297
Common logarithms, 205
Composition of functions, 24, 173
Compound amount, 230, 231
Compound interest, 230
Concave down, 106
Concave up, 106
Concentration of a drug, 104, 246
Constant function, 16, 56
Constant-multiple rule, 80, 83
Constrained optimization, 319ff
Constraint equation, 135, 341
Consumers' surplus, 281–82
Continuous annuity, 468

Continuous function, 75
Continuously compounded interest, 233, 282
Continuous random variable, 486ff
Continuous stream of income, 287, 413–14
Convergent improper integral, 420, 421
Cosecant, 377
Cosine, 359
   alternate definition of, 361
   derivate of, 368, 369
   graph of, 364
   properties of, 361–62
   table of values for, A7
Cost:
   average, 168
   fixed, 40
   marginal, 93, 158, 168
Cost functions, 40, 55, 93, 150, 151–52, 255
Cotangent, 377
Coughing, 144
Crime rate, 349
Cumulative distribution function, 490
   relation to probability density function, 491
Curve sketching, 119ff
Curve sketching, summary, 130
Cylinder, 288

D

Decay:
   exponential, 181, 223
   radioactive, 223–26
   sales, 226
Decay constant, 223
Decreasing function, 104
Definite integral(s), 259
   approximation of, 277
   as area, 261
   change of limits rule, 401
   properties, 266
Demand curve, 152, 281
Demand function, 152, 307–308, 312, 342
Derivative(s), 55
   chain rule for, 174, 175
   of a constant function, 56

   constant-multiple rule for, 80, 83
   of cosine functions, 368, 369
   of exponential functions, 187–89
   general power rule for, 80, 82
   of the natural logarithm, 208
   notation for, 55, 61, 88
   partial (*see* Partial derivatives)
   of a power funtion, 57
   of a product, 166, 169
   of a quotient, 167, 169
   second, 87
   of sine functions, 368, 369
   of a sum, 80, 81, 83
   of tangent functions, 380
   third, 91
Differentiable, 68

Differential, total, 330ff
Differential equation, 194, 427ff
  autonomous, 447
  initial condition, 430
  qualitative theory of, 447ff
Differentiation, 55
Differentiation, logarithmic, 214

Diffusion of information by mass media,
  237–38
Discrete random variables, 477ff
Divergent improper integral, 421
Double integral, 343ff
Drug time-concentration curve, 13–14,
  104–105, 206

E

Ebbinghaus model for forgetting, 245
E. Coli infection, 228
Ecological range, 341
Effective annual rate of interest, 231
Elimination time (of a drug), 246
Enzyme kinetics, 20
Epidemic, 92
Epidemic model, 241–43
Euler's method, 443
Evan's price adjustment model, 468
e^x, 189
  (*see also* Exponential function(s))
Expected value (mean) of a random variable,
  481, 497
Exponential decay, 181, 223–26
Exponential density function, 504

Exponential function(s), 185, 189, 195, 204
  chain rule for, 192
  derivative of, 187–89
  graph of, 186, 189
  integration of, 252
  properties of, 195
  table of values for, A2
Exponential growth, 181, 220–23, 284
Exponentially distributed, 505
Exponents:
  fractional, 182
  laws of, 183
  negative, 183
Extreme point, 104
Extreme points, locating, 119, 310

F

Factoring polynomials, 29ff
First derivative rule, 113
First derivative test for extrema, 310, 316
Fixed costs, 40
Fluid flow, 145
Food, consumption of, 308
Forgetting, Ebbinghaus model for, 245
Function(s), 5
  composition of, 24, 173
  constant, 16
  domain of, 7

  graph of, 8
  linear, 15
  polynomial, 18
  power, 18
  quadratic, 17
  rational, 18
  of several variables, 293ff
  value at x, 5
  zero of, 27, 107
Fundamental Theorem of Calculus, 261

G

Gene frequency, 464
General power rule, 80, 82
Glucose infusion, 239, 245
Gompertz growth curve or equation, 197, 440,
  456
Graphs:
  of equations, 11

  of functions, 8
  of functions of several variables, 294–96
Growth:
  bacteria, 220–21
  exponential, 220–23
  of fish population, 241
  of fruit flies, 222

Growth (cont.)
  logistic, 239ff, 459–61
  population, 284

H

Half-life, 223
  carbon-14, 224
  cobalt-60, 227
  iodine-131, 225
  radioactive potassium, 227

I

Improper integral, 418ff
  convergent, 420, 421
  divergent, 421
Increasing function, 104
Indefinite integral, 252
Inflection point, 106
Inflection points, locating, 122
Infinity, 70
Information, diffusion by mass media, 237–38
Integral, 249ff
  definite, 259
  double, 343ff
  improper, 418ff
  indefinite, 252
  iterated, 344
Integral sign, 252
Integration:

K

Kidney function, 335–336

L

Lagrange multipliers, 319ff
Learning curve, 237
Least squares, method of, 336ff
Least squares error, 337–38
Limit:
  criterion for continuity, 76
  of a function, 65
  at infinity, 70
  theorems, 66
Line(s):
  equations of, 15–16, 38, 42
  parallel, 42
  perpendicular, 42

restricted organic, 239–41
Growth constant, 220

  radium, 286
  strontium-90, 224
Heat loss, 296, 303, 312, 325–27
Horizontal asymptote, 107

  and antiderivatives, 252
  of exponential functions, 252
  by parts, 395ff, 402–403
  of power functions, 252, 253
  by subsitution, 389ff, 400–402
Intercept, 107
Interest:
  compound, 230
  compounded continuously, 233, 282
Intramuscular injection of drug, 206
Intravenous infusion of glucose, 239, 245
Intrinisic rate of growth, 461
Inventory costs, minimizing, 142–44
Investment, capital, 468
Iodine-131, 224
IQ's, distribution of, 409
Interated integral, 344

  slope of, 38
  slope properties of, 41ff
  tangent, 49–50
  tangent to a semicircle, 45
  y-intercept of, 38
Linear function, 15
ln x, 202
  (see Natural logarithm(s))
Logarithmic differentiation, 214
Logarithm to the base 10, 205
Logistic curve, 239ff
Lung cancer, 339

M

Manufacturing costs, 297, 320
Marginal cost function, 55, 93, 158, 168, 255, 260, 264, 273
Marginal heat loss, 304, 327
Marginal productivity of capital, 304, 324
Marginal productivity of labor, 304, 324
Marginal productivity of money, 324, 334
Marginal profit function, 55
Marginal revenue function, 55, 152, 158, 255
Mathematical model, 103, 458
Maxima and minima:
　constrained, 319ff
　definitions of, 104, 309

first derivative test for, 310, 316
　for functions of one variable, 133
　for functions of several variables, 309ff
　second derivate test for, 314
Maximum point, 104, 309
Mean (expected value) of a random variable, 481, 497
Median of a random variable, 503
Memoryless property of an exponential random variable, 516
Minimum point, 104, 309
Monopoly, 152, 155–58, 312
Multiplier, effect, 541–42

N

National debt, 105
Natural logarithm(s), 202–203
　applications of 219–47
　derivative of, 208
　graph of, 202
　integration involving, 253
　properties of, 203, 212
　table of values, A4
Net change, 258
Net primary production of nutrients, 17

Newton-Raphson algorithm, 532ff
Newton's law of cooling, 430, 432
Nominal rate of interest, 231
Normal curve, 409, 422, 508, 509
Normal density function, 508
Normally distributed, 508
Normal random variable, 508
Notation for derivatives, 55, 61, 88
Numerical solution of differential equations, 441ff

O

Objective equation, 135, 319
Oil, demand for, 250, 254, 259, 272–73
One-compartment problems, 463
Optimal design, principle of, 326

Optimal reorder quantity, 142–44, 148
Optimization problems, 133ff, 319ff
Optimization problems, procedures for solving, 138

P

Parabola, 17
Paraboloid, 238
Parallel lines, 42
Pareto distribution, 518
Partial derivatives, 299ff
　interpretation of, 302–303
　notation for, 299, 302, 304–305
　second, 304–305
Parts, integration by, 395ff, 402–403
Periodic functions, 353, 364, 372, 472
Perpendicular lines, 42
Point-slope formula, 42
Poisson random variable, 551
Polynomial function, 18
Population, U.S., 105

Population, world, 284
Population density function, 416
Population genetics, 464–66
Population growth, 220–23, 241, 284
Position function, 94
Power function, 18
Power rule, 57
Predator-prey models, 372 470ff
Present value, 234
Present value of an income stream, 287, 413–15
Price discrimination, 312
Principal amount, 230
Probability, 479, 487
Probability density function, 487
　beta, 489, 501, 502

Probability density function (cont.)
 chi-square, 518
 exponential, 504
 gamma, 520
 normal, 508
 relation to cumulative distribution function,
  491
 uniform, 498
Probability density histogram, 482

Probability table, 479
Producers' surplus, 285
Production, maximization of, 322
Production functions, 297, 304, 322, 333–34
Production possibilities curve, 329
Production schedule, 329
Product rule, 166, 169
Profit functions, 55, 150, 155ff, 318
Proportional, 430–31, 458

Q

Quadratic formula, 27
Quadratic function, 17

Qualitative theory of differential equations
 447ff
Quotient rule, 167, 169

R

Radian, 354
Radian measure, 354
Radioactive dating, 225–26
Radioactive decay, 223–26
Radioactive fallout, 224
Random variables:
 continuous, 486ff
 discrete, 477ff
 exponential, 505
 normal, 508
 Poisson, 551
 standard normal, 509
 uniform, 498

Rate of change, 42, 91ff, 302, 430
Rate of learning, 237
Rate of net investment, 432
Rational function, 18
Rectangle rule, 277
Reflection (through a line), 202
Relative frequency histogram, 478
Reliability function, 514
Renal clearance, 335–36
Restricted organic growth, 239–41
Revenue functions, 55, 150, 152–55, 255
Riemann sum, 276–77, 344
Rumor, spread of, 37–38, 243

S

Saddle-shaped graph, 315
Sales decay curve, 226
Secant-line calculation of derivative, 58ff
Second derivative, 87
Second derivative rule, 114
Second derivative test for extrema, 314
Second partial derivatives, 304–305
Separation of variables, 432ff
Sine, 359
 alternate definition of, 361
 derivative of, 368, 369
 graph of, 363–64
 properties of, 361–62
 table of values for, A7
Slope:
 of a curve, 49–50

 of a line, 38
 properties, 41ff
Solid of revolution, 279–80
Sphere, 287
Standard deviation, 478
 of a random variable, 481
Standard normal random variable, 509
Standard position of angles, 356
Straight line (see Line(s))
Strontium-90, 224
Substitution:
 in definite integrals, 400–402
 integration by, 389ff
Sum rule, 80, 81, 83
Supply and demand, 285
Survival function, 514

T

Tangent function, 377
  derivative of, 380
  graph of, 381
  table of values for, A7
Tangent line, 44, 49–50
Tax, imposition of, 157
Terminal velocity (of a skydiver), 236
Theory of the firm, 150

Total differentials, 330ff
Trachea, contraction of, 144
Trapezoidal rule, 405ff
Trigonometric functions, 378
  definitions of, 359, 377
  table of values for, A7
Truncated cone, 288

U

Uniform random variable, 498

V

Variance:
  alternate formula for, 500
  of a probability table, 480
  of a random variable, 481, 497
Velocity, 95, 254, 260
Velocity constant of elimination, 239

Verhulst, P., 241
Vertical asymptote, 108
Vertical line test, 10
Volume of a solid of revolution, 279–80
Voting model, 6, 171

W

Wage per unit of capital, labor, 308
War fever, 468

Water pollution, 16
World population, 284

X

x-intercept, 16, 107

Y

y-intercept, 16, 38, 107

Z

Zero of a function, 27, 107